Gas Phase NMR

New Developments in NMR

Editor-in-Chief:
Professor William S. Price, *University of Western Sydney, Australia*

Series Editors:
Professor Bruce Balcom, *University of New Brunswick, Canada*
Professor István Furó, *Industrial NMR Centre at KTH, Sweden*
Professor Masatsune Kainosho, *Tokyo Metropolitan University, Japan*
Professor Maili Liu, *Chinese Academy of Sciences, Wuhan, China*

Titles in the Series:
1: Contemporary Computer-Assisted Approaches to Molecular Structure Elucidation
2: New Applications of NMR in Drug Discovery and Development
3: Advances in Biological Solid-State NMR
4: Hyperpolarized Xenon-129 Magnetic Resonance: Concepts, Production, Techniques and Applications
5: Mobile NMR and MRI: Developments and Applications
6: Gas Phase NMR

How to obtain future titles on publication:
A standing order plan is available for this series. A standing order will bring delivery of each new volume immediately on publication.

For further information please contact:
Book Sales Department, Royal Society of Chemistry, Thomas Graham House, Science Park, Milton Road, Cambridge, CB4 0WF, UK
Telephone: +44 (0)1223 420066, Fax: +44 (0)1223 420247
Email: booksales@rsc.org
Visit our website at www.rsc.org/books

Gas Phase NMR

Edited by

Karol Jackowski
University of Warsaw, Poland
Email: kjack@chem.uw.edu.pl

and

Michał Jaszuński
Polish Academy of Sciences, Warsaw, Poland
Email: michal.jaszunski@icho.edu.pl

ROYAL SOCIETY OF CHEMISTRY

New Developments in NMR No. 6
Print ISBN: 978-1-78262-161-4
PDF eISBN: 978-1-78262-381-6
EPUB eISBN: 978-1-78262-722-7
ISSN: 2044-253X

A catalogue record for this book is available from the British Library

Published by The Royal Society of Chemistry,
Thomas Graham House, Science Park, Milton Road,
Cambridge CB4 0WF, UK

Registered Charity Number 207890

For further information see our website at www.rsc.org

Printed in the United Kingdom by CPI Group (UK) Ltd, Croydon, CR0 4YY, UK

Preface

Nuclear magnetic resonance (NMR) is a very special kind of molecular spectroscopy. It is extensively used by chemists, primarily for the investigation of organic molecules in liquids and solids. NMR experiments in the gas phase are somewhat less popular, even though the first systematic studies of gases have been started more than 50 years ago. This is presumably connected with the time-consuming procedure of sample preparation, because the NMR observation of gaseous compounds is fairly easy – almost all the new experimental techniques known and applied for liquids can be successfully adopted for the study of gases. Gas phase NMR has certainly its own areas of unique applications in science and for this reason the investigation of gases is emerging as an important tool, involving many new applications in the fields of physics, chemistry, biology, and medicine. It is especially important that gas phase NMR experiments deliver the values of spectral parameters which are free from intermolecular interaction effects and therefore suitable for direct comparison with quantum chemical calculations, usually performed for isolated molecules. The combined use of experimental and theoretical methods in this area gives rise to a new outlook on magnetic properties of molecules and on multinuclear NMR experiments themselves; the gas phase studies are enormously enriched when they are connected with the calculation of spectral parameters.

This book includes eleven chapters, discussing various aspects of NMR spectroscopy; only some of the numerous covered topics are listed below. The starting point is a general overview of problems and challenges encountered in the gas phase. It shows how the dependence of NMR parameters (nuclear magnetic shielding and spin–spin coupling) on gas density and temperature gives insight into the theories of intermolecular effects and intramolecular motion. Microwave spectroscopy and molecular beam resonance methods, described next, provide valuable information

New Developments in NMR No. 6
Gas Phase NMR
Edited by Karol Jackowski and Michał Jaszuński

Published by the Royal Society of Chemistry, www.rsc.org

about NMR parameters. For instance, the molecular beam resonance methods yield indirect nuclear spin–spin coupling tensors, while the nuclear spin-rotation tensors are especially important for the semi-empirical determination of absolute shielding in molecules.

Having the absolute shielding values in small molecules and the accurate magnetic moment of the proton makes it possible to extend our knowledge to the magnetic moments of other nuclei. The more accurate values of nuclear magnetic moments thus obtained are essential for NMR spectroscopy; moreover they can be applied for the measurement of shielding in routine research work. In addition, such measurements may be used to determine primary isotope effects in shielding. Molecules in the gas phase are of interest as objects of chemical analysis and modern NMR spectrometers are so fast that two-dimensional spectra can be successfully used to monitor the progress of chemical reactions. As an illustration, the gaseous decomposition of di-*tert*-butyl peroxide is elucidated in detail. As mentioned above, accurate values of the shielding in small molecules are needed for further applications, *e.g.* for the comparison with calculated shielding parameters or the determination of nuclear magnetic moments. This is important in the reviewed ^{17}O and ^{33}S studies, which are especially difficult because they require advanced NMR techniques for the detection of gaseous compounds when the concentration of molecules is low and in addition the natural abundances of oxygen-17 and sulfur-33 nuclei are low.

The discussion of the theoretical studies of NMR parameters begins with a short explanatory account of the methods used to determine accurate shielding constants. Particular attention is paid to the hierarchy of *ab initio* methods, because their use permits to improve systematically the results and to estimate the error bars. Next, the development of efficient theoretical methods needed to calculate zero-point vibration and temperature effects and the magnitude of these contributions to NMR parameters are discussed. The nuclear motion effects, related to the rotation and vibration of the molecule, have to be considered when the experimental data are compared with computed shielding or spin–spin coupling constants. Relativistic methods are required to determine reliable values of all the NMR parameters when there is a heavy atom in the molecule. Significant progress made in the last few years in the development and implementation of two- and four-component approaches which yield the relativistic values of NMR parameters, reflected by increasing accuracy of the results, is next reviewed.

Molecules are usually observed exploring the thermal equilibrium of nuclear magnetic moments, but the sensitivity of NMR spectroscopy is enormously increased after gas hyperpolarization. The hyperpolarization can be achieved by parahydrogen induced polarization (PHIP) for molecular hydrogen or by the optical pumping methods for noble gases like ^{3}He or ^{129}Xe. The hyperpolarized gases are utilized in a large variety of NMR experimental studies, with the most spectacular application in magnetic resonance imaging (MRI) of the lung. This non-invasive diagnostic method could be an excellent tool of modern medicine; unfortunately, there is a shortage of helium-3 on the market and for medical purposes the application of alternative gases like xenon-129 and fluorinated compounds must be also examined. The use of fluorinated gases in MRI can be accomplished without the hyperpolarization process.

We are pleased to present the first book which covers so many different aspects of gas phase NMR spectroscopy. We hope that this book will be helpful for everyone familiar with NMR methods, giving a better understanding of spectral parameters and more knowledge about the role and possible applications of gas phase NMR experiments.

Karol Jackowski and Michał Jaszuński

Contents

Chapter 1 Fundamental Intramolecular and Intermolecular Information from NMR in the Gas Phase 1
Cynthia J. Jameson

1.1 Why Should One Do Gas Phase NMR Measurements? 1
1.2 The Effect of Intermolecular Interactions on NMR Chemical Shifts 3
1.2.1 The Density Coefficient of the Chemical Shift 4
1.2.2 The Intermolecular Shielding Function 6
1.2.3 Contact Shifts in the Gas Phase 8
1.3 The Intramolecular Effects on Shielding 8
1.3.1 The Temperature Dependence of Chemical Shift in the Zero-density Limit 9
1.3.2 The Intramolecular Shielding Surface for Diatomic and Polyatomic Molecules 10
1.3.3 Rovibrational Averaging. The Connection between the Temperature Dependence in the Zero-density Limit and Isotope Shifts 13
1.3.4 Absolute Shielding Scales and Comparison with State-of-the-Art Quantum Calculations 17
1.4 The Spin–Spin Coupling in the Gas Phase 18
1.4.1 The Density Coefficient of the Spin–Spin Coupling 19
1.4.2 The *J* Surface and the Effects of Rovibrational Averaging 19

New Developments in NMR No. 6
Gas Phase NMR
Edited by Karol Jackowski and Michał Jaszuński

Published by the Royal Society of Chemistry, www.rsc.org

1.5 Spin–Lattice Relaxation in the Gas Phase. Cross-sections for Angular Momentum Transfer and Molecular Reorientation in the Binary Collision Limit 23
1.5.1 Temperature-dependent Classical Cross-Sections from Gas Phase Studies 23
1.5.2 Spin-rotation Mechanism 27
1.5.3 Quadrupolar Mechanism 28
1.5.4 Intramolecular Dipole–Dipole Mechanism 29
1.5.5 Chemical Shift Anisotropy Mechanism 29
1.5.6 Relaxation Rates Add When Two or More Mechanisms are Operative 30
1.5.7 Intermolecular Dipolar, Quadrupolar, Spin-rotation, and Chemical Shift Anisotropy Mechanism 30
1.5.8 Intermolecular Nuclear Spin Dipole Electron Spin Dipole Mechanism, Spin Relaxation in the Presence of O_2 33
1.5.9 Classical Trajectory Calculations of Relaxation Cross-sections 35
1.5.10 The Special Case of Hydrogen Molecule 37
1.6 Conformational Dynamics in the Gas Phase 38
List of Abbreviations 41
References 42

Chapter 2 Obtaining Gas Phase NMR Parameters from Molecular Beam and High-resolution Microwave Spectroscopy **52**
Alexandra Faucher and Roderick E. Wasylishen

2.1 Introduction 52
2.2 The Hyperfine Hamiltonian 54
2.3 Nuclear Spin Rotation 54
2.4 Nuclear Magnetic Shielding 57
2.5 The Ramsey–Flygare Method 60
2.5.1 Linear Molecules 61
2.5.2 Non-linear Molecules 63
2.5.3 Relativistic Methods 66
2.6 The Quadrupolar Interaction 73
2.6.1 Applications of Quadrupolar Tensors from Molecular Spectroscopy 75

2.7 Nuclear Spin–Spin Coupling 78
2.7.1 Characterization of Indirect Spin–Spin Coupling Tensors 80
2.8 Conclusions 84
Appendix A: The Measurement of Nuclear Magnetic Moments 85
References 88

Chapter 3 Nuclear Magnetic Moments and NMR Measurements of Shielding 95
Karol Jackowski and Piotr Garbacz

3.1 Introduction 95
3.2 NMR Experimental Methods in the Gas Phase 96
3.2.1 Gas Samples 96
3.2.2 High-pressure Techniques 97
3.3 Resonance Frequency in an Isolated Molecule 98
3.4 Nuclear Magnetic Moments 101
3.4.1 The Magnetic Moment of the Proton 102
3.4.2 Nuclear Magnetic Moments from Gas Phase NMR Experiments 105
3.5 Direct Measurements of Shielding 109
3.5.1 Referencing of Shielding Measurements 111
3.5.2 External and Internal Referencing of Shielding 115
3.6 Applications of Shielding Measurements 115
3.6.1 Standardization of NMR Spectra 115
3.6.2 Verification of Shielding Calculations 118
3.6.3 Primary Isotope Effects in Shielding 119
3.6.4 ^{13}C Shielding Scale for NMR Measurements in Solids 119
3.6.5 Adsorbed Gases 120
3.7 Conclusions 120
Acknowledgements 121
References 122

Chapter 4 Gas Phase NMR for the Study of Chemical Reactions: Kinetics and Product Identification 126
Alexander A. Marchione and Breanna Conklin

4.1 Introduction 126

4.2 Experimental Considerations – Concentration, Pressure, Temperature, Vessel Design 127
4.2.1 Concentration and Pressure 127
4.2.2 Temperature and Sample Temperature Calibration 129
4.2.3 Vessel Design and Material-of-construction 130
4.3 Spectroscopic Considerations – Probe Design, Phase and Frequency Drift, Spectral Acquisition Schedule 133
4.3.1 Probe Design 133
4.3.2 Phase and Frequency Drift 134
4.3.3 Acquisition Parameters 134
4.4 Survey of Published Studies 135
4.5 Current Example 139
4.6 Characterization of Reaction Products – 2D Correlation Experiments and DOSY 143
4.6.1 Gas Phase Correlation Experiments 144
4.6.2 Gas Phase DOSY 147
4.7 Conclusions and Outlook 148
Acknowledgements 149
References 149

Chapter 5 ^{17}O and ^{33}S NMR Spectroscopy of Small Molecules in the Gas Phase 152
Włodzimierz Makulski

5.1 Introduction 152
5.2 Background 153
5.2.1 Oxygen and Sulfur in Chemistry of Small Molecules 153
5.2.2 NMR Parameters of ^{17}O and ^{33}S Nuclei 154
5.2.3 ^{17}O and ^{33}S-labelled Compounds 156
5.3 NMR Experiments in Gas Phase 157
5.3.1 Experimental Approach and Problems 157
5.3.2 Gas Phase Experimental Characteristic 158
5.3.3 Absolute Shielding 158
5.3.4 Spin–Spin Coupling 162
5.4 ^{17}O and ^{33}S Shielding from Gas Phase Measurements 163
5.4.1 C^{17}O Molecule as Reference of Oxygen Shielding 163
5.4.2 The "Isolated" Water Molecule 164
5.4.3 ^{17}O Magnetic Shielding of Small Molecules 164
5.4.4 CO^{33}S as the Reference of Sulfur Shielding 168

5.4.5 Uniqueness of the SF_6 Molecule 168
5.4.6 Other Sulfur Containing Compounds 169
5.4.7 Intermolecular Interactions 170
5.5 Effects of Condensation 171
5.6 Isotope Effects on Chemical Shifts and Spin–Spin Coupling 173
5.6.1 Isotope Effects Observed on ^{17}O and ^{33}S Nuclei 173
5.6.2 Isotope Effects Observed on Other Nuclei 174
5.7 Spin–Spin Coupling Involving ^{17}O or ^{33}S Nuclei 175
5.8 Summary 180
Acknowledgements 181
References 182

Chapter 6 Accurate Non-relativistic Calculations of NMR Shielding Constants 186
Andrej Antušek and Michał Jaszuński

6.1 Introduction 186
6.2 Non-relativistic Theory of NMR Parameters 187
6.3 Analysis of the Shielding Constants within *Ab Initio* Electronic Structure Methods 190
6.3.1 Basis Sets in the Calculation of NMR Shielding Constants 191
6.3.2 Electron Correlation Effects 192
6.3.3 Relativistic Effects 193
6.3.4 Zero-point Vibrational and Temperature Effects 194
6.3.5 Intermolecular Interactions 195
6.4 Applications 196
6.4.1 Approaching Accurate NMR Shielding Constants: Two Examples 196
6.4.2 Basis Sets Effects 198
6.4.3 Electron Correlation Effects 198
6.4.4 Relativistic Effects 205
6.4.5 Zero-point Vibrational and Temperature Effects 207
6.4.6 Intermolecular Interactions 208
6.4.7 Determining the Nuclear Magnetic Dipole Moments 210
6.4.8 Available Software Packages 210
6.5 Conclusions 211
Acknowledgements 212
References 212

Chapter 7 Rovibrational and Temperature Effects in Theoretical Studies of NMR Parameters **218**
Rasmus Faber, Jakub Kaminsky and Stephan P. A. Sauer

7.1 Methods for Calculation of Rovibrational Corrections 218
7.1.1 Perturbation Theory Approach 219
7.1.2 Vibrational Corrections 221
7.1.3 The Effective Geometry Approach 223
7.1.4 Rotational Contributions 224
7.1.5 Temperature Averaging 225
7.1.6 Secondary Isotope Effects 226
7.1.7 Alternative Perturbation Expansions 227
7.1.8 Calculation of the Required Parameters 228
7.2 Examples of Vibrational Corrections to Shieldings 229
7.2.1 Vibrational Corrections to Hydrogen Shieldings 230
7.2.2 Vibrational Corrections to Carbon Shieldings 231
7.2.3 Vibrational Corrections to Nitrogen Shieldings 233
7.2.4 Vibrational Corrections to Oxygen Shieldings 234
7.2.5 Vibrational Corrections to Fluorine Shieldings 235
7.2.6 Vibrational Corrections to Phosphorus and Transition Metal Shieldings 236
7.2.7 Transferability 237
7.2.8 Methodological Aspects 238
7.2.9 Temperature Effects and Isotopic Shifts 245
7.2.10 Solvent Effects 246
7.2.11 Practical Aspects of ZPVC Calculations 247
7.3 Examples of Vibrational Corrections to Coupling Constants 247
7.3.1 High-level Wavefunction Calculations on Small Molecules 249
7.3.2 DFT Calculations 255
7.3.3 Systems with Relativistic Effects 260
7.3.4 Isotope Effects 260
7.3.5 General Trends 261
References 262

Chapter 8 Relativistic Calculations of Nuclear Magnetic Resonance Parameters **267**
Michal Repisky, Stanislav Komorovsky, Radovan Bast and Kenneth Ruud

8.1 Introduction 267
8.2 Basic Theoretical Models of Relativistic Quantum Chemistry 269
8.2.1 Relativistic Four-component Hamiltonians 269
8.2.2 Relativistic Two-component Hamiltonians 274
8.3 Relativistic Quantum Chemical Models for NMR Parameters 279
8.3.1 External Field-dependent Unitary Transformation (EFUT) 284
8.3.2 Restricted Magnetic Balance (RMB) 284
8.3.3 Simple Magnetic Balance (sMB) 285
8.3.4 Other Methods for Solving the Magnetic Balance Problem 286
8.4 Examples of Relativistic Effects on NMR Parameters 288
8.4.1 Nuclear Magnetic Shielding Constants and Chemical Shifts 289
8.4.2 Indirect Nuclear Spin–Spin Coupling Constants 295
8.5 Concluding Remarks 297
Acknowledgements 298
References 298

Chapter 9 High-resolution Spectra in PHIP **304**
Rodolfo H. Acosta, Ignacio Prina and Lisandro Buljubasich

9.1 Introduction 304
9.2 Parahydrogen Induced Polarization (PHIP) 306
9.2.1 Brief Description of p-H_2 307
9.2.2 ALTADENA and PASADENA 307
9.2.3 Hydrogenation 309
9.3 *J*-Spectroscopy 310
9.3.1 Theoretical Background 310
9.3.2 Partial *J*-Spectra 317
9.3.3 Technical Considerations 317
9.4 *J*-Spectroscopy in PHIP (PhD-PHIP) 319
9.4.1 Theoretical Basis 319
9.4.2 Experimental Results 324

9.5 PhD-PHIP in Gases 327
9.6 Summary 331
Acknowledgements 332
References 332

Chapter 10 Optical Hyperpolarization of Noble Gases for Medical Imaging **336**
Tadeusz Pałasz and Bogusław Tomanek

10.1 Introduction 336
10.2 Boltzmann Equilibrium Polarization and Hyperpolarization 337
10.3 Spin Exchange Optical Pumping of ^{3}He and ^{129}Xe 339
10.3.1 Optical Pumping of Alkali Metal Atoms 341
10.3.2 Spin Exchange between Optically Pumped Alkali Metal Atoms and Noble Gas Nuclei 345
10.3.3 Relaxation Processes 349
10.4 Metastability Exchange Optical Pumping of ^{3}He 352
10.4.1 Optical Pumping of ^{3}He and Metastability Exchange 352
10.4.2 Compression of Polarized ^{3}He 355
10.4.3 MEOP at High Magnetic Field and Elevated Pressures 356
10.5 Summary 359
References 360

Chapter 11 Medical Applications of Hyperpolarized and Inert Gases in MR Imaging and NMR Spectroscopy **364**
Marcus J. Couch, Matthew S. Fox, Barbara Blasiak, Alexei V. Ouriadov, Krista M. Dowhos, Boguslaw Tomanek and Mitchell S. Albert

11.1 Introduction 364
11.2 Hyperpolarized ^{3}He and ^{129}Xe Lung MRI 366
11.2.1 Overview of HP Gas MRI 366
11.2.2 Static Breath-hold Imaging 367
11.2.3 Diffusion Imaging 371
11.2.4 Probing Dissolved-phase ^{129}Xe 372

11.3 ^{19}F Lung Imaging 373
11.4 ^{129}Xe MRI of the Brain 379
11.5 Hyperpolarized ^{129}Xe Biosensors 382
11.6 Conclusions 384
Acknowledgements 385
References 385

Subject Index **392**

CHAPTER 1

Fundamental Intramolecular and Intermolecular Information from NMR in the Gas Phase

CYNTHIA J. JAMESON

University of Illinois at Chicago, USA
Email: cjjames@uic.edu

1.1 Why Should One Do Gas Phase NMR Measurements?

In the gas phase we have a well-defined homogeneous physical system, and the theory for dilute gas behavior is in an advanced stage. In dilute gases, we can expand the molecular electronic property (*e.g.*, nuclear magnetic shielding, J coupling, nuclear quadrupole coupling) in a virial expansion, in which the property virial coefficients can be expressed theoretically in closed form and can be obtained unequivocally experimentally in the binary interaction limit. These experimentally measured quantities depend on two quantum-mechanical mathematical surfaces: the shielding, or J, or electric field gradient (efg) at the nucleus as a function of intermolecular nuclear coordinates and the weak intermolecular interaction potential energy surfaces that are also a function of the same intermolecular nuclear coordinates. Furthermore, we can extrapolate the measured NMR data (shielding, J, efg) to the zero-density limit to obtain these electronic properties for the

New Developments in NMR No. 6
Gas Phase NMR
Edited by Karol Jackowski and Michał Jaszuński

Published by the Royal Society of Chemistry, www.rsc.org

isolated molecule, that are much more closely related to quantum-mechanical calculations than quantities measured in condensed phases. To validate theoretical methods, it is always preferable to benchmark the results by comparing them with available experimental data, preferably for isolated molecules. Extrapolation to this limit is only possible for gas phase measurements. Here too, the temperature dependence of the electronic property at the zero-density limit is a function of two quantum-mechanical mathematical surfaces: the shielding (or *J* or efg) as a function of intramolecular nuclear coordinates and the intramolecular potential energy surface that are also a function of the same coordinates. The latter is commonly characterized by specifying the derivatives at the equilibrium intramolecular configuration, namely the quadratic, cubic, quartic force constants. The shielding is particularly sensitive to the anharmonicity of the intramolecular potential surface. Thus, gas phase NMR data for shielding, *J*, and efg provide stringent tests of theoretical descriptions of both the quantum-mechanical electronic property surfaces and also the potential energy surfaces over which they are averaged, to yield the temperature-dependent experimental data (property virial coefficients and zero-density limiting values) that are available only in the dilute gas phase. In addition to temperature, another variable, isotopic masses of neighboring (and observed) nuclei, can affect the measured data, given the same electronic property surfaces and the same potential energy surfaces; thus, isotope effects provide an independent test of these quantum-mechanical surfaces. While these observations and their interpretation are of specific interest to NMR spectroscopists, they are of more general interest as prototypes of rovibrational averaging and intermolecular effects on molecular electronic properties. Fortunately, it is possible in NMR spectroscopy to make very precise measurements of quantities that are very sensitive to changes in electronic environment, nuclear magnetic shielding and *J*, molecular electronic properties that are sensitive indices of the chemical bond and that vary with nuclear displacements from the equilibrium molecular configuration, leading to changes in resonance frequencies that are amenable to highly precise measurements under precisely controlled constant temperature conditions over a wide range of temperatures. Thus, gas phase measurements in NMR provide valuable tests of quantum-mechanically calculated molecular electronic property surfaces. Indeed, the dihedral-angle dependence of three-bond *J* coupling by Martin Karplus (known to NMR spectroscopists as the Karplus equation) was the earliest (1959) example of an experimentally testable quantum-mechanically calculated property surface.[1] An important disadvantage of gas phase NMR, however, is that only the isotropic values of the NMR tensor quantities can be obtained.

For the same reasons, NMR spectra of dilute gases provide thermodynamic and kinetic information that are important from a theoretical point of view. The gas phase allows the separation of intramolecular and environmental effects on the energy requirements for molecular processes. Gas phase NMR data provide the free energy barriers for conformational changes, from which torsional parameters for molecular dynamics (MD) force fields are obtained. Furthermore, pressure can be used as an experimental variable in gas phase studies; rate constants are both temperature- and pressure-dependent. Use of dynamic gas phase NMR techniques permits the complete characterization of rate processes within both temperature and pressure ranges, allowing the kinetics of chemical rate processes to be investigated in both the unimolecular and bimolecular regimes. Information about internal vibrational redistribution and collisional energy transfer in kinetic processes is obtained from these NMR studies. Thus, conformational dynamics can be characterized under well-defined limiting conditions in the gas phase, free energy barriers can be obtained, and theoretical interpretation of results using well-established methods can provide detailed interpretation. A collateral experimental advantage is the rapid spin–lattice relaxation that facilitates multiple acquisitions; ^{13}C relaxation times are at least two orders of magnitude shorter in the gas phase for some systems than in condensed phases. In the gas phase, we can measure spin–lattice relaxation rates that are of fundamental interest in their own right. The rates are resolvable into well-defined mechanisms *via* measurements as a function of field, of temperature, of density. Furthermore, in the gas phase, each relaxation mechanism is capable of being theoretically calculated *via* classical trajectory calculations in the binary collision limit, yielding well-defined relaxation cross-sections that are well-established descriptions of fundamental dynamic molecular events, such as transfer of rotational angular momentum and molecular reorientation, that provide valuable stringent tests of the anharmonicity of intermolecular potential surfaces.

Reviews of gas phase NMR studies include some of these measurements of shielding and spin–spin coupling,[2–5] spin-relaxation studies,[3] and conformational changes,[4,6,7] that provide more detailed information and references to original literature not included in the present overview.

1.2 The Effect of Intermolecular Interactions on NMR Chemical Shifts

Buckingham and Pople proposed in 1956 that any electromagnetic properties of gases be expanded in a virial expansion.[8] For nuclear magnetic shielding in a pure gas

$$\sigma(T, \rho) = \sigma_0(T) + \sigma_1(T)\rho + \sigma_2(T)\rho^2 + \sigma_3(T)\rho^3 + \cdots \quad (1.1)$$

For a nucleus X in molecule A in a dilute mixture of gases A and B,

$$\sigma^{\mathrm{X\,in\,A}}(T, \rho_{\mathrm{A}}, \rho_{\mathrm{B}}) = \sigma_0^{\mathrm{X\,in\,A}}(T) + \sigma_{1\mathrm{AA}}(T)\rho_{\mathrm{A}} + \sigma_{1\mathrm{AB}}(T)\rho_{\mathrm{B}} + \cdots \quad (1.2)$$

In the gas of pure A, this expansion permits the study of the intermolecular contributions by investigating the temperature dependence of the density coefficient of nuclear shielding $\sigma_{1\mathrm{AA}}(T)$, *i.e.*, the slope of $\sigma^{\mathrm{X\,in\,A}}(T, \rho_{\mathrm{A}})$ as a function of density ρ_{A}, in the limit of linear behavior. At the same time this permits the study of the intramolecular contributions, $\sigma_0^{\mathrm{X\,in\,A}}(T)$, by investigating the temperature dependence of nuclear shielding in the limit of zero gas density. This term arises from the variation with temperature of the populations of the rotational and vibrational states, each one of which has a different characteristic average shielding. This is mathematically equivalent to extrapolation to a pressure that is low enough that collisional deformation of the molecules no longer contributes to σ, however there are still sufficient collisions to provide averaging over the rovibrational states of the molecule. The quantity $\sigma_0^{\mathrm{X\,in\,A}}(T)$ is the shielding in a molecule free of intermolecular interactions and therefore equivalent to an isolated molecule. From experimental measurements in the linear-density regime, each observed shift can be corrected for the intermolecular contributions $\sigma_1(T)\rho$ so that the remainder, $[\sigma_0(T) - \sigma_0(300\ \mathrm{K})]$, is obtained.

In gas mixtures with low mole fraction of A in B, subtraction of the accurately determined AA contributions permits the determination of $\sigma_{1\mathrm{AB}}(T)$. The quantity $\sigma_{1\mathrm{AA}}(T)$ is a measure of the effects on nuclear magnetic shielding of X in molecule A from binary collisions of A with another A molecule and $\sigma_{1\mathrm{AB}}(T)$ is a measure of the effects from binary collisions of A with molecule B. The excess intermolecular property, $\sigma_2(T)\rho^2 + \sigma_3(T)\rho^3 + \cdots$, has been investigated in some cases, for example for ^{129}Xe in Xe gas, where collectively this has been found to be opposite in sign to $\sigma_1(T)\rho$.[9] There are experimental indications that this is true for other nuclei in other gases as well, for example, ^{19}F in $H_2C{=}CF_2$.[10] Our main focus in this section is on the density coefficient of NMR properties in the limit of zero density, that is, the second virial coefficient of shielding. There is an experimental quantity that also has a linear density dependence, the bulk susceptibility contribution to the observed chemical shift,[11] that is the same amount for all nuclei in the sample, that is an artifact of the sample shape and vanishes for spherical samples. It is understood in this section that experimental values of the second virial coefficient will have been corrected for the sample shape contribution, since we are interested in the true shielding response that arises from binary intermolecular interactions. This susceptibility correction limits the precision of experimental values, but is of consequence only in those cases where the true second virial coefficient of shielding is smaller than this correction.

1.2.1 The Density Coefficient of the Chemical Shift

The first observation of the density coefficient of the chemical shift in a gas was by Streever and Carr in 1961 for ^{129}Xe in xenon gas,[12] followed soon thereafter by Gordon and Dailey for ^{1}H in CH_4 and C_2H_6,[13] and in 1962 by Raynes, Buckingham, and Bernstein for ^{1}H in H_2S, CH_4, and C_2H_6 and for HCl in various gas mixtures.[14] Measurements of the second virial coefficient of nuclear magnetic shielding have been carried out for a variety of nuclei; the largest values are those for ^{129}Xe in Xe atom interacting with another rare gas atom or molecule.[9,15–19] Second virial coefficients of shielding of other nuclei, for example, of ^{1}H in HCl, HBr,[20] HCN,[21] NH_3,[22] C_2H_4, C_2H_6, C_3H_8,[23] of ^{11}B in BF_3,[24] of ^{13}C in CO,[20] CO_2,[25] HCN,[21] CH_4,[23,24] of ^{15}N in N_2,[26] NNO,[25] NH_3,[22] of ^{17}O in CO, CO_2, NNO, OCS,[27] of ^{31}P in PH_3,[28] PF_3, PF_5, POF_3,[29] of ^{19}F in a large number of molecules (ref. 30, 31 and references therein) have been investigated as a function of temperature. Recent additions to these include all the nuclei in propene,[32] in cyclopropane,[33] in CHF_3,[34] in CH_2F_2,[35] in SO_2 and SO_3,[36] in SiF_4,[37] in CH_3OH,[38] in $(CH_3)_2O$,[39] in CH_3NH_2,[40] in CH_3CN,[41,42] and in $(CH_3)_4Sn$.[43] The linear-density coefficient of ^{13}C in benzene, acetylene, and CH_3Br,[44–46] and of ^{33}S in SF_6[47] have also been studied.

An advantage of gas phase studies in the linear-density regime is that the intermolecular effects on shielding can be expressed in closed mathematical form, just as derived in general for any electromagnetic molecular property by Buckingham and Pople.[8] For Xe interacting with CF_4, for example,

$$\sigma_1(T) = \iiint \{\sigma(R,\theta,\phi) - \sigma(\infty)\} \mathrm{e}^{\left[\frac{-V(R,\theta,\phi)}{k_B T}\right]} R^2 \mathrm{d}R \sin\theta \, \mathrm{d}\theta \, d\phi \tag{1.3}$$

The theoretically expected behavior of $\sigma_1(T)$ in rare gas systems over a wide temperature range has been shown to be negative (deshielding with increasing density), increasing in magnitude with increasing temperature, then switching over and decreasing in magnitude with increasing temperature (see Figure 6 in Ref. 48). For rare gas atoms, it has been found that the sign of $\sigma_1(T)$ is indeed negative at all temperatures, that is, the nucleus becomes more deshielded with increasing density. For nuclei of end atoms in a molecule, the sign of σ_1 is generally negative at all temperatures. Known exceptions are $\sigma_1(^{15}\mathrm{N})$ in CH_3CN and HCN,[21,41] in which intermolecular interactions involve the lone pair and thereby affect $n \rightarrow \pi^*$ contributions to the ^{15}N shielding toward less deshielding (such as that which accompanies a blue shift in the $n \rightarrow \pi^*$ transition energy). For more centrally located nuclei in a molecule, the general behavior, sign, or temperature dependence has not been calculated, but magnitudes are expected to be smaller than for end atoms. This behavior is not generalizable since, unlike the end atoms which experience intermolecular effects directly, a nucleus in centrally located atoms (except in linear molecules) can only experience intermolecular effects indirectly through chemical bonds, hence is dependent on the specific molecular structure surrounding the observed nucleus. In those cases where the temperature dependence has been measured over a wide range of temperatures, it has been found that the magnitude of $\sigma_1(T)$ generally decreases with increasing temperature; exceptions are ^{129}Xe in CO and ^{129}Xe in N_2,[19] which are not anomalous behavior since $\sigma_1(T)$ has been theoretically predicted to turn around to decreasing magnitudes at much lower temperatures, in general.[48] The magnitudes and signs of very small σ_1 for less exposed nuclei such as ^{33}S in SF_6, ^{13}C in CH_4, ^{29}Si in SiF_4[37,47,49] are difficult to obtain accurately in experiments because they are generally small and the experimental density coefficient is thus dominated by bulk susceptibility contributions for non-spherical sample shapes.

There is clear evidence of a nuclear site effect in the same molecule, with more exposed nuclei having larger values of σ_1 than less exposed ones.[50] In a clear example, the three chemically inequivalent F nuclei in XFC=CF$_2$, F_{gem}, F_{trans}, F_{cis}, have different distances from the center of mass of the molecule, and these relative distances change as X goes from light to heavy mass. Within the freely rotating molecule, the relative exposure of each of the three ^{19}F sites to intermolecular interactions are reflected in their respective values of σ_1 in XFC=CF$_2$ molecules (X = H, F, Cl, Br, I), completely consistent with the relative distance of each of F_{gem}, F_{trans}, F_{cis} from the center of mass as the latter changes systematically from X = H, to F, to Cl, to Br, to I. A more transparent and elegant example of the nuclear site effect was demonstrated experimentally by Beckett and Carr in the density dependence (in HD gas of varying density with a small amount of D_2) of the isotope shift $[\sigma(D_2) - \sigma(HD)] = a + b\rho$.[51] Here, $a = [\sigma_0(D_2) - \sigma_0(HD)]$ is the isotope shift extrapolated to the zero-density limit. Their observation that the density coefficient of the isotope effect, $b < 0$ means that $|\sigma_1(D_2)| > |\sigma_1(HD)|$ since all σ_1 is known to be negative. The greater magnitude of $|\sigma_1(D_2)|$arises from the more exposed deuterium nucleus in D_2 ($R/2$ from the center of mass in this isotopomer) compared to HD where the D is $R/3$ from the center of mass. The density coefficient b can be calculated from the site factors, as shown in ref. 50. Thus, the nuclear site effect gives rise to the observed density dependence of the isotope shift in the Beckett and Carr experiments.

In some cases, intermolecular effects on shielding have been measured from very low density gas to the liquid phase in a single experiment. It is especially interesting when both gas and liquid are observed in the same sample tube as a function of temperature. The difference in chemical shift between the liquid and the overhead vapor should approach zero in the limit of the critical temperature. Indeed, this behavior has been observed for ^{19}F in a large number of compounds (see for example ref. 50).

1.2.2 The Intermolecular Shielding Function

The first *ab initio* shielding function calculated for a rare gas pair $[\sigma(R) - \sigma(\infty)]$ was that for Ar–Ar;[48] these were restricted Hartree–Fock (RHF) calculations of $\sigma(R)$ from large separations all the way to an internuclear distance of 1 Å, or 0.30 times r_0(Ar–Ar). In this specific case the united atom in the correlation diagram of the two Ar atoms is a closed shell ground state (Kr) so that it is possible to extrapolate the shielding function all the way to the united atom, and thereby observe the general shape of an intermolecular shielding function. The intermolecular shielding function for Xe interacting with rare gas atoms Ne, Ar, Kr, Xe has been calculated with very large basis sets at various levels of accuracy.[52,53] With the inclusion of relativistic corrections,[53] the final agreement with the experimental temperature dependence is almost within experimental error. The shielding functions all change steeply in the vicinity of r_0 of the potential energy surface (PES).

It turns out that scaling is possible for rare gas pairs using the shielding sensitivity that is proportional to $\langle a_0{}^3/r^3\rangle$ for the atom whose shielding is being calculated and using the corresponding-states type of factors for the strength of intermolecular interactions, in terms of electric dipole polarizabilities α and ionization potentials *IP*:[54]

$$\frac{\alpha_{\mathrm{Xe}}(0)}{\alpha_{\mathrm{Rg}}(0)}\frac{IP_{\mathrm{Xe}}}{IP_{\mathrm{Rg}}}\frac{IP_{\mathrm{Xe}}+IP_{\mathrm{Rg}}}{2IP_{\mathrm{Xe}}}$$

This has been shown both at the RHF and DFT (Density Functional Theory) level by comparing *ab initio* calculated shielding functions for various rare gas pairs with scaled shielding functions based on the Ar–Ar pair.[52]

What is the range of interaction measured by the intermolecular chemical shift, *i.e.*, which distances make the major contributions to the observed density coefficient of the chemical shift? We answered this question using ^{129}Xe shielding in rare gas pairs.[52] This is clearly shown by the reduced function that compares all three rare gas pairs at their corresponding states. When the integrands in $\sigma_1(T)=4\pi\int_0^{\infty}\{\sigma(r)-\sigma(\infty)\}\,\exp[-V(r)/k_BT]r^2\,\mathrm{d}r$ for Xe-Rg, Rg = Xe, Kr, Ar, Ne are scaled using the scaling factor $\{[\alpha_{\mathrm{Xe}}(0)/\alpha_{\mathrm{Rg}}(0)]\cdot[IP_{\mathrm{Xe}}/IP_{\mathrm{Rg}}]\cdot[(IP_{\mathrm{Xe}}+IP_{\mathrm{Rg}})/2\mathrm{IP}_{\mathrm{Xe}}]\}$ and $r^*=r/r_0$ then at the reduced temperature $T/T_c=1$ all the integrand functions superimpose into one curve when plotted *vs.* r^*. From this curve we find that $\sigma_1(T)$ is nearly entirely accounted for by the sum over the range $0.90r_0$ to $1.5r_0$, with $r=0.96r_0$ to $1.24r_0$ providing approximately 80% of the observed density dependence of the nuclear shielding for rare gas pairs.[52] This is the range of interaction measured by the intermolecular chemical shift when only van der Waals (vdW) interactions are involved. *Ab initio* calculations of Xe shielding for Xe atom interacting with small molecules such as N_2, CO_2, CO, CH_4 and CF_4[55,56] and in cages such as C_{60} and $(H_2O)_n$[57] indicate that the sharp deshielding that is observed in the Xe-Rg shielding function for interatomic distances shorter than the r_0 of the potential function is also observed in these later examples. State-of-the-art coupled-cluster calculations of the intermolecular shielding surface for hetero and homo rare gas pairs among the set Ar, Ne, and He[58] exhibit the same general behavior of $[\sigma(R)-\sigma(\infty)]$ and of $\sigma_1(T)$ for all rare gas pairs, as was already described above, found earlier with RHF calculations for these systems.[48]

The observed scaling discussed above permits us to predict that $\sigma_1(T)$ for ^{1}H will be small so that the density coefficient will be dominated by the sample-shape-defined bulk susceptibility contribution, and $\sigma_1(T)$ for other nuclei in end atoms in molecules will scale with $\langle a_0{}^3/r^3\rangle$ for the atom, just as the chemical shift ranges for different nuclei do.[59] Of course, when the observed nucleus is in a molecule with structure, any secondary effects such as changes in torsion angles resulting from the intermolecular interactions may contribute significantly to the intermolecular shifts observed for the nucleus in question. Also, where hydrogen bonding is involved, the shielding of the donor atom, the acceptor atom and the proton should be affected significantly, though not in the same monotonic way as for rare gas pairs. To our knowledge, there has been no equivalent investigation to determine the range of interaction measured by the intermolecular chemical shift in a hydrogen-bonded system.

1.2.3 Contact Shifts in the Gas Phase

For interactions with a paramagnetic gas, in addition to the $\sigma_1(T)$ that would be present for the diamagnetic gas, there is a contribution coming from the electron spin density at the position of the observed nucleus. For example, for Xe interacting with O_2,

$$\sigma_1(T)_{\text{hyperfine}} = -\left(\frac{16\pi g_e^2 \mu_B^2}{9k_B T}\right) 2\pi \iint \left[\rho_{\text{spin}}(R,\theta) - \rho_{\text{spin}}(\infty)\right] \mathrm{e}^{\left[\frac{-V(R,\theta)}{k_B T}\right]} R^2 \mathrm{d}R \sin\theta \, \mathrm{d}\theta \tag{1.4}$$

and, of course, the molar paramagnetic bulk susceptibility of the paramagnetic gas provides a large but predictable sample-shape-specific contribution. In eqn (1.4), the sign of the electron spin density at the Xe nucleus, $[\rho_{\text{spin}}(R,\theta) - \rho_{\text{spin}}(\infty)]$, is negative when the probability density of the β spin dominates over the α at the Xe nuclear position (R,θ). The linear-density coefficient for Xe in O_2 and NO have been obtained as a function of temperature.[60–62] The calculation of the ^{129}Xe hyperfine tensor of the Xe@O_2 molecular system permits comparison with the experimental data.[63] At low temperatures, the explicit T^{-1} dependence in the hyperfine contribution dominates over the weaker temperature dependence in the intermolecular weighting factor. Thus, a sample of Xe in O_2 can serve as a very sensitive thermometer in NMR measurements.

1.3 The Intramolecular Effects on Shielding

Experimental intramolecular effects on shielding have been reviewed earlier,[2,64,65] and also as a subset of the more general concept of rovibrational averaging of molecular electronic properties.[66] These manifest themselves as isotope shifts and temperature dependence in the isolated molecule. In 1952 Norman Ramsey considered the vibrational and rotational averages of shielding and spin–spin coupling for the case of diatomic molecules, using hydrogen molecule and its isotopomers as specific examples.[67] Although temperature-dependent shieldings due to rovibrational averaging in isolated molecules were already predicted by Buckingham in 1962,[68] our first observations of the temperature dependence of shielding in isolated molecules occurred in 1977 in the determination of $\sigma_1(T)$ for ^{19}F in the small molecules F_2 and ClF,[20] followed by ^{19}F in BF_3, CF_4, SiF_4, and SF_6.[69] We explored ^{19}F rather than ^{1}H chemical shifts in studying molecules in the gas phase because of the known large chemical shift range for ^{19}F nucleus, in addition to its being an end atom more exposed to intermolecular interactions. Extrapolation of our measured ^{19}F chemical shifts to zero density revealed intercepts that were strongly temperature-dependent, thus leading to our investigation of the temperature dependence in the limit of the isolated molecule, not only for ^{19}F in a large number of molecules, but also for other nuclei such as ^{31}P, ^{15}N, and others. We explored the concept of the shielding surface for a nucleus in a molecule, the dependence of the nuclear magnetic shielding on bond distances, bond angles, and other internal degrees of freedom in a polyatomic molecule, by starting from the observed temperature dependence and extracting information about derivatives of the intramolecular shielding surface. Raynes, on the other hand, from 1971 on, started from the theoretical shielding surface and tried to predict the magnitudes of the temperature dependence and the mass dependence, so he was inclined to start with ^1H in H_2,[70,71] and H_2O,[72,73] and ^1H and ^{13}C in CH_4,[74–76] since the accuracy of quantum calculations of shielding were limited by the number of electrons in the molecule and the size of the set of basis functions. On the other hand, for ^1H, the temperature dependence of shielding in the isolated molecule is small and, thus, difficult to accurately measure experimentally. H_2O is particularly difficult to observe as an isolated molecule.

1.3.1 The Temperature Dependence of Chemical Shift in the Zero-density Limit

We measured the temperature dependence of the shieldings for ^{13}C in CO and CO_2,[26,25] ^{15}N in N_2 and NNO,[26,25] ^{31}P in PH_3,[28] ^{11}B in BF_3 and ^{13}C in CH_4,[24] ^{31}P in PF_3 and POF_3,[29] ^{15}N and ^1H in NH_3,[22] and ^{19}F in a large number of fluorine-containing molecules (references for individual ^{19}F sites are given in ref. 30 and 31). For most cases, the temperature dependence of shielding in the isolated molecule is non-linear and has a negative $d\sigma_0/dT$ throughout the temperature range, that is, more deshielding with increasing temperature. The largest temperature dependence we observed was for ^{19}F in F_2 molecule.[20] Here, as well as in all other ^{19}F $\sigma_0(T)$ that we reported for a large number of fluorohydrocarbons, the slopes $d\sigma_0/dT$ are negative and the curvature is in a uniform direction, more pronouncedly negative with increasing temperature. This we also observed for ^{19}F in SF_6, SeF_6, TeF_6, and WF_6.[77] There is an interesting correlation between the absolute shielding in the isolated molecule σ_0 and the temperature coefficient $d\sigma_0/dT$ for ^{19}F; the more deshielded ^{19}F sites also have the largest temperature coefficients.[78,79] The R_e that minimizes the molecular potential energy at the equilibrium geometry of a diatomic molecule, for example, is correlated with the quantum-mechanical behavior of the electrons which determines the steepness of the shielding function at that position, as will be seen in Section 1.3.2. The curvature of the $\sigma_0(T)$ function is an expected natural consequence of the non-linear dependence on temperature of the dynamically averaged displacements of the bond length from its equilibrium value. Notable exceptions are $\sigma_0(T)$ functions for ^{13}C in CO and ^{15}N in N_2,[26] which appear to be linear. When centrifugal distortion dominates the dynamically averaged displacement of the bond length from its equilibrium value, $(\langle R\rangle^T - R_e)$, and when the shielding surface is nearly linear at the equilibrium geometry for the molecule, then a linear $\sigma_0(T)$ may be expected over a wide temperature range. Raynes *et al.* reported linear temperature dependences for the ^1H and ^{13}C in CH_4, C_2H_6, C_2H_4, C_3H_8 and other alkanes, with predominantly negative values of $d\sigma_0/dT$ for ^1H, and either sign for ^{13}C.[23] The predominantly negative values of $d\sigma_0/dT$ observed for nuclei of end atoms could be rationalized in terms of dominant local bond anharmonic stretching that leads to average bond lengths that increase with increasing temperature and implies $(\partial\sigma/\partial r)_e < 0$ for those systems. We shall see in Section 1.3.2 that this is not always the case. Negative values for $d\sigma_0/dT$ have also been observed for central atoms, such as ^{13}C in CH_4,[23,24] ^{77}Se in SeF_6 and ^{126}Te in TeF_6,[77] where the totally symmetric breathing mode is largely responsible for the observed $d\sigma_0/dT$. On the other hand, the ^{13}C nuclear sites in ethane and higher alkanes are involved in many vibrational modes of the appropriate symmetry, so the sign of $d\sigma_0/dT$ is less easily predicted.

1.3.2 The Intramolecular Shielding Surface for Diatomic and Polyatomic Molecules

The first complete intramolecular shielding surface was calculated in 1979 by Hegstrom for the entire range ($R = 0$ to ∞) for $\sigma_\perp$ and $\sigma_{||}$ for H_2^+ molecule ion:[80] At $R = 0$ the shielding is characteristic of the diamagnetic shielding of He nucleus in free He^+ ion (the united atom for this system) and at $R = \infty$ the smaller diamagnetic shielding of an isolated H atom, passing through a minimum at R longer than R_e, so that at $R = R_e$, the derivative of the shielding function is negative. Earlier, some calculations of nuclear shielding in the immediate vicinity of the equilibrium bond length had been reported for a few diatomic molecules, *e.g.*, ^1H and Li in LiH by Stevens and Lipscomb (1964),[81] in H_2 by Raynes *et al.* (1971),[70] so that the first and second derivatives of the shielding with respect to bond length could be determined at the equilibrium geometry of the molecule. Later, Chesnut calculated first derivatives of the isotropic shielding at the equilibrium geometry for a number of nuclei in small molecules (including all hydrides across the periodic table from LiH to FH and NaH to ClH) and found that all derivatives for ^1H shielding in these molecules were negative, that is, at the equilibrium geometry, the ^1H shielding surface is becoming deshielded with slight increase in bond length. On the other hand, while most shielding derivatives were negative for the heavier nucleus in the hydride molecules, some were positive.[82] We obtained the same shape of the shielding surface for ^{23}Na in NaH between 0.5 and 5.0 Å[54] as Hegstrom obtained for H_2^+ molecule ion, except that the ^{23}Na shielding surface has a minimum at R shorter than R_e, so that at $R = R_e$, the first derivative of the shielding function is positive (likewise for Li in LiH), just as found by Chesnut. On the other hand, for ^{19}F in F_2 and ^{35}Cl in ClF and for ^{19}F in HF and ^{35}Cl in HCl, the shielding functions are decreasing with distance in the vicinity of R_e, that is, $[d\sigma/d(R-R_e)]_e<0$. The general behavior of $[d\sigma^X/d(R-R_e)]_e$ across the periodic table for XH_n hydrides going from LiH to FH and going from NaH to ClH is to smoothly vary with R_e from positive to negative across a row. Furthermore, as a description of the general behavior across the periodic table, we found that all 14 cases could be placed on one curve, if $[d\sigma^X/d(R-R_e)]_e$ were scaled by $\langle a_0{}^3/r^3\rangle_X$ and R_e.[54]

The highest level of theory is to have a full treatment of electron correlation, *i.e.*, full configuration interaction (FCI) calculations in the basis set limit. This is rarely used except for the smallest systems. To include electron correlation at all, we go beyond self-consistent field (SCF), *i.e.*, Hartree–Fock calculations. The most accurate of the single reference methods, yet still practically feasible, treats electron correlation at a level of coupled clusters singles, doubles and a perturbation correction for triple excitations, CCSD(T). When a single reference calculation is insufficient, multi-configurational methods are used, for example MCSCF (multi-configuration self-consistent field), RASSCF (restricted active space SCF), CASSCF (complete active space SCF). Also often used is Møller–Plesset perturbation theory at various orders (MP2, MP3, MP4). The second-order polarization propagator approximation (SOPPA) is the method of choice for those using the polarization propagator formalism as an alternative approach to study atomic and molecular properties within both regimes, relativistic and non-relativistic. Finally, there is density functional theory (DFT), which is a very popular method for including electron correlation, particularly for very large systems, but is still in a stage of development of improving the exchange-correlation functionals. These are some of the methods which have been used for calculations of NMR quantities, including intramolecular and intermolecular property surfaces.

The most accurate *ab initio* shielding surfaces for simple molecules have been calculated by Gauss *et al.* using CCSD(T). For example, using large basis sets, they have calculated shieldings at seven different bond distances for diatomic molecules H_2, HF, N_2, CO, and F_2 to find the first and second derivatives of shielding for all nuclei.[83,84] Their results agree with the trends found in the earlier RHF calculations, and their results provide better agreement with experimental temperature dependence and isotope shifts.

Shielding surfaces for more complicated molecules include the variation of the shielding with respect to bond angles and dihedral angles in addition to bond lengths. Shielding surfaces for polyatomic molecules are best expressed in terms of the symmetry coordinates that are the symmetry-adapted linear combinations of bond stretches, bond angles, and dihedral angles, rather than local modes. This is especially advantageous to use for small molecules such as H_2O, NH_3, PH_3, and CH_4, where there is a small number of symmetry coordinates. The first complete analysis of the shielding surface of a polyatomic molecule and the accompanying rovibrational averaging was carried out for ^1H and ^{17}O in H_2O molecule by Fowler and Raynes.[72,73] Later, Raynes *et al.* carried out the quantum-mechanical calculations and complete analysis of the ^{13}C and ^{1}H shielding surfaces for CH_4 and also the rovibrational averaging.[74–76] We did the same for NH_3[85] and PH_3,[86] including solving numerically for the highly anharmonic vibrational wavefunctions for the umbrella inversion coordinate. A comparison of the shielding surfaces for the central atom in these molecules (H_2O, NH_3, PH_3, CH_4,) with respect to the symmetry coordinates reveals similar signs and curvatures. There is a marked difference in the X shielding change with respect to HXH bond angle; however; the minimum in this trace on the shielding surface in H_2O is at the equilibrium bond angle, while the minimum occurs at slightly smaller and markedly larger bond angles than equilibrium for NH_3 and PH_3, respectively,[86] that lead to different overall temperature dependences for the X nuclei when rovibrational averaging is carried out, as described in the following section. These shielding calculations were done at the RHF level. Subsequent correlated calculations reveal very similar shapes of the shielding surfaces.

Correlated calculations for shielding surfaces for all nuclei in H_2O have been carried out by Fukui *et al.* using finite field MP3,[87] and by Vaara *et al.*[88] and Raynes *et al.* using MCSCF.[89] The most accurate shielding surface calculations for H_2O have been carried out using CCSD(T) by Gauss *et al.*[90] Fukui *et al.* also carried out finite-field MP3 calculations of the shielding surfaces of NH_3 and PH_3, CH_4 and SiH_4, as well as H_2S, in terms of the symmetry coordinates,[87] but these still do not constitute accurate calculations since their rovibrationally corrected anisotropies do not agree with experimental values. The most accurate shielding surface calculations for H_2S have been carried out using CCSD(T) by Gauss *et al.*;[91] they did the same calculations for SO_2 and OCS molecules at the same level of theory. The full shielding surfaces with respect to seven symmetry coordinates have been calculated for HC≡CH at the MCSCF level.[92]

The shielding surfaces for ^{13}C and ^{77}Se in CSe_2 have been calculated using MCSCF and DFT.[93] Of interest is the result that the derivatives $(\partial\sigma^{Se}/\partial r)_e$ and $(\partial\sigma^{Se}/\partial r')_e$ are roughly equal, which the authors found somewhat surprising, as it means that the effect of change in the bond not directly attached to the observed nucleus may be as important as that in the directly attached bond. This was also found in the case of ^{13}C shieldings in HC≡CH molecule,[92] but not for ^{1}H in this molecule, nor for ^{1}H in the di- or tri-hydrides of the first and second row in the periodic table.[82] The dependence of the shielding on torsion angles (that involve the observed nucleus) has been shown experimentally and theoretically to be the primary determining factor for the dispersion of the ^{13}C chemical shifts in proteins. The ^{13}C chemical shifts of the alanine residues, for example, in a folded protein differ from those of the random coil version of the protein, largely because the torsion angles of the various alanine residues in the folded protein are determined by the secondary and tertiary structure of the folded protein. Theoretical calculations of this torsion angle dependence led to the realization that ^{13}C chemical shifts in proteins are robust indicators of protein structure.[94]

Early reviews of intramolecular shielding surfaces are given in ref. 95 and 96, with direct comparison of the shapes of intermolecular and intramolecular shielding surfaces, and including shielding surfaces for ^{13}C and ^{15}N in amino acids and proteins.[96] For current annual reviews, consult Section 2.2 Shielding Surfaces and Rovibrational Averaging in the Specialist Periodical Reports on Nuclear Magnetic Resonance published by the Royal Society of Chemistry.[97]

The large errors in DFT calculations of shielding surfaces has been demonstrated in comparison to CCSD(T) calculations of the non-relativistic part in XeF_2 molecule.[98] For ^{19}F shielding, DFT drastically overestimates the correlation effects in this molecule. For molecules containing heavy atoms, there are relativistic contributions to shielding. Here the relativistic corrections to the absolute shielding and shielding anisotropy are very important for both nuclei. For ^{19}F in XeF_2, both the non-relativistic and the relativistic terms are deshielding with increasing bond length, thus relativistic effects further enhance the decrease in shielding compared to that found for ^{19}F in fluorohydrocarbons. On the other hand, for the central atom, Xe, almost full mutual cancellation of the negative non-relativistic and positive relativistic contributions results in small derivatives of the shielding for Xe. The greater sensitivity of the spin–orbit (SO) contribution relative to the scalar relativistic correction to bond stretch in the Xe and F shielding surfaces in XeF_2 has also been noted. In another example, it has been found that the spin–orbit contribution to the ^{1}H shielding in HI has an opposite distance dependence to the non-relativistic contributions; SO shielding increases with increasing bond length, opposite to the trend for ^{19}F in XeF_2, while the non-relativistic contributions behave as usual for ^{1}H, decreasing with bond length.[99] Similarly, SO shielding for ^{13}C in CTe_2 increases with increasing bond length and the second derivatives with bond stretch and bending are also positive, but the mixed second derivative with respect to the two bond distances is negative.[100] There has been no systematic study of the signs and magnitudes of the bond-length dependence of the relativistic scalar and SO contributions to shielding as has been done for the non-relativistic shielding.

1.3.3 Rovibrational Averaging. The Connection between the Temperature Dependence in the Zero-density Limit and Isotope Shifts

Rovibrational averaging of shielding is interesting in its own right and is perhaps the most precisely measured among molecular electronic properties. A general discussion of the theoretical and observed effects of rovibrational averaging reveals the ways in which the observed rovibrational effects on all these molecular properties can be understood for individual nearly isolated molecules using the same theoretical framework.[66] In all cases we need the intramolecular potential energy surface in the vicinity of the equilibrium geometry. The PES can be used to solve for the anharmonic vibrational wavefunctions or else its derivatives (force constants up to cubic, even quartic, may be necessary) may be available either from theoretical calculations or from vibrational spectroscopy. In some cases the quality of the quantum-mechanically calculated potential surface can be good enough to carry out the calculations of the anharmonic vibrational wavefunctions. Then we also need the electronic property surface. This is sometimes close to linear with respect to a displacement coordinate at the equilibrium geometry. In this case, it is the anharmonicity of the PES that provides most of the temperature dependence. The temperature dependence of the dynamic averages of nuclear displacement coordinates coupled with the derivatives of the shielding surface with respect to these coordinates permit the observed temperature dependence to be understood quantitatively. Sometimes, the minimum of the potential energy surface corresponds to a region in the property surface with significant curvature (non-negligible $(\partial^2 P/\partial r^2)_e$, $(\partial^2 P/\partial \theta^2)_e, \ldots$). In this case even using harmonic vibrational wavefunctions can lead to significant temperature dependence. On the other hand, the zero-point vibrational (ZPV) correction to the property may be largely reproduced by harmonic terms only. The isotope effect on molecular electronic properties is treated in the same theoretical framework. The mass dependence of the dynamic averages of nuclear displacement coordinates coupled with the derivatives of the shielding surface combine to provide isotope shifts. The observed temperature dependence of the shielding provides a stringent test of theoretical shielding surface calculations, as do the observed isotope shifts. Reviews with particular emphasis on isotope shifts are given in ref. 101 and 102; for current annual reviews, consult Section 2.3 Isotope Shifts in the Specialist Periodical Reports on Nuclear Magnetic Resonance published by the Royal Society of Chemistry.[97]

Any molecular electronic property P that is a function of nuclear configuration may be expressed as an expansion in terms of the dimensionless normal coordinates

$$P = P_e + \sum_{si} \left(\frac{\partial P}{\partial q_{si}}\right)_e q_{si} + \frac{1}{2}\sum_{si}\sum_{s'i'} \left(\frac{\partial^2 P}{\partial q_{si}\partial q_{s'i'}}\right)_e q_{si}\, q_{s'i'} + \cdots \tag{1.5}$$

where the subscript e designates the value at the equilibrium configuration, s denotes the sth vibrational mode, and i classifies each of the degenerate vibrations. The observed shielding at a given temperature may thus be written in terms of derivatives of nuclear shielding. The expectation values of the dimensionless normal coordinates should be calculated to at least first order using anharmonic vibrational wavefunctions while it is sufficient to use the zeroth order vibrational wavefunctions to calculate the expectation values of $q_{si}\, q_{s'i'}$. These vibrational state averages can then be weighted according to the populations of these states at a given temperature. Early work expressed the expectation values in terms of force constants up to cubic, and then, rather than a proper statistical weighting by populations, instead used an approximate sum over harmonic states in the high-temperature limit, since this sum could be expressed in closed form with the coth function.[103]

Complete analysis of high-resolution rotational–vibrational spectra of the molecule provides the required molecular constants needed in calculating the expectation values and populations, including the rotational portion (centrifugal distortion).

The observed large temperature dependence of ^{19}F shielding in diatomic molecules F_2 and ClF[103] and CF_4, SiF_4, and BF_3 served as first examples in this analysis.[104] A molecule-dependent mass-dependent transformation relates the dimensionless normal coordinates to the curvilinear internal displacement coordinates such as $(R\text{-}R_e)_{bond}$ and angle deformations. Symmetry dictates that only totally symmetric vibrational modes can contribute to the linear term in eqn (1.5). For T_d AX_4, D_{3h} AX_3, and O_h AX_6, $D_{\infty h}$ AX_2 molecules, there is only one totally symmetric mode; for $C_{\infty v}$ ABX molecules, there are two, involving the two distinct $(R\text{-}R_e)_{bond}$, and for C_{2v} AX_2 and C_{3v} AX_3 molecules, there are two, one involving bond stretches, the other involving angle deformations.[104–106] Consideration of the mass dependence of the derived expressions in diatomic molecules, and in those cases where only bond stretches contribute to the totally symmetric vibrational mode, permitted a derivation of isotope shifts in terms of the same constants as those for the temperature dependence of shielding for polyatomic molecules. This led to an approximate expression that explicitly relates the isotope shift to the fractional change in mass $(m' - m)/m'$,[77,107] and also provided the theoretical basis for the many observed trends in experimental isotope shifts.[108–110] On the other hand, the calculated temperature dependence of A in C_{3v} AX_3 or C_{2v} AX_2 molecules is found to depend on angle deformations as well. A complete treatment of the rovibrational averaging for H_2O and its isotopomers was carried out by Fowler and Raynes,[73] using the *ab initio* shielding surfaces they had calculated at the RHF level. Improved calculations for rovibrational averaging of shielding in H_2O were later carried out using correlated shielding calculations by Vaara *et al.*[88] and by Raynes *et al.*,[89] using RASSCF and MCSCF methods, respectively. The temperature dependence of ^{15}N in NH_3 was found to be very small, and this was due to the opposite temperature dependence arising from the umbrella inversion mode compared to all other modes.[85] Except for the inversion, the shielding surfaces of ^{15}N in NH_3 and ^{31}P in PH_3 are remarkably similar and do scale to one another. But for ^{31}P in PH_3, the temperature dependence contributions coming from various terms reinforce rather than oppose each other, leading to an overall negative temperature dependence (deshielding with increasing temperature).[86]

For diatomic molecules eqn (1.5) reduces to a very simple form and much of our understanding and physical insight about isotope shifts and temperature dependence of shielding was developed by studies of F_2, ClF, N_2, and CO.[103] What are the relative contributions to the temperature dependence observed for shielding of the isolated polyatomic molecule in the gas phase? If we start out with the *ab initio* calculations of the shielding surface, then, to answer this question, the surface calculations and the averaging of the dynamic variables is best carried out in terms of the symmetry coordinates and normal modes, as in eqn (1.5). However, thinking in terms of local bond stretching and other local displacements involving the observed nucleus permits extension of understanding gained from diatomic molecules and small symmetric molecules to much bigger molecular systems, without carrying out the full surface calculations and rovibrational averaging indicated in eqn (1.5). For nuclei of end atoms such as ^{19}F and ^{1}H in a polyatomic molecule, the local bond stretching internal coordinate turns out to dominate the temperature dependence. This is still true at the highest level of calculations using CCSD(T) for both shielding and potential functions.[83,84,92] This is still the case for ^{19}F in XeF_2, even though there are relativistic contributions that make the analysis more complicated.[98] The results for the ^{77}Se on the end atoms in CSe_2 are as expected, but $(\partial\sigma^{Se}/\partial r')_e$ is not much smaller than $(\partial\sigma^{Se}/\partial r)_e$.[93] It had been proposed early on that two-bond isotope shifts had two important contributions: (a) the shielding change with respect to the stretch of the remote bond coupled with the mass-dependence of the remote bond length upon isotopic substitution of one of the atoms participating in the bond, (b) the shielding change with respect to the stretch of the local bond coupled with the dependence of this average bond length on a remote mass change, and that (a) was likely more important than (b).[101,102] In other words, observations of 2- and 3-bond isotope shifts are by themselves experimental indicators of significant change in shielding upon stretch of a bond 2 or 3 bonds away from the observed NMR nucleus in the molecule. The $^{74-82}Se$-induced ^{77}Se isotope shifts in CSe_2 are well-reproduced by using all first and second derivatives.[93]

The case for centrally located nuclei is more ambiguous. We have already mentioned the various contributions in the cases of ^{17}O, ^{15}N, and ^{31}P in H_2O, NH_3, and PH_3 molecules. For the ^{13}C shielding in HC≡CH, the dominant nuclear motion contribution comes from the bending at "the other" carbon atom with the combined stretching contributions being only 20% of those from bending.[92] The relative importance of first and second derivatives of ^{13}C shielding to the zero-point vibrational contributions and isotope shifts in substituted methanes CF_nH_{4-n} has been investigated by Bour *et al.* but this is probably not the last word on these systems since they are unable to reproduce the experimental zero density ^{13}C shifts relative to CH_4.[111] The results for ^{13}C in CSe_2 are likewise ambiguous, so that the Se mass effects on the ^{13}C spectrum are not as well reproduced as those in the ^{77}Se spectrum.[93] For ^{129}Xe in XeF_2, there is a practically negligible (less than 0.1%), slightly negative ZPV correction to σ^{Xe}, that slightly increases only up to a few ppm due to finite temperature contributions. This is partly due to the almost full cancellation of the non-relativistic and relativistic contributions to $\langle\sigma^{Xe}{}_{r}\rangle^{T}$ (the first-order term in the rovibrational contributions arising from the stretch of the bond to the nucleus in question) that results in almost constant vibrational contribution in the whole temperature range. In addition, heavy cancellation takes place between the second-order terms $\langle\sigma^{Xe}{}_{rr}\rangle^{T}$ and $\langle\sigma^{Xe}{}_{rr'}\rangle^{T}$, thus the $\langle\sigma^{Xe}{}_{\theta\theta}\rangle^{T}$ term is mostly responsible for the Xe temperature dependence (the indices correspond to the contributions to the vibrational corrections from first and second shielding derivatives).[98]

It has been found that it is more efficient to solve directly for the vibrational wavefunctions from quantum-mechanically calculated potential surfaces. This method has been applied to finding zero-point vibrational corrections to the shielding and other properties in HF, H_2O, NH_3, and CH_4,[112] ^{19}F in 24 fluorohydrocarbons,[113] and also applied to all nuclei in H_2, HF, and H_2O.[114]

1.3.4 Absolute Shielding Scales and Comparison with State-of-the-Art Quantum Calculations

For molecules with larger numbers of atoms, DFT is still the most efficient method of introducing electron correlation, but the absolute errors in DFT calculations cannot be revealed by comparing against experimental chemical shifts because of cancellation of computational errors in taking shielding differences, especially between similar types of molecules. Comparing different functionals against one another in this way actually may lead to wrong conclusions about the usefulness of particular functionals for shielding estimates that may be applied to structural assignments of NMR spectra in condensed phase; we have seen already the quantitative measure of intermolecular effects on shielding that only gas phase experiments reveal. Thus, the practice of comparing a set of calculated shieldings against the chemical shifts measured for the same set of molecules relative to some standard reference does not provide a true test of the quality of the theoretical results. For comparisons of very accurate *ab initio* calculations with experiment (thermal average for the isolated molecule at 300 K), it is necessary to include the zero-point vibrational corrections to obtain $\langle\sigma\rangle^{0K}$ and the additional thermal corrections to obtain $\langle\sigma\rangle^{300K}$. Then one will have the absolute shielding for the nucleus in that specific isolated molecule. These corrections have been carried out by Gauss *et al.* for several nuclear sites in a variety of molecules (molecules with lighter atoms where the relativistic corrections are expected to be small), so as to test the quality of various levels of theory, including DFT, using a variety of functionals, against the gold standard, CCSD(T), and against experiments in the gas phase extrapolated to zero density at 300 K.[115] By doing so, the deficiencies of the DFT methods for shielding calculations have been revealed. In future, incremental improvements in exchange-correlation functionals can be tested by using these large numbers of molecular systems for which the theoretical absolute shieldings $\langle\sigma\rangle^{300K}$ have been ultimately checked against experimental values of $\langle\sigma\rangle^{300K}$. Experimental absolute shieldings for small molecules are obtained by measurements of chemical shifts in gas phase experiments extrapolated to the isolated molecule limit, coupled with the determination of $\langle\sigma\rangle^{300K}$ in at least one standard reference molecule containing the nucleus of interest, derived from high-resolution microwave measurements of the spin-rotation constant. The method of determining from an experimental spin-rotation constant measured for a particular rovibrational state, the paramagnetic part of the shielding at the equilibrium geometry and adding to this the calculated diamagnetic term at the equilibrium geometry and then making the rovibrational corrections to obtain $\langle\sigma\rangle^{300K}$ for that standard reference molecule has been in use for some time.[30] It continues to be used in current work, on ^{17}O, for example.[88,90] Thus, $\langle\sigma\rangle^{300K}$ data are available for sets of molecules for ^{1}H,[116] for ^{13}C,[117] for ^{15}N,[118] for ^{17}O,[119,120] for ^{19}F,[30,31] for ^{29}Si,[121,122] for ^{31}P,[123] for ^{33}S,[124,125] for ^{77}Se and ^{125}Te.[126] Since accurate chemical shifts between molecules all extrapolated to the zero density limit have been reported for these cases, in future, more accurate values of spin-rotation constants for the standard reference can always be used to improve the reported absolute $\langle\sigma\rangle^{300K}$ values. For example, more accurate values for the standard ^{17}O reference H_2O,[90] ^{19}F reference HF,[83] ^{31}P reference PH_3,[127] ^{33}S reference H_2S and SO_2,[91] have become available. Gauss *et al.* have used experimental values for $\langle\sigma\rangle^{300K}$, a set of values that have been measured in the gas phase in the isolated molecule limit based on the spin-rotation constant of one specific standard molecule among the set, to compare accurately calculated shieldings at the CCSD(T) level and large basis sets and including rovibrational corrections for ^{13}C,[128] for ^{17}O,[129] for ^{19}F,[130] and for ^{15}N and ^{31}P,[131] and finally for benchmarking theoretical calculations against CCSD(T) for these nuclei and also ^{7}Li, ^{27}Al, and ^{33}S in selected molecules.[115]

As reviewed earlier[132] and in Chapter 3 by Jackowski and Garbacz in this book, it is also possible to measure absolute shieldings directly without using a spin-rotation standard reference by using the ratio of resonance frequencies for two nuclei in the same sample, and ultimately a suggestion to use a single standard reference for all nuclei, ^{3}He in the He atom.[133] The proposed new method of shielding measurements neither removes nor solves the problem of bulk susceptibility correction when the helium sample is used as the external standard. On the other hand, any NMR experiment performed for a gaseous compound with the extrapolation of results to the zero-density at 300 K gives immediately the exact value of the shielding constant when the reference used is ^{3}He. This latter method has been applied to the determination of the absolute shieldings $\langle\sigma\rangle^{300K}$ for ^{1}H in several molecules,[116] and for $^{35/37}$Cl in HCl.[134]

1.4 The Spin–Spin Coupling in the Gas Phase

There are alternative measurements of J in isolated molecules: hyperfine data obtained from high resolution molecular beam and microwave spectroscopies, in particular, the parameters c_3 and c_4 yield the complete experimental indirect spin–spin coupling tensor for an isolated molecule in the gas phase,[135,136] but here we consider only the results from gas phase NMR experiments. The general approaches used as described above for shielding apply equally well to other molecular electronic properties, in particular the spin–spin coupling J.[66] This NMR quantity does not appear to have the very marked density and temperature dependence in the gas phase that has been found for shielding. Nevertheless, in 1966, Carr *et al.* observed both a small increase in J(HD) in the proton spectrum with increasing temperature at constant density and a small decrease with increasing density at constant temperature.[137] They were able to account for the temperature dependence of J(HD) in terms of the centrifugal stretching of the HD bond when rotational states are excited as the temperature increased. Theoretical calculations of J as a function of geometry are more difficult than those for shielding since four parts (Fermi-contact, spin-dipolar, orbital diamagnetic, and orbital paramagnetic mechanisms) contribute in the gas phase, each having a different dependence on internal coordinates.

1.4.1 The Density Coefficient of the Spin–Spin Coupling

The first measured second virial coefficient of J coupling in a polyatomic molecule was reported in 1971 for the one-bond $J(^{29}Si\text{-}^{19}F)$ in SiF_4 gas.[138] Since then, only a few values of the second virial coefficient J_1 (analogous to σ_1) have been reported. Examples are J_1 for the one-bond couplings $J(^{11}B\text{-}^{19}F)$ in BF_3,[24,139] $J(^{13}C\text{-}^{19}F)$ in CD_3F,[140] in CH_2F_2,[35] and in CHF_3,[34] $J(^{13}C\text{-}^{1}H)$ in CH_2F_2,[35] in CHF_3,[34] in CH_3Br,[46] in CH_3I,[141] and in benzene,[44] $J(^{13}C\text{–}^{13}C)$ in $HC{\equiv}CH$,[45] $J(^{13}C\text{-}^{15}N)$ in CH_3CN,[41] and J(Si-F) in $(SiF_3)_2O$.[37] A few J_1 have been reported for two-bond couplings J(HCF) in CHF_3,[34] and CH_2F_2,[35] J(DCF) in CD_3F,[140] J(HCC) in CH_3CN,[41] and for three-bond coupling, J_1 has been reported for J(HCCN) in CH_3CN.[41] Jackowski has reviewed gas phase studies of spin–spin coupling.[142]

1.4.2 The J Surface and the Effects of Rovibrational Averaging

The earliest (1959) theoretical calculation of a J surface is that for the dependence of the 3-bond coupling constant on the dihedral angle by Karplus,[1] that has turned out to be an extremely useful result that applies to the general coupling path J(X-B-C-Y), with nuclei X and Y throughout the periodic table and any intervening atoms B and C. The universal form of the Karplus equation is:

$$J(\phi) = a\cos(2\phi) + b\cos\phi + c \tag{1.6}$$

with the parameters a, b, c depending on the four atoms in the coupling path. The original work used various rotated ethanic fragments and the valence-bond method, and assumed that the Fermi-contact (FC) mechanism dominated the coupling. It is worth noting that the Karplus equation has remained valid after more than five decades. When the four parts of the coupling rather than just the Fermi contact have been calculated for rotated ethanes,[143] the common assumption that the Fermi-contact term is totally dominant has been confirmed. The derivatives of the orbital paramagnetic and orbital diamagnetic terms are significant but opposite in sign for this case. It is found that the coefficients in the Karplus equation, when electron correlation is included (*e.g.*, using SOPPA(CCSD) or MCSCF), are in good agreement with coefficients derived from experimental coupling constant data. It is further observed that extending the Fourier series in the Karplus equation to include $\cos(3\phi)$ and $\cos(4\phi)$ terms neither significantly improves the quality of the fit nor significantly changes the values of the other coefficients.

The reduced coupling constant $K(\mathrm{AB}) = J(\mathrm{AB})\ 4\pi^2[h\gamma_A\gamma_B]^{-1}$, the nuclear moment-independent electronic part of the J coupling, is usually considered instead of J itself, so as to be able to compare the sign of the coupling between nuclear pairs throughout the periodic table,[144,145] to compare relative sensitivities to bond displacements from one pair of nuclear sites to another in a variety of molecular systems, or to discuss isotope effects resulting from vibrational averaging.[102,146] The dependence of the one-bond spin–spin coupling on bond length is manifest in observed isotope effects on K, for example $K(\mathrm{DF}) \neq K(\mathrm{HF})$ for hydrogen fluoride molecule. This is called a primary isotope effect $= |K(\mathrm{DF})| - |K(\mathrm{HF})|$, arising from isotopic substitution of one of the coupled nuclei. Note that this definition involves the difference between the absolute magnitude of the coupling constant for the heavy minus the light isotopomer. Secondary isotope effects are defined similarly, except that they arise from isotopic substitution of other than the coupled nuclei in the molecule. Occasionally, the magnitudes of secondary isotope effects on K can be larger than the primary isotope effects. This, too, is a manifestation of the K mathematical surface in terms of internal coordinates of the molecule.

A simple example system is of course a diatomic molecule, of which the simplest is the HD molecule. The temperature dependence of J(HD) had been measured from proton resonance studies in the gas phase over a 250-degree temperature range by Beckett and Carr.[147] By fitting these data to the thermal average of eqn (1.5), Raynes and Panteli obtained the first derivative of J with respect to bond extension by neglecting the smaller temperature dependence of the mean square displacement.[148] For the isotopomers of HD, Raynes *et al.* found that the term in the first derivative is mostly responsible for the primary isotope effect on the coupling, with the opposite signed term in the second derivative making a small contribution.[149]

Although there were some earlier calculations at various levels with and without electron correlation, for the HD molecule the highest level of theory, full configuration interaction calculations in the basis set limit, can actually be used. With FCI, an equilibrium value of $J_e(\mathrm{HD}) = 41.22$ Hz is obtained in the basis set limit.[150] Adding a calculated zero-point vibrational correction of 1.89 Hz and a temperature correction of 0.20 Hz at 300 K leads to a total calculated spin–spin coupling constant $\langle J_0(\mathrm{HD})\rangle_{\mathrm{FCI}}{}^{300\mathrm{K}} = 43.31(5)$ Hz, which is within the error bars of the experimental gas phase NMR value, $\langle J_0(\mathrm{HD})\rangle_{\mathrm{EXPT}}{}^{300\mathrm{K}} = 43.26(6)$ Hz, obtained by extrapolating values measured in HD–He mixtures to zero density. These results are the ultimate as far as experiments and theoretical calculations are concerned, but are not very different from the earlier calculations by Oddershede *et al.* using SOPPA(CCSD) at 11 R values, that led to a vibrational correction of 1.81 Hz.[151] or the MCSCF calculations that led to a thermally averaged value of 43.15 Hz.[152]

For the HF molecule, CASSCF calculations provide a ZPV correction of 25 Hz and the thermal average $\langle J_0(\mathrm{HF})\rangle^{300\mathrm{K}} = 510(10)$ Hz[153] that compares well to the molecular beam results $\langle J(\mathrm{HF})\rangle_{\nu=0} = 500(20)$ Hz.[154] The value at the equilibrium geometry, J_e(HF), and also the bond length dependence are dominated by the Fermi-contact term. On the other hand, in N_2 and CO, the nuclear spin–spin coupling bond length dependence is sharp and largely due to the Fermi-contact term. While the other mechanisms do not show any appreciable geometry dependence, the FC term even changes sign near the equilibrium geometry. This feature, a sharply varying FC term with an inflection point close to equilibrium, gives an explanation for the apparently great importance of non-contact mechanisms for CO and N_2 molecules and possibly also for other multiply bonded systems.[155]

Raynes *et al.* have comprehensively investigated the rovibrational effects on the J couplings in CH_4, using SOPPA(CCSD)[156–158] and MCSCF,[159] to obtain the coupling surfaces in terms of the symmetry coordinates, to obtain vibrational averages of both J(CH) and J(HCH), and to calculate the isotope effects at various temperatures for these spin–spin couplings.[158] They have carried out similarly comprehensive studies for SiH_4.[159,160] There is parallel behavior between respective surfaces of SiH_4 and CH_4 in the reduced couplings. It is not surprising that K values for $^{29}SiH_4$ are generally twice those for $^{13}CH_4$. This is consistent with the observation that the one-bond K(XF) for X in analogous compounds exhibits the same periodicity across the periodic table when plotted against atomic number as does the $|\Psi_{ns}(r=0)|^2$ for the X atom.[161] Furthermore, bending is also relatively more important in SiH_4 than in CH_4. In the total nuclear motion effects, first-order stretching is dominant but there are significant contributions also from the second-order terms in SiH_4.

Raynes *et al.* have also investigated the J surfaces of H_2O[162] and HC≡CH,[163] at the SOPPA(CCSD) level, calculating all nuclear motion effects. Other calculations on the water molecule surfaces, using MCSCF,[164] find only small differences when compared to the SOPPA(CCSD). All the ZPV corrections discussed above were calculated assuming small-amplitude nuclear motions. For large-amplitude nuclear motions, other approaches, such as statistical averaging over conformational isomers (for 3-bond coupling across a dihedral angle, for example) and molecular dynamics, are needed for a meaningful comparison with experimental measurements. For NH_3, it is found that the umbrella inversion mode has significant contributions to the dynamic averaging of the coupling constants,[165] just as had been found for the dynamic averaging of ^{15}N shielding in this molecule.[85]

Stanton *et al.* have carried out benchmark calculations of all J surfaces (for all one-, two-, and three-bond couplings) in HC≡CH, $H_2C{=}CH_2$, CH_3CH_3, and cyclopropane, including all four mechanisms for each, and using coupled cluster theory to CCSD level with large uncontracted basis sets for accurate description of the Fermi-contact contribution that dominates in these molecules.[166] They also carried out vibrational averaging to second order in the normal coordinates and evaluated the zero-point vibrational average using CCSD for the anharmonic potential surface. They found that two-bond CCH couplings are the most affected by vibrational averaging. The relative importance of first and second derivatives of one- and two-bond J couplings to the zero-point vibrational contributions and isotope effects in substituted methanes CF_nH_{4-n} have been investigated by Bour *et al.*[111] The uniformly positive values of the first derivatives of the one-bond J(CH) with respect to the CH bond stretch in CH_4 and $H_2C{=}CH_2$ likewise accounts for the generally negative CH primary isotope effect in these molecules, just as found in alkyl sites in general.[146]

The performance of various exchange-correlation functionals for the Fermi-contact contribution to J varies from one molecular system to another in the same series,[167] therefore DFT would not be a method of choice for J calculations. Nevertheless, DFT calculations of the dependence of one-bond and two-bond J couplings on normal coordinates have been carried out,[168] particularly calculations of one-bond coupling including relativistic corrections, for example in the dependence of $J(^{31}P\text{–}^{31}P)$ in $H_2P\text{-}PH_2$ and $H_2P\text{-}PF_2$ molecules on the dihedral angle between the bisectors of the two ∠HPH, or the ∠HPH and the ∠FPF.[169] Since biphosphines are not fixed in a particular conformation, observations of $J(^{31}P\text{–}^{31}P)_{iso}$ represent conformationally averaged values. It has been suggested that one practical strategy for J calculations would be for J at the equilibrium geometry to be calculated accurately using coupled cluster theory, then use DFT to calculate the vibrational corrections.[170]

A study of the general trends in primary and secondary isotope effects on spin–spin coupling in small molecules can reveal interesting information about general trends in spin coupling surfaces.[102,146] Primary isotope effects on reduced one-bond coupling, *e.g.*, $|K(D_2)| - |K(HD)|$ is negative for hydrogen molecule, are generally negative for CH in alkyl sites and SiH in silyl sites, negative for SnH in $SnH_{4-n}D_n$, for SnH in $[SnH_{3-n}D_n]^+$, for PH in $[PH_{4-n}D_n]^+$, for $P^{(V)}$H in $H_2P(O)OH$ and for other similar P sites.[102] In these molecules the main electronic factor that is responsible for the primary isotope effect on K is the first derivative $(\partial K/\partial r)_e$. Since the mean bond length in the heavy isotopomer is shorter than that in the light isotopomer, the negative isotope effects on K in these cases correspond to positive $(\partial K/\partial r)_e$. In other words, the reduced spin–spin coupling increases with increasing bond length in all these systems. On the other hand, the primary isotope effect is positive for HF, and positive for NH in $NH_{3-n}D_n$, for $P^{(III)}$H in $PH_{3-n}D_n$, for $[PH_{2-n}D_n]^-$, for SeH in $SeH_{2-n}D_n$ and for SnH in $[SnH_{3-n}D_n]^-$. The positive isotope effects on K in these cases correspond to negative $(\partial K/\partial r)_e$ where one of the coupled nuclei is an atom with one or more lone pairs. The lone pair on A is known to be responsible for negative contributions to the reduced coupling K(AH); apparently it is also the lone pair which is responsible for the greater sensitivity of the reduced coupling to bond extension and for the negative sign of $(\partial K/\partial r)_e$. See Table 3 in ref. 102 for contrasting examples with and without lone pairs and ref. 146 for the rationalization of these trends in terms of the Fermi-contact term.

1.5 Spin–Lattice Relaxation in the Gas Phase. Cross-sections for Angular Momentum Transfer and Molecular Reorientation in the Binary Collision Limit

Spin lattice relaxation in the gas phase provides qualitatively different information from that in condensed fluid phases in many ways; different relaxation mechanisms dominate and interpretation of the experimental relaxation times provide different types of information about the system. Since the intermolecular dynamics in the gas phase can be modeled more accurately (using well-established mathematical theory of non-uniform gases) than in the liquid phase, the gas phase provides critical tests of relaxation theories, permits quantitative separation of two or more contributing mechanisms, and provides a direct connection to intermolecular potential functions.[3] Collisions that do not reorient the molecule containing the nuclear spins make no contribution to relaxation of spin magnetizations. This is the reason that T_1 measurements inherently provide a valuable source of information on the anisotropy of the intermolecular interaction or serve as a sensitive test of anisotropy of proposed *ab initio* or semi-empirical potential energy surfaces.

As a function of gas density, ρ, T_1 is long at very low densities for which the collision frequency is very low (where T_1 is inversely proportional to the gas density, the reciprocal density regime), passes through a characteristic minimum corresponding to a matching between the spin-precession frequency and the collision frequency, then passes into a regime in which T_1 increases linearly with gas density. Early gas phase studies, particularly in the vicinity of the T_1 minimum, are reviewed in ref. 171 and 172. For the purposes of determining classical cross-sections in a pure gas, the regime of densities that is appropriate to study is that for which T_1 is proportional to the density of the gas, sometimes called the "extreme narrowing limit" (see also ref. 3). In the following sections, we restrict our discussion to this linear-density regime. For a discussion of the lower density regions where different relaxation rates for different nuclear spin symmetry species may be expected and systems where a quantum scattering treatment is required, see Armstrong's review in ref. 172.

1.5.1 Temperature-dependent Classical Cross-Sections from Gas Phase Studies

In 1966, Roy Gordon developed a kinetic theory for nuclear spin relaxation in dilute gases and mixtures of gases,[173] employing classical mechanics for the molecular translational and rotational motion, assuming binary collisions, and no correlation between the effects of successive binary collisions, resulting in expressions relating the measured spin relaxation times in the linear-density regime to two cross-sections that could be calculated for any intermolecular potential having angle-dependent terms. The spin relaxation times in a dilute gas are found to depend only on the net changes produced by collisions in the molecular rotational angular momentum vector but not on the details of the trajectory during a collision. This permits the description of relaxation in the dilute gas in terms of cross-sections. For the quadrupolar relaxation mechanism,

$$T_{1Q} = \frac{160I^2(2I-1)}{3(2I+3)}\left(\frac{\hbar}{eqQ}\right)^2 \rho\bar{v}\sigma_{\theta,2} \tag{1.7}$$

$eqQ/\hbar$ is the nuclear quadrupole coupling constant, ρ is the number density of the collision partner measured in amagat (2.687×10^{25} molecules m^{-3}), and $\bar{v}$ is the mean relative speed that is given by $(8k_BT/\pi\mu)^{1/2}$ with μ the reduced mass of the colliding pair. The electronic coupling affects the populations of the nuclear magnetic spin states of a nucleus with $I>1/2$ since the magnetic moment of the nucleus is directed along the axis of the nuclear electric charge distribution of this nucleus. The subscript in the cross-section signifies the connection to the $P_2(\cos\theta)$ autocorrelation function, where θ is the angle between the molecular rotational angular momentum vector of A before and after a collision with B.

$$\int_0^\infty \langle P_2[\mathbf{u}(0)\cdot\mathbf{u}(t)]\rangle \mathrm{d}t = \frac{1}{4}\left[\rho\bar{v}\sigma_{\theta,2}\right]^{-1} \tag{1.8}$$

The same cross-section $\sigma_{\theta,2}$ is involved in the dipole–dipole (DD) relaxation mechanism (for like spins),[173]

$$T_{1DD} = \frac{2}{\gamma^4\hbar^2 I(I+1)\langle r^{-3}\rangle^2}\rho\bar{v}\sigma_{\theta,2} \tag{1.9}$$

(For the relaxation of the I spin by dipolar coupling to the unlike I' spin, we replace γ^4 by $(\gamma'\gamma)^2$ and 2 by 3.)

Nuclear spin relaxation can also be affected by the molecular rotation when a magnetic coupling exists between the nuclear magnetic moment and the magnetic moment associated with the molecular rotation. The spin-rotation (SR) mechanism is important for nuclei with a spherical charge distribution (spin $I=1/2$). The spin-rotation relaxation mechanism involves a different cross-section, σ_J.[173]

$$T_{1,\mathrm{SR}} = \frac{3}{2C_{\mathrm{eff}}^2\langle J(J+1)\rangle}\rho\bar{v}\sigma_J \tag{1.10}$$

These relations can be applied when the following assumptions hold: (a) The Larmor frequency is small compared with the collision frequency. (b) The duration of a collision is short compared with the average time between collisions. (c) The interactions among the collision partners do not significantly influence their collisions with the observed molecule. (d) Bound states between the observed molecule and the collision partner have no significant effect on the spin relaxation.

These two relaxation cross-sections are among the 45 collision cross-sections that can be defined in a general formalism and calculated from a known or proposed intermolecular potential function. In order to facilitate a comparison among related information that can be obtained from NMR, depolarized Rayleigh light scattering, microwave non-resonant absorption, transport properties, transport coefficients in the presence of electric or magnetic fields, it is necessary to have an unambiguous yet physically meaningful definition of a collision cross-section for the process being studied. For example, the cross-sections obtained by studying the pressure broadening of the depolarized Rayleigh light scattering spectrum turns out to be the same cross-section as that which can be obtained from NMR relaxation by the quadrupolar mechanism. McCourt and co-workers provide a detailed derivation of the collision cross-sections related to the transport and relaxation properties that are currently used in testing non-reactive potential surfaces.[174,175] From the Boltzmann equation, the Chapman–Enskog procedure[176] can be used to obtain classical definitions of the kinetic theory cross-sections. The nomenclature used to label the collision cross-sections specifies the nature of the collisional process that contributes to the phenomenon. In general, a collision or effective cross-section is written in terms of indices that represent the pre-collisional and post-collisional tensorial ranks or powers of the microscopic polarizations that are coupled and also label which collision partner the polarizations belong to. When the pre- and post-collisional values are identical and changes in only one partner are relevant (in NMR relaxation we observe only one of the collision partners at a time), the cross-section can be abbreviated; $\mathfrak{S}(pqst|\mathrm{A})_{\mathrm{AB}}$, for example, is the cross-section for molecule A in the collision of A with B. The index p denotes the p-fold tensor product of the reduced peculiar velocity $\mathbf{W} = (m/2k_\mathrm{B}T)^{1/2}\mathbf{v}$ of molecule A, q denotes the tensorial rank in the molecular angular momentum $\boldsymbol{J}$ for molecule A. The s and t indices denote the scalar dependencies of the cross-section on the translational and reduced rotational energy of molecule A. For example, the cross-section for diffusion is $\mathfrak{S}(1000|\mathrm{A})_{\mathrm{AB}}$ and the shear viscosity cross-section is $\mathfrak{S}(2000|\mathrm{A})_{\mathrm{AB}}$. For spin–lattice relaxation T_1, $p = 0$, $s = 0$, $t = 0$ since the relaxation cross-sections have a dependence only on the molecular rotational angular momentum $\boldsymbol{J}$. Liu and McCourt demonstrated the connection between the reorientation collision cross-section expressions arising in NMR relaxation in the gas phase when described from the point of view of kinetic theory and from the point of view of traditional correlation function theory.[177] Thus, the following are identified:

$$\mathfrak{S}(0100|\mathrm{A})_{\mathrm{AB}} \equiv \sigma_\mathrm{J} \quad \text{and} \quad \mathfrak{S}'(0\hat{2}00|\mathrm{A})_{\mathrm{AB}} \equiv \sigma_{\theta,2} \tag{1.11}$$

Intramolecular dipole–dipole, chemical shift anisotropy, and electric quadrupolar relaxation rates obtained in the gas phase all provide the cross-section $\mathfrak{S}'(0\hat{2}00|\mathrm{A})_{\mathrm{AB}}$. The hat symbol over the tensorial rank 2 indicates the use of normalized angular momentum, just as in Gordon's derivation. The prime means this is the "self-only" part which has no dependence upon the collision partner except as introduced *via* the intermolecular potential and *via* the number density of the collision partner. Molecular reorientation in classical language corresponds to a change in the quantum number M_J without a change in the quantum number J. When the spacing between J levels is large (as in H_2 molecule) quantum scattering is the appropriate description of the collision events, but the classical limit for rotation is easily satisfied for most molecular systems. Spin-rotation relaxation rates in the gas phase provide the cross-section for changes in the molecular rotational angular momentum quantum number J of the observed molecule A upon collisions (*i.e.*, in classical terms, a change in the molecule's rotational energy, *i.e.*, molecular rotational energy transfer). Thus, the cross-section $\mathfrak{S}(0100|\mathrm{A})_{\mathrm{AB}}$ is also known as σ_J. Thus, eqn (1.10) can also be written as

$$T_1 = \frac{3}{2C_{\mathrm{eff}}^2\langle J(J+1)\rangle}\rho\bar{v}\mathfrak{S}(0100|\mathrm{A})_{\mathrm{AB}} \tag{1.12}$$

For a linear molecule C_{eff} is the perpendicular component of the spin-rotation tensor,[173]

$$C_{\mathrm{eff}}^2 = C_\perp^2 \quad \text{and} \quad \langle J(J+1)\rangle = \frac{2I_0}{k_B T} \tag{1.13}$$

For a nucleus, say ^{19}F or ^{1}H, in a spherical top such as CF_4, CH_4, SiF_4, SF_6, SeF_6, TeF_6,[178]

$$C_{\mathrm{eff}}^2 = \left[\frac{1}{3}\left(C_\parallel + 2C_\perp\right)\right]^2 + \frac{4}{45}\left[C_\parallel - C_\perp\right]^2 \quad \text{and} \quad \langle J(J+1)\rangle = \frac{3I_0}{k_B T} \tag{1.14}$$

By using more than one isotope it is possible to determine both cross-sections. For example, the ^{15}N spin in the $^{15}N_2$ molecule in the gas phase is completely dominated by the spin-rotation mechanism, so the measurements of T_1 as a function of temperature for $^{15}N_2$ in a mixture of $^{15}N_2$ and Kr can provide the cross-section $\mathfrak{S}(0100|N_2)_{\mathrm{N2\text{-}Kr}}$. On the other hand, the ^{14}N spin in the $^{14}N_2$ molecule in the gas phase is completely dominated by the nuclear quadrupolar mechanism, so the measurements of T_1 for $^{14}N_2$ in a mixture of $^{14}N_2$ and Kr can provide the cross-section $\mathfrak{S}'(0200|N_2)_{\mathrm{N2\text{-}Kr}}$ as a function of temperature.

When two or more relaxation mechanisms contribute to T_1, it is necessary to separately determine the individual relaxation rates and analyze them individually so as to obtain the relaxation cross-sections that may be compared with those from classical trajectory calculations. In what follows, we will consider each relaxation mechanism in turn. Fortunately, there are many experimental examples where one particular mechanism dominates over all others in the range of gas densities and temperatures studied. Thus, we may characterize the temperature and density dependence of the relaxation rate arising from that specific mechanism with the same precision as the original data.

1.5.2 Spin-rotation Mechanism

We restrict our discussion to the density regime in which T_1 is proportional to the density of the gas. For several pure gases in which relaxation has been found to be dominated by the spin-rotation mechanism, (T_1/ρ) is found experimentally to be consistent with the power law

$$(T_1/\rho)_{\mathrm{lin}} \propto T^n \tag{1.15}$$

where n is negative; that may be written as

$$\left(\frac{T_1}{\rho}\right)_{\mathrm{lin},T} = \left(\frac{T_1}{\rho}\right)_{\mathrm{lin},300\,\mathrm{K}} \left(\frac{T}{300}\right)^n \tag{1.16}$$

Thus, all measured T_1 values in the pure gas can be characterized by two quantities, $(T_1/\rho)_{\mathrm{lin}}$ at some reference temperature, say 300 K, and the power n; *i.e.*, these two numbers will reproduce the results of every experimental T_1 measurement in the linear-density regime at any temperature in the range of temperatures for which n was fitted. In Table 1.1 we provide these quantities in eqn (1.16) for various pure gases.[179,180–186] In addition ^{19}F spin-rotation relaxation in WF_6, MoF_6, and UF_6 gases have been studied.[187] It is said that the signature of the spin-rotation mechanism is that T_1 has a temperature dependence close to $T^{-1.5}$. We can see in Table 1.1 for the pure gases that this is indeed the case. In eqn (1.10), we note that the explicit T^{-1} arising from the average rotational angular momentum square gives the spin-rotation relaxation rate a steep temperature dependence, while $\bar{v} = (8k_BT/\pi\mu)^{1/2}$ appears in expressions for all T_1 mechanisms in the linear density regime, and the cross-section σ_J itself has a temperature dependence with a power close to -1.

Table 1.1 Characteristics of spin-rotation relaxation for various nuclei in linear and spherical top molecules.

Nucleus	Pure gas	$(T_1/\rho)_{\mathrm{lin,300\ K}}$ ms amagat^{-1}	n	T range, K	Ref.
^{13}C	CO	1.231(30)	−1.32(3)	230–420	179
^{13}C	CO_2	21.6(5)	−1.51(5)	290–400	180
^{13}C	CH_4	10.2(5)	−1.43(2)	230–400	181
^{15}N	$^{15}N_2$	2.23(6)	−1.20(3)	215–400	182
^{15}N	^{15}N^{15}NO end	100.5(19)	−1.417(14)	240–400	183
^{15}N	^{15}N^{15}NO cen	54.0(11)	−1.417(14)	240–400	183
^{19}F	SF_6	2.132(23)	−1.75(2)	290–400	184
^{19}F	SeF_6	3.21(7)	−1.97(5)	310–400	185
^{19}F	TeF_6	10.03(10)	−1.82(2)	310–400	185
^{19}F	CF_4	1.948(39)	−1.41(2)	210–400	186
^{1}H	CH_4	20.2(4)	−1.40(3)	230–400	181

In a mixture of gases A and B, the spin-rotation relaxation times of a nucleus in molecule A in the extreme narrowing limit are additive.[173]

$$T_1(A) = \left(\frac{T_1}{\rho}\right)_{\text{lin},A-A} \rho_A + \left(\frac{T_1}{\rho}\right)_{\text{lin},A-B} \rho_B \tag{1.17}$$

This additivity of course depends on the neglect of correlations between the effects of successive collisions, as well as on the assumption of binary collisions.

The values characterizing the relaxation of the spin in the molecule (listed in columns 3–4 of Table 1.1) infinitely dilute in a buffer gas (including the following gases: Ar, Kr, Xe, N_2, CO, CO_2, HCl, CH_4, CF_4, SF_6) are also provided in the same references as given in Table 1.1. The cross-section for rotational angular momentum transfer for an observed target molecule by a collision partner is uniquely determined by the details of the anisotropy of the intermolecular potential. Nevertheless, there are general trends in the observations at room temperature across the large number of collision pairs included in ref. 179–186. A physically intuitive simple model provides a comparison between the cross-sections for this wide range of buffer gases and proposes that the efficiency for rotational angular momentum transfer for a target molecule upon collision with various molecules may be thought of as a product of three factors: the anisotropy of the shape of the target molecule, the electronic factors that depend largely on electric polarizabilities and electronic moments of the target and projectile molecules, and a kinematic factor involving molecular diameters, moments of inertia and reduced mass.[188]

1.5.3 Quadrupolar Mechanism

When the nucleus has spin $I > 1/2$, then the quadrupolar mechanism may dominate the relaxation. We consider here the case when the nucleus is in a molecule so that the electric field gradient at the nucleus is an intrinsic molecular electronic property. (In Section 1.5.7 we will consider a transient quadrupole coupling arising from the binary collision itself, an electric field gradient arising from the intermolecular interaction as in the case of ^{131}Xe or ^{83}Kr in the rare gas.) The ^{14}N in $^{14}N_2$ and the end ^{14}N in NNO have been found to relax nearly entirely by the quadrupolar mechanism,[189,190] which permits the characterization of quadrupolar relaxation cross-sections $\mathfrak{S}'(0\hat{2}00|\mathrm{A})_{\mathrm{AB}}$ or $\sigma_{\theta,2}$ in the gas phase. The ^{17}O relaxation has been studied and is likewise dominated by the quadrupolar mechanism.[191] Just as for spin-rotation relaxation, we find experimentally that the temperature dependence for quadrupolar relaxation can be described in the form of a power law. Table 1.2 shows the examples for ^{14}N; the quantities are the ones analogous to the quantities in eqn (1.16).

Once again, in a mixture of gases A and B, the quadrupolar relaxation times of a nucleus in molecule A in the extreme narrowing limit are additive,[173] just as in eqn (1.17), so that $(T_1/\rho)_{\mathrm{lin,A-B}}$ provides the cross-sections

Table 1.2 Characteristics of quadrupolar relaxation for ^{14}N nuclei in linear molecules.

Nucleus	Pure gas	$(T_1/\rho)_{lin}$ at 300 K ms amagat^{-1}	n	T range, K	Ref.
^{14}N	$^{14}N_2$	0.050(3)	−0.17(2)	225–405	189
^{14}N	$^{14}N^{14}NO$ (end)	4.92(12)	−0.35(4)	265–400	190

for molecule A in collision with B. The cross-section $\sigma_{\theta,2}$ ranges from 29.6(9) Å^2 to 73(2) Å^2 and has a temperature dependence close to −1, ranging from −0.63(4) to −0.91(6) for $^{14}N_2$ with the 10 different collision partners.[189] From the end ^{14}N nucleus in NNO, we obtain the cross-section that ranges from 43.0(19) Å^2 to 99.4(28) Å^2 and has a temperature dependence close to −1, ranging from −0.66(6) to −0.95(4), for NNO with 10 different collision partners.[190] Of course, the cross-section is a property of the molecule in a collision pair and does not depend on which of its nuclei has been used for the relaxation measurement. In the case of NNO, the end nitrogen provides a more precise determination of the cross-section since the middle nitrogen has a smaller quadrupole coupling constant and therefore the quadrupolar relaxation of the center ^{14}N does not dominate the relaxation rate.

1.5.4 Intramolecular Dipole–Dipole Mechanism

The intramolecular dipolar mechanism is a very significant relaxation mechanism in the liquid phase, but is not so important in the gas phase. Even for ^1H in CH_4 gas, with a short C–H bond, the intramolecular dipolar mechanism is only a very minor contributor; the ^1H T_1 has a temperature dependence of $T^{-1.40(3)}$, typical of spin-rotation relaxation, whereas the intramolecular dipolar mechanism is expected to behave roughly as $T^{-0.5}$. Although the intramolecular dipolar relaxation rate may become important at very low temperatures, at temperatures close to room temperature it contributes very little. Therefore, when attempting to determine cross-sections for molecular reorientation, $\sigma_{\theta,2}$, it is better to use a quadrupolar nucleus and obtain the cross-section from its T_1. The intramolecular dipolar mechanism is important for H_2 molecule, but H_2 relaxation cannot be treated classically (see Section 1.5.10); it is also important for ^1H relaxation in HCl molecule with Ar, which has to be treated at least semi-classically.[192]

1.5.5 Chemical Shift Anisotropy Mechanism

The relaxation rate for the chemical shift anisotropy (CSA) mechanism is proportional to the square of the magnetic field strength and the chemical shift anisotropy. In an axial case,

$$T_{1,\text{CSA}} = \frac{15}{2\gamma^2 B_0^2 (\sigma_{\parallel} - \sigma_{\perp})^2} \rho \bar{v} \sigma_{\theta,2} \tag{1.18}$$

This mechanism could become dominant in nuclei with large chemical shift ranges, and in bonding situations that produce large shielding anisotropies, and of course at high fields. For ^{77}Se in CSe_2 gas, both the spin-rotation and the chemical shift anisotropy should be significant relaxation mechanisms, similarly for ^{129}Xe in XeF_2 gas, but not for ^{13}C in ^{13}CO where the CSA is not large enough, so spin-rotation mechanism dominates in the latter. The dependence of the CSA relaxation rate on B_0^2 means that multiple field studies will allow its determination even when it is not the dominant mechanism, as described below.

1.5.6 Relaxation Rates Add When Two or More Mechanisms are Operative

When the electric quadrupole coupling constant is not very large, then the quadrupolar mechanism is no longer dominant. Competing spin-rotation and quadrupolar mechanisms have been found for the ^2D in CD_4[193] and for the middle ^{14}N in NNO.[190] In these cases, the relaxation rates add

$$\left(\frac{\rho}{T_1}\right)_{\text{lin}} = \left(\frac{\rho}{T_1}\right)_{\text{lin,SR}} + \left(\frac{\rho}{T_1}\right)_{\text{lin,Q}} \tag{1.19}$$

Other less important mechanisms for these gases are the chemical shift anisotropy, intramolecular dipolar, and intermolecular dipolar mechanisms. For ^2H in CD_4, the spin-rotation mechanism is found to comprise an average of 7% of the total relaxation rate and ranges from 6 to 8% for individual buffer gases.[193] For the middle ^{14}N in NNO the spin-rotation is competitive with the quadrupolar relaxation due to the smaller electric field gradient for the middle N in this molecule, so that any errors in the subtracted SR relaxation rate leave errors in the deduced quadrupolar relaxation rate.[190] We consider other cases of competing relaxation mechanisms in Sections 1.5.8 and 1.5.10.

1.5.7 Intermolecular Dipolar, Quadrupolar, Spin-rotation, and Chemical Shift Anisotropy Mechanism

The primary relaxation mechanisms for rare gas pairs are intermolecular. Whereas relaxation mechanisms we have discussed in the preceding sections depend on intramolecular quantities (r, q, $\Delta\sigma$, or C), in the case of the rare gases the collision pair generates the corresponding intermolecular quantities: the dipole–dipole interaction is between the pair undergoing binary collision. An electric field gradient is induced by the intermolecular interaction during a collision, thereby producing a transient electric quadrupole coupling. The intermolecular shielding in a rare gas pair is an anisotropic tensor. A spin-rotation coupling is generated by an intermolecular pair. For ^3He relaxation the intermolecular dipolar mechanism dominates since the other intermolecular mechanisms depend on electronic quantities that correlate with the electric polarizability of the rare gas atoms which, for ^{3}He atoms, is too small to generate a large enough spin-rotation coupling, or intermolecular anisotropic shielding, to compete favorably with the intermolecular dipolar mechanism. On the other hand, ^{129}Xe has a well-characterized distance-dependent shielding anisotropy and spin-rotation tensor in the Xe–Xe collision pair that makes the spin-rotation and chemical shift anisotropy mechanism significant in pure ^{129}Xe gas. For quadrupolar nuclei ^{83}Kr and ^{131}Xe, the transient electric field gradient created by the collision pair can provide a quadrupolar relaxation mechanism. Since, by their nature, these intermolecular relaxation mechanisms are less effective than intramolecular ones, collisions with the walls can become relatively important, especially at low densities. Thus, for applications that depend on maintaining the hyperpolarization of rare gases (such as ^{3}He, ^{83}Kr, ^{129}Xe) for long times, considerable effort has been expended in preparing surface coatings that render the surface collisions less effective for spin relaxation (such as by eliminating paramagnetic sites from the container surface, which we will consider no further). For those applications, very long relaxation times are desirable since they allow users to polarize the rare gas sample prior to and in a different location than the actual experiment. Excluding surface effects, relaxation rates by any of the intermolecular mechanisms mentioned above should increase linearly with the number density.

The thermally averaged ^{3}He intermolecular dipolar relaxation rate for a pair of colliding fermions such as for ^{3}He in He gas has been derived by Happer *et al.*[194] The expression is the same form as was derived in 1973 by Shizgal[195] and also by Richards *et al.*,[196] although arrived at *via* different routes. Numerical calculations for temperatures from 0.1 to 550 K give a relaxation time increasing with temperature; T_{1DD} is 74.4 h for a He density of 10 amagat at room temperature.[194]

The relaxation rate of ^{129}Xe in xenon gas has been investigated precisely and comprehensively by Moudrakovski *et al.*,[197] under various conditions of density, temperature, and magnetic field strengths. The density dependence of the relaxation rate is linear with density up to 160 amagat, as expected for any intermolecular mechanism. However, it begins to exhibit a different behavior at lower densities, particularly below 20 amagat. The authors attribute this to wall effects beginning to be competitive and ultimately becoming dominant at 3 amagat (more about this low-density regime below). By studying different isotopic compositions (natural abundance and ^{131}Xe-depleted xenon gas) they established experimentally that the scalar relaxation of the second kind arising from the collisions of ^{129}Xe with ^{131}Xe (the latter relaxing *via* an intermolecular quadrupolar mechanism) was not significant. By using different B_0 fields, the authors established that $[T_1\rho]^{-1}$ gives a straight-line plot against $B_0{}^2$ with an intercept that is field-independent. Both the field-dependent part and the field-independent part are found to be directly proportional to density, using only samples that are

40 amagat or greater. The authors arrived at the conclusion that at least two intermolecular mechanisms are very significant: spin-rotation (originally suggested by Torrey in 1963)[198] and chemical shift anisotropy. The spin-rotation mechanism is field-independent and the CSA mechanism goes as B_0^2; they are expected to have very different temperature dependences. The authors were able to characterize each of these relaxation mechanisms separately. They also carried out a theoretical calculation of the spin-rotation relaxation rate $[T_1\rho]^{-1}$ as a function of temperature, which gave good agreement with the experimental temperature dependence of the field-independent part.

The low-density regime in ^{129}Xe relaxation was studied further by Walker *et al.*[199] and by Saam *et al.*[200,201] These authors find that at Xe densities below 14 amagat, not only wall mechanisms are responsible for ^{129}Xe relaxation. They propose that persistent (as opposed to transient) Xe_2 dimers, or van der Waals molecules (constituting about 1–3% of the xenon in 1 amagat of pure xenon gas at room temperature in two different estimates[200,202]), contribute to the relaxation. The Xe_2 dimer is a well-known molecular species, has a well depth of 282 K, deep enough to contain 25 or 26 vibrational levels and many rotational states that provide high-resolution lines in vacuum ultraviolet spectra observed for transitions including vibrational quantum numbers $\nu = 0$ to 9 in the ground electronic state.[203] They investigated this mechanism by means of introducing other buffer gases (He, Ar, or N_2) that provide third-body collisions that can cause the break-up of the Xe_2 dimer. The behavior of the relaxation as a function of concentration of the buffer gas supports this model. In other words, ^{129}Xe spin relaxation in Xe_2 molecules that persist at low densities (fewer collisions with third bodies) is responsible for some of the relaxation previously attributed entirely to wall effects. To calculate the field-dependent CSA relaxation rate over the entire density range for which experimental data are available ($\sim$1 amagat and >20 amagat), Vaara *et al.*[202] used their *ab initio* Xe–Xe shielding function that had given a good account of the shielding second virial quantity $\sigma_1(T)$,[53] and assumed pairwise additivity for the instantaneous Xe clusters in MD simulations. The simulations provide time-correlation functions from which spectral density functions could be obtained and then $(T_1)^{-1}$. Vaara *et al.* find that the relativistic effects on the calculated shielding anisotropy lead to much steeper change with distance at R values below the equilibrium Xe–Xe distance than their non-relativistic counterparts. Good agreement was found with the results of Moudrakovski *et al.* for all densities >20 amagat[197] and Saam *et al.* for $\sim$1 amagat[201] for pure Xe. They did not carry out MD simulations for gases containing third bodies like N_2 or Ar at the intermediate low-density regimes $1 < \rho < 14$ amagat for which experiments also exist.

The intermolecular quadrupolar mechanism in the gas phase was first observed by Brinkmann *et al.* for ^{131}Xe in 1962 at 298 K and 0.76 T magnetic field strength,[204] and much later, also for ^{83}Kr for gas densities 25–156 amagat at 300 K.[205] They find that $[T_1\rho]^{-1}$ was 0.0392 and

0.00213 (s amagat)$^{-1}$ for ^{131}Xe and ^{83}Kr in the pure gases, respectively. The intermolecular relaxation rate can be written as

$$\frac{1}{T_1} = \frac{3(2I+3)}{160I^2(2I-1)}\left(\frac{eqQ}{\hbar}\right)^2 \rho \frac{d^4}{\bar{v}} F\left(\frac{V}{k_B T}\right) \tag{1.20}$$

where d is a characteristic length of the interaction and $F(V/k_BT)\cdot\pi d^2$ is an effective cross-section, where $F(V/k_BT)$ is the collision efficiency that depends on the intermolecular potential function V and temperature, of course. Parallel to the case of ^{129}Xe, the high-density mechanism is the transient electric field gradient created during binary collisions, but later studies including much lower densities by Meersmann *et al.* suggest that in lower densities third-body break-up of stable dimer molecules possessing electric field gradients characteristic of Xe_2 or Kr_2 diatomic molecules is the likely operative mechanism, in addition to wall effects.[206,207] Meersmann *et al.* also confirm in the presence of buffer gases the additivity of $[T_1]^{-1}$,[207]

$$[T_1(A)]^{-1} = \left(\frac{1}{T_1\rho}\right)_{\mathrm{lin},A-A} \rho_A + \left(\frac{1}{T_1\rho}\right)_{\mathrm{lin},A-B} \rho_B + \left(\frac{1}{T_1}\right)_{\mathrm{int}} \tag{1.21}$$

and the additional density-independent intercept (third term in eqn (1.21)) that results from a combination of Kr-surface interactions and the formation of $^{83}Kr_2$ van der Waals dimers.[207,208]

1.5.8 Intermolecular Nuclear Spin Dipole Electron Spin Dipole Mechanism, Spin Relaxation in the Presence of O_2

An instance in which the intermolecular dipolar mechanism could become dominant for a nucleus in a molecule is in the case of a nuclear spin dipole interacting with an electron spin dipole on the collision partner. This is indeed the case when the collision partner is an O_2 or an NO molecule. The theoretical limit for a hard sphere potential at the high translational energy limit, in the zero-magnetic field limit ($\omega = 0$) is known from earlier work,[209,210]

$$\left(\frac{1}{T_1^{DD}}\right)_{\text{theor limit}} = \frac{16}{3} S(S+1)\gamma_1^2\gamma_S^2 \frac{\hbar^2}{d^2}\left(\frac{\pi\mu}{8k_B T}\right)^{1/2} N_S \tag{1.22}$$

where πd^2 is the hard sphere cross-section, as before, $\bar{v}$ is the mean relative speed that is given by $(8k_BT/\pi\mu)^{1/2}$ and $\langle S(S+1)\rangle$ is taken to be a constant of the motion for O_2 molecule. The experimental intermolecular dipole–dipole relaxation rate is analyzed using the following equation:[211]

$$\left(\frac{1}{T_1^{DD}}\right)_{\text{inter}} = \left(\frac{1}{T_1^{DD}}\right)_{\text{theor limit}} \cdot F\left(\frac{V}{k_B T}\right) \cdot \left\{1 - f(T)\omega^{1/2}\right\} \tag{1.23}$$

where the magnetic field dependence appears as the low-frequency limiting form that applies when the nuclear-spin bearing molecule suffers several collisions during one Larmor precession of the nucleus, and $F(V/k_BT)$ is a function that is a measure of the collision efficiency that goes to 1 for a hard-sphere spherical potential, that is, the actual effective cross-section is $F(V/k_BT)\cdot\pi d^2$. For ^{19}F in SF_6 in a mixture of SF_6 and O_2, the spin-rotation mechanism, which had been shown to dominate the relaxation rate in pure SF_6, still applies for SF_6–SF_6 collisions, but the intermolecular ^{19}F-electron spin dipole interaction is also operating. Because of the large γ_S, this is a very significant relaxation mechanism. A multiple magnetic field study permits the separation of the field-independent intramolecular $[T_{1SR}]^{-1}$ and the field-dependent $[T_{1DD,\mathrm{inter}}]^{-1}$ relaxation rates. The functional form of eqn (1.23) was validated and the temperature dependence of the various parts of eqn (1.23) were found for SF_6 in O_2; further studies were carried out for ^{19}F in CF_4 and SiF_4 in O_2,[212] as well as ^{19}F in SeF_6 and TeF_6 in O_2, and for ^1H in CH_4 in CH_4-O_2 mixtures.[213] By varying the density of O_2 in the samples, it is possible to include a wide range of relative contributions to the relaxation rate. For example, for ^{19}F in CF_4 in O_2, in mixtures such that the relaxation rates are 10%DD/90%SR up to 80%DD/20%SR, we successfully determined the density, temperature, and magnetic field dependence of the intermolecular DD relaxation rate since the dependences of the rates of the two mechanisms on these three factors are opposite, *i.e.*, ρ *vs.* $1/\rho$, $\sim T^{-1}$ *vs.* $T^{+3/2}$, 1 *vs.* $[1-f(T)\omega_F{}^{1/2}]$. We find that the experimental $f(T)$ function is reasonably close to our theoretical estimate in eqn (1.24),

$$f(T)=\frac{1}{24}\left(\frac{d}{\bar{v}}\right)^{1/2}\left[3+7\left(\frac{\gamma_S}{\gamma_I}\right)^{1/2}\right] \tag{1.24}$$

where the temperature dependence appears only in the $\bar{v}$. At 300 K the theoretical estimate given by eqn (1.24) is between 92% (CF_4 in O_2) and 108% (TeF_6 in O_2) of the experimental values for the six different systems mentioned above.

The temperature-dependent experimental cross-section for intermolecular dipolar interaction with the electron spin of O_2 is $F(V/k_BT)\cdot\pi d^2$. The efficiency $F(V/k_BT)$ is found to increase dramatically in the order $CH_4 < CF_4 < SF_6 < SiF_4 < SeF_6 < TeF_6$ at 300 K,[212] magnitudes that are about 2–4 times as large as for a square well potential counterpart of the best empirical estimates of the PES for the interaction of these molecules with O_2. The temperature dependence of the experimental cross-section is more pronounced than for a square well potential, and likewise becomes more pronounced in the same relative order. It is quite clear that the experimental cross-sections indicate significant long-range contributions.

We also investigated the relaxation of ^{129}Xe in the presence of O_2 gas.[214] In this case the other intermolecular relaxation rates available to ^{129}Xe (such as those described in Section 1.5.7) are much too weak to compete with the relaxation due to ^{129}Xe-electron-spin–dipole interactions during binary collisions when the oxygen densities are significant (mole fraction of O_2 in the Xe–O_2 mixtures >0.005). The temperature-dependent effective cross-section

$F(V/k_BT)\cdot\pi d^2$ found experimentally is $250(T/300\ \text{K})^{-0.27}\ \text{Å}^2$. The relaxation of ^{3}He in the presence of O_2 has been investigated by Saam *et al.*[215] and ^{3}He in the presence of NO by Hayden *et al.*[216] The analysis of the latter is much more complicated in that the effective magnetic moment of the NO is a result of both electron spin and orbital angular momentum that couple and leads to a temperature-dependent effective magnetic moment, in contrast with O_2 in which S is a good quantum number.

1.5.9 Classical Trajectory Calculations of Relaxation Cross-sections

Precise determination of the intermolecular potential energy surface remains one of the most important problems in chemical physics. The accurate and consistent calculation of potential-energy surfaces (PES) for van der Waals complexes, from short through intermediate to large intermolecular separations, remains a severe technical challenge for *ab initio* quantum mechanics. Only a few small systems have been subjected to CCSD(T) level calculations in the limit of complete basis sets. There is thus a need to validate the PES, where possible, against available experimental data, including second interaction virial coefficient, transport properties, and relaxation phenomena data for binary mixtures, crossed-beam total differential and total integral scattering, as well as microwave and infrared spectra for the vdW complex. All these data are required to validate a PES because each is sensitive to a different portion of the PES. For example, the bound state properties associated with microwave and infrared spectra of the vdW complex are fairly well defined for inter-species distances less than 8 Å and the microwave spectrum primarily provides the moment of inertia of the vdW complex, thus the distances and angles close to the global minimum of the PES. The vdW infrared spectra are mainly sensitive to the shape (anharmonicity) of the van der Waals well. Properties such as the virial coefficients and transport and relaxation phenomena require an accurate representation of the long-range part of the PES. Crossed-beam experiments provide a post-collision angular distribution; the dependence of the cross-section on scattering angle is quantified by the differential cross-section. The integral of the differential cross-section (DCS) over scattering angles gives the total or integral cross-section. The supernumerary rainbows arising from interference effects provide information about the range and shape of the potential near the minimum. Total differential and total integral scattering data are therefore sensitive to both the attractive and repulsive parts of the potential, particularly sensitive to the anisotropy (angle-dependence) about the repulsive wall. The NMR relaxation cross-sections $\mathfrak{S}'(0\hat{2}00|\text{A})_{\text{AB}}$ and $\mathfrak{S}(0100|\text{A})_{\text{AB}}$ are particularly sensitive to the anisotropy of the PES for all intermolecular distances because only anisotropy can cause molecular reorientation or rotational angular momentum change, unlike the interaction virial coefficients that can be accounted for by an isotropic potential, with only minor corrections arising from anisotropy.

A fully classical trajectory code has been developed by Dickinson *et al.* for atom collisions with a diatomic molecule,[217] and for pure gases of linear molecules;[218] finally, the theory has been extended to rigid molecules of arbitrary structure, *i.e.*, asymmetric tops.[219] Since symmetric tops and spherical tops can be considered as special cases of asymmetric tops, this last development allows transport and relaxation properties of important molecules such as benzene, methane, and sulfur hexafluoride to be calculated (but not any molecules that have internal rotation, such as ethane). Given a particular PES function, all the temperature-dependent cross-sections can be calculated using this code, including the cross-sections that are identified with many gas phase thermophysical properties such as binary diffusion coefficients, mixture viscosities, mixture thermal conductivity, mole fraction dependencies of the interaction second virial coefficient, the binary diffusion coefficient, the interaction viscosity, the mixture shear viscosity and thermal conductivity coefficients, field effects on these properties, as well as the cross-sections associated with spin relaxation.

For N_2-Ar, the NMR relaxation cross-sections[182,189] have the tightest experimental uncertainties, and are available over a more extended temperature range than are the effective cross-sections extracted from any other relaxation phenomenon. Among the eight N_2-Ar potential energy surfaces tested by McCourt *et al.*[220] using classical trajectory calculations, one PES, denoted as XC(fit), gives consistently better agreement with the values of $\mathfrak{S}'(0\hat{2}00|\mathrm{A})_{\mathrm{AB}}$ determined from the NMR measurements[189] than does any other of the potential-energy surfaces, including the previously "best" MMSV (Morse–Morse–Spline–van der Waals) PES that had been fitted to previously known thermophysical and crossed beam data. The XC(fit) PES does provide distinctly altogether better agreement with these NMR relaxation experimental results than do any of the other four new N_2-Ar potential-energy surfaces the authors considered, and is for the moment the "best" intermolecular potential for N_2-Ar.

We had carried out classical trajectories on six simple model potentials for N_2-Kr;[221] these include a previously published empirical surface derived from fits to molecular beam experiments and various model potentials of the Tang and Toennies type that differ in the set of dispersion coefficients employed. Forty-five effective cross-sections that determine the bulk transport and relaxation phenomena were calculated by classical trajectories for temperatures ranging from 100 to 800 K for each of the six PES. The sensitivity of the NMR-derived cross-sections to the various characteristics of the anisotropy of the potential (such as the anisotropy in the well depth, in the high repulsive wall, in the low repulsive wall, and at $V = 0$) are examined. It is found that both the radial anisotropy and the anisotropy in well depth contribute to $\mathfrak{S}(0100|\mathrm{N}_2)_{\mathrm{N2\text{-}Kr}}$ or σ_J and $\mathfrak{S}'(0\hat{2}00|\mathrm{N}_2)_{\mathrm{N2\text{-}Kr}}$ or $\sigma_{\theta,2}$. The often-assumed empirical power law dependence of the NMR cross-sections on temperature within a 200 degree range (200–400 K) is found to be consistent with the results of the classical trajectory calculations of these cross-sections for all of the six potentials considered, although a more complex temperature dependence would be necessary to describe a much greater temperature range. It was found that better overall agreement with experimental data previously considered could be obtained by slight modification of one of the previously used potentials.[222] A modification of this surface to include a recent *ab initio* determination of the C_6 dispersion coefficient, and to bring in the virial and microwave data, gives a new potential surface that is in good agreement with all available experimental data and thus could be considered as the current best N_2-Kr PES.

We subsequently carried out a similar study of the CO_2–Ar potential.[223] Twelve potential energy surfaces that have been proposed for the CO_2–Ar interaction were considered in detail. The anisotropies of these surfaces are compared and their ability to predict the interaction second virial coefficient as a function of temperature has been examined. Intermolecular bending and stretching quadratic force constants predicted by each and the mean square torque calculated for each are compared with the experimental values. Quantum diffusion Monte Carlo simulations provide the average rotational constants and geometry for the ground vibrational state as well as the dissociation energy in each case. These are compared with the experimental values. Classical trajectory calculations were carried out to obtain 45 types of thermal average cross-sections for six of these surfaces. Various thermophysical properties calculated from these cross-sections and the NMR relaxation cross-sections are compared with experimental data. It is found that the spectroscopic constants define the depth and shape of the well at the global minimum, whereas the NMR cross-sections and mean square torque probe the anisotropy in a broader sense. The thermophysical properties (viscosity, diffusion coefficient, and thermal conductivity) are not strongly discriminating between the surfaces, whereas the temperature dependence of the second virial coefficient detects the weaknesses in the low and upper repulsive walls of those surfaces that were modified specifically to improve greatly the shape of the well so as to reproduce the spectroscopic constants.[223] A more recent *ab initio* PES for CO_2–Ar has been calculated and tested only against the infrared spectra of the vdW complex.[224] It remains to be seen whether this one can reproduce the NMR relaxation data and the thermophysical properties. We also carried out trajectory calculations for NNO-Kr and NNO-Ar.[225] The available PES are unable to reproduce both NMR cross-sections accurately. Similar studies have been carried out by Dickinson *et al.* for N_2–N_2 and CO_2–CO_2, using the NMR relaxation cross-sections and thermophysical properties in a multi-property analysis of the PES.[226,227]

1.5.10 The Special Case of Hydrogen Molecule

Some of the earliest theoretical treatment and experimental T_1 studies in the gas phase involved the hydrogen molecule.[228,229] Experimental studies of H_2, HD, and D_2 in collisions with rare gas atoms were carried out by Armstrong, McCourt, and co-workers in He and Ne,[230–233] and by McCourt *et al.* in

Ar.[234–237] These experiments were accompanied by theoretical calculations that permitted the testing of various *ab initio* and semi-empirical potential surfaces for H_2, HD, or D_2 interacting with He, Ne, or Ar. A study of the relaxation times of all the isotopomers of hydrogen in Ar gas have been carried out by McCourt *et al.* over the entire density range from the reciprocal density regime through the T_1 minimum all the way to the linear-density regime,[236,237] and then they carried out a full critical test of a new PES for this system.[235] For hydrogen molecule, a classical treatment does not apply because the rotational energy levels are very widely spaced, so only a quantum-mechanical approach can be used. They used a potential function that Bissonnette *et al.* determined by modifying a starting exchange-Coulomb type PES fitted to highly accurate spectroscopic data for H_2-Ar, D_2-Ar, and HD-Ar van der Waals molecules, plus interaction second virial coefficient data and Raman collisional shift data for H_2–Ar binary mixtures.[238] Utilization of this particular set of data ensures that both the isotropic component of the PES and the anisotropy in the potential well are determined very accurately. Indeed, this PES provides excellent agreement with bulk transport and relaxation data not utilized in the determination of the original PES. Using quantum-mechanical close-coupled computations based on the H_2–Ar potential energy surface obtained by Bissonnette *et al.* and using all the ^1H and ^2H relaxation data, McCourt *et al.* concluded that significant differences found between the experimental and theoretical results indicate that the short-range anisotropy of the proposed PES is too weak. The reciprocal density regime (densities below the T_1 minimum) is shown to have a much higher sensitivity to changes in the anisotropic component of the intermolecular potential energy surface than the linear-density regime, and therefore, for H_2–Ar, the discrepancy between the calculated and experimental cross-sections is a much more stringent test of the PES.

1.6 Conformational Dynamics in the Gas Phase

N. S. True and co-workers have developed the use of pressure-dependent gas phase NMR spectroscopy as a probe of conformational dynamics in the gas phase. Their present capabilities allow spectral acquisition at sample pressures as low as 0.1 torr. With these capabilities they are able to pursue studies that address both the accumulation and the disposal of intramolecular and intermolecular vibrational energy in simple molecules undergoing structural exchange. Gas phase NMR studies have also been carried out on several other systems such as PF_5, ethers and alkyl nitrites, and cyclic amines. Since, for many of these systems, it is possible to obtain rate data in the bimolecular kinetic region, they can probe intermolecular energy transfer efficiencies accompanying these processes. They also can determine temperature-dependent equilibrium constants, and activation thermodynamic quantities for chemical exchange processes, that can provide stringent tests of high-level *ab initio* calculations of transition states and activation energies. Temperature-dependent equilibrium constants for conformational equilibria that have been measured by the True group in the gas phase include the keto $\leftrightarrow$ enol tautomerism of acetyl acetone,[239] the syn $\leftrightarrow$ anti conformational equilibrium of methyl nitrite,[240] and the larger nitrites *n*-propyl, *n*-butyl, and isobutyl nitrite.[241,242]

The rate constants for chemical exchange processes in the gas phase are pressure-dependent due to competition between bimolecular deactivation and reaction of energized molecules. Three pressure regions can in principle be observed for a gas phase chemical exchange process, namely unimolecular at high pressure, fall-off at intermediate pressure, and bimolecular at low pressure. Internal rotation has been studied for several symmetrically substituted amides at or near the unimolecular limit,[243–246] and also for thioamides.[247] In each case, exchange-broadened ^{1}H spectra were obtained for samples containing the amide at its vapor pressure and several atmospheres of an inert gas; measurements were made at several pressures to ensure that the rate constants were at the unimolecular limit. ^{13}C spectra in isotopically enriched samples have also been used, for example, to measure the rate constants in *N*,*N*-dimethylformamide.[248] Gibbs activation energies are 5–10 kJ mol^{-1} lower in the gas phase than those in solution. Activation energies for ring inversions in various molecules have been studied, for example cyclohexane,[249] cyclohexene,[250] tetrahydropyran,[251] *N*,*N*-dimethylpiperazine,[252] *N*-methylpiperazine,[253] *N*-methylpiperidine,[254] and *N*-ethylmorpholine.[255] The degenerate Cope rearrangement of bullvalene [tricyclo(3.3.2.0)deca-2,7,9-triene] was observed in ^{1}H NMR in the gas phase in samples with 1 torr of bullvalene in a 6 up to 2580 torr of a bath gas (SF_6).[256] The measured pressure-dependence of the rate constants at 356 K are in the unimolecular and fall-off kinetic regions. For the bullvalene rearrangement the bimolecular kinetic region occurs at pressures considerably below 5 torr. Unimolecular rate constants obtained for the rearrangement are *ca.* 15% lower than those observed in solutions of bullvalene in CS_2(liq) at the same temperatures.

Data on pressure-dependent rate constants of unimolecular processes provide tests of statistical kinetic theories such as RRKM (Rice–Ramsperger–Kassel–Marcus).[257,258] RRKM theory is the method of choice for practical predictions of gas phase dissociation and isomerization rate coefficients. Account is taken of the way in which the different normal-mode vibrations and rotations contribute to reaction, and allowance is made for the zero-point energies. The total internal energy is partitioned into active and inactive components, such that only the active component can flow freely among the internal modes and thus contribute to reaction. The assumed equilibrium ratio of active-to-inactive components is evaluated using partition functions. In applications where gas-phase collisions are important, the rates of activation and deactivation take into account their energy dependence, and in the high-pressure limit the transition state is in equilibrium with non-activated reactants and RRKM reduces to conventional transition-state theory. RRKM and other statistical kinetic theories assume that the rate constant for intramolecular energy redistribution in critically energized molecules is rapid (compared with the energy-dependent rate constant) and ergodic. Statistical theories of chemical reactions are therefore most applicable to large molecules undergoing processes at high activation energies. However, at the low activation energies required for conformational processes, critically energized molecules have sparse density of states and the anharmonic coupling constants among vibrational states are small. Under these conditions, statistical kinetic theories may not provide an adequate description of these processes. Conformational processes of the small molecules studied in the gas phase by the True group are just the type of systems that may challenge the assumptions of RRKM theory.

When the process can be modeled with RRKM theory, it is valid to apply transition state theory to the high-pressure rate constants. When this is the case, then accurate *ab initio* calculations of ground- and transition-state structures and vibrational frequencies can be used to predict gas-phase activation parameters ($\Delta G^{\neq}{}_{298}$, $\Delta H^{\neq}{}_{298}$, $\Delta S^{\neq}{}_{298}$) for internal rotation (or pseudorotation or ring inversion) that can then be directly compared to the corresponding activation parameters that are experimentally obtained from the temperature-dependent kinetic data from ^{1}H (or ^{13}C or ^{19}F) NMR in the gas phase. Note that all this is possible only for dilute gas phase results, but not for solution phase data. Only data obtained in the dilute gas phase can be used to test quantum calculations of ground state structures, transition state structures, vibrational frequencies, and activation barriers since in these solvent-free systems it is actually possible to use the highest levels of quantum-mechanical theory and not to have to resort to DFT or MD simulations with empirical force fields that are the typical theoretical approaches to condensed phase kinetics. We illustrate with two examples from work in the dilute gas phase by True *et al.*

The True group has carried out ^{1}H experiments to observe the chemical exchange spectra for 2 torr of ^{15}*N*-trifluoroacetamide in 600 torr of bath gas (SF_6) and for 1 torr of the molecule in 300 torr of the SF_6.[259] They had previously shown that the internal rotation process for this molecule was statistical. From the total line shape analysis they obtained the activation parameters for internal rotation of trifluoroacetamide. Their *ab initio* MP2 calculations of these activation parameters agree with experiment within experimental errors, whereas DFT calculations (with the B3PW91 functional) using the same basis set did not.

Another example is the Berry pseudorotation in SF_4, which concertedly exchanges the magnetically inequivalent sets of axial and equatorial F atoms. This is one of the smallest molecules that undergo an intramolecular rearrangement that has rate constants accessible to NMR measurement.[260] The True group found that the strong collision RRKM model as applied to Berry pseudorotation of SF_4 reproduces very well their experimental pressure-dependent rate constants at 335 K. The curvature and displacement of the experimental fall-off curve are not significantly perturbed by effects of weak collisions and non-statistical intramolecular vibrational energy redistribution. Since the process can be modeled with RRKM theory, it is valid to apply transition state theory to the high-pressure (at 7.9 atm) rate constants. From analysis of the exchange-broadened ^{19}F NMR spectra, they obtain the temperature-dependent rate constants characterized by $E^{\infty} = 11.9(0.2)$ kcal mol^{-1}, $A^{\infty} = 3.56(1.09)\times10^{12}$ s^{-1}, and the activation parameters $\Delta G^{+}{}_{298} = 12.2(0.1)$ kcal mol^{-1}, $\Delta H^{+}{}_{298} = 11.3(0.4)$ kcal mol^{-1}, and $\Delta S^{+}{}_{298} = -3.3(0.4)$ cal mol^{-1} K^{-1}. Quantum calculations at the MP4 level predict $\Delta H^{+}{}_{298} = 11.55$ kcal mol^{-1} and $\Delta S^{+}{}_{298} = -3.91$ cal mol^{-1} K^{-1}, in excellent agreement with their experiment, whereas DFT calculations (using hybrid functionals B3LYP and B3PW91) provide considerably less accurate results.

Reviews of the experimental and theoretical work in these areas provide an overview.[6,7]

List of Abbreviations

CASSCF	Complete active space self-consistent-field method
CCSD	Coupled-cluster singles and doubles method
CCSD(T)	CCSD model augmented by perturbative corrections for triple excitations
CSA	Chemical shift anisotropy
DCS	Differential cross-section
DD	Dipole–dipole
DFT	Density functional theory
efg	Electric field gradient
FC	Fermi contact
FCI	Full configuration interaction
MCSCF	Multi-configuration self-consistent-field method
MD	Molecular dynamics
MMSV	Morse–Morse–Spline–van der Waals potential function
MP2, MP3, MP4	Møller–Plesset perturbation theory (second, third, fourth order)
NMR	Nuclear magnetic resonance
PES	Potential energy surface
RASSCF	Restricted active space self-consistent-field method
RHF	Restricted Hartree–Fock
SO	Spin–orbit
SOPPA	Second-order polarization propagator approximation
SR	Spin-rotation
UV	Ultraviolet
vdW	van der Waals
ZPV	Zero-point vibration

References

1. M. Karplus, *J. Chem. Phys.*, 1959, **30**, 11–15.
2. C. J. Jameson, *Chem. Rev.*, 1991, **91**, 1375–1395.
3. C. J. Jameson, Gas phase studies of intermolecular interactions and relaxation, in *Encyclopedia of Nuclear Magnetic Resonance*, ed. D. M. Grant and R. K. Harris, John Wiley, London, 1996, vol. 4, pp. 2179–2189.
4. C. B. LeMaster, *Prog. Nucl. Magn. Reson. Spectrosc.*, 1997, **31**, 119–154.
5. K. Jackowski, *Appl. Magn. Reson.*, 2003, **24**, 379–385.
6. N. S. True, Gas phase studies of chemical exchange processes, in *Encyclopedia of NMR*, ed. D. M. Grant and R. K. Harris, John Wiley, London, 1996, pp. 2173–2178. EMagRes, DOI: 10.1002/9780470034590.emrstm0184.
7. N. S. True and C. Suarez, Gas phase NMR studies of conformational processes, in *Advances in Molecular Structure Research*, ed. I. Hargittai and M. Hargittai, JAI Press, Greenwich, CT, 1995, ch. 4, vol. 1, pp. 115–155.
8. A. D. Buckingham and J. A. Pople, *Discuss. Faraday Soc.*, 1956, **22**, 17–21.
9. A. K. Jameson, C. J. Jameson and H. S. Gutowsky, *J. Chem. Phys.*, 1970, **53**, 2310–2321.
10. C. J. Jameson, A. K. Jameson and D. Oppusunggu, *J. Chem. Phys.*, 1984, **81**, 2571–2573.
11. A. D. Buckingham, T. Schaefer and W. G. Schneider, *J. Chem. Phys.*, 1960, **32**, 1227–1233.
12. R. L. Streever and H. Y. Carr, *Phys. Rev*, 1961, **121**, 20–25.
13. S. Gordon and B. P. Dailey, *J. Chem. Phys.*, 1961, **34**, 1084–1085.
14. W. T. Raynes, A. D. Buckingham and H. J. Bernstein, *J. Chem. Phys.*, 1962, **36**, 3481–3488.
15. C. J. Jameson, A. K. Jameson and S. M. Cohen, *J. Chem. Phys.*, 1973, **59**, 4540–4546.
16. C. J. Jameson, A. K. Jameson and S. M. Cohen, *J. Chem. Phys.*, 1975, **62**, 4224–4226.
17. C. J. Jameson, A. K. Jameson and S. M. Cohen, *J. Chem. Phys.*, 1976, **65**, 3401–3406.
18. C. J. Jameson, A. K. Jameson and S. M. Cohen, *J. Chem. Phys.*, 1977, **66**, 5226–5230.
19. C. J. Jameson, A. K. Jameson and H. Parker, *J. Chem. Phys.*, 1978, **68**, 3943–3944.
20. A. K. Jameson, K. Schuett, C. J. Jameson, S. M. Cohen and H. Parker, *J. Chem. Phys.*, 1977, **67**, 2821–2824.
21. C. J. Jameson, A. K. Jameson, D. Oppusunggu and S. Wille, *J. Chem. Phys.*, 1982, **76**, 152–162.
22. C. J. Jameson, A. K. Jameson, S. M. Cohen, H. Parker, D. Oppusunggu, P. M. Burrell and S. Wille, *J. Chem. Phys.*, 1981, **74**, 1608–1612.

23. B. Bennett and W. T. Raynes, *Magn. Reson. Chem.*, 1991, **29**, 946–954.
24. A. K. Jameson, J. Moyer and C. J. Jameson, *J. Chem. Phys.*, 1978, **68**, 2873–2877.
25. C. J. Jameson, A. K. Jameson, H. Parker, S. M. Cohen and C. L. Lee, *J. Chem. Phys.*, 1978, **68**, 2861–2867.
26. C. J. Jameson, A. K. Jameson, S. Wille and P. M. Burrell, *J. Chem. Phys.*, 1981, **74**, 853–856.
27. W. Makulski and K. Jackowski, *Chem. Phys. Lett.*, 2001, **341**, 369–372.
28. C. J. Jameson, A. K. Jameson and H. Parker, *J. Chem. Phys.*, 1978, **68**, 2868–2872.
29. C. J. Jameson, A. K. Jameson and S. Wille, *J. Phys. Chem.*, 1979, **83**, 3372–3378.
30. C. J. Jameson, A. K. Jameson and P. M. Burrell, *J. Chem. Phys.*, 1980, **73**, 6013–6020.
31. C. J. Jameson, A. K. Jameson and J. Honarbakhsh, *J. Chem. Phys.*, 1984, **81**, 5266–5267.
32. M. Kubiszewski and W. Makulski, *Pol. J. Chem.*, 2008, **82**, 1457–1462.
33. W. Makulski and M. Wilczek, *Pol. J. Chem.*, 2006, **80**, 1055–1060.
34. M. Kubiszewski, W. Makulski and K. Jackowski, *J. Mol. Struct.*, 2005, **737**, 7–10.
35. M. Kubiszewski, W. Makulski and K. Jackowski, *J. Mol. Struct.*, 2004, **704**, 211–214.
36. W. Makulski and K. Jackowski, *J. Mol. Struct.*, 2004, **704**, 219–222.
37. W. Makulski, *J. Mol. Struct.*, 2013, **1036**, 168–173.
38. W. Makulski, *J. Mol. Struct.*, 2008, **872**, 81–86.
39. W. Makulski, *J. Mol. Struct.*, 2005, **744**, 439–446.
40. E. Wielogorska, W. Makulski, W. Kozminski and K. Jackowski, *J. Mol. Struct.*, 2004, **704**, 305–309.
41. M. Wilczek, W. Kozminski and K. Jackowski, *Chem. Phys. Lett.*, 2002, **358**, 263–270.
42. K. Jackowski and M. Wilczek, *J. Mol. Struct.*, 2003, **651**, 259–263.
43. W. Makulski, *J. Mol. Struct.*, 2012, **1017**, 45–50.
44. K. Jackowski, E. Maciaga and M. Wilczek, *J. Mol. Struct.*, 2005, **744**, 101–105.
45. K. Jackowski, M. Wilczek, M. Pecul and J. Sadlej, *J. Phys. Chem. A*, 2000, **104**, 5955–5958.
46. K. Jackowski, M. Kubiszewski and M. Wilczek, *Chem. Phys. Lett.*, 2007, **440**, 176–179.
47. K. Jackowski, M. Wilczek, W. Makulski and W. Kozminski, *J. Phys. Chem. A*, 2002, **106**, 2829–2832.
48. C. J. Jameson and A. C. de Dios, *J. Chem. Phys.*, 1992, **97**, 417–434.
49. K. Jackowski and W. T. Raynes, *Mol. Phys.*, 1977, **34**, 465–475.
50. C. J. Jameson, A. K. Jameson and D. Oppusunggu, *J. Chem. Phys.*, 1984, **81**, 2313–2317.
51. J. Beckett and H. Y. Carr, *Phys. Rev. A: At., Mol., Opt. Phys.*, 1981, **24**, 144–148.

52. C. J. Jameson, D. N. Sears and A. C. de Dios, *J. Chem. Phys.*, 2003, **118**, 2575–2580.
53. M. Hanni, P. Lantto, N. Runeberg, J. Jokisaari and J. Vaara, *J. Chem. Phys.*, 2004, **121**, 5908–5919.
54. C. J. Jameson and A. C. de Dios, *J. Chem. Phys.*, 1993, **98**, 2208–2217.
55. A. C. de Dios and C. J. Jameson, *J. Chem. Phys.*, 1997, **107**, 4253–4270.
56. D. N. Sears and C. J. Jameson, *J. Chem. Phys.*, 2004, **121**, 2151–2157.
57. D. Stueber and C. J. Jameson, *J. Chem. Phys.*, 2004, **120**, 1560–1571.
58. A. Antusek, M. Jaszunski and A. Rizzo, *J. Chem. Phys.*, 2007, **126**, 074303.
59. C. J. Jameson and H. S. Gutowsky, *J. Chem. Phys.*, 1964, **40**, 1714–1724.
60. C. J. Jameson and A. K. Jameson, *Mol. Phys.*, 1971, **20**, 957–959.
61. C. J. Jameson, A. K. Jameson and S. M. Cohen, *Mol. Phys.*, 1975, **29**, 1919–1927.
62. C. J. Jameson, A. K. Jameson and S. M. Cohen, *J. Chem. Phys.*, 1976, **65**, 3397–3400.
63. L. Vukovic, C. J. Jameson and D. N. Sears, *Mol. Phys.*, 2006, **104**, 1217–1225.
64. C. J. Jameson, *Bull. Magn. Reson.*, 1980, **3**, 3–28.
65. C. J. Jameson, *Annu. Rev. Phys. Chem.*, 1996, **47**, 135–169.
66. C. J. Jameson, Rovibrational averaging of molecular electronic properties, in *Theoretical Models of Chemical Bonding*, ed. Z. B. Maksic, Part 3. Molecular Spectroscopy, Electronic Structure, and Intramolecular Interactions, Springer-Verlag, Berlin, 1991, pp. 457–519.
67. N. F. Ramsey, *Phys. Rev.*, 1952, **87**, 1075–1079.
68. A. D. Buckingham, *J. Chem. Phys.*, 1962, **36**, 3096.
69. C. J. Jameson, A. K. Jameson and S. M. Cohen, *J. Chem. Phys.*, 1977, **67**, 2771–2774.
70. D. B. Cook, A. M. Davies and W. T. Raynes, *Mol. Phys.*, 1971, **21**, 113–122.
71. W. T. Raynes, A. M. Davies and D. B. Cook, *Mol. Phys.*, 1971, **21**, 123–133.
72. P. W. Fowler, G. Riley and W. T. Raynes, *Mol. Phys.*, 1981, **42**, 1463–1481.
73. P. W. Fowler and W. T. Raynes, *Mol. Phys.*, 1981, **43**, 65–82.
74. W. T. Raynes, P. Lazzeretti and R. Zanasi, *Mol. Phys.*, 1987, **61**, 1415–1421.
75. P. Lazzeretti, R. Zanasi, A. J. Sadlej and W. T. Raynes, *Mol. Phys.*, 1987, **62**, 605–616.
76. W. T. Raynes, P. W. Fowler, P. Lazzeretti, R. Zanasi and M. Grayson, *Mol. Phys.*, 1988, **64**, 143–162.
77. C. J. Jameson and A. K. Jameson, *J. Chem. Phys.*, 1986, **85**, 5484–5492.
78. C. J. Jameson, *Mol. Phys.*, 1985, **54**, 73–79.
79. C. J. Jameson, A. K. Jameson and D. Oppusunggu, *J. Chem. Phys.*, 1986, **85**, 5480–5483.
80. R. A. Hegstrom, *Phys. Rev. A: At., Mol., Opt. Phys.*, 1979, **19**, 17–30.
81. R. M. Stevens and W. N. Lipscomb, *J. Chem. Phys.*, 1964, **40**, 2238–2247.

82. D. B. Chesnut, *Chem. Phys.*, 1986, **110**, 415–420.
83. D. Sundholm, J. Gauss and A. Schafer, *J. Chem. Phys.*, 1996, **105**, 11051–11059.
84. D. Sundholm and J. Gauss, *Mol. Phys.*, 1997, **92**, 1007–1014.
85. C. J. Jameson, A. C. de Dios and A. K. Jameson, *J. Chem. Phys.*, 1991, **95**, 1069–1079.
86. C. J. Jameson, A. C. de Dios and A. K. Jameson, *J. Chem. Phys.*, 1991, **95**, 9042–9053.
87. H. Fukui, T. Baba, J. Narumi, H. Inomata, K. Miura and H. Matsuda, *J. Chem. Phys.*, 1996, **105**, 4692–4699.
88. J. Vaara, J. Lounila, K. Ruud and T. Helgaker, *J. Chem. Phys.*, 1998, **109**, 8388–8397.
89. R. D. Wigglesworth, W. T. Raynes, S. P. A. Sauer and J. Oddershede, *Mol. Phys.*, 1999, **96**, 1595–1607.
90. C. Puzzarini, G. Cazzoli, M. E. Harding, J. Vázquez and J. Gauss, *J. Chem. Phys.*, 2009, **131**, 234304.
91. T. Helgaker, J. Gauss, G. Cazzoli and C. Puzzarini, *J. Chem. Phys.*, 2013, **139**, 244308.
92. R. D. Wigglesworth, W. T. Raynes, S. Kirpekar, J. Oddershede and S. P. A. Sauer, *J. Chem. Phys.*, 2000, **12**, 736–746.
93. J. Lounila, J. Vaara, Y. Hiltunen, A. Pulkkinen, J. Jokisaari, M. Ala-Korpela and K. Ruud, *J. Chem. Phys.*, 1997, **107**, 1350–1361.
94. A. C. de Dios, J. Pearson and E. Oldfield, *Science*, 1993, **260**, 1491–1496.
95. C. J. Jameson and A. C. de Dios, The nuclear shielding surface: The shielding as a function of molecular geometry and intermolecular separation, in *Nuclear Magnetic Shielding and Molecular Structure*, ed. J. A. Tossell, Kluwer Academic Publishers, Dordrecht, 1993, pp. 95–116.
96. A. C. de Dios and C. J. Jameson, *Annu. Rep. NMR Spectrosc*, 1994, **29**, 1–69.
97. C. J. Jameson and A. C. de Dios, Theoretical and Physical Aspects of Nuclear Shielding, in *A Specialist Periodical Report, Nuclear Magnetic Resonance*, ed. K. Kamienska-Trela and J. Wojcik, RSC Publishing, Royal Society of Chemistry, UK, 2014, vol. 43, pp. 49–80.
98. P. Lantto, S. Kangasvieri and J. Vaara, *J. Chem. Phys.*, 2012, **137**, 214309.
99. B. Minaev, J. Vaara, K. Ruud, O. Vahtras and H. Agren, *Chem. Phys. Lett.*, 1998, **295**, 455–461.
100. P. Lantto, J. Vaara, A. M. Kantola, V.-V. Telkki, B. Schimmelpfennig, K. Ruud and J. Jokisaari, *J. Am. Chem. Soc.*, 2002, **124**, 2762–2771.
101. C. J. Jameson and H. J. Osten, *Annu. Rep. NMR Spectrosc.*, 1986, **17**, 1–78.
102. C. J. Jameson, The dynamic and electronic factors in isotope effects on NMR parameters, in *Isotopes in the Physical and Biomedical Sciences*, ed. E. Buncel and J. R. Jones, Isotopic Applications in NMR Studies, Elsevier, Amsterdam, 1991, ch. 1, vol. 2, pp. 1–54.
103. C. J. Jameson, *J. Chem. Phys.*, 1977, **66**, 4977–4982.
104. C. J. Jameson, *J. Chem. Phys.*, 1977, **67**, 2814–2820.
105. H. J. Osten and C. J. Jameson, *J. Chem. Phys.*, 1985, **82**, 4595–4606.

106. C. J. Jameson and H. J. Osten, *J. Chem. Phys.*, 1984, **81**, 2556–2561.
107. C. J. Jameson, *J. Chem. Phys.*, 1977, **66**, 4983–4988.
108. C. J. Jameson and H. J. Osten, *J. Chem. Phys.*, 1984, **81**, 4293–4299.
109. C. J. Jameson and H. J. Osten, *J. Chem. Phys.*, 1984, **81**, 4300–4305, erratum, ibid, 1985, 83, 915.
110. C. J. Jameson and H. J. Osten, *J. Am. Chem. Soc.*, 1985, **107**, 4158–4161.
111. M. Dracinsky, J. Kaminsky and P. Bour, *J. Chem. Phys.*, 2009, **130**, 094106.
112. K. Ruud, P. O. Astrand and P. R. Taylor, *J. Chem. Phys.*, 2000, **112**, 2668–2683.
113. P. O. Astrand and K. Ruud, *Phys. Chem. Chem. Phys.*, 2003, **5**, 5015–5020.
114. H. Sabzyan and B. Buzari, *Chem. Phys.*, 2008, **352**, 297–305.
115. A. M. Teale, O. B. Lutnæs, T. Helgaker, D. J. Tozer and J. Gauss, *J. Chem. Phys.*, 2013, **138**, 024111.
116. P. Garbacz, K. Jackowski, W. Makulski and R. E. Wasylishen, *J. Phys. Chem. A*, 2012, **116**, 11896–11904.
117. A. K. Jameson and C. J. Jameson, *Chem. Phys. Lett.*, 1987, **134**, 461–466.
118. C. J. Jameson, A. K. Jameson, D. Oppusunggu, S. Wille, P. M. Burrell and J. Mason, *J. Chem. Phys.*, 1981, **74**, 81–88.
119. R. E. Wasylishen and D. L. Bryce, *J. Chem. Phys.*, 2002, **117**, 10061–10066.
120. W. Makulski and K. Jackowski, *J. Mol. Struct.*, 2003, **651**, 265–269.
121. C. J. Jameson and A. K. Jameson, *Chem. Phys. Lett.*, 1988, **149**, 300–305.
122. W. Makulski, K. Jackowski, A. Antusek and M. Jaszunski, *J. Phys. Chem. A*, 2006, **110**, 11462–11466.
123. C. J. Jameson, A. Dedios and A. K. Jameson, *Chem. Phys. Lett.*, 1990, **167**, 575–582.
124. R. E. Wasylishen, C. Connor and J. O. Friedrich, *Can. J. Chem.*, 1984, **62**, 981–985.
125. K. Jackowski, W. Makulski and W. Kozminski, *Magn. Reson. Chem.*, 2002, **40**, 563–565.
126. C. J. Jameson and A. K. Jameson, *Chem. Phys. Lett.*, 1987, **135**, 254–259.
127. P. Lantto, K. Jackowski, W. Makulski, M. Olejniczak and M. Jaszunski, *J. Phys. Chem. A*, 2011, **115**, 10617–10623.
128. A. A. Auer, J. Gauss and J. F. Stanton, *J. Chem. Phys.*, 2003, **118**, 10407–10417.
129. A. A. Auer, *J. Chem. Phys.*, 2009, **131**, 024116.
130. M. E. Harding, M. Lenhart, A. A. Auer and J. Gauss, *J. Chem. Phys.*, 2008, **128**, 244111.
131. E. Prochnow and A. A. Auer, *J. Chem. Phys.*, 2010, **132**, 064109.
132. M. Jaszunski, A. Antusek, P. Garbacz, K. Jackowski, W. Makulski and M. Wilczek, *Prog. Nucl. Magn. Reson. Spectrosc.*, 2012, **67**, 49–63.
133. K. Jackowski, M. Jaszunski and M. Wilczek, *J. Phys. Chem. A*, 2010, **114**, 2471–2475.

134. M. Jaszunski, M. Repisky, T. B. Demissie, S. Komorovsky, E. Malkin, K. Ruud, P. Garbacz, K. Jackowski and W. Makulski, *J. Chem. Phys.*, 2013, **139**, 234302.
135. D. L. Bryce and R. E. Wasylishen, *J. Am. Chem. Soc.*, 2000, **122**, 3197–3205.
136. D. L. Bryce and R. E. Wasylishen, *Acc. Chem. Res.*, 2003, **36**, 327–334.
137. D. K. Green, A. Hartland and H. Y. Carr, *Bull. Am. Phys. Soc.*, 1966, **11**, 311.
138. A. K. Jameson and J. P. Reger, *J. Phys. Chem.*, 1971, **75**, 437–439.
139. K. Jackowski, W. Makulski, A. Szyprowska, A. Antusek and M. Jaszunski, *Magn. Reson. Chem.*, 2009, **47**, 857–861.
140. K. Jackowski, M. Kubiszewski and W. Makulski, *J. Mol. Struct.*, 2002, **614**, 267–272.
141. M. Wilczek, M. Kubiszewski and K. Jackowski, *J. Mol. Struct.*, 2004, **704**, 311–313.
142. K. Jackowski, *Int. J. Mol. Sci.*, 2003, **4**, 125–142.
143. M. Grayson and S. P. A. Sauer, *Mol. Phys.*, 2000, **98**, 1981–1990.
144. C. J. Jameson and H. S. Gutowsky, *J. Chem. Phys.*, 1969, **51**, 2790–2803.
145. C. J. Jameson, Theoretical considerations: Spin-spin coupling, in *^{31}P NMR Spectroscopy in Stereochemical Analysis: Organic Compounds and Metal Complexes*, ed. J. G. Verkade and L. D. Quin, Verlag Chemie, Weinheim, 1987, pp. 205–230.
146. C. J. Jameson and H. J. Osten, *J. Am. Chem. Soc.*, 1986, **108**, 2497–2503.
147. J. Beckett, Ph.D. Thesis, Rutgers University, 1979.
148. W. T. Raynes and N. Panteli, *Chem. Phys. Lett.*, 1983, **94**, 558–560.
149. W. T. Raynes and J. P. Riley, *Mol. Phys.*, 1974, **27**, 337–343.
150. T. Helgaker, M. Jaszuński, P. Garbacz and K. Jackowski, *Mol. Phys.*, 2012, **110**, 2611–2617.
151. J. Oddershede, J. Geertsen and G. E. Scuseria, *J. Phys. Chem.*, 1988, **92**, 3056–3059.
152. O. Vahtras, H. Ågren, P. Jørgensen, H. J. A. Jensen, S. B. Padkjær and T. Helgaker, *J. Chem. Phys.*, 1992, **96**, 6120–6125.
153. P.-O. Astrand, K. Ruud, K. V. Mikkelsen and T. Helgaker, *J. Chem. Phys.*, 1999, **110**, 9463–9468.
154. S. M. Bass, R. L. De Leon and J. S. Muenter, *J. Chem. Phys.*, 1987, **86**, 4305–4312.
155. O. Vahtras, H. Ågren, P. Jørgensen, T. Helgaker and H. J. Aa Jensen, *Chem. Phys. Lett.*, 1993, **209**, 201–206.
156. J. Geertsen, J. Oddershede, W. T. Raynes and G. E. Scuseria, *J. Magn. Reson.*, 1991, **93**, 458.
157. W. T. Raynes, J. Geertsen and J. Oddershede, *Int. J. Quantum Chem.*, 1994, **52**, 153–163.
158. R. D. Wigglesworth, W. T. Raynes, S. P. A. Sauer and J. Oddershede, *Mol. Phys.*, 1997, **92**, 77–88.

159. S. Kirpekar, T. Enevoldsen, J. Oddershede and W. T. Raynes, *Mol. Phys.*, 1997, **91**, 897–907.
160. S. P. A. Sauer, W. T. Raynes and R. A. Nicholls, *J. Chem. Phys.*, 2001, **115**, 5994–6006.
161. C. J. Jameson, Spin-spin Coupling in *Multinuclear NMR*, ed. J. Mason, Plenum Press, New York, 1987, ch. 4. pp. 89–132.
162. R. D. Wigglesworth, W. T. Raynes, S. P. A. Sauer and J. Oddershede, *Mol. Phys.*, 1998, **94**, 851–862.
163. R. D. Wigglesworth, W. T. Raynes, S. Kirpekar, J. Oddershede and S. P. A. Sauer, *J. Chem. Phys.*, 2000, **112**, 3735–3746.
164. J. Casanueva, J. San Fabián, E. Díez and A. L. Esteban, *J. Mol. Struct.*, 2001, **565–566**, 449–454.
165. A. Yachmenev, S. N. Yurchenko, I. Paidarová, P. Jensen, W. Thiel and S. P. A. Sauer, *J. Chem. Phys.*, 2010, **132**, 114305.
166. K. Sneskov and J. F. Stanton, *Mol. Phys.*, 2012, **110**, 2321–2327.
167. C. I. Oprea, Z. Rinkevicius, O. Vahtras, H. Ågren and K. Ruud, *J. Chem. Phys.*, 2005, **123**, 014101.
168. T. A. Ruden, O. B. Lutnæs and T. Helgaker, *J. Chem. Phys.*, 2003, **118**, 9572.
169. M. A. M. Forgeron, M. Gee and R. E. Wasylishen, *J. Phys. Chem. A*, 2004, **108**, 4895–4908.
170. T. A. Ruden, T. Helgaker and M. Jaszuński, *Chem. Phys.*, 2004, **296**, 53–62.
171. M. Bloom and I. Oppenheim, Intermolecular Forces Determined by Nuclear Magnetic Resonance, in *Intermolecular Forces*, ed. J. O. Hirschfelder, Interscience, New York, 1967, ch. 10, pp. 549–600.
172. R. L. Armstrong, Longitudinal Nuclear Spin Relaxation Time Measurements in Molecular Gases, in *NMR Basic Principles and Progress*, Springer-Verlag, Berlin, 1976, vol. 13, pp. 71–95.
173. R. G. Gordon, *J. Chem. Phys.*, 1965, **44**, 228–234.
174. F. R. W. McCourt, J. J. M. Beenakker, W. E. Kohler and I. Kuscer, *Nonequilibrium Phenomena in Polyatomic Gases Part I*, The Dilute Gas, Oxford University, Oxford, 1990, vol. 1.
175. F. R. W. McCourt, J. J. M. Beenakker, W. E. Kohler and I. Kuscer, *Nonequilibrium Phenomena in Polyatomic Gases Part II*, Cross-sections, Scattering and Rarefied Gases, Oxford University, Oxford, 1990, vol. 2.
176. S. Chapman and T. G. Cowling, *Mathematical Theory of Nonuniform Gases*, Cambridge University, Cambridge, 3rd edn, 1970.
177. W. K. Liu and F. R. McCourt, *Chem. Phys.*, 1977, **19**, 137–146.
178. J. A. Courtney and R. L. Armstrong, *Can. J. Phys.*, 1972, **50**, 1252–1261.
179. C. J. Jameson, A. K. Jameson and K. Buchi, *J. Chem. Phys.*, 1986, **85**, 697–700.
180. C. J. Jameson, N. C. Smith and K. Jackowski, *J. Chem. Phys.*, 1987, **86**, 2717–2722.
181. C. J. Jameson, A. K. Jameson, N. C. Smith, J. K. Hwang and T. Zia, *J. Phys. Chem.*, 1991, **95**, 1092–1098.

182. C. J. Jameson, A. K. Jameson and N. C. Smith, *J. Chem. Phys.*, 1987, **86**, 6833–6838.
183. C. J. Jameson, A. K. Jameson, J. K. Hwang and N. C. Smith, *J. Chem. Phys.*, 1988, **89**, 5642–5649.
184. C. J. Jameson and A. K. Jameson, *J. Chem. Phys.*, 1988, **88**, 7448–7452.
185. C. J. Jameson, A. K. Jameson and R. Terry, *J. Phys. Chem.*, 1991, **95**, 2982–2988.
186. C. J. Jameson and A. K. Jameson, *J. Chem. Phys.*, 1988, **89**, 866–870.
187. I. Ursu, M. Bogdan, P. Fitori, A. Darabont and D. E. Demco, *Mol. Phys.*, 1985, **56**, 297–302.
188. C. J. Jameson and A. K. Jameson, *J. Chem. Phys.*, 1990, **93**, 3237–3244.
189. C. J. Jameson, A. K. Jameson and M. A. ter Horst, *J. Chem. Phys.*, 1991, **95**, 5799–5808.
190. C. J. Jameson, M. A. ter Horst and A. K. Jameson, *J. Chem. Phys.*, 1998, **109**, 10227–10237.
191. A. Brown and R. E. Wasylishen, *J. Phys. Chem. A*, 2012, **116**, 9769–9776.
192. W. B. Neilsen and R. G. Gordon, *J. Chem. Phys.*, 1973, **58**, 4149–4170.
193. M. A. ter Horst, C. J. Jameson and A. K. Jameson, *Magn. Reson. Chem.*, 2006, **44**, 241–248.
194. N. R. Newbury, A. S. Barton, G. D. Cates, W. Happer and H. Middleton, *Phys. Rev. A: At., Mol., Opt. Phys.*, 1993, **48**, 4411–4420.
195. B. Shizgal, *J. Chem. Phys.*, 1973, **58**, 3424–3431.
196. W. J. Mullin, F. Laloë and M. G. Richards, *J. Low Temp. Phys.*, 1990, **80**, 1–13.
197. I. L. Moudrakovski, S. R. Breeze, B. Simard, C. I. Ratcliffe, J. A. Ripmeester, T. Seideman, J. S. Tse and G. Santyr, *J. Chem. Phys.*, 2001, **114**, 2173–2181.
198. H. C. Torrey, *Phys. Rev.*, 1963, **130**, 2306–2312.
199. B. Chann, I. A. Nelson, L. W. Anderson, B. Driehuys and T. G. Walker, *Phys. Rev. Lett.*, 2002, **88**, 113201.
200. B. N. Berry-Pusey, B. C. Anger, G. Laicher and B. Saam, *Phys. Rev. A: At., Mol., Opt. Phys.*, 2006, **74**, 063408.
201. B. C. Anger, G. Schrank, A. Schoeck, K. A. Butler, M. S. Solum, R. J. Pugmire and B. Saam, *Phys. Rev. A: At., Mol., Opt. Phys.*, 2008, **78**, 043406.
202. M. Hanni, P. Lantto and J. Vaara, *Phys. Chem. Chem. Phys.*, 2011, **13**, 13704–13708.
203. D. E. Freeman, K. Yoshino and Y. Tanaka, *J. Chem. Phys.*, 1974, **61**, 4880–4889.
204. D. Brinkmann, E. Brun and H. H. Staub, *Helvet. Phys. Acta*, 1962, **35**, 431–436.
205. D. Brinkmann and D. Kuhn, *Phys. Rev. A: At., Mol., Opt. Phys.*, 1980, **21**, 163–167.
206. K. F. Stupic, Z. I. Cleveland, G. E. Pavlovskaya and T. Meersmann, *J. Magn. Reson.*, 2011, **208**, 58–69.
207. Z. I. Cleveland and T. Meersmann, *J. Chem. Phys.*, 2008, **129**, 244304.

208. Z. I. Cleveland and T. Meersmann, *ChemPhysChem*, 2008, **9**, 1375–1379.
209. I. Oppenheim and M. Bloom, *Can. J. Phys.*, 1961, **39**, 845–869.
210. F. M. Chen and R. F. Snider, *J. Chem. Phys.*, 1967, **46**, 3937–3940.
211. C. J. Jameson, A. K. Jameson, J. K. Hwang and D. Dabkowski, *J. Phys. Chem.*, 1988, **92**, 5937–5941.
212. C. J. Jameson, A. K. Jameson and J. K. Hwang, *J. Chem. Phys.*, 1991, **94**, 172–178.
213. C. J. Jameson, A. K. Jameson and J. K. Hwang, *J. Phys. Chem.*, 1989, **93**, 634–638.
214. C. J. Jameson, A. K. Jameson and J. K. Hwang, *J. Chem. Phys.*, 1988, **89**, 4074–4081.
215. B. Saam, W. Happer and H. Middleton, *Phys. Rev. A: At., Mol., Opt. Phys.*, 1995, **52**, 862–865.
216. G. Archibald, E. Brief, C. Lei, T. Pausak and M. E. Hayden, *Phys. Rev. A: At., Mol., Opt. Phys.*, 2006, **73**, 022721.
217. A. S. Dickinson and M. S. Lee, *J. Phys. B: At. Mol. Phys.*, 1985, **18**, 3987–4000.
218. E. L. Heck and A. S. Dickinson, *Comput. Phys. Commun*, 1996, **95**, 190–220.
219. A. S. Dickinson, R. Hellmann, E. Bich and E. Vogel, *Phys. Chem. Chem. Phys.*, 2007, **9**, 2836–2843.
220. A. K. Dham, W. J. Meath, J. W. Jechow and F. R. W. McCourt, *J. Chem. Phys.*, 2006, **124**, 034308.
221. M. A. ter Horst and C. J. Jameson, *J. Chem. Phys.*, 1995, **102**, 4431–4446.
222. F. R. W. McCourt, M. A. ter Horst and C. J. Jameson, *J. Chem. Phys.*, 1995, **102**, 5752–5760.
223. M. A. ter Horst and C. J. Jameson, *J. Chem. Phys.*, 1996, **105**, 6787–6806.
224. Y. Cui, H. Ran and D. Xie, *J. Chem. Phys.*, 2009, **130**, 224311.
225. M. A. ter Horst and C. J. Jameson, *J. Chem. Phys.*, 1998, **109**, 10238–10243.
226. D. Cappelletti, F. Vecchiocattivi, F. Pirani, E. L. Heck and A. S. Dickinson, *Mol. Phys.*, 1998, **93**, 485–499.
227. S. Bock, E. Bich, E. Vogel, A. S. Dickinson and V. Vesovic, *J. Chem. Phys.*, 2004, **121**, 4117–4122.
228. N. Bloembergen, E. M. Purcell and R. V. Pound, *Phys. Rev.*, 1948, **73**, 679–712.
229. E. M. Purcell, R. V. Pound and N. Bloembergen, *Phys. Rev.*, 1946, **70**, 986–987.
230. C. Lemaire, R. L. Armstrong and F. R. W. McCourt, *J. Chem. Phys.*, 1984, **81**, 5275–5280.
231. R. S. Wagner, R. L. Armstrong, E. C. Bissonnette and F. R. W. McCourt, *J. Chem. Phys.*, 1990, **92**, 5907–5918.
232. R. L. Armstrong, M. Bogdan, K. R. Jeffrey, C. Bissonnette and F. R. W. McCourt, *J. Chem. Phys.*, 1993, **99**, 5754–5761.
233. R. S. Wagner, R. L. Armstrong, C. Lemaire and F. R. W. McCourt, *J. Chem. Phys.*, 1986, **84**, 1137–1141.

234. C. Lemaire, R. L. Armstrong and F. R. W. McCourt, *J. Chem. Phys.*, 1987, **87**, 6499–6501.
235. H. Sabzyan, F. R. W. McCourt and W. P. Power, *J. Chem. Phys.*, 2004, **120**, 4306–4315.
236. H. Sabzyan, W. P. Power and F. R. W. McCourt, *J. Chem. Phys.*, 1998, **108**, 6170–6184.
237. H. Sabzyan, W. P. Power and F. R. W. McCourt, *J. Chem. Phys.*, 1998, **106**, 2361–2374.
238. C. Bissonnette, C. E. Chauqui, K. G. Crowell, R. J. Le Roy, R. J. Wheatley and W. J. Meath, *J. Chem. Phys.*, 1996, **105**, 2639–2653.
239. M. M. Folkendt, B. E. Weiss-Lopez, J. P. Chauvel Jr. and N. S. True, *J. Phys. Chem.*, 1985, **89**, 3347–3352.
240. J. P. Chauvel Jr. and N. S. True, *J. Phys. Chem.*, 1983, **87**, 1622–1625.
241. P. O. Moreno, N. S. True and C. B. Lemaster, *J. Mol. Struct.*, 1991, **243**, 373–384.
242. P. O. Moreno, N. S. True and C. B. Lemaster, *J. Phys. Chem.*, 1990, **94**, 8780–8787.
243. C. Suarez, M. Tafazzoli, N. S. True, S. Gerrard, C. B. LeMaster and C. L. LeMaster, *J. Phys. Chem.*, 1995, **99**, 8170–8176.
244. A. N. Taha, S. M. Neugebauer-Crawford and N. S. True, *J. Phys. Chem. A*, 1998, **102**, 1425–1430.
245. A. N. Taha, S. M. Neugebauer-Crawford and N. S. True, *J. Am. Chem. Soc.*, 1998, **120**, 1934–1935.
246. A. N. Taha and N. S. True, *J. Phys. Chem. A*, 2000, **104**, 8609–8616.
247. S. M. Neugebauer-Crawford, A. N. Taha, N. S. True and C. B. LeMaster, *J. Phys. Chem. A*, 1977, **101**, 4699–4706.
248. B. D. Ross and N. S. True, *J. Am. Chem. Soc.*, 1984, **106**, 2451–2452.
249. P. O. Moreno and N. S. True, *J. Phys. Chem.*, 1991, **95**, 57–62.
250. C. Suarez, M. Tafazzoli and N. S. True, *J. Mol. Struct.*, 1992, **71**, 89–94.
251. P. S. Chu and N. S. True, *J. Phys. Chem.*, 1985, **89**, 2625–2630.
252. C. B. Lemaster, C. L. Lemaster, C. Suarez, M. Tafazzoli and N. S. True, *J. Phys. Chem.*, 1989, **93**, 3993–3996.
253. M. Tafazzoli, C. Suarez, N. S. True, C. B. Lemaster and C. L. Lemaster, *J. Mol. Struct.*, 1994, **317**, 137–145.
254. M. Tafazzoli, C. Suarez, N. S. True, C. B. Lemaster and C. L. Lemaster, *J. Phys. Chem.*, 1992, **96**, 10201–10205.
255. M. Tafazzoli, S. Gerrard, N. S. True and C. B. LeMaster, *J. Mol. Struct.*, 1994, **317**, 131–136.
256. P. O. Moreno, C. Suarez, M. Tafazzoli, N. S. True and C. B. Lemaster, *J. Phys. Chem.*, 1992, **96**, 10206–10212.
257. R. A. Marcus and O. K. Rice, *J. Phys. Colloid Chem.*, 1951, **55**, 894–908.
258. R. A. Marcus, *J. Chem. Phys.*, 1952, **20**, 359–364.
259. C. B. Lemaster, C. L. Lemaster and N. S. True, *J. Am. Chem. Soc.*, 1999, **121**, 4478–4484.
260. A. N. Taha, N. S. True, C. B. LeMaster, C. L. LeMaster and S. M. Neugebauer-Crawford, *J. Phys. Chem. A*, 2000, **104**, 3341–3348.

CHAPTER 2

Obtaining Gas Phase NMR Parameters from Molecular Beam and High-resolution Microwave Spectroscopy

ALEXANDRA FAUCHER AND RODERICK E. WASYLISHEN*

Department of Chemistry, Gunning-Lemieux Chemistry Centre, University of Alberta, Edmonton, AB, Canada T6G 2G2
*Email: roderick.wasylishen@ualberta.ca

2.1 Introduction

This chapter is concerned with the connection between parameters that NMR spectroscopists typically measure and those that can be measured by high-resolution rotational spectroscopy[1–6] or molecular beam resonance techniques,[7–9] all of which ultimately have as their goal the determination of molecular structure. The objective of this chapter is to illustrate these connections and show how NMR spectroscopists can use data from microwave and molecular beam spectroscopy. After a brief discussion of the important and relevant Hamiltonians that connect NMR and molecular spectroscopy, we will illustrate how spin-rotation tensors can be used to establish absolute magnetic shielding scales, provide a better understanding of the relationship between molecular structure and magnetic shielding, and allow theoreticians to rigorously test quantum chemistry computations. Several representative examples will be presented, however, we wish to indicate that

New Developments in NMR No. 6
Gas Phase NMR
Edited by Karol Jackowski and Michał Jaszuński

Published by the Royal Society of Chemistry, www.rsc.org

this is not meant to be a comprehensive review of the many shielding scales that have been proposed in the literature (the annual reviews of Jameson and De Dios[10] are highly recommended). The discussion of spin-rotation and magnetic shielding tensors is followed by a discussion of how accurate measurements of nuclear electric quadrupolar coupling tensors *via* microwave spectroscopy or molecular beam techniques can be combined with quantum chemistry computations to provide nuclear electric quadrupole moments, or eQ values. Finally, we will discuss how spin–spin coupling tensors can be characterized using molecular beam resonance techniques and their importance in understanding the mechanisms of indirect spin–spin coupling. The advantage of measuring NMR parameters in the gas phase is that one can eliminate intermolecular effects on NMR parameters.[11] Measurement of the temperature and density dependence of NMR parameters *via* gas phase NMR spectroscopy also allows one to quantify effects of rotational-vibrational averaging.[12,13]

Both NMR spectroscopy and microwave spectroscopy began to flourish in the early 1950s, approximately five years after the Second World War. In fact, progress in these two fields was often reviewed together because of their close connection.[14] Important in the early development of NMR spectroscopy was the interpretation of magnetic shielding, responsible for chemical shifts, and the interpretation of indirect spin–spin coupling (also known as J-coupling), responsible for valuable line splitting observed in NMR spectra.[15] N. F. Ramsey played a leading role in providing the theoretical foundation for interpretation of these parameters.[16–18] As is often the case in the development of molecular quantum mechanics and spectroscopy, hydrogen (H_2, HD, and D_2) has played a critical role in this development.[19–25] Ramsey's theory of magnetic shielding in molecules consisted of two terms, a diamagnetic term, σ^d, that depends only on the ground electronic state of the molecule, and a paramagnetic term, σ^p, that arises from mixing of some excited electronic states with the ground state.[16] The latter term was a challenge to calculate from first principles, even for diatomic molecules; however, Ramsey recognized that it was related to the spin-rotation constant, C_I. This led to his predictions of the isotropic magnetic shielding constant in H_2; 26.8 ppm in 1950,[16] 26.2(4) ppm in 1956,[7] and 26.43(60) ppm in 1966,[26] close to the present accepted value of 26.293(5) ppm at 300 K.[25] While spin-rotation constants for ^{1}H and ^{19}F in diatomic molecules are on the order of 10–100 kHz, indirect spin–spin interactions are typically several orders magnitude smaller and difficult to measure using microwave and molecular beam techniques (one early exception was thallium fluoride,[27] where $^1J(^{205}\text{Tl}, ^{19}\text{F}) = -13.3 \pm 0.7$ kHz). However, the indirect spin–spin coupling constant in HD was readily measured *via* NMR by Carr and Purcell,[28] with $^1J(^2\text{H}, ^1\text{H}) = 43.5 \pm 1$ Hz; see Garbacz[24] for a more recent and accurate value, 43.140(10) Hz. Typical spectral resolution in NMR was routinely on the order of ± 0.1 Hz, even in the 1960s. With the exception of TlF, the first observation of indirect spin–spin coupling *via* molecular beam methods was in 1970 for hydrogen fluoride, where

$^1J(^{19}F,\ ^1H) = +529(23)$ Hz.[29] Ramsey recognized much earlier that not only was the isotropic indirect spin–spin coupling available from molecular beam measurements but also the direct dipolar coupling constant and anisotropy in the indirect spin–spin coupling, ΔJ.[7,8] Though these early developments highlight the complementary nature of NMR and microwave spectroscopy, in our experience this connection is little known amongst modern-day spectroscopists. We hope that this chapter will provide the NMR community with a relevant introduction to the microwave and molecular beam spectroscopy literature.

2.2 The Hyperfine Hamiltonian

Before discussing the individual interactions measurable by molecular beam and microwave spectroscopy, and their relationships to those nuclear interactions of interest to NMR spectroscopists, we briefly summarize the hyperfine Hamiltonian relevant to rotational spectroscopy for a diatomic molecule. The total hyperfine Hamiltonian for a diatomic molecule including one spin-1/2 (I) and one quadrupolar (S) nucleus in molecular beam or microwave spectroscopy is[6–9,30]

$$\hat{\mathscr{H}}_{\mathrm{HF}} = \hat{\mathscr{H}}_{\mathrm{Q,S}} + \hat{\mathscr{H}}_{\mathrm{SR,I}} + \hat{\mathscr{H}}_{\mathrm{SR,S}} + \hat{\mathscr{H}}_{\mathrm{DD,IS}} + \hat{\mathscr{H}}_{\mathrm{J,IS}} \tag{2.1}$$

$$\hat{\mathscr{H}}_{\mathrm{HF}} = \bar{\bar{V}}_{\mathrm{S}}{:}\bar{\bar{Q}}_{\mathrm{S}} + C_{\mathrm{I}}\bar{I}\cdot\bar{J} + C_{\mathrm{S}}\bar{S}\cdot\bar{J} + c_3\bar{I}\cdot\bar{\bar{d}}_{\mathrm{T}}\cdot\bar{S} + c_4\bar{I}\cdot\bar{S} \tag{2.2}$$

where $\hat{\mathscr{H}}_{\mathrm{Q}}$, $\hat{\mathscr{H}}_{\mathrm{SR}}$, $\hat{\mathscr{H}}_{\mathrm{DD}}$, and $\hat{\mathscr{H}}_{\mathrm{J}}$ refer to the quadrupolar, spin-rotation, direct dipolar, and indirect spin–spin coupling interactions, respectively. Here, $\bar{I}$ and $\bar{S}$ represent the nuclear spin angular momentum vectors for the I and S nuclei, respectively, $\bar{J}$ represents the rotational angular momentum due to molecular rotation, $\bar{\bar{Q}}_{\mathrm{S}}$ is the nuclear electric quadrupole moment tensor, $\bar{\bar{V}}_{\mathrm{S}}$ is the electric field gradient tensor, C_{I} is the spin-rotation constant, and the constants c_3 and c_4 determined in molecular beam/microwave experiments are equal to R_{eff} and J_{iso}, the effective dipolar coupling constant and the isotropic indirect spin–spin coupling constant, respectively (*vide infra*). The tensor $\bar{\bar{d}}_{\mathrm{T}}$ contains contributions from both direct and indirect spin–spin coupling. Here we point out that in reading the literature one finds that there appears to be no accepted convention for reporting the sign of the spin-rotation tensor. For example, in eqn (2.2), the second term is often written with a negative sign, $-C_{\mathrm{I}}\bar{I}\cdot\bar{J}$ (see p. 208 of ref. 7). Throughout this chapter we have attempted to report the signs as found in the original literature.

2.3 Nuclear Spin Rotation

The nuclear spin-rotation interaction is the coupling of the nuclear spin angular momentum with the rotational angular momentum of a molecule due to intramolecular magnetic fields produced by molecular rotation. The spin-rotation interaction Hamiltonian is

$$\hat{\mathscr{H}}_{\mathrm{SR}} = (\hat{I}_x, \hat{I}_y, \hat{I}_z) \begin{bmatrix} C_{xx} & C_{xy} & C_{xz} \\ C_{yx} & C_{yy} & C_{yz} \\ C_{zx} & C_{zy} & C_{zz} \end{bmatrix} \begin{pmatrix} \hat{J}_x \\ \hat{J}_y \\ \hat{J}_z \end{pmatrix} \tag{2.3}$$

where $\bar{I} = (\hat{I}_x, \hat{I}_y, \hat{I}_z)$ are the nuclear spin angular momentum operators for the observed nucleus, $\bar{\bar{C}}$ is the nuclear spin-rotation tensor, and $\bar{J} = (\hat{J}_x, \hat{J}_y, \hat{J}_z)$ are the angular momentum operators corresponding to molecular rotation. Within the vector model, the total angular momentum of the molecule is represented by $\bar{F}$, a sum of the nuclear spin angular momentum and the molecular rotational angular momentum. Details are outlined in several excellent texts,[1,4,6] and will not be repeated here. In microwave experiments, one determines the nuclear spin-rotation tensor in the inertial principal axis system, where the diagonal elements of $\bar{\bar{C}}$ are denoted C_{aa}, C_{bb}, and C_{cc}, after the inertial principal directions. For asymmetric tops, the principal components of the moment of inertia tensor are $I_{\mathrm{a}} \neq I_{\mathrm{b}} \neq I_{\mathrm{c}} \neq 0$. This complicates the rotational spectrum, but allows for the determination of each diagonal component of $\bar{\bar{C}}$. For symmetric tops, where $I_{\mathrm{a}} = I_{\mathrm{b}} < I_{\mathrm{c}}$ or $I_{\mathrm{a}} < I_{\mathrm{b}} = I_{\mathrm{c}}$, the rotational spectrum is simplified and it is more difficult to obtain the three diagonal components of $\bar{\bar{C}}$. For linear molecules, the principal axis system for the nuclear spin-rotation tensor is coincident with the inertial principal axis system, and $\bar{\bar{C}}$ consists of two unique elements, $C_{\parallel}$ and $C_{\perp}$ ($C_{\parallel}$ is zero). The spin rotation constant C_{I} is equal to $C_{\perp}$. The Hamiltonian is thus simplified to

$$\hat{\mathscr{H}}_{\mathrm{SR}} = C_{\mathrm{I}} \left(\hat{I}_x, \hat{I}_y, \hat{I}_z\right) \begin{pmatrix} \hat{J}_x \\ \hat{J}_y \\ \hat{J}_z \end{pmatrix}. \tag{2.4}$$

The nuclear spin rotation constant depends on the intramolecular magnetic field created by other nuclei and electrons, and can be partitioned as follows, where $C_{\mathrm{I}}^{\mathrm{rel.}}$ accounts for relativistic effects.

$$C_{\mathrm{I}} = C_{\mathrm{I}}^{\mathrm{nucl.}} + C_{\mathrm{I}}^{\mathrm{elec.}} + C_{\mathrm{I}}^{\mathrm{rel.}} \tag{2.5}$$

The electronic contribution, following Brown and Carrington,[6] is

$$C_{\mathrm{I}}^{\mathrm{elec.}} = -\frac{2e}{m_{\mathrm{e}}} \gamma_{\mathrm{I}} \hbar^3 \sum_{n>0} \frac{\left\langle 0 \left| \sum_i \bar{l}_{i\mathrm{I}} r_{i\mathrm{I}}^{-3} \right| n \right\rangle \langle n | \bar{L}_{\mathrm{I}} | 0 \rangle}{E_0 - E_n} \cdot \bar{\bar{I}}_{\mathrm{eff}}^{-1} \tag{2.6}$$

where e is the elementary charge, m_{e} is the electron rest mass, γ_{I} is the magnetogyric ratio for the observed nucleus I, $r_{i\mathrm{I}}$ and $\bar{l}_{i\mathrm{I}}$ are the position vectors and angular momentum operators for the ith electron with respect to nucleus I, $\bar{L}_{\mathrm{I}}$ is the total electronic angular momentum operator with the origin at nucleus I, and $\bar{\bar{I}}_{\mathrm{eff}}^{-1}$ is the inverse of the effective moment of inertia tensor. Note eqn (2.6) is valid in the limit where relativistic effects are ignored.

The simplest case involving the spin-rotation interaction occurs for a diatomic molecule with one spin-1/2 isotope and one magnetically inactive ($I=0$) isotope. In conventional microwave spectroscopy, one observes rotational transitions which obey the selection rule $\Delta J = \pm 1$. The approximate rotational energy and transition frequency between energy levels are described below.

$$E_J = B_\nu J(J+1) - D_\nu [J(J+1)]^2 \tag{2.7}$$

$$\nu_{(J+1)\leftarrow J} = 2B_\nu (J+1) - 4D_\nu (J+1)^3 \tag{2.8}$$

Here,

$$B_\nu = \frac{h}{8\pi^2 I} \tag{2.9}$$

and

$$D_\nu \approx \frac{4B^3}{\omega^2} \tag{2.10}$$

where B_ν is the rotational constant for vibrational level ν, and D_ν is the corresponding centrifugal distortion constant. The latter is always positive for diatomic molecules. The rotational energy levels are modified by the spin-rotation interaction according to

$$E_{\text{Total}} = E_J + E_{\text{SR}} \tag{2.11}$$

$$E_{\text{SR}} = \frac{C_I}{2}[F(F+1) - I(I+1) - J(J+1)] \tag{2.12}$$

where

$$F = J+I, \quad J+I-1, \quad J+I-2, \ldots, \quad |J-I| \tag{2.13}$$

(see Figure 2.1). Thus the nuclear spin-rotation interaction causes splittings and shifts in the rotational energy levels.

Note that in the case of quadrupolar nuclei, *e.g.*, oxygen-17 in $^{12}C^{17}O$, the microwave spectrum is further complicated by the quadrupolar interaction. The total electronic energy in this case is

$$E_{\text{Total}} = E_J + E_Q + E_{\text{SR}} \tag{2.14}$$

where

$$E_Q = -C_Q \frac{\frac{3}{4}C(C+1) - I(I+1)J(J+1)}{2(2J-1)(2J+3)I(2I-1)} \tag{2.15}$$

$$C = F(F+1) - J(J+1) - I(I+1) \tag{2.16}$$

(see Figure 2.2).

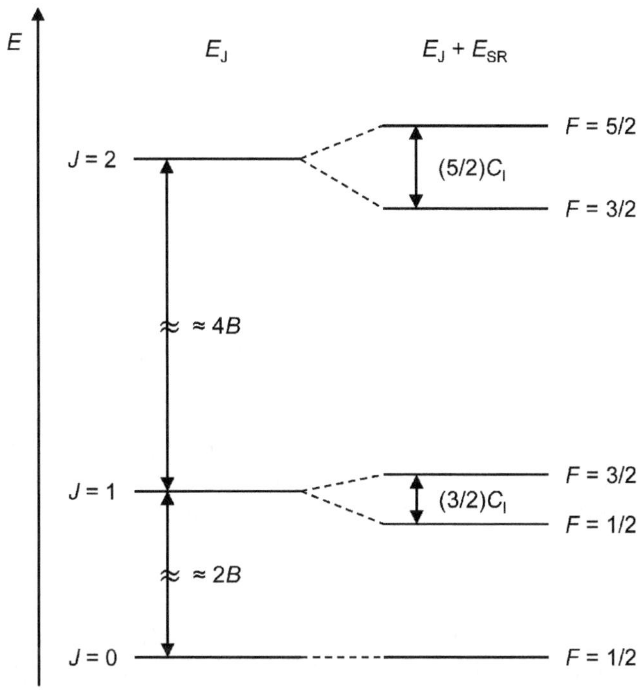

Figure 2.1 Splitting in rotational energy levels of a diatomic molecule due to a nucleus with $I = 1/2$ as a result of the nuclear spin-rotation interaction, in the absence of an applied electric or magnetic field. Splittings due to nuclear spin-rotation are greatly exaggerated.

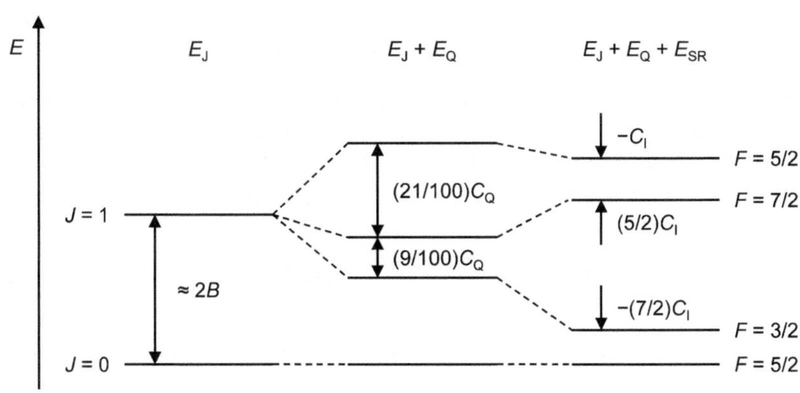

Figure 2.2 Splitting of rotational energy levels of a diatomic molecule due to a nucleus with $I = 5/2$ as a result of the nuclear electric quadrupolar and spin-rotation interactions. Splittings are greatly exaggerated.

2.4 Nuclear Magnetic Shielding

The primary reason why NMR spectroscopists have interest in the nuclear spin rotation tensor is its connection with the nuclear magnetic shielding tensor, quantified in NMR spectroscopy by means of the chemical shift. In the Cartesian laboratory frame, the magnetic shielding Hamiltonian in NMR spectroscopy is[31–33]

$$\hat{\mathcal{H}}_{\mathrm{MS}} = \gamma_{\mathrm{I}}\left(\hat{I}_x, \hat{I}_y, \hat{I}_z\right)\begin{bmatrix} \sigma_{xx} & \sigma_{xy} & \sigma_{xz} \\ \sigma_{yx} & \sigma_{yy} & \sigma_{yz} \\ \sigma_{zx} & \sigma_{zy} & \sigma_{zz} \end{bmatrix}\begin{pmatrix} 0 \\ 0 \\ B_0 \end{pmatrix} \tag{2.17}$$

$$\hat{\mathcal{H}}_{\mathrm{MS}} = \gamma_{\mathrm{I}}[\hat{I}_x\sigma_{xz}B_0 + \hat{I}_y\sigma_{yz}B_0 + \hat{I}_z\sigma_{zz}B_0] \tag{2.18}$$

where γ_{I} is the gyromagnetic ratio for the I nuclei, $\hat{I}_x$, $\hat{I}_y$, and $\hat{I}_z$ are the nuclear spin angular momentum operators for the observed nucleus, B_0 is the applied magnetic field strength (assumed to be along the laboratory z-axis), and σ_{ij} $(i,j = x,y,z)$ are elements of the magnetic shielding tensor. In NMR spectroscopy one is concerned with the secular terms of this Hamiltonian, which commute with $\hat{I}_z$. This simplifies the shielding Hamiltonian to

$$\hat{\mathcal{H}}_{\mathrm{MS}} = \gamma_{\mathrm{I}}\sigma_{zz}B_0\hat{I}_z \tag{2.19}$$

where σ_{zz} is the magnetic shielding constant. The magnetic shielding interaction thus describes the coupling of the nuclear spin angular momentum with the applied magnetic field, by means of the electrons surrounding the observed nuclei. In the principal axis system, the symmetric part of the magnetic shielding tensor, $\bar{\bar{\sigma}}$, is diagonal.

$$\bar{\bar{\sigma}}^{\mathrm{PAS}} = \begin{bmatrix} \sigma_{11} & 0 & 0 \\ 0 & \sigma_{22} & 0 \\ 0 & 0 & \sigma_{33} \end{bmatrix} \tag{2.20}$$

Note that the total magnetic shielding tensor also possesses an antisymmetric component, however, it has a negligible effect on observed NMR spectra.[32–37] The observed shielding constant can be related to the magnetic shielding tensor in the principal axis system as follows[38]

$$\sigma_{zz}(\theta) = \frac{1}{3}\mathrm{Tr}\bar{\bar{\sigma}}^{\mathrm{PAS}} + \frac{1}{3}\sum_{j=1}^{3}\left(3\cos^2\theta_j - 1\right)\sigma_{jj} \tag{2.21}$$

where $\mathrm{Tr}\bar{\bar{\sigma}}^{\mathrm{PAS}}$ is the trace of the magnetic shielding tensor in the principal axis system, σ_{jj} $(j = 1, 2, 3)$ are the elements of this tensor, and θ is the angle between the principal axes and the applied magnetic field direction. Note that in a solution or in the gas phase, rapid molecular tumbling eliminates NMR spectral effects from the second term in eqn (2.21). The chemical shift, δ, is a magnetic field-independent parameter measured in an NMR experiment, and is related to the nuclear resonance frequency by

$$\delta = \frac{\nu_{\mathrm{sample}} - \nu_{\mathrm{ref}}}{\nu_{\mathrm{ref}}} \tag{2.22}$$

where

$$\nu = \frac{\gamma_I B_0}{2\pi}(1-\sigma). \tag{2.23}$$

Thus, the nuclear magnetic shielding constant is related to the NMR chemical shift by

$$\delta = \frac{\sigma_{ref} - \sigma}{1 - \sigma_{ref}} \approx \sigma_{ref} - \sigma \tag{2.24}$$

where the approximation on the right holds for nuclei with small σ_{ref} values. The nuclear resonance frequency, ν, can in general be measured very accurately. In a solution NMR experiment, the precision of measured ν values may be on the order of tenths of Hz or better. The issue with determining precise values of magnetic shielding constants, necessary for relating the results of quantum chemistry calculations with experiment, is that they are not readily available from NMR experiments as the magnetic field strength, B_0, and most magnetogyric ratios, γ, are not known accurately.

Following Ramsey,[16] the nuclear magnetic shielding constant can be divided into a "diamagnetic" and "paramagnetic" contribution (note the "zz" subscript is dropped). Note that this is a non-relativistic theory.[1,16,18]

$$\sigma = \sigma^{d} + \sigma^{p} \tag{2.25}$$

The diamagnetic contribution to the nuclear magnetic shielding for a nucleus, I, depends on the ground state electron configuration of the molecule and can be calculated accurately.[1,39–41] The paramagnetic contribution to the nuclear magnetic shielding depends on electronic excited states and is thus more difficult to calculate. The paramagnetic contribution can be written as[6]

$$\sigma^{p} = \frac{e^2\hbar^2}{2m_e^2}\sum_{n>0}\frac{\left\langle 0\left|\sum_i \bar{l}_{iI} r_{iI}^{-3}\right|n\right\rangle\langle n|\bar{L}_I|0\rangle + \text{c.c.}}{E_0 - E_n} \tag{2.26}$$

where "c.c." denotes the complex conjugate of the preceding term. This expression is derived for non-relativistic electrons and nuclei. This expression along with the corresponding expression given above (eqn (2.6)) for the nuclear spin rotation constant are presented to highlight the similarities between the electronic contribution to the nuclear spin-rotation tensor and the paramagnetic part of the nuclear magnetic shielding tensor. As outlined below, this was recognized by Ramsey many years ago. Note that shielding and spin-rotation tensors are no longer calculated using the above equations, but instead are calculated as derivatives of the total electronic energy.[10,41]

The "average" diamagnetic and paramagnetic contributions to the nuclear magnetic shielding discussed above are the average of the tensor components in the principal axis system, which, for a linear molecule, are oriented within the molecular reference frame as parallel to and perpendicular to the internuclear bond axis. In a linear molecule, the value of $\sigma_{\parallel}^{p}$ is zero in the non-relativistic limit, while both $\sigma_{\parallel}^{d}$ and $\sigma_{\perp}^{d}$ are non-zero.[42] The values of both σ_{av}^{d} and σ_{av}^{p} depend on the rotational and vibrational state of the molecule (*vide infra*).

2.5 The Ramsey–Flygare Method

Due to the difficulties in experimentally measuring magnetic shielding constants, it is necessary to employ alternative means by which absolute shielding scales can be established. Other methods for obtaining σ are not widely used, but include NMR measurements of ratios of relaxation rates,[43] as well as the variable pressure measurement of nuclear resonance frequencies of gases, where the shielding constant of a nucleus in a particular molecule or atom is known. This latter technique requires that one knows γ_N of the isotope. These methods have been discussed previously.[40] Recently, the latter has been used to determine accurate values for $\langle\sigma_0\rangle_{300K}$ for several hydrogen-containing species, as well as establish the currently accepted ^{1}H absolute shielding scale, based on the magnetic shielding of the proton in H_2(g), $\langle\sigma_0\rangle_{300K} = 26.293(5)$ ppm.[25] However, the majority of experimental absolute shielding tensors documented in the NMR literature have been determined from nuclear spin-rotation tensors using the Ramsey–Flygare method. The Ramsey–Flygare method exploits similarities between the paramagnetic part of the nuclear magnetic shielding tensor and the electronic part of the spin-rotation tensor to obtain values for magnetic shielding tensor components. These tensors are more accurately regarded as semi-experimental, as they are based on experimental spin-rotation tensors and on calculated values for the diamagnetic contribution to σ. Furthermore, they are only valid in the non-relativistic limit. As mentioned, Ramsey was the first to develop the theory behind this relationship, however, Flygare was largely responsible for extending Ramsey's theory beyond linear molecules.[44,45] In general, the total magnetic shielding, σ_{av}, for a molecule can be determined *via* the expression[42,45]

$$\sigma_{av} = \sigma^{d}(\text{free atom}) - \left(\frac{\mu_0}{4\pi}\right)\frac{e^2}{3m_e}\sum_{N'} R_{N'} \cdot \frac{\langle\rho\rangle_{N'}}{R_{N'}^3} + \frac{m_p}{2m_e g_I}\left(\frac{1}{3}\right)\sum\frac{C_{gg}}{B_{gg}} \tag{2.27}$$

which can be approximated by

$$\sigma_{av} \approx \frac{m_p}{2m_e g_I}\left(\frac{1}{3}\right)\sum\frac{C_{gg}}{B_{gg}} + \sigma^{d}(\text{free atom}) \tag{2.28}$$

where m_e is the electron mass, m_p is the mass of the proton, μ_0 is the magnetic constant, e is the elementary charge, g_I is the nuclear g-factor for the observed nucleus (see Appendix A for definition), C_{gg} are the diagonal elements of the nuclear spin-rotation tensor, B_{gg} are the corresponding molecular rotational constants, N′ denotes all nuclei other than the observed nucleus, $R_{N'}$ is the internuclear bond distance, and $\rho_{N'}$ is the charge displacement distance if the nuclear charge is not centered on N′.

2.5.1 Linear Molecules

For an isolated diatomic molecule at equilibrium, the paramagnetic contribution to the magnetic shielding is described by

$$\sigma_{\perp}^{\mathrm{p}} = \frac{3}{2}\sigma^{\mathrm{p}} = \frac{m_{\mathrm{p}}C_{\mathrm{I}}}{2m_{\mathrm{e}}g_{\mathrm{I}}B} - \left(\frac{3}{2}\right)\left(\frac{\mu_0}{4\pi}\right)\frac{e^2Z}{3m_{\mathrm{e}}r} \tag{2.29}$$

where σ^{p} is the isotropic paramagnetic shielding, Z is the atomic number of the other nucleus, and r is the internuclear distance. The average diamagnetic component can be approximated using the free atom value for the diamagnetic shielding.

$$\sigma^{\mathrm{d}} \approx \sigma^{\mathrm{d}}(\text{free atom}) + \left(\frac{\mu_0}{4\pi}\right)\frac{e^2Z}{3m_{\mathrm{e}}r} \tag{2.30}$$

Note that the second term in both equations, the nuclear contribution to the magnetic shielding, need not be computed if one wants only the total magnetic shielding. Experimentally, one measures spin-rotation constants in a given vibrational and rotational state, thus vibrational effects must be removed before using eqn (2.27)–(2.29). The equilibrium value for the spin rotation constant for a diatomic molecule is calculated by[46]

$$\begin{aligned} C_{\mathrm{I}} = \langle C_{\mathrm{I}}\rangle_{\mathrm{v,J}} - \left(v+\frac{1}{2}\right)\frac{B_{\mathrm{e}}}{\omega_{\mathrm{e}}}\left[\left(\frac{\partial^2 C_{\mathrm{I}}}{\partial\xi^2}\right)_{\xi=0} - 3a\left(\frac{\partial C_{\mathrm{I}}}{\partial\xi}\right)_{\xi=0}\right] \\ - 4(J^2+J)\left(\frac{B_{\mathrm{e}}}{\omega_{\mathrm{e}}}\right)^2\left(\frac{\partial C_{\mathrm{I}}}{\partial\xi}\right)_{\xi=0} \end{aligned} \tag{2.31}$$

where ξ is the bond displacement and $\xi = 0$ is the equilibrium bond length,

$$\xi = \frac{(r - r_{\mathrm{e}})}{r_{\mathrm{e}}} \tag{2.32}$$

and

$$a = -\left(1 + \frac{\alpha_{\mathrm{e}}\omega_{\mathrm{e}}}{6B_{\mathrm{e}}^2}\right). \tag{2.33}$$

The value of B_{e} can be calculated from rotational constants measured for molecules in two or more vibrational states (*e.g.*, B_0 and B_1), and ω_{e} is obtained from vibrational spectroscopy experiments. After vibrational corrections are carried out and the spin-rotation constant is used to calculate the paramagnetic part of the magnetic shielding, the magnetic shielding constant, σ, can be calculated using the following equation.

$$\sigma = \frac{\sigma_{\parallel}^{\mathrm{d}} + 2\left(\sigma_{\perp}^{\mathrm{d}} + \sigma_{\perp}^{\mathrm{p}}\right)}{3} \tag{2.34}$$

Note that these same magnetic shielding tensor components can also be used to calculate the span of the shielding tensor,

$$\Omega = \sigma_{\parallel} - \sigma_{\perp}. \tag{2.35}$$

As the diamagnetic components of the magnetic shielding tensor are not typically experimentally available, these values can be readily calculated using quantum chemistry methods. Once the equilibrium value for the magnetic shielding constant is obtained, rotational-vibrational effects are reintroduced,[47]

$$\langle \sigma_0 \rangle_{\mathrm{T}} = \sigma + \frac{B_{\mathrm{e}}}{\omega_{\mathrm{e}}}\left[\left(\frac{\partial^2 \sigma}{\partial \xi^2}\right)_{\xi=0} - 3a\left(\frac{\partial \sigma}{\partial \xi}\right)_{\xi=0}\right]\left[\frac{1}{2} + \frac{\mathrm{e}^{-hc\omega_{\mathrm{e}}/kT}}{1 - \mathrm{e}^{-hc\omega_{\mathrm{e}}/kT}}\right] + \frac{4kT}{hcB_{\mathrm{e}}}\left(\frac{B_{\mathrm{e}}}{\omega_{\mathrm{e}}}\right)^2\left(\frac{\partial \sigma}{\partial \xi}\right)_{\xi=0} \tag{2.36}$$

to yield the final semi-experimental value of the magnetic shielding for the isolated molecule, $\langle \sigma_0 \rangle_{\mathrm{T}}$. The values of $\left(\frac{\partial \sigma}{\partial \xi}\right)_{\xi=0}$ and $\left(\frac{\partial^2 \sigma}{\partial \xi^2}\right)_{\xi=0}$ can be estimated from quantum chemistry calculations, or experimentally by measuring isotope effects on magnetic shielding values.[48] Note that vibrational averaging is also necessary to properly compare any calculated spectroscopic parameters to those determined by experiment. Corrections for magnetic susceptibility, gas pressure, or gas-to-liquid shifts are also carried out as necessary (see Figure 2.3).

A classic example of a magnetic shielding scale developed using the Ramsey–Flygare method is that of ^{13}C, based on the spin-rotation constant of the carbon nucleus in ^{13}C^{16}O. The constants $B_0(^{13}\mathrm{C}^{16}\mathrm{O})$ and $D_0(^{13}\mathrm{C}^{16}\mathrm{O})$ have been determined several times (summarized in 2002),[49] with the most recent by Cazzoli *et al.*,[50] where $B_0 = 55.101$ GHz. The ^{13}C spin-rotation constant in ^{13}C^{16}O was first reported in 1968 by Ozier, Crapo, and Ramsey,[51] who used the molecular beam magnetic resonance technique and found $C_{\mathrm{I}}(^{13}\mathrm{C}) = -32.59 \pm 0.15$ kHz. Meerts *et al.* later determined $C_{\mathrm{I}}(^{13}\mathrm{C}) = -32.70(12)$ kHz *via* molecular beam electric resonance spectroscopy.[52] These data were used by Jameson and Jameson in 1987 to refine the then-current ^{13}C absolute shielding scale, with $\sigma_{300\mathrm{K}}(^{13}\mathrm{C}) = 1.0 \pm 1.2$ ppm.[53] The equilibrium value of the ^{13}C magnetic shielding determined by Jameson and Jameson was 3.0 ± 0.9 ppm, *i.e.*, the correction due to rotational–vibrational averaging was ≈ 2 ppm at 300 K. Raynes *et al.* later re-examined the data and proposed

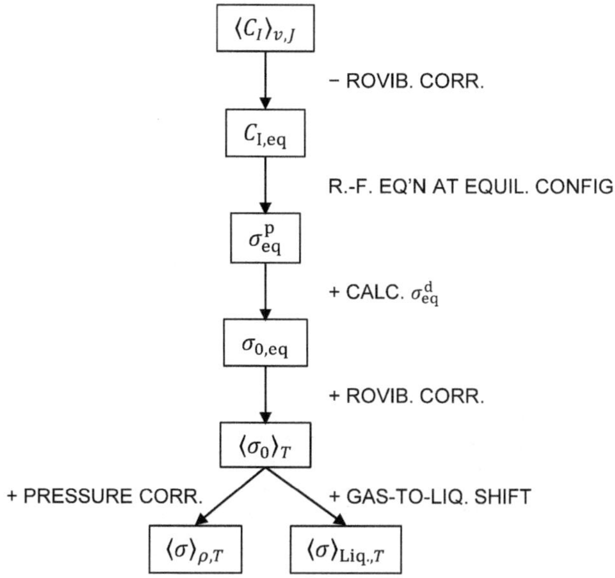

Figure 2.3 Procedure for deriving semi-experimental magnetic shielding constants from experimentally measured spin-rotation constants. Subscript "0" refers to the isolated molecule.
Figure adapted from ref. 42.

$\sigma_{300\mathrm{K}}(^{13}\mathrm{C}) = 0.6 \pm 0.9$ ppm.[54] This was followed by Sundholm *et al.*,[55] who also used the value of $C_{\mathrm{I}}(^{13}\mathrm{C}) = -32.70(12)$ kHz, and found $\sigma_{300\mathrm{K}}(^{13}\mathrm{C}) = 0.9 \pm 0.9$ ppm. CCSD(T) calculations reported in 2003[56] were in good agreement with these three experimental values.

2.5.2 Non-linear Molecules

Several modern magnetic shielding scales are now based on microwave studies of non-linear molecules, *e.g.*, the ^{17}O and ^{33}S magnetic shielding scales discussed below, based on $H_2{}^{17}O$ and $H_2{}^{33}S$. Here we will briefly outline the approach to determining magnetic shielding tensors in non-linear molecules.

Consider sulfur difluoride, SF_2, a rather unstable triatomic molecule with C_{2v} point-group symmetry. SF_2 and various other sulfur fluorides (*e.g.*, SFSF, SF_3SF, SF_2SF_2, SF_6, SF_4, *etc.*) that are present in preparations of SF_2 were studied by gas phase ^{19}F NMR by Gombler *et al.*[57] A peak at $\delta = -167.0$ ppm with respect to CCl_3F was assigned to SF_2. Note the absolute fluorine shielding of CCl_3F(g) is 195.7 ppm,[58] which tentatively puts SF_2 at $\sigma = 362.7$ ppm (see Figure 2.4 and Table 2.1). Although the structure of SF_2 was determined in 1969 by microwave spectroscopy,[59] ^{19}F hyperfine structure was first reported by Gerry and co-workers in 1997.[60] Since SF_2 is an asymmetric rotor, three components of the spin-rotation tensor were obtained from analysis of the high-resolution microwave spectrum.

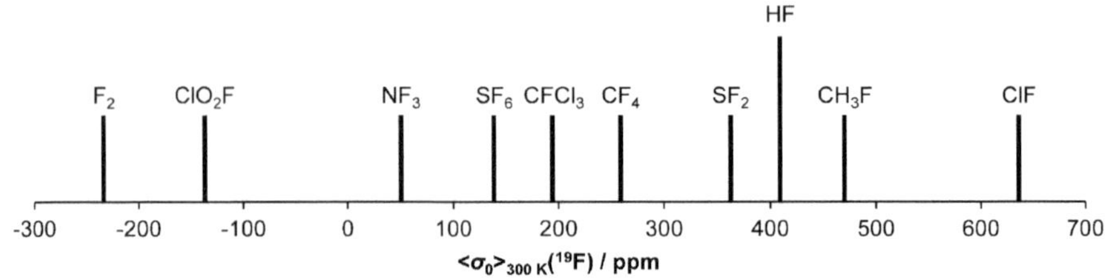

Figure 2.4 The ^{19}F absolute shielding scale for isolated molecules in the gas phase, including select molecules, based on $\langle\sigma_0\rangle_{300\,K}(^{19}F)$ in HF. Magnetic shielding values are from ref. 40.

Table 2.1 Fluorine-19 nuclear magnetic shielding parameters (in ppm) for SF_2, as determined by Fourier-transform microwave spectroscopy (FTMW), NMR spectroscopy, and relativistic (ZORA) DFT calculations carried out for the equilibrium structure using the ADF program package.[178–180]

	FTMW Ref. 61	*ZORA/QZ4P* This work	*NMR* Ref. 57
σ_{av}	363(6)	328	362.7
σ_{11}		−9.4	
σ_{22}		39.2	
σ_{33}		954	
σ_{aa}	525(8)	503	
σ_{bb}	516(11)	490	
σ_{cc}	47(11)	−9.4	

In rotational spectroscopy one measures any internal interaction tensor in the inertial principal axis system. These components are therefore the diagonal elements of the spin-rotation tensor in the inertial axis system, $C_{aa} = -11.66(270)$ kHz, $C_{bb} = -5.88(296)$ kHz, and $C_{cc} = 13.49(150)$ kHz. Note that the molecule lies in the *ab*-plane with the *b*-axis coincident with the C_2 axis. From these C_{gg} values, the authors obtain σ_{gg} values (*i.e.*, the diagonal components of the shielding tensor in the moment of inertia principal axis system) *via* a method similar to that outlined above for diatomic molecules.[60] For σ_{av} the authors obtained 419(71) ppm, within experimental error of that estimated from the gas phase NMR data (note that the trace of the magnetic shielding tensor and therefore σ_{av} is independent of the axis system). In 2001, Brupbacher-Gatehouse extended her study of the hyperfine structure in SF_2 and obtained somewhat different values for C_{aa}, C_{bb}, and C_{cc}: −6.96(87), −2.31(48) and 16.20(37) kHz, respectively.[61] In the non-linear case, the paramagnetic component of the magnetic shielding can be obtained *via*

$$\sigma^{p} = \frac{m_p}{2m_e g_I}\left(\frac{1}{3}\sum\frac{C_{gg}}{B_{gg}}\right) - \left(\frac{\mu_0}{4\pi}\right)\frac{e^2}{3m_e}\sum_{N'}\frac{Z_{N'}}{r_{N'}} \tag{2.37}$$

Analysis of these data leads to $\sigma_{aa} = 525(8)$ ppm, $\sigma_{bb} = 516(11)$ ppm, and $\sigma_{cc} = 47(11)$ ppm, and correspondingly $\sigma_{av} = 363(6)$ ppm. This compares well with the experimental ^{19}F gas phase NMR value, 362.7 ppm, confirming the assignment of Gombler *et al.* Furthermore, one can compare these results with those obtained from quantum chemical computations by transforming the computed principal components of the shielding tensor to the moment of inertia tensor axis system.[62] In order to do this, one requires information about the orientation of both tensors. Gerry *et al.*[60] assumed one component of the ^{19}F magnetic shielding tensor lies along the S–F bond and indeed our calculations support this notion, with the σ_{22} component oriented approximately 4.6° from the S–F bond direction (Faucher and Wasylishen, unpublished). Calculated magnetic shielding tensor components reported by Gerry *et al.* are $\sigma_{aa} = 518$ ppm, $\sigma_{bb} = 618$ ppm, and $\sigma_{cc} = 76$ ppm. *Ab initio* calculations have also been reported by Schindler in 1988,[63] and Chan and Eckert in 2001.[64] The latter authors did not transform their calculated shielding tensor to the moment of inertia principal axis system, invalidating their assessment of the agreement between theory and experiment. Note the analysis used in the above-mentioned papers on SF_2 and FSCl essentially follows that used by Flygare in his early investigation of OF_2.[65] Very recently Teale *et al.*[66] compared Flygare's experimental spin-rotation tensor values for OF_2 with those calculated using various DFT functionals and found reasonable agreement, impressive considering the experimental values were obtained in 1965.

For our next example we consider ammonia, an oblate symmetric top which has been used to establish the nitrogen magnetic shielding scale.[67] Ammonia has played a central role in the development of molecular spectroscopy,[1,4] the MASER,[68] quantum mechanics,[69] and nuclear magnetic resonance.[70–72] Ammonia was the first polyatomic molecule detected in interstellar space and continues to play an important role in the study of interstellar medium.[73] An excellent early introductory discussion of the physics and spectroscopy of ammonia is given by J. P. Gordon[74] as well as in the review by Ho and Townes.[73] The rotational energy of NH_3 is described by two quantum numbers, J and K, which correspond to the total angular momentum and its projection along the symmetry axis (the C_3 axis), respectively. Two distinct isomers of NH_3 exist, one where all three ^{1}H spins are parallel (ortho) and the other where the three spins are not aligned (para); as expected, transitions between these forms are forbidden. One important characteristic of ammonia is that it has a tunneling vibration that splits the J and K states (for $K \neq 0$) into doublets (see Ho and Townes[73] for details). Ammonia is an oblate symmetric rotor, and thus $I_c > I_b = I_a$. Microwave spectroscopy provides B_0 (*i.e.*, rotation of the symmetry axis ($I_b = I_a \approx 2.8147 \times 10^{-47}$ kg m^2)) and high-resolution vibration-rotation spectra provide C_0 ($I_c = 4.5164 \times 10^{-47}$ kg m^2). See David[75] for an introductory discussion of the vibrational–rotational spectrum of ammonia. Determining magnetic shielding constants in this symmetric top is a similar process to that described above for SF_2. Using the principal components of the ^{14}N spin-rotation tensor, $C_{cc} = -6.695 \pm 0.005$ kHz, and $C_{bb} = C_{aa} = -6.764 \pm 0.005$ kHz,[76] the average paramagnetic component of the magnetic shielding, $\sigma^{\mathrm{p}}_{\mathrm{av}}$, is calculated to be −90.06 ppm. The rotational constants $B_0 = 9.9443\ \mathrm{cm}^{-1}$ (298.15 GHz) and $C_0 = 6.196\ \mathrm{cm}^{-1}$ (185.75 GHz) are experimentally available,[77] and the value of $\sigma^{\mathrm{d}} = 354.56$ ppm was determined from *ab initio* calculations,[78] making 264.54 ± 0.05 ppm the final value for σ(^{14}N in NH_3).[67] For the sake of comparison, suppose we consider hydrogen cyanide, HCN, where the paramagnetic term is much larger in magnitude. Here, $C_{^{14}\mathrm{N}} = 10.4 \pm 0.3$ kHz and $B_0 = 44.316$ GHz.[79] In this case, $\sigma^{\mathrm{p}}_{\mathrm{av}}$ is −408.98 ppm,[67] thus $\sigma_0 = -31.4$ ppm. Jameson *et al.* use the measured ^{15}N chemical shift of HCN relative to ammonia to obtain −20.4 ppm for $\langle\sigma_0\rangle_{300\mathrm{K}}$.[67] These numbers are in line with what computational chemists provide, *e.g.*, Gauss and Stanton[80] obtained CCSD values of $\sigma_{0,\mathrm{eq}} = 269.7$ and −16.7 ppm for ammonia and HCN, respectively. More recently, Sun *et al.* obtained 269.5 ppm and −14.1 ppm at the optimized equilibrium bond length.[81]

2.5.3 Relativistic Methods

The importance of relativistic effects in chemistry and in computations of magnetic resonance parameters has been extensively documented and the reader is referred to some recent reviews,[82,83] as well as the text edited by M. Kaupp *et al.*,[41] which has contributions from many experts in the field. The Ramsey–Flygare relationship between spin-rotation and nuclear magnetic shielding constants maps the experimental (necessarily relativistic) value for C_{I} to a non-relativistic value for σ^{p}. Even with the inclusion of relativistic calculations of σ^{d}, the errors incurred from this non-relativistic mapping can be significant, particularly for heavy nuclei. Thus the Ramsey–Flygare method is most useful for light nuclei where relativistic contributions to C_{I} and σ can be ignored. Indeed, for molecules containing only first-row elements it has generally been accepted that relativistic effects are negligible compared to errors in spin-rotation principal components, intermolecular effects, susceptibility effects, *etc.*, *i.e.*, on the order of 1 to 5 ppm. Recent calculations support this contention (*vide infra*). That being said, many absolute shielding constants for heavier nuclei determined using the Ramsey–Flygare method have been reported as it was not until recently that a relativistic theory describing the nuclear spin-rotation tensor was developed.[84–89] This development allowed conversion between nuclear spin rotation tensors and nuclear magnetic shielding tensors with the inclusion of relativistic effects.

In 2007, Xiao *et al.*[84] examined several existing four-component exact and approximate (*i.e.*, without consideration of the negative energy solutions to the Dirac equation) relativistic methods by which magnetic shielding tensors could be computed. Shielding constants for He to Rn and No were calculated. Performance of the approximate methods was based on comparison with the external field-dependent unitary transformation (EFUT) method. The exact methods (*i.e.*, the full field-dependent unitary transformation at matrix level (FFUTm), the external field-dependent unitary transformation at operator level, and the orbital decomposition approach at matrix level

(ODA)) gave consistent results and were therefore evaluated to perform equally well. This research was followed in 2013 by the description of a relativistic molecular Hamiltonian in a body-fixed reference frame for both linear[86] and non-linear[85] molecules. This Hamiltonian was used to develop a description of the nuclear spin–rotation interaction which includes a fully relativistic treatment of electrons and quasi-relativistic treatment of nuclei. That is, the first-order relativistic correction for the nuclear kinetic energy and nucleus–electron interactions is considered. The authors explicitly present an equation for the relativistic "mapping" of nuclear spin-rotation to nuclear magnetic shielding tensors, essentially a relativistic correction to the Ramsey–Flygare equation. Following Xiao *et al.*,[85] the true relativistically mapped value of the magnetic shielding tensor can be obtained by the equation below,

$$\sigma^{I}_{uv}(\text{Expt}) = -\frac{I^{0}_{vv}}{2g_{\text{I}}\mu_{n}}\left[M^{\text{I}}_{uv}(\text{Expt}) - M^{\text{I,d}}_{uv}\right] + \Delta^{\text{I,P}}_{e,uv}(\text{EFUT}) + \sigma^{\text{I,d}}_{uv}(\text{EFUT}) \tag{2.38}$$

for the uv component of the magnetic shielding tensor of nucleus I, where I^{0}_{vv} is the moment of inertia tensor, g_{I} is the nuclear g-factor for nucleus I, μ_{n} is the nuclear magneton, $M^{\text{I}}_{uv}(\text{Expt})$ is the experimentally measured nuclear spin-rotation tensor, $M^{\text{I,d}}_{uv}$ is the calculated diamagnetic spin-rotation term, $\sigma^{\text{I,d}}_{uv}(\text{EFUT})$ is the calculated diamagnetic component of the magnetic shielding, and $\Delta^{\text{I,P}}_{e,uv}(\text{EFUT})$ is the relativistic correction to the mapping of the paramagnetic magnetic shielding term. All calculations are performed using the external field-dependent unitary transformation (EFUT) approach. The relativistic correction amounts to the difference in the contribution of relativistic effects to the computations of the nuclear magnetic shielding and the nuclear spin rotation tensors. In the non-relativistic limit, this expression for relativistic mapping between these two tensors reduces to the Ramsey–Flygare equation. This equation was later applied to developing relativistic, semi-experimental absolute shielding scales based on ^{1}H, ^{19}F, ^{35}Cl, ^{79}Br, and ^{127}I magnetic shielding constants in HX molecules.[90] The general procedure for the relativistic mapping between nuclear spin rotation and nuclear magnetic shielding tensors is outlined by Xiao *et al.*[86] and is very similar to that outlined for the Ramsey–Flygare method above.

In 2012, Aucar *et al.* independently developed a relativistic method by which molecular computations of nuclear magnetic shielding and nuclear spin rotation tensors could be performed.[87] Aucar *et al.* use a perturbation theory approach to introduce relativistic effects into the molecular Hamiltonian, where the motions of nuclei as well as magnetic nucleus–nucleus interactions are treated in a non-relativistic manner whilst electrons are treated relativistically. This work also showed that the relativistic contribution to the two interaction tensors differs, and that the spin-rotation tensor is in general less affected by relativistic effects. Based on this work, Aucar *et al.* published calculated spin-rotation constants for H and X in HX (X = F, Cl, Br, I).[88,89] Jaszuński *et al.* also reported calculations of the spin-rotation and magnetic shielding constants in HCl, with $\sigma_{300K}(^{35}Cl) = 976.202$ ppm,[91] using the method of Aucar *et al.*,[87] to be compared with the non-relativistic value of 946.3 ± 0.9 ppm. Note that the theory for relativistic calculations of shielding constants in paramagnetic systems has also been developed.[92]

In light of the recent advances with respect to the computation of accurate, relativistic nuclear spin-rotation and magnetic shielding tensors,[84–89] we present some literature highlighting the main results and implications of these new methods. Since the relationship between rotational and NMR spectroscopy has been highlighted in a 2003 review,[93] and as updates in absolute shielding scales are reviewed annually,[10,94,95] we will focus only on a few key examples which demonstrate the importance of relativistic effects and highlight where they must be considered.

2.5.3.1 Experimentally Derived Nuclear Magnetic Shielding Scales

Up until recently, light nuclei for which relativistic effects could be ignored were the main focus of the development of absolute shielding scales. As such, many of the absolute shielding scales for light nuclei have undergone multiple revisions over the past several decades, revisions which are often cross-disciplinary in terms of computational chemistry, NMR spectroscopy, and rotational spectroscopy. For our first example, we consider one of the most well-established absolute shielding scales, that for ^{19}F, which is based on the historically relevant molecule hydrogen fluoride. This example illustrates the interplay between the fields of NMR spectroscopy and rotational spectroscopy over the course of the ^{19}F nuclear magnetic shielding scale's history, and is an example of a system for which relativistic effects are not very important.[90] The literature discussed in this section is by no means a comprehensive compilation of the many papers on magnetic shielding in HF, though they are some of the more important experimental contributions to this topic. The nuclear spin-rotation constants and effective dipolar coupling constant for HF were reported to be $|C_I\,(^1H)| = 71 \pm 3$ kHz, $|C_I\,(^{19}F)| = 305 \pm 2$ kHz, and $c_3 = 57 \pm 2$ kHz in 1961.[96] These data were obtained *via* molecular beam magnetic resonance experiments carried out under the supervision of Norman Ramsey. Further results from molecular beam experiments were reported by Ramsey and coworkers in 1964,[97] and an absolute shielding scale for ^{19}F was derived based on $|C_I\,(^{19}F)|$ values for HF and F_2. The obtained shielding value, σ, for ^{19}F in HF $= 414.9 \pm 1.4$ ppm and $\sigma(^{19}F)$ in $F_2 = -210 \pm 8$ ppm. Using these derived magnetic shielding constants, the nuclear magnetic moment of ^{19}F was calculated to be $+2.628\,353 \pm 0.000\,005$ nuclear magnetons. A classic paper by Hindermann and Cornwell[46] documents gas phase ^{19}F NMR data for several small molecules. The authors detail a vibrational correction for the fluorine shielding in HF and obtain $\sigma_{300K}(^{19}F)$ in HF $= 410 \pm 6$ ppm and $\sigma_{300K}(^1H) = 28.8 \pm 0.5$ ppm for the ground vibrational state of HF at zero pressure. For the non-vibrating monomer, $\sigma_0(^1H)$ in HF = 29.2 ± 0.5 ppm. The radio frequency spectra of HF and DF were measured by the molecular beam electric resonance method in 1970.[29] The measurements were carried out with molecules in the lowest vibrational state ($\nu = 0$) and first rotational state ($J = 1$), and the spin-rotation constants obtained were $C_F = 307.637(20)$ kHz and $C_H = -71.128(24)$ kHz. In 1973, radiofrequency spectra of HF and $H^{35}Cl$ in an external electric and magnetic field measured using a high-resolution molecular beam electric resonance spectrometer were reported.[98] This research provided values for the anisotropy in the magnetic shielding of both the ^1H and ^{19}F nuclei in HF, with $\Delta\sigma(^1H) = 24(9)$ ppm and $\Delta\sigma(^{19}F) = 108(9)$ ppm. Gas phase ^{19}F NMR measurements reported by Jameson *et al.* in 1980 yielded resonance frequencies at the zero-density limit and at 300 K for 20 different gases.[99] This research was also extended in 1984.[58] Note, the absolute fluorine shielding constants given in this paper are based on $\sigma_{300\,K}(^{19}F)$ in HF = 410.0 ppm from Hindermann and Cornwell,[46] a value which seems to have withstood the test of time as high-level computations still give results close to 410.0 ppm. (Computational research such as Harding *et al.*[100] and Sun *et al.*[81] illustrate the importance of having access to spin-rotation data.) Note the equilibrium value of Hindermann and Cornwell is 419.7 ± 0.3 ppm, which compares well to the non-relativistic, semi-experimental equilibrium value of Xiao *et al.*,[90] 420.59 ppm. The relativistic correction for ^{19}F in HF, according to Xiao *et al.*, is 4.5 ppm, putting the relativistic semi-experimental equilibrium shielding value at 425.09 ppm. Note that though the nuclear magnetic shielding scale of ^{19}F is based on shielding values for $^1H^{19}F$, chemists use $CFCl_3$ as the ^{19}F chemical shift reference. The absolute shielding of $CFCl_3$ was also determined by Jameson *et al.*[58] for an isolated molecule in the gas phase, $\langle\sigma_0\rangle_{300K}(^{19}F) = 195.7$ ppm, again, based on $\sigma_{300K}(^{19}F) = 410.0$ ppm for HF.[46]

The shielding scales of ^{17}O and ^{33}S, which have undergone several revisions in the recent past, highlight the role of relativistic effects for light nuclei. In 2009, Puzzarini and co-workers proposed an update to the ^{17}O absolute shielding scale based on the non-relativistic semi-experimental value for $\sigma_{300K}(^{17}O)$ in $H_2{}^{17}O$ of 325.3(3) ppm.[101] Note that this value is for an isolated water molecule, and the gas-to-liquid shift for water is known to be large, *ca.* −36 ppm according to experiment.[102] The spin-rotation tensor used to calculate this value was determined *via* microwave experiments using the Lamb-dip technique.[101] Their spin-rotation tensor and absolute shielding tensor values were compared with high-level quantum chemical calculations (*i.e.*, CCSD(T) calculations using several series of correlation-consistent basis sets) and were found to agree well. Puzzarini *et al.* also proposed an updated ^{33}S absolute shielding scale in 2013,[103] on the basis of their microwave measurements of the spin-rotation tensor in $H_2{}^{33}S$. Their value of $\sigma_{300K}(^{33}S) = 716(5)$ ppm agreed with $\sigma_{300K}(^{33}S) = 719$ ppm obtained from high-level calculations. Note that the previous absolute shielding scale was based on $\sigma_0(^{33}S) = 817(12)$ ppm in the OCS molecule.[104] The ^{17}O and ^{33}S absolute shielding scales were revised again in 2015 by Komorovsky *et al.*[105]

The difference in this approach *versus* that of Puzzarini *et al.* a few years earlier is that these new semi-experimental absolute shielding values include a relativistic correction to the Ramsey–Flygare relationship, which was determined from four-component relativistic DFT calculations of both the magnetic shielding and spin-rotation tensors. The new ^{17}O scale was derived from spin-rotation tensors obtained for $H_2{}^{17}O$[101] and $C^{17}O$[106] and is based on absolute shielding values of $\sigma_{300K}(^{17}O) = 328.4(3)$ ppm and $-59.05(59)$ ppm, respectively. The calculated relativistic contribution to the magnetic shielding for ^{17}O is on the order of what was observed for ^{19}F in HF (discussed above), approximately 3 ppm for $H_2{}^{17}O$ and 2 ppm for $C^{17}O$.[105] Moving down the group, Komorovsky *et al.*[105] demonstrate that relativistic effects are surprisingly important for determining absolute shielding values for sulfur. They determined $\sigma_{300K}(^{33}S)$ to be 742.9(4.6) ppm based on microwave data for H_2S, with relativistic effects amounting to *ca.* 25 ppm. For comparison, note the chemical shift range for ^{33}S is approximately 1000 ppm.

Finally, we highlight the first results reported for the method of direct relativistic mapping between nuclear spin-rotation and nuclear magnetic shielding constants proposed by Xiao *et al.*[84–86,90] This method was implemented by the same research group, who reported experimentally derived magnetic shielding constants for both H and X in a series of hydrogen halides (*i.e.*, HX, X = F, Cl, Br, I) in 2014.[90] Values for $\sigma_{eq}(X)$ are 425.09, 995.61, 2961.03, and 5829.97 ppm for HF, HCl, HBr, and HI, respectively. Some of their reported data are shown in Figure 2.5, where it is clear not only

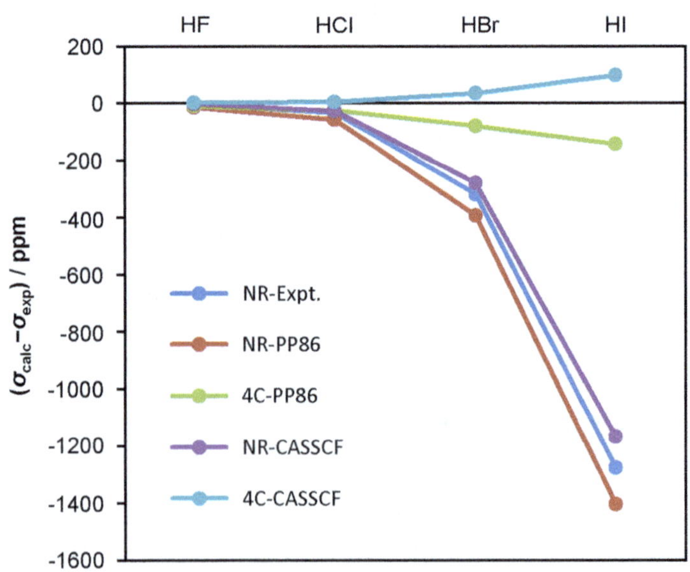

Figure 2.5 Deviations of theoretical and non-relativistic experimentally derived nuclear magnetic shielding constants from the relativistically mapped semi-experimental nuclear magnetic shielding constants for X in HX (X = F, Cl, Br, I).
Adapted from ref. 90.

that the non-relativistic computational methods fail to reproduce the relativistically mapped value, but that the non-relativistic Ramsey–Flygare relationship breaks down fairly quickly for heavier atoms as well.

2.5.3.2 Computations of Spin-rotation and Magnetic Shielding Tensors

The accuracy of quantum chemistry computations of both C_I and σ values has increased significantly over the past several decades. In light of these advances, some have opted to determine absolute shielding values by means of calculation only. This can be advantageous to performing experiments to obtain C_I values for a number of reasons, particularly when experimentally derived magnetic shielding constants for an isotope are not available due to experimental difficulties. Accurate experimental values for these tensors are useful for benchmarking quantum chemistry methods and both experimental and theoretical data provide a means by which to evaluate and revise experimental nuclear magnetic dipole moments, which are important for the accurate calculation of NMR parameters such as indirect spin–spin coupling tensors. Some examples of recently reported calculated spin-rotation and magnetic shielding tensor components are highlighted below. Note that in each case, experimental spin-rotation tensors are used to verify the accuracy of quantum chemistry calculations.

Recent computations by Malkin *et al.*[107] indicate that the ^{119}Sn absolute shielding scale, based on gas phase ^{119}Sn NMR spectroscopy measurements of $Sn(CH_3)_4$,[108] is in need of revision. Based on the work of Aucar *et al.*,[87] Malkin's four-component relativistic calculations indicate a roughly 1000 ppm discrepancy between calculated relativistic $\sigma^P(^{119}Sn)$ values and non-relativistic $\sigma^P(^{119}Sn)$ values determined using the Ramsey–Flygare method from both experimental and calculated spin-rotation constants for SnH_4, $Sn(CH_3)_4$, and $SnCl_4$. For example, the calculated $\sigma_{300K}(^{119}Sn) = 3199$ ppm *versus* the semi-experimental value[108] of $\sigma_{300K}(^{119}Sn) = 2172 \pm 200$ ppm in $Sn(CH_3)_4$. This calculated value of Malkin *et al.* agrees well with the value of $\sigma_0(^{119}Sn) = 3467$ ppm in $Sn(CH_3)_4$ expected based on previous[109] calculations, which included relativistic corrections. The relativistic calculations by Malkin *et al.* accurately reproduced the experimental spin-rotation constants for all three molecules, and adequately reproduced the relative difference between the semi-experimental $\sigma(^{119}Sn)$ values (*i.e.*, within a few hundred ppm). The 1000 ppm difference is significant, even when compared to the moderate chemical shift range of ^{119}Sn, approximately 5000 ppm. Using this consistent difference between calculated relativistic shielding constants and non-relativistic, experimentally derived shielding constants, Malkin *et al.* tentatively redefine the experimental ^{119}Sn absolute shielding constant in SnH_4 to be 3661 ± 132 ppm (or 3299 ± 286 ppm for liquid $Sn(CH_3)_4$). They also propose a revised nuclear magnetic dipole moment for ^{119}Sn, -1.0447773 μ_N.

Calculated magnetic shielding constants for group 13 elements have been recently reported in a series of diatomics by Jaszuński *et al.*[110] The theoretical values were determined based on high-level coupled-cluster and four-component relativistic DFT calculations of both X and F nuclei in XF (X = ^{11}B, ^{27}Al, ^{69}Ga, ^{115}In, ^{205}Tl) molecules. Calculated spin-rotation constants were in good agreement with experimentally determined C_I values, *i.e.*, typically within 1–2 kHz, supporting the accuracy of the calculated magnetic shielding values. The absolute shielding values determined for the X nuclei in XF were 83.48, 583.20, 2197.99, 4763.35, and 12195.32 ppm for X = B, Al, Ga, In, and Tl, respectively. In this work the importance of the inclusion of relativistic effects is stressed, as the relativistic corrections to σ(X) were up to 3669 ppm (*i.e.*, for Tl), not surprisingly increasing as one moves down the group. Relativistic corrections to spin-rotation constants displayed similar trends, *e.g.*, in AlF the relativistic correction was −0.07 kHz compared to the total spin-rotation constant, −8.69 kHz, whereas the total calculated C_{Tl} value was −152.71 kHz after a relativistic correction of −103.69 kHz. In lieu of accurate absolute shielding constants determined semi-experimentally for these nuclei, these values computed by Jaszuński *et al.* are expected to be of considerable interest to computational chemists. Of course, these theoretical shielding scales may be of less value to the NMR experimentalist since one cannot readily carry out NMR measurements on these molecules to obtain the corresponding values of δ_{iso}.

The same research group published a series of coupled cluster and four-component relativistic DFT calculations of spin-rotation and magnetic shielding constants in XF_6 molecules (X = S, Se, Te, Mo, W).[111] Like their work on the group 13 hydrides, these calculations indicated that relativistic corrections are increasingly important for determining the magnetic shielding constants of heavier nuclei. Note that the only experimentally available C_I values for the X nuclei were determined from NMR relaxation data,[112] and are believed to be less reliable than values determined from microwave or molecular beam experiments. Ruud *et al.*[111] calculated σ_0(X) values of 392.6, 1512.8, and 3554.1 ppm for XF_6, where X = ^{33}S, ^{77}Se, and ^{125}Te, respectively, with relativistic corrections of 42.6, 265.4, and 852.0 ppm.

Finally, we see another example of the importance of relativistic effects for light nuclei with the ^{31}P absolute shielding scale, revised in 2011. Lantto *et al.*[113] reported a calculated, relativistically corrected value of $\sigma_{300K}(^{31}P) = 614.7$ ppm in PH_3, an approximately 24 ppm difference when compared with the semi-experimental value. Though this difference is relatively small when compared with the ^{31}P chemical shift range of approximately 2000 ppm, it is clear that even for relatively light nuclei, relativistic corrections to shielding constants can be significant. These calculations, in conjunction with data from gas phase NMR experiments, were used to propose a new ^{31}P nuclear magnetic dipole moment, $-1.1309246(50)\ \mu_N$.

2.5.3.3 *Implications for Experimentally Derived Absolute Shielding Scales*

Relativistic corrections for first-row elements such as ^{13}C, ^{17}O, and ^{19}F in molecules containing only light atoms have been reported to be on the order of a few ppm, though in practice it is unlikely to significantly affect semi-experimental shielding scales. Relativistic effects in general played a smaller role in the calculation of spin-rotation constants.[88] For second-row elements relativistic corrections on the order of tens of ppm (up to a maximum of $\approx$30 ppm for chlorine) have been reported, and thus for these and heavier elements it is clear that the Ramsey–Flygare relationship, in use since the 1950s, is now somewhat antiquated. Relativistic effects should henceforth be included in the establishment of experimentally derived absolute shielding scales. Of even greater concern is the derivation of semi-experimental shielding constants of light nuclei in the presence of heavy nuclei, for which relativistic effects have been shown to be significant. For example, relativistic corrections to ^{1}H shielding values in hydrogen halides are up to 2 ppm (*i.e.*, for HI), according to Xiao *et al.*[90] Calculated $\sigma(^{19}\text{F})$ values in the series of group 13 fluorides (XF, X = B, Al, Ga, In, Tl), and in the hexafluorides (*i.e.*, XF_6, X = S, Se, Te, Mo, W) also display these trends. Relativistic effects accounted for 3.4 ppm of the theoretical ^{19}F shielding constant in BF, increased to −76.4 ppm in TlF and 67.7 ppm in WF_6. Relativistic contributions to spin-rotation and magnetic shielding constants were also seen to vary for the same light nucleus from molecule to molecule,[105] and though this amounted to only a few ppm it may be of greater concern for heavier nuclei with large chemical shift ranges. It is thus always preferable to develop absolute shielding scales using C_I values from several different molecules in order to ensure accuracy. It is apparent from recent calculations that many nuclear magnetic shielding scales will need to be revised, even those for relatively light nuclei. As the accuracy of calculations improves, it is also reasonable to expect many absolute shielding scales will be based on purely calculated values of σ, and for this to expedite the development of shielding scales for less routinely studied NMR-active nuclei.

2.6 The Quadrupolar Interaction

Nuclei with spin $I \geq 1$ possess a nuclear quadrupole moment, eQ, because the nuclear charge distribution is non-spherical. If the shape of the charge distribution is that of a prolate spheroid, eQ is positive, and if it is oblate, eQ is negative. The nuclear quadrupole moment interacts with the electric field gradient (EFG) at the nucleus, represented by the quadrupolar Hamiltonian,

$$\hat{\mathscr{H}}_\text{Q} = \frac{\text{eQ}}{2I(2I-1)\hbar}(\hat{S}_x, \hat{S}_y, \hat{S}_z)\begin{bmatrix} V_{xx} & V_{xy} & V_{xz} \\ V_{yx} & V_{yy} & V_{yz} \\ V_{zx} & V_{zy} & V_{zz} \end{bmatrix}\begin{pmatrix} \hat{S}_x \\ \hat{S}_y \\ \hat{S}_z \end{pmatrix}. \tag{2.39}$$

The EFG is described by a symmetric, traceless second-rank tensor with five independent components, $V_{\alpha\beta}$ $(\alpha,\beta=x,y,z)$ where $V_{\alpha\beta}=V_{\beta\alpha}$. In its principal axis system, the EFG tensor can be characterized by two independent parameters: the largest principal component,

$$V_{ZZ}=eq_{ZZ} \tag{2.40}$$

and the asymmetry parameter,

$$\eta=\frac{V_{XX}-V_{YY}}{V_{ZZ}} \tag{2.41}$$

with the convention that $|V_{ZZ}|\geq|V_{YY}|\geq|V_{XX}|$. The product of the nuclear quadrupole moment and the largest component of the EFG tensor is known as the quadrupolar coupling constant,

$$C_Q=\frac{eQV_{ZZ}}{h}. \tag{2.42}$$

As with spin-1/2 nuclei such as ^{13}C discussed above, the rotational spectra of molecules containing quadrupolar nuclei exhibit fine structure due to the interaction of the nuclear spin angular momentum, $\bar{I}$, with the molecular rotational angular momentum, $\bar{J}$. Due to the coupling between the nuclear spin and rotation angular moments, analysis of high-resolution rotational spectra of quadrupolar nuclei provides both the nuclear spin-rotation tensors and nuclear quadrupolar tensors. In linear molecules, there is only one non-zero component of the spin-rotation tensor and the EFG tensor is characterized only by $C_{Q,}$ as η_Q is zero.

It is important to recognize that from microwave spectroscopy (or molecular beam techniques) one obtains the EFG tensor in the moment of inertia principal axis system. For example, in relatively simple molecules such as HO^2H, methylene chloride, and $CH_2{=}CH^{35}Cl$, the PAS of the EFG tensor at ^2H and ^{35}Cl nuclei do not have symmetry requirements to be either along the $O{-}^2H$[114] or the $C{-}^{35}Cl$ bond.[115] To familiarize oneself with nuclear electric quadrupolar coupling, we have found it useful to calculate the EFG and $C_Q(^2H)$ at a bare deuterium nucleus located 1 Å from a proton, assuming the nuclei lie along the z-axis.[116] First, the potential, V at the deuterium nucleus is

$$V=\frac{e}{4\pi\varepsilon_0 r} \tag{2.43}$$

where ε_0 is the permittivity of vacuum, 8.8542×10^{-12} C^2 m^{-1} J^{-1}, and $r=1.0\times10^{-10}$ m. The electric field, E, is the first derivative of the potential with respect to r,

$$E=\frac{\partial V}{\partial r}=-\frac{e}{4\pi\varepsilon_0 r^2} \tag{2.44}$$

and the electric field gradient is the first derivative of E with respect to r,

$$\mathrm{EFG}=eq_{zz}=\frac{\partial E}{\partial r}=\frac{2e}{4\pi\varepsilon_0 r^3} \tag{2.45}$$

which in our example is 2.880×10^{21} J C^{-1} m^{-2}. The nuclear quadrupole coupling constant, C_Q, in frequency units is

$$C_Q(^2\text{H}) = \frac{eq_{zz}eQ(^2\text{H})}{h} \tag{2.46}$$

and the value of $Q(^2\text{H})$ is $+0.2860$ fm^2,[117,118] making

$$C_Q(^2\text{H}) = \frac{(2.880\times10^{21}\,\text{J}\,\text{C}^{-1}\,\text{m}^{-2})(1.6022\times10^{-19}\,\text{C})(0.2860\times10^{-30}\,\text{m}^2)}{6.6261\times10^{-34}\,\text{J}\,\text{s}}$$
$$= 199.2\,\text{kHz}. \tag{2.47}$$

Note, experimentally, $C_Q(^2\text{H}) = 224.54$ kHz for HD,[119] which is on the same order of the magnitude as that calculated here. If one carries out quantum chemistry calculations of electric field gradients the results are typically given in atomic units. One atomic unit is 9.717362×10^{21} V m^{-2}.[117,120] Quadrupolar tensors can also be characterized by NQR and NMR spectroscopy, however, using these techniques one usually cannot characterize isolated molecules.[121]

2.6.1 Applications of Quadrupolar Tensors from Molecular Spectroscopy

The numerous applications of electric field gradient tensor information is discussed in previous texts and reviews,[2,4,122,123] thus we will limit our discussion to only a few. The relationship between experimental C_Q values (*i.e.*, EFG tensors) and molecular structure has been widely studied. An early example is the proposed relationship between C_Q values and the ionic character of diatomic molecules.[124,125] This idea was discussed in several texts.[2,4,126] Microwave and molecular beam spectroscopic experiments have also been used to study hydrogen bonding and weak intermolecular interactions (*e.g.*, van der Waals molecules).[127] These experimental studies are particularly powerful when used in combination with state-of-the-art computational chemistry techniques.[128,129] Knowledge of C_Q values is useful for the interpretation of NMR relaxation data, particularly for gases.[43,130] Note also that C_Q values are routinely obtained from high-resolution microwave spectroscopy experiments, and the line splitting caused by the quadrupolar interaction is directly dependent on this parameter (Figure 2.2). This gives spectroscopists access to both the magnitude and sign of C_Q, the latter rarely available from NMR experiments. Nuclei and molecules that are difficult to study *via* NMR spectroscopy are also accessible *via* microwave and molecular beam spectroscopy, including those containing heavy nuclides (which often also have large nuclear quadrupole moments), important for testing the accuracy of relativistic calculations (see Table 2.2).

Table 2.2 Select experimental nuclear quadrupolar coupling constants obtained *via* molecular beam and microwave spectroscopy. Values are for the lowest vibrational state ($\nu = 0$).

Molecule	C_Q/MHz	Year	Ref.
^{35}ClF	−145.782 11(9)	1977	181
^{75}AsP	−247.949 5(46)	2006	132
^{121}Sb^{14}N	644.500 0(55)	2004	182
^{121}Sb^{15}N	644.662 0(67)	2004	
^{123}Sb^{14}N	821.648 3(63)	2004	
^{123}Sb^{15}N	821.822 2(10)	2004	
^{121}SbF	−588.851 3(7)	2005	183
^{123}SbF	−750.728 0(8)	2005	
^{121}SbP	617.4417(47)	2004	182
^{123}SbP	787.148 6(53)	2004	
^{121}Sb^{35}Cl	−516.361 1(3)	2005	183
^{121}Sb^{37}Cl	−516.338 5(7)	2005	
^{121}Sb**^{35}Cl**	−41.476 1(5)	2005	
^{121}Sb**^{37}Cl**	−32.686 0(11)	2005	
^{175}LuF	−4957.538 64(77)	2005	184
^{176}LuF	−6994.149 5(12)	2005	
^{175}Lu^{16}O	−4671.66(46)	2011	185
^{175}Lu^{35}Cl	−4290.655 60(42)	2005	184
^{175}Lu^{37}Cl	−4290.853 27(86)	2005	
^{176}Lu^{35}Cl	−6053.794(60)	2005	
^{177}Hf^{32}S	−5158.5235(57)	2002	186
^{179}Hf^{32}S	−5829.3481(63)	2002	
^{129}Xe**^{197}Au**F	−527.637(79)	2004	187
^{131}Xe**^{197}Au**F	−527.45(13)	2004	
^{197}AuF	−53.2344(67)	2000	188
^{209}Bi^{14}N	894.5607(69)	2004	189
^{209}Bi^{15}N	894.8811(109)	2004	
^{209}BiP	898.2172(46)	2004	

One important application of measurements of EFG tensor information is the ability to, in conjunction with accurate quantum chemistry calculations, determine values for nuclear electric quadrupole moments. Having reliable values of Q is critical for being able to reliably execute quantum chemistry calculations involving EFG parameters, either for the prediction of C_Q values of molecules based on experimental structures (*e.g.*, determined from X-ray crystallography or other spectroscopic methods) or for the evaluation of the accuracy of computational procedures, *e.g.*, methods and/or basis sets. Accepted Q values are tabulated by IUPAC,[117] and more recently determined Q values have been summarized by Pyykkö.[131] To determine eQ for a nucleus, experimental C_Q values in conjunction with high-level quantum chemistry calculations of the electric field gradient at the relevant nuclei (*i.e.*, eq_{ZZ} values) are used, according to eqn (2.42). Microwave and molecular beam data for small molecules is the most common source of these C_Q values. A recent example is the ^{75}As nucleus, little-studied *via* NMR spectroscopy due primarily to its large nuclear quadrupole moment, which was determined to be $Q(^{75}\text{As}) = 31.1(2)$ fm^2

based on rotational data and coupled-cluster calculations on the diatomic molecule AsP.[132,133]

Another noteworthy example is the recent determination of eQ(^{63}Cu), which was reported in 2014 by Santiago *et al.*[134] The electric field gradient at the copper nucleus in diatomic molecules is notoriously difficult to calculate,[135] particularly for DFT,[136] with many previous attempts resulting in EFG tensor components with the wrong magnitude and/or sign. The work of Santiago *et al.*[134] is of particular interest due to the careful approach of the authors, who used experimental C_Q(^{63}Cu) values for 16 small linear molecules obtained from high-resolution microwave spectroscopy along with four-component relativistic calculations of the electric field gradient at copper (Figure 2.6) to determine Q(^{63}Cu). Calculation of Q(^{63}Cu) was performed using the equilibrium values of C_Q(^{63}Cu) for the diatomic molecules, which were determined from the experimental values and a vibrational correction. EFG calculations were carried out at the experimental equilibrium geometries using relativistic adapted Gaussian basis sets (RAGBS) for the Cu atom and relatively large correlation-consistent or Dyall basis sets for atoms bound to copper. Several methods were used to calculate the ^{63}Cu EFG tensors, including the Hartree–Fock, DFT, and coupled cluster approach. The authors acknowledge that the absolute values of the calculated EFGs are not well described by the employed computational methods, despite their sophistication, and nuclear electric quadrupole moments

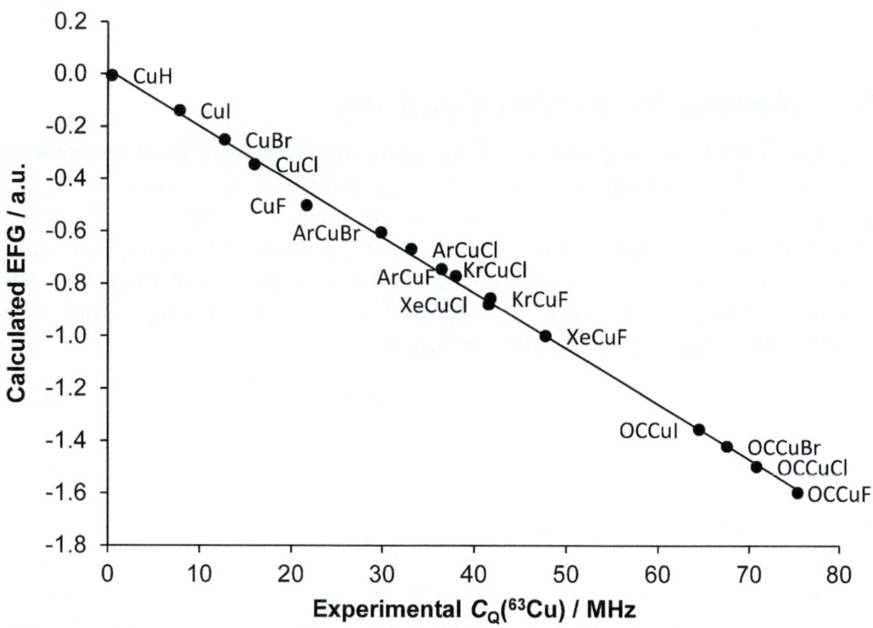

Figure 2.6 Determination of the nuclear electric quadrupole moment for the ^{63}Cu nucleus.
Adapted from ref. 134.

determined by the direct approach (*i.e.*, by use of eqn (2.42) on an individual basis) varied from molecule to molecule and between computational methods. In light of this, the authors use an indirect approach to calculate the final value of the ^{63}Cu nuclear quadrupole moment. The indirect approach uses the variation in EFG and C_Q values, *i.e.*, the difference between EFG and C_Q values when compared with appropriate reference values. As such, it is a reliable method when one possesses data on several molecules, and it eliminates effects from systematic errors in the calculation of the EFGs. The indirect approach, according to Santiago *et al.*,[134] is used by means of linear regression to determine Q according to the equation below.

$$Q(\mathrm{X}) = \frac{\Delta\nu_Q(\mathrm{X})}{234.9647\Delta q(\mathrm{X})} \tag{2.48}$$

Note that $\nu_Q = C_Q/2$ for nuclei with $I = 3/2$, as $\mathrm{X} = {}^{63}\mathrm{Cu}$ in this case. The indirect approach gave much more consistent results across the computational methods used, with the results from the highest level of theory, Dirac–Coulomb coupled cluster calculations with single, double, and triple excitations included, ultimately showing the highest accuracy, *i.e.*, the small intercept values obtained in the linear regression when DC-CCSD(T) and DC-CCSDT EFG values are used suggests that the calculated value of $Q(^{63}\mathrm{Cu})$ is accurate. The final reported value of $Q(^{63}\mathrm{Cu})$ was based on the highest level of theory, the DC-CCSDT calculations, and is -19.8 ± 1.0 fm^2. The previously accepted value of $Q(^{63}\mathrm{Cu})$ was $-22.0(15)$ fm^2, obtained by the muonic method.[137]

2.7 Nuclear Spin–Spin Coupling

The basics of direct and indirect spin–spin coupling have been extensively discussed.[138,139] Direct nuclear spin–spin coupling is a source of line splitting in NMR spectra of solids and an important mechanism of nuclear relaxation, whilst indirect spin–spin coupling is used extensively for elucidating molecular structure and conformation. As they both deal with the coupling of neighboring nuclear spins, the direct and indirect spin–spin coupling Hamiltonians have similar forms.

$$\hat{\mathscr{H}}_{\mathrm{DD}} = R_{\mathrm{DD}}(\bar{I} \cdot \bar{\bar{D}} \cdot \bar{S}) \tag{2.49}$$

$$\hat{\mathscr{H}}_{\mathrm{J}} = (\bar{I} \cdot \bar{\bar{J}} \cdot \bar{S}) \tag{2.50}$$

The 2nd rank tensors $\bar{\bar{D}}$ and $\bar{\bar{J}}$ have some differences. $\bar{\bar{D}}$ is symmetric, has a trace of zero, and in the absence of vibrational averaging it is axially symmetric, *i.e.*, direct-dipolar interactions average to zero in liquids and gas phase NMR studies, though in principle they do influence high-resolution rotational microwave spectra. Of course, they are also observed in NMR studies of solids and partially ordered liquids (*i.e.*, solutions of liquid crystals), as well as through the nuclear Overhauser effect in solutions. $\bar{\bar{J}}$, on the other hand, is in general not symmetric, and generally has a non-zero trace, though, like magnetic shielding, the antisymmetric component of $\bar{\bar{J}}$ only affects the NMR spectrum in special circumstances.[34,36,140] As is well known,[93,141] it is not possible, using any spectroscopic method, to experimentally distinguish between the direct dipolar coupling, R_{DD}, and anisotropy in the indirect spin–spin coupling, ΔJ. Instead, one obtains an experimentally measured effective dipolar coupling constant, R_{eff}, which consists of both a direct and an indirect spin–spin coupling component.

$$R_{eff} = R_{DD} - \frac{\Delta J}{3} \tag{2.51}$$

Note that the dipolar coupling constant, R_{DD}, is defined as

$$R_{DD} = \frac{\mu_0 \gamma_I \gamma_S \hbar}{8\pi^2} \left\langle \frac{1}{r_{IS}^3} \right\rangle \tag{2.52}$$

where r_{IS} is the internuclear separation. Whilst one can theoretically calculate R_{DD} from the molecular structure, the indirect spin–spin coupling is a second-order parameter that cannot easily be calculated. For a linear molecule the anisotropy of $\bar{\bar{J}}$ is defined as

$$\Delta J = J_{\parallel} - J_{\perp} \tag{2.53}$$

where $J_{\parallel}$ and $J_{\perp}$ are the indirect spin–spin coupling parallel and perpendicular to the internuclear axis. Since the value of J depends on the magnetogyric ratios for both nuclei, the reduced spin–spin coupling constant, K, is sometimes reported instead. For coupling between nuclei I and S, the reduced coupling constant is written below.

$$K_{IS} = \frac{4\pi^2 J_{IS}}{h\gamma_I \gamma_S} \tag{2.54}$$

The magnitude of the reduced coupling constant is independent of the magnetogyric ratios for the coupled nuclei, thus K values are important in allowing one to establish periodic trends.

Indirect spin–spin coupling is a two-stage process. The nuclear spin of one nucleus perturbs the electrons in its vicinity, and the resulting perturbation is transferred *via* the electronic framework of the molecule to electrons in the vicinity of the second nucleus, thus coupling the two nuclear spins. The total values of indirect spin–spin coupling tensor components are sums of contributions arising from three different electron–nuclear interactions, the spin–orbital (SO), spin–dipolar (SD), and Fermi-contact (FC) mechanisms, as well as the spin–dipolar Fermi-contact cross-term (SD×FC).[142,143] The most well known, as it is the major contributor to isotropic J-coupling constants in organic molecules, is the Fermi-contact mechanism. This mechanism only contributes to J_{iso} and not to ΔJ. Conversely, the SD×FD term only contributes to ΔJ. The sign of J_{iso} can be either positive or negative, though NMR experiments typically only provide the absolute value of J_{iso}. If the indirect spin–spin coupling interaction stabilizes the antiparallel arrangement of the nuclear spins, then J is positive.

2.7.1 Characterization of Indirect Spin–Spin Coupling Tensors

Many important relationships between the isotropic indirect spin–spin coupling constant, J_{iso}, and molecular structure have been established. Though isotropic spin–spin coupling constants are routinely measured in liquid- and solid-state NMR experiments, J tensors are difficult to characterize experimentally *via* NMR spectroscopy and very few have been measured with any degree of confidence.[139,141,144] For coupling between light atoms, contributions to R_{eff} from ΔJ are often ignored. This is usually a good approximation, and can now be verified by quantum chemistry computations.

2.7.1.1 Experimentally Determined Indirect Spin–Spin Coupling Tensors

The ability to experimentally measure indirect spin–spin coupling tensors *via* microwave spectroscopy is fairly limited due to the typically small values of J_{iso} and ΔJ.[145] Typical resolution in a microwave spectroscopy experiment is on the order of kHz to hundreds of Hz, making J-coupling constants on the order of 500 Hz or less very difficult if not impossible to characterize. On the other hand, molecular beam resonance techniques can provide one with experimental values for constants c_3 and c_4, which are equal to R_{eff} and J_{iso}, respectively.[8,9,93] The molecular beam electric resonance spectrometer built by Ramsey at Harvard in the 1960s (which has subsequently been moved and used at St. Olaf College in Northfield, Minnesota, for three decades, before finding its current location at Southern Polytechnic State University in Marietta, Georgia)[146] is capable of resolving isotropic J-coupling constants with values less than 100 Hz, with reported accuracies on the order of 1 Hz. The research group of James Cederberg has compiled an impressive collection of accurate spin–spin coupling tensors for alkali metal halides such as LiF, KF, CsF, and RbF, as well as many of the analogous molecules with heavier halides. Cederberg *et al.* have, to date, reported molecular beam experiments for complete sets of alkali metal halides, including KF–KI[147–149] and RbF–RbI.[146,150–152] What is impressive about this research is that one obtains not only the sign and magnitude of J_{iso}, but also ΔJ, a parameter that often eludes conventional NMR measurements.[153] For a list of some of Cederberg's recently reported c_3 and c_4 values, see Table 2.3. For a list of values of c_3 and c_4 measured using molecular beam and microwave spectroscopy prior to year 2000, see Vaara *et al.*[153] Note that out of those values previously reported, small molecules such as LiH and HF were of particular interest historically as model compounds for quantum chemistry calculations.

Table 2.3 Experimental R_{eff} (c_3) and J_{iso} (c_4) values obtained from molecular beam spectroscopy reported between years 2000 and 2015.

Molecule	c_3/kHz	c_4/kHz	Year	Ref.
^{6}LiI	$0.23805(58) - 0.0030(11)(v+1/2) + 0.00035(42)(v+1/2)^2$	$0.02485(45) - 0.00138(85)(v+1/2) + 0.00021(33)(v+1/2)^2$	2005	190
^{7}LiI	$0.6287(15) - 0.0075(28)(v+1/2) + 0.00080(96)(v+1/2)^2$	$0.0656(12) - 0.0034(21)(v+1/2) + 0.00047(75)(v+1/2)^2$	2005	
^{85}RbF	$0.79757(29) - 0.00754(40)(v+1/2) + 0.000111(97)(v+1/2)^2$	$0.23708(24) - 0.00200(32)(v+1/2) - 0.000057(75)(v+1/2)^2$	2006	150
^{87}RbF	$2.70292(99) - 0.0255(14)(v+1/2) + 0.00037(33)(v+1/2)^2$	$0.80346(80) - 0.0068(11)(v+1/2) - 0.00019(25)(v+1/2)^2$	2006	
^{85}Rb^{35}Cl	$0.0330(11) - 0.0007(23)(v+1/2) - 0.00017(76)(v+1/2)^2$	$0.02651(63) - 0.0013(13)(v+1/2) - 0.00061(41)(v+1/2)^2$	2006	151
^{87}Rb^{35}Cl	$0.1120(36) - 0.0022(76)(v+1/2) - 0.0006(26)(v+1/2)^2$	$0.0898(21) - 0.0045(42)(v+1/2) - 0.0021(14)(v+1/2)^2$	2006	
^{85}Rb^{37}Cl	$0.02751(89) - 0.0005(18)(v+1/2) - 0.00014(61)(v+1/2)^2$	$0.02207(52) - 0.0011(10)(v+1/2) + 0.00049(33)(v+1/2)^2$	2006	
^{87}Rb^{37}Cl	$0.0932(30) - 0.0018(62)(v+1/2) - 0.0005(20)(v+1/2)^2$	$0.0748(18) - 0.0037(34)(v+1/2) - 0.0016(11)(v+1/2)^2$	2006	
^{39}K^{79}Br	$0.03749(40) + 0.00013(41)(v+1/2) - 0.000042(96)(v+1/2)^2 + 0.000006(21)[J(J+1)]$	$0.02189(18)$	2008	149
^{39}K^{81}Br	$0.04041(44) + 0.00014(44)(v+1/2) - 0.00004(10)(v+1/2)^2 + 0.000007(23)[J(J+1)]$	$0.02359(19)$	2008	
^{39}K^{127}I	$0.01092(56) - 0.00020(36)(v+1/2)$	$0.02409(27) + 0.00002(19)(v+1/2)$	2008	
^{41}K^{127}I	$0.00599(31) - 0.00011(20)(v+1/2)$	$0.01322(15) + 0.00001(11)(v+1/2)$	2008	
^{23}Na^{19}F	$4.042(11) - 0.174(29)(v+1/2) + 0.087(20)(v+1/2)^2 - 0.0181(41)(v+1/2)^3 + 0.00084(73)[J(J+1)]$	$0.1479(81) + 0.084(21)(v+1/2) - 0.063(14)(v+1/2)^2 + 0.0135(29)(v+1/2)^3 - 0.00071(58)[J(J+1)]$	2010	191
^{85}Rb^{127}I	$-0.04404(26) + 0.00058(14)(v+1/2)$	$0.10344(35) + 0.00047(23)(v+1/2)$	2011	152
^{87}Rb^{127}I	$-0.14926(90) + 0.00197(48)(v+1/2)$	$0.3505(12) + 0.00157(79)(v+1/2)$	2011	
^{85}Rb^{79}Br	$0.0183(28) - 0.0008(17)(v+1/2)$	$0.09398(98) + 0.0016(16)(v+1/2) - 0.00090(43)(v+1/2)^2$	2014	146
^{85}Rb^{81}Br	$0.0197(30) - 0.0008(18)(v+1/2)$	$0.1013(11) + 0.0017(17)(v+1/2) - 0.00095(45)(v+1/2)^2$	2014	

Consider indirect spin–spin coupling in HF. The classic NMR spectroscopy paper by Solomon and Bloembergen[154] is concerned with relaxation of ^{1}H and ^{19}F nuclei in anhydrous hydrofluoric acid. While splittings in the NMR spectra of HF are not observed directly, the authors used relaxation data and the nuclear Overhauser effect to deduce that A/h or $J(^{19}\mathrm{F}, {}^{1}\mathrm{H}) = 615 \pm 50$ Hz. A variable-pressure gas phase NMR study reported by Hindermann and Cornwell in 1968 highlights the dynamic behavior of pure HF gas and HF–HD gas mixtures.[155] In this research, splittings due to indirect spin–spin coupling were not observed, instead the spectrum consisted of a single averaged ^{19}F resonance. This was evidence for rapid monomer exchange processes in the gas phase, *i.e.*, the breaking and re-forming of HF/HD polymers. The molecular beam electric resonance spectra of HF and DF, reported in 1970 by Muenter and Klemperer,[29] indicated that $J_{\mathrm{HF}} = 0.529(23)$ kHz. Finally, in 1974, Martin and Fujiwara[156] were able to observe splittings due to $J(^{19}\mathrm{F}, {}^{1}\mathrm{H})$ for HF in aprotic solvents, *e.g.*, splittings of 479 ± 4 Hz were observed in acetonitrile. In a 1977 review[157] dealing with the calculation of nuclear spin–spin coupling, Kowalewski presented a table showing that there was some controversy on the sign of $^{1}J(^{19}\mathrm{F}, {}^{1}\mathrm{H})$ until Muenter and Klemperer confirmed that it was positive (*i.e.*, quantum chemistry calculations performed earlier than 1975 were inconsistent).

2.7.1.2 Calculations of Indirect Spin–Spin Coupling Tensors

Ab initio calculations of J-tensors were not carried out extensively until the late 20th century, as J-tensors were difficult to calculate, and early calculations were often unreliable. Considerable progress has been made in computing J-tensors in the past 15 years.[41,141,153,158,159] High-resolution spin–spin coupling data (*i.e.*, J_{iso} and, in particular, ΔJ values) obtained from molecular beam experiments have been extremely valuable for testing theoretical approaches to calculating indirect spin–spin coupling tensors.[153,160] Knowledge of experimental J-tensors, including calculations thereof, has been used to propose several periodic trends for J_{iso} and ΔJ values.[93,158] As well, a breakdown of the individual contributions to indirect spin–spin coupling tensors is only available through quantum chemistry calculations, in which case experimental values for the isotropic J-coupling and anisotropy in J (*i.e.*, c_4 and the indirect contribution to c_3) are used to verify the accuracy of the overall calculations. Data from molecular beam experiments are especially valuable for this purpose as they are typically carried out on isolated molecules, and are thus free from intermolecular effects.

One of the first comprehensive reports of *ab initio* indirect spin–spin coupling tensor calculations which utilized microwave and molecular beam data was that of Bryce and Wasylishen in 2000.[141] This research compared experimental values for J_{iso} and ΔJ obtained from high-resolution microwave and molecular beam spectroscopy to the results from multi-configurational SCF calculations in order to reliably establish a method for computing the entire J tensor. This was followed up in 2002 with ZORA DFT calculations of

J-tensors in diatomic molecules,[158] with the purpose of establishing periodic trends for the entire tensor, and in 2009 an updated study on *J*-tensors was reported by Bryce and Autschbach,[160] which used relativistic (*i.e.*, ZORA) hybrid DFT calculations to compute *J* tensors for a series of diatomic molecules. These studies revealed several trends in both K_{iso} and ΔK as well as the contribution from the mechanisms behind each. K_{iso} and ΔK absolute values tend to increase as one moves down and across the periodic table and, in general, there is a correlation between the product of the atomic numbers, Z_1Z_2, and the magnitude of both K_{iso} and ΔK. The latter is thought to be due to the dependence of the spin–dipolar and spin–orbit coupling mechanisms on the inverse cube of the electron–nucleus distance, $\frac{1}{\langle r^3 \rangle}$, in the non-relativistic limit, as first described by Ramsey.[17] Note that ΔK values are negative for the interhalogen molecules, indicating that the largest component of the *J*-tensor is oriented perpendicular to the bond axis. The Fermi-contact mechanism was long thought to be the most important *J*-coupling mechanism. Importantly, these calculations revealed that the Fermi-contact mechanism, while it plays the dominant role in the isotropic spin–spin coupling of alkali metal diatomics, plays a relatively minor role in contributing to K_{iso} for any homonuclear or heteronuclear interhalogen diatomic molecule. In fact, calculations suggest that the paramagnetic spin–orbit coupling mechanism is responsible for approximately 70 to 80% of the K_{iso} value for these molecules (see Figure 2.7). In general, calculations suggest

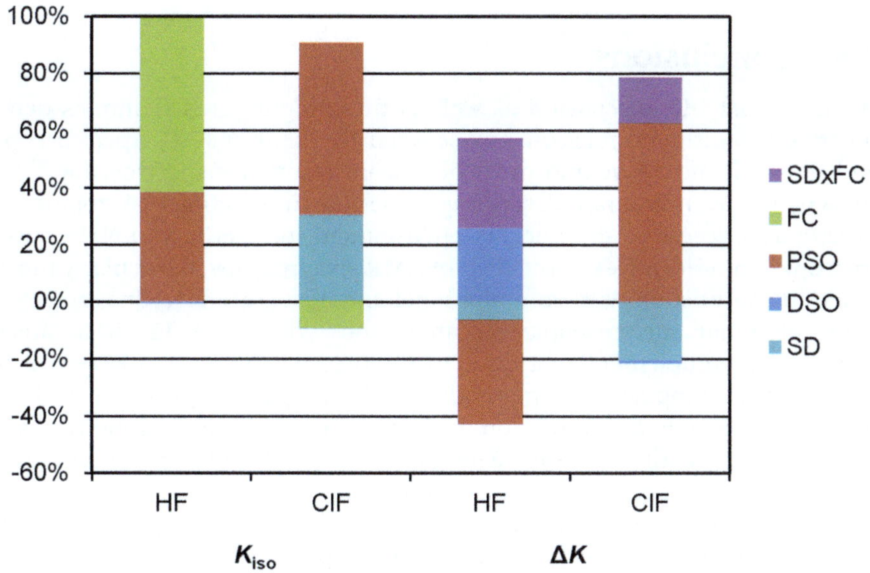

Figure 2.7 Contribution of spin–spin coupling mechanisms to the isotropic and anisotropic reduced spin–spin coupling of HF and ClF as determined by MCSCF calculations. Note the Fermi-contact mechanism contributes little to the reduced isotropic spin–spin coupling constant for ClF. Data from ref. 141.

Table 2.4 Observed and calculated values of K_{iso} and ΔK for the thallium halides.[a]

	K_{iso} (obs.)	K_{iso} (calc.)	ΔK (obs.)	ΔK (calc.)	Expt. Ref.
TlF	−202	−248	173	200	27
TlCl	−224	−259	262	251	192
TlBr	−361	−399	448	484	193
TlI	−474	−521	664	728	194

[a]Most of the experimental data are accurate to two significant figures. The reader should consult the original paper(s) for details.

that as one moves from left to right across the periodic table, the Fermi-contact mechanism becomes less important for mediating indirect spin–spin coupling between nuclei. Likewise, the Fermi-contact/spin dipolar cross-term contributes substantially to the anisotropy in K for the alkali metal halides and, again, the paramagnetic spin–orbit term dominates for interhalogen molecules. Calculations also suggest that any mechanism for spin–spin coupling can in general contribute substantially to both K_{iso} (*i.e.*, FC, PSO, DSO, SD) or ΔK (*i.e.*, FCxSD, PSO, DSO, SD).

Finally, we include the experimental indirect spin–spin coupling data for the thallium halides[153] as well as the results of two-component relativistic hybrid density functional computations.[161,162] These data are important for testing the accuracy of relativistic calculations, though the experimental data may not have high precision. Furthermore, they illustrate the general trend that K_{iso} and ΔK increase in magnitude as one moves down any group in the periodic table, *e.g.*, $\Delta K(\text{TlF}) < \Delta K(\text{TlCl}) < \Delta K(\text{TlBr}) < \Delta K(\text{TlI})$ (see Table 2.4).

2.8 Conclusions

We hope that this discussion as well as these highlighted examples demonstrate the intimate relationship between the field of NMR spectroscopy and those of microwave and molecular beam resonance spectroscopy.[93,122] Indeed, the Hamiltonian describing hyperfine interactions in rotational spectra also describes the nucleus–environment interactions in NMR spectroscopy. Many advances in the field of NMR spectroscopy, particularly those made in conjunction with computational chemistry, are directly connected with rotational spectroscopic research, and work done in these areas continues to supplement that done in the field of NMR spectroscopy. It is stressed that computational methods that NMR spectroscopists, as well as other physical chemists, rely on to determine molecular properties are developed using the high-precision data obtained from microwave and molecular beam spectroscopy as benchmarks.

Many of our absolute shielding scales, connecting theory with experimental values for chemical shifts, were made possible by molecular beam/microwave spectroscopic research in which spin-rotation constants were measured. The majority of these are based on the relationship recognized by Ramsey, which connects the paramagnetic part of the magnetic shielding to the nuclear spin-rotation constant from rotational spectroscopy. Though the non-relativistic expression used historically is now somewhat antiquated, we hope this overview is a valuable introduction to this topic, as well as the older literature in which absolute shielding scales were developed. There now exist increasingly reliable and relatively easily implemented computational methods for estimating the nuclear spin-rotation and magnetic shielding tensors with the inclusion of electron relativistic effects, as well as a method by which spin-rotation tensors can be directly "relativistically mapped" to nuclear magnetic shielding tensors. Several new absolute shielding scales have been proposed for both light and heavy nuclei based on these methods. In justifying a "theoretical" shielding scale we strongly urge that high-level (or high-quality) computations be tested on as many molecules as possible. That is, magnetic shielding scales should be based on calculations and experimental data of more than one molecule. For example, in the case of oxygen, accurate ^{17}O spin-rotation data are available for both CO and H_2O, and gas phase ^{17}O NMR data are available for several small molecules,[163,164] so to confirm the internal consistency of a magnetic shielding scale accurate computations of several molecules may be checked.

A significant fraction of current and historical nuclear electric quadrupole moments were also derived using precise nuclear quadrupolar coupling constants determined from rotational spectra, in combination with calculations of eq_{ZZ}. Another significant contribution to the field of NMR spectroscopy is the ability to characterize R_{eff} and, by association, ΔJ values, a parameter that is elusive in most NMR experiments. Spin–spin coupling interactions are generally small in magnitude compared to quadrupolar or spin–rotation interactions, thus in terms of rotational spectroscopy indirect spin–spin coupling is generally only observed in molecular beam experiments where the resolution is on the order of Hz or tens of Hz. Together with quantum chemistry calculations, the determination of experimental ΔJ values has increased our understanding of fundamental spin–spin interactions, allowing spectroscopists to propose several periodic trends as well as gain insight into the mechanisms of spin–spin coupling. Clearly, the interplay of experimental NMR spectroscopy, molecular spectroscopy, and quantum chemistry computations will continue to advance one's understanding of molecular structure.

Appendix A: The Measurement of Nuclear Magnetic Moments

It is instructive to review how one can use gas phase NMR, molecular beam, or microwave spectroscopic data to determine nuclear magnetic moments. Early experimental techniques to measure nuclear magnetic moments were summarized by Ramsey in 1953.[165] As well, this text provided an early compilation of nuclear magnetic moments. Accounts of the history associated with measuring the proton nuclear magnetic moment are given by J. S. Rigden,[166] and two classic papers concerning the measurement of the proton magnetic moment were published by research groups led by O. Stern[167] and by I. I. Rabi.[168] Otto Stern received the 1943 Nobel Prize in Physics, in part for his contribution to measuring the magnetic moment of the proton, and Isidor I. Rabi won the 1944 Nobel Prize in Physics for his research in measuring nuclear magnetic moments *via* molecular beam methods.

Nuclear magnetic moments are generally represented as

$$\mu_{\mathrm{I}} = \gamma_{\mathrm{I}}\hbar[I(I+1)]^{1/2} \tag{2.55}$$

where the magnitude of the magnetic moment depends on the total length of the nuclear spin angular momentum vector. However, magnetic moments are also often written as

$$\mu_{\mathrm{I}} = \gamma_{\mathrm{I}}\hbar I \tag{2.56}$$

where the magnitude of the magnetic moment depends on the largest projection of the nuclear spin angular momentum on the z-axis. They are generally listed in units of nuclear magnetons, μ_{N}, where the value of $\mu_{\mathrm{N}} = 5.050\ 783\ 53 \times 10^{-27}$ J T^{-1}.[120] For many applications, scientists often prefer to simply report the magnetogyric ratio, γ_{I}, or the nuclear "g_{I}" value,

$$g_{\mathrm{I}} = \frac{\gamma_{\mathrm{I}}\hbar}{\mu_{\mathrm{N}}}. \tag{2.57}$$

It is important to recognize that values of γ_{I} and g_{I} pertain to the "bare" unshielded nucleus. The most accurate magnetic dipole moment values and their associated errors can be found on the NIST website under fundamental constants.[120] Tables of currently accepted magnetic moments,[169] as well as earlier IUPAC recommendations,[170] are available, however, many of these values should be used with caution. Older tables of magnetic moments used, by today's standards, crude estimates of shielding simply because of the problem associated with calculating σ. Typically only a diamagnetic correction to the magnetic shielding was performed. We first noticed that there were problems with available tables of magnetic moments in 1995 while attempting to establish an absolute shielding scale for tin, for which several dissimilar values for the ^{119}Sn magnetic moment were available.[171] The general problem with experimental values for nuclear magnetic moments was also described in more detail by Gustavsson and Mårtensson-Pendrill.[172] At this point it is useful to remind readers of the most general and currently used method of determining nuclear magnetic dipole moments. Values of nuclear magnetic moments, μ_{I}, are generally referenced or established based on accepted values for the proton, the helion (bare helium nucleus) or the shielded helion (*i.e.*, the helium atom). As was already mentioned, magnetic moments may also be derived *via* nuclear shielding constants determined from nuclear spin-rotation constants.

We will briefly work through an example, the magnetic moment of ^{83}Kr, recently reported by W. Makulski.[173] In this research, the ^{3}He and ^{83}Kr NMR resonance frequencies of mixtures of helium–krypton gas were measured and extrapolated to zero gas density to eliminate interatomic interactions. At an applied magnetic field of approximately 11.75 T, $\nu_0(^3\text{He}) = 381.356\ 606\ 62(7)\,\text{MHz}$ and $\nu_0(^{83}\text{Kr}) = 19.264\ 669\ 87(12)\,\text{MHz}$. The resonance frequencies are related to the magnetogyric ratios and the magnetic field, B_0, as follows.

$$\nu_0(^3\text{He}) = \frac{B_0\gamma_{\text{He}}}{2\pi}(1 - \sigma_{0,\text{He}}) \tag{2.58}$$

and

$$\nu_0(^{83}\text{Kr}) = \frac{B_0\gamma_{\text{Kr}}}{2\pi}(1 - \sigma_{0,\text{Kr}}) \tag{2.59}$$

therefore

$$\frac{\nu_0(^{83}\text{Kr})}{\nu_0(^3\text{He})} = \frac{[\gamma_{\text{Kr}}(1 - \sigma_{0,\text{Kr}})]}{[\gamma_{\text{He}}(1 - \sigma_{0,\text{He}})]} = 0.050\ 516\ 156\ 1(3). \tag{2.60}$$

The shielded helion magnetogyric ratio over 2π is $32.434\ 100\ 84(81)\,\text{MHz T}^{-1}$. Therefore,

$$\gamma_{\text{Kr}} = \left(\frac{\nu_0(^{83}\text{Kr})}{\nu_0(^3\text{He})}\right)\frac{[\gamma_{\text{He}}(1 - \sigma_{0,\text{He}})]}{(1 - \sigma_{0,\text{Kr}})} \tag{2.61}$$

$$\gamma_{\text{Kr}} = (0.050\,516\,156\,1\,(3))\frac{2\pi(32.434\,100\,84(81)\times 10^6\ \text{Hz T}^{-1})}{(1 - \sigma_{0,\text{Kr}})} \tag{2.62}$$

$$\gamma_{\text{Kr}} = \frac{1.029\,466\,05(17)\times 10^7\ \text{rad s}^{-1}\,\text{T}^{-1}}{(1 - \sigma_{0,\text{Kr}})} \tag{2.63}$$

Notice that in order to determine γ_{Kr} we need to know the absolute magnetic shielding constant for krypton at zero density. The situation is the same regardless of which magnetic moment we wish to determine. As outlined in this chapter, measured spin-rotation tensors provide an experimental method of obtaining shielding tensors and this is an important source of accurate magnetic shielding tensors. In recent years quantum chemistry computations have served as complementary methods providing shielding information and in some cases are the only reasonable source of shielding. In the case of the krypton atom, one can expect that quantum computations of the shielding can be accurately predicted, certainly to ± 50 ppm. Following Makulski, we use $\sigma_0(^{83}\text{Kr}) = 3577.3(33)$ ppm.[174] Thus,

$$\gamma_{\text{Kr}} = 1.0331619(34)\times 10^7\ \text{rad s}^{-1}\ \text{T}^{-1} \tag{2.64}$$

or

$$\mu_{\text{I}} = \gamma_{\text{I}}\hbar I = (1.0331619(34)\times 10^7\ \text{rad s}^{-1}\ \text{T}^{-1})(9/2)\hbar \tag{2.65}$$

$$\mu_{\text{I}} = 4.902945(16)\times 10^{-27}\ \text{J T}^{-1} = 0.9707297(32)\ \mu_{\text{N}} \tag{2.66}$$

The NMR experiment described here does not give the sign of the nuclear magnetic moment, which is negative for ^{83}Kr. The previously accepted value was −0.967 221(12) μ_N.[175] Note in the IUPAC tables of Harris *et al.*[170] the value for ^{83}Kr is listed as −1.07311 μ_N as these tables give $\mu_I = \gamma_I \hbar [I(I+1)]^{1/2}$.

We would like to indicate that the research groups made up of M. Jaszuński, K. Jackowski, W. Makulski and co-workers have devoted considerable effort into improving values of nuclear magnetic dipole moments. These have been summarized recently.[176,177] Finally, there seems to be a disconnect between the nuclear physics community (*e.g.*, N. J. Stone)[169] and the chemical physics community.

References

1. W. H. Flygare, *Molecular Structure and Dynamics*, Prentice-Hall, Englewood Cliffs, 1978.
2. C. H. Townes and A. L. Schawlow, *Microwave Spectroscopy*, McGraw-Hill, New York, 1955, pp. 225–247.
3. H. W. Kroto, *Molecular Rotation Spectra*, John Wiley & Sons, London, 1975.
4. W. Gordy and R. L. Cook, *Microwave Molecular Spectra*, Wiley, New York, 1984.
5. J. D. Graybeal, *Molecular Spectroscopy*, McGraw-Hill, rev 1st edn, 1988.
6. J. Brown and A. Carrington, *Rotational Spectroscopy of Diatomic Molecules*, Cambridge University Press, Cambridge, 2003.
7. N. F. Ramsey, *Molecular Beams*, Oxford University Press, 1956.
8. N. F. Ramsey, *Am. Sci.*, 1961, **49**, 509.
9. T. R. Dyke and J. S. Muenter, *Molecular Structure and Properties, Physical Chemistry Series Two*, ed. A. D. Buckingham, Butterworth & Co., London, 1975, vol. 2, pp. 27–92.
10. C. J. Jameson and A. C. De Dios, *Nuclear Magnetic Resonance*, ed. K. Kamieńska-Trela, RSC Publishing, Cambridge, 2015, ch. 2, vol. 44.
11. W. T. Raynes, A. D. Buckingham and H. J. Bernstein, *J. Chem. Phys.*, 1962, **36**, 3481.
12. C. J. Jameson, *Chem. Rev.*, 1991, **91**, 1375.
13. K. Jackowski, *J. Mol. Struct.*, 2006, **786**, 215.
14. B. P. Dailey, *Annu. Rev. Phys. Chem.*, 1953, **4**, 425.
15. A. Abragam, *Principles of Nuclear Magnetism*, Oxford University Press, New York, 1961.
16. N. F. Ramsey, *Phys. Rev.*, 1950, **78**, 699.
17. N. F. Ramsey, *Phys. Rev.*, 1953, **91**, 303.
18. P. Pyykkö, *Theor. Chem. Acc.*, 2000, **103**, 214.
19. W. T. Raynes and N. Panteli, *Mol. Phys.*, 1983, **48**, 439.
20. W. T. Raynes and N. Panteli, *Chem. Phys. Lett.*, 1983, **94**, 558.
21. J. Oddershede, J. Geertsen and G. E. Scuseria, *J. Phys. Chem.*, 1988, **92**, 3056.
22. G. B. Bacskay, *Chem. Phys. Lett.*, 1995, **242**, 507.

23. M. Jaszuński, G. Łach and K. Strasburger, *Theor. Chem. Acc.*, 2011, **129**, 325.
24. P. Garbacz, *Chem. Phys.*, 2014, **443**, 1.
25. P. Garbacz, K. Jackowski, W. Makulski and R. E. Wasylishen, *J. Phys. Chem. A*, 2012, **116**, 11896.
26. T. Myint, D. Kleppner, N. F. Ramsey and H. G. Robinson, *Phys. Rev. Lett.*, 1966, **17**, 405.
27. R. v. Boeckh, G. Gräff and R. Ley, *Z. Phys.*, 1964, **179**, 285.
28. H. Y. Carr and E. M. Purcell, *Phys. Rev.*, 1952, **88**, 415.
29. J. S. Muenter and W. Klemperer, *J. Chem. Phys.*, 1970, **52**, 6033.
30. T. C. English and J. C. Zorn, *Molecular Physics*, ed. D. Williams, Academic Press, New York and London, 2nd edn, 1974, vol. 3, p. 669.
31. M. Mehring, *Principles of High Resolution NMR in Solids*, Springer-Verlag, New York, 2nd edn, 1983.
32. U. Haeberlen, *High Resolution NMR in Solids: Selective Averaging*, Academic Press, New York, 1976.
33. W. S. Veeman, *Prog. Nucl. Magn. Reson. Spectrosc.*, 1984, **16**, 193.
34. R. F. Schneider, *J. Chem. Phys.*, 1968, **48**, 4905.
35. F. A. L. Anet and D. J. O'Leary, *Concepts Magn. Reson.*, 1991, **3**, 193.
36. J. B. Robert and L. Wiesenfeld, *Phys. Rep.*, 1982, **86**, 363.
37. S. W. Ding and C. H. Ye, *Chem. Phys. Lett.*, 1990, **170**, 277.
38. D. C. Apperley, R. K. Harris and P. Hodgkinson, *Solid State NMR Basic Principles & Practice*, Momentum Press, New York, 2012.
39. W. N. Lipscomb, *Advances in Magnetic Resonance*, ed. J. S. Waugh, Academic Press, New York, 1966, vol. 2, pp. 137–176.
40. *Multinuclear NMR*, ed. J. Mason, Plenum Press, New York, 1987.
41. *Calculation of NMR and EPR Parameters: Theory and Applications*, ed. M. Kaupp, M. Bühl and V. G. Malkin, Wiley-VCH, Weinheim, 2004.
42. C. J. Jameson, *Encyclopedia of NMR*, ed. R. K. Harris and R. E. Wasylishen, Wiley, Chichester, 2012, vol. 1, pp. 493–502.
43. C. J. Jameson, *Chem. Rev.*, 1991, **91**, 1375.
44. W. H. Flygare, *J. Chem. Phys.*, 1964, **41**, 793.
45. W. H. Flygare, *Chem. Rev.*, 1974, **74**, 653.
46. D. K. Hindermann and C. D. Cornwell, *J. Chem. Phys.*, 1968, **48**, 4148.
47. A. D. Buckingham, *J. Chem. Phys.*, 1962, **36**, 3096.
48. R. E. Wasylishen, J. O. Friedrich, S. Mooibroek and J. B. Macdonald, *J. Chem. Phys.*, 1985, **83**, 548.
49. G. Cazzoli, L. Dore, L. Cludi, C. Puzzarini and S. Beninati, *J. Mol. Spectrosc.*, 2002, **215**, 160.
50. G. Cazzoli, C. Puzzarini and A. V. Lapinov, *Astrophys. J.*, 2004, **611**, 615.
51. I. Ozier, L. M. Crapo and N. F. Ramsey, *J. Chem. Phys.*, 1968, **49**, 2314.
52. W. L. Meerts, F. H. De Leeuw and A. Dymanus, *Chem. Phys.*, 1977, **22**, 319.
53. A. K. Jameson and C. J. Jameson, *Chem. Phys. Lett.*, 1987, **134**, 461.
54. W. T. Raynes, R. McVay and S. J. Wright, *J. Chem. Soc., Faraday Trans. 2*, 1989, **85**, 759.

55. D. Sundholm, J. Gauss and A. Schäfer, *J. Chem. Phys.*, 1996, **105**, 11051.
56. A. A. Auer and J. Gauss, *J. Chem. Phys.*, 2003, **118**, 10407.
57. W. Gombler, J. Schaebs and H. Willner, *Inorg. Chem.*, 1990, **29**, 2697.
58. C. J. Jameson, A. K. Jameson and J. Honarbakhsh, *J. Chem. Phys.*, 1984, **81**, 5266.
59. D. R. Johnson and F. X. Powell, *Science*, 1969, **164**, 950.
60. B. Gatehouse, H. S. P. Müller and M. C. L. Gerry, *J. Chem. Phys.*, 1997, **106**, 6916.
61. B. Brupbacher-Gatehouse, *J. Mol. Struct.*, 2001, **599**, 51.
62. J. Preusser and M. C. L. Gerry, *J. Chem. Phys.*, 1997, **106**, 10037.
63. M. Schindler, *J. Chem. Phys.*, 1988, **88**, 7638.
64. J. C. C. Chan and H. Eckert, *J. Mol. Struct.*, 2001, **535**, 1.
65. W. H. Flygare, *J. Chem. Phys.*, 1965, **42**, 1157.
66. A. M. Teale, O. B. Lutnæs, T. Helgaker, D. J. Tozer and J. Gauss, *J. Chem. Phys.*, 2013, **138**, 024111.
67. C. J. Jameson, A. K. Jameson, D. Oppusunggu, S. Wille, P. M. Burrell and J. Mason, *J. Chem. Phys.*, 1981, **74**, 81.
68. W. H. Culver, *Science*, 1957, **126**, 810.
69. C. Puzzarini, *Theor. Chem. Acc.*, 2008, **121**, 1.
70. J. Oddershede, I. Paidarová and V. Špirko, *J. Mol. Spectrosc.*, 1992, **152**, 342.
71. C. J. Jameson, A. K. Jameson, S. M. Cohen, H. Parker, D. Oppusunggu, P. M. Burrell and S. Wille, *J. Chem. Phys.*, 1981, **74**, 1608.
72. C. J. Jameson, A. C. de Dios and A. K. Jameson, *J. Chem. Phys.*, 1991, **95**, 1069.
73. P. T. P. Ho and C. H. Townes, *Annu. Rev. Astron. Astrophys.*, 1983, **21**, 239.
74. J. P. Gordon, *Phys. Rev.*, 1955, **99**, 1253.
75. C. W. David, *J. Chem. Educ.*, 1996, **73**, 46.
76. S. G. Kukolich, *J. Am. Chem. Soc.*, 1975, **97**, 5704.
77. W. S. Benedict and E. K. Plyler, *Can. J. Phys.*, 1957, **35**, 1235.
78. E. A. Laws, R. M. Stevens and W. N. Lipscomb, *J. Chem. Phys.*, 1972, **56**, 2029.
79. F. DeLucia and W. Gordy, *Phys. Rev.*, 1969, **187**, 58.
80. J. Gauss and J. F. Stanton, *J. Chem. Phys.*, 1995, **103**, 3561.
81. M. Sun, I. Y. Zhang, A. Wu and X. Xu, *J. Chem. Phys.*, 2013, **138**, 124113.
82. M. Iliaš, V. Kellö and M. Urban, *Acta Phys. Slovaca*, 2010, **60**, 259.
83. J. Autschbach, *Philos. Trans. R. Soc., A*, 2014, **372**, 20120489.
84. Y. Xiao, W. Liu, L. Cheng and D. Peng, *J. Chem. Phys.*, 2007, **126**, 214101.
85. Y. Xiao and W. Liu, *J. Chem. Phys.*, 2013, **138**, 134104.
86. Y. Xiao and W. Liu, *J. Chem. Phys.*, 2013, **139**, 034113.
87. I. A. Aucar, S. S. Gómez, M. C. Ruiz de Azúa and C. G. Giribet, *J. Chem. Phys.*, 2012, **136**, 204119.
88. I. A. Aucar, S. S. Gómez, J. I. Melo, C. G. Giribet and M. C. Ruiz de Azúa, *J. Chem. Phys.*, 2013, **138**, 134107.

89. I. A. Aucar, S. S. Gómez, C. G. Giribet and M. C. Ruiz de Azúa, *J. Chem. Phys.*, 2013, **139**, 094112.
90. Y. Xiao, Y. Zhang and W. Liu, *J. Chem. Theory Comput.*, 2014, **10**, 600.
91. M. Jaszuński, M. Repisky, T. B. Demissie, S. Komorovsky, E. Malkin, K. Ruud, P. Garbacz, K. Jackowski and W. Makulski, *J. Chem. Phys.*, 2013, **139**, 234302.
92. S. Komorovsky, M. Repisky, K. Ruud, O. L. Malkina and V. G. Malkin, *J. Phys. Chem. A*, 2013, **117**, 14209–14219.
93. D. L. Bryce and R. E. Wasylishen, *Acc. Chem. Res.*, 2003, **36**, 327.
94. C. J. Jameson and A. C. de Dios, *Nuclear Magnetic Resonance*, ed. K. Kamieńska-Trela, RSC Publishing, Cambridge, 2014, ch. 2, vol. 43, pp. 49–80.
95. C. J. Jameson and A. C. de Dios, *Nuclear Magnetic Resonance*, ed. K. Kamieńska-Trela and J. Wójcik, RSC Publishing, Cambridge, 2013, ch. 2, vol. 42, pp. 45–77.
96. M. R. Baker, H. M. Nelson, J. A. Leavitt and N. F. Ramsey, *Phys. Rev.*, 1961, **121**, 807.
97. M. R. Baker, C. H. Anderson and N. F. Ramsey, *Phys. Rev. A: At., Mol., Opt. Phys.*, 1964, **133**, 1533.
98. F. H. De Leluw and A. Dymanus, *J. Mol. Spectrosc.*, 1973, **48**, 427.
99. C. J. Jameson, A. K. Jameson and P. M. Burrell, *J. Chem. Phys.*, 1980, **73**, 6013.
100. M. E. Harding, M. Lenhart, A. A. Auer and J. Gauss, *J. Chem. Phys.*, 2008, **128**, 244111.
101. C. Puzzarini, G. Cazzoli, M. E. Harding, J. Vázquez and J. Gauss, *J. Chem. Phys.*, 2009, **131**, 234304.
102. T. M. Nymand and P.-O. Åstrand, *J. Chem. Phys.*, 1997, **106**, 8332.
103. T. Helgaker, J. Gauss, G. Cazzoli and C. Puzzarini, *J. Chem. Phys.*, 2013, **139**, 244308.
104. K. Jackowski, W. Makulski and W. Koźmiński, *Magn. Reson. Chem.*, 2002, **40**, 563.
105. S. Komorovsky, M. Repisky, E. Malkin, K. Ruud and J. Gauss, *J. Chem. Phys.*, 2015, **142**, 091102.
106. G. Cazzoli, L. Dore, C. Puzzarini and S. Beninati, *Phys. Chem. Chem. Phys.*, 2002, **4**, 3575.
107. E. Malkin, S. Komorovsky, M. Repisky, T. B. Demissie and K. Ruud, *J. Phys. Chem. Lett.*, 2013, **4**, 459.
108. W. Makulski, *J. Mol. Struct.*, 2012, **1017**, 45.
109. W. Makulski, K. Jackowski, A. Antušek and M. Jaszuński, *J. Phys. Chem. A*, 2006, **110**, 11462.
110. M. Jaszuński, T. B. Demissie and K. Ruud, *J. Phys. Chem. A*, 2014, **118**, 9588.
111. K. Ruud, T. B. Demissie and M. Jaszuński, *J. Chem. Phys.*, 2014, **140**, 194308.
112. C. J. Jameson and A. K. Jameson, *Chem. Phys. Lett.*, 1987, **135**, 254.

113. P. Lantto, K. Jackowski, W. Makulski and M. Olejniczak, *J. Phys. Chem. A*, 2011, **115**, 10617.
114. J. Verhoeven and A. Dymanus, *J. Chem. Phys.*, 1970, **52**, 3222.
115. *E.g.*, Z. Kisiel, E. Białkowska-Jaworska and L. Pszczółkowski, *J. Chem. Phys.*, 1998, **109**, 10263.
116. A re-wording of a problem in Chapter 3 of A. Carrington and A. D. McLachlan, *Introduction to Magnetic Resonance*, Harper and Row, 1967.
117. E. R. Cohen, T. Cvitaš, J. G. Frey, B. Holmström, K. Kuchitsu, R. Marquardt, I. Mills, F. Pavese, M. Quack, J. Stohner, H. L. Strauss, M. Takami and A. J. Thor, *Quantities, Units and Symbols in Physical Chemistry*, RSC Publishing, Cambridge, 3rd edn, 2007.
118. D. M. Bishop and L. M. Cheung, *Phys. Rev. A: At., Mol., Opt. Phys.*, 1979, **20**, 381.
119. W. E. Quinn, J. M. Baker, J. T. LaTourrette and N. F. Ramsey, *Phys. Rev.*, 1958, **112**, 1929.
120. The NIST Reference on Constants, Units, and Uncertainty. http://physics.nist.gov/cuu/Constants/index.html (accessed May 12, 2015).
121. *NMR of Quadrupolar Nuclei in Solid Materials*, ed. R. E. Wasylishen, S. E. Ashbrook and S. Wimperis, Wiley, Chichester, 2012.
122. C. Puzzarini, J. F. Stanton and J. Gauss, *Int. Rev. Phys. Chem.*, 2010, **29**, 273.
123. Y. Xu and W. Jäger, *Applications of Physical Methods to Inorganic and Bioinorganic Chemistry, Encyclopedia of Inorganic Chemistry*, ed. R. A. Scott, Wiley, 2008.
124. B. P. Dailey and C. H. Townes, *J. Chem. Phys.*, 1955, **23**, 118.
125. A. H. Barrett and M. Mandel, *Phys. Rev.*, 1958, **109**, 1572.
126. M. Karplus and R. N. Porter, *Atoms & Molecules: An Introduction For Students of Physical Chemistry*, W. A. Benjamin, Inc., New York, 1970.
127. J. M. Hutson, *Annu. Rev. Phys. Chem.*, 1990, **41**, 123.
128. *E.g.*, S. P. Dempster, O. Sukhorukov, Q.-Y. Lei and W. Jäger, *J. Chem. Phys.*, 2012, **137**, 174303.
129. *E.g.*, C. J. Knapp, Y. Xu and W. Jäger, *J. Mol. Spectrosc.*, 2011, **268**, 130.
130. A. Brown and R. E. Wasylishen, *J. Phys. Chem. A*, 2012, **116**, 9769.
131. P. Pyykkö, *Mol. Phys.*, 2008, **106**, 1965.
132. F. Leung, S. A. Cooke and M. C. L. Gerry, *J. Mol. Spectrosc.*, 2006, **238**, 36.
133. L. Demovič, V. Kellö and A. J. Sadlej, *Chem. Phys. Lett.*, 2010, **498**, 10.
134. R. T. Santiago, T. Q. Teodoro and R. L. A. Haiduke, *Phys. Chem. Chem. Phys.*, 2014, **16**, 11590.
135. M. Srebro and J. Autschbach, *J. Phys. Chem. Lett.*, 2012, **3**, 576.
136. P. Schwerdtfeger, M. Pernpointner and W. Nazarewicz, *Calculation of NMR and EPR Parameters. Theory and Application*, ed. M. Kaupp, M. Bühl and V. G. Malkin, Wiley-VCH, Weinheim, 2004, p. 288.
137. B. Effenberger, W. Kunold, W. Oesterle, M. Schneider and L. M. Simons, *Z. Phys. A: At. Nucl.*, 1982, **309**, 77.

138. R. E. Wasylishen, *NMR Crystallography*, ed. R. K. Harris, R. E. Wasylishen and M. J. Duer, Wiley, Chichester, 2009, pp. 127–136.
139. R. E. Wasylishen, *Encyclopedia of NMR*, ed. R. K. Harris and R. E. Wasylishen, Wiley, Chichester, 2012, vol. 4, pp. 2074–2083.
140. K. J. Harris, D. L. Bryce and R. E. Wasylishen, *Can. J. Chem.*, 2009, **87**, 1338.
141. D. L. Bryce and R. E. Wasylishen, *J. Am. Chem. Soc.*, 2000, **122**, 3197.
142. C. J. Jameson, *Multinuclear NMR*, ed. J. Mason, Plenum Press, New York, 1987, pp. 89–131.
143. J. N. Murrell, *Prog. Nucl. Magn. Reson. Spectrosc.*, 1971, **6**, 1.
144. *E.g.*, see K. Eichele, G. Wu, R. E. Wasylishen and J. F. Britten, *J. Phys. Chem.*, 1995, **99**, 1030.
145. *E.g.*, see R. E. Wasylishen, D. L. Bryce, C. J. Evans and M. C. L. Gerry, *J. Mol. Spectrosc.*, 2000, **204**, 184.
146. J. Cederberg, *J. Mol. Spectrosc.*, 2014, **299**, 22.
147. G. Paquette, A. Kotz, J. Cederberg, D. Nitz, A. Kolan, D. Olson, K. Gunderson, S. Lindaas and S. Wick, *J. Mol. Struct.*, 1988, **190**, 143.
148. D. Nitz, J. Cederberg, A. Kotz, K. Hetzler, T. Aakre and T. Walhout, *J. Mol. Spectrosc.*, 1984, **108**, 6.
149. J. Cederberg, J. Randolph, B. McDonald, B. Paulson and C. McEachern, *J. Mol. Spectrosc.*, 2008, **250**, 114.
150. J. Cederberg, E. Frodermann, H. Tollerud, K. Huber, M. Bongard, J. Randolph and D. Nitz, *J. Chem. Phys.*, 2006, **124**, 244304.
151. J. Cederberg, S. Fortman, B. Porter, M. Etten, M. Feig, M. Bongard and L. Langer, *J. Chem. Phys.*, 2006, **124**, 244305.
152. J. Cederberg, B. Paulson and C. Conklin, *J. Mol. Spectrosc.*, 2011, **265**, 92.
153. J. Vaara, J. Jokisaari, R. E. Wasylishen and D. L. Bryce, *Prog. Nucl. Magn. Reson. Spectrosc.*, 2002, **41**, 233.
154. I. Solomon and N. Bloembergen, *J. Chem. Phys.*, 1956, **25**, 261.
155. D. K. Hindermann and C. D. Cornwell, *J. Chem. Phys.*, 1968, **48**, 2017.
156. J. S. Martin and F. Y. Fujiwara, *J. Am. Chem. Soc.*, 1974, **96**, 7632.
157. J. Kowalewski, *Prog. Nucl. Magn. Reson. Spectrosc.*, 1977, **11**, 1.
158. D. L. Bryce, R. E. Wasylishen, J. Autschbach and T. Ziegler, *J. Am. Chem. Soc.*, 2002, **124**, 4894.
159. T. Helgaker, M. Jaszuński and M. Pecul, *Prog. Nucl. Magn. Reson. Spectrosc.*, 2008, **53**, 249.
160. D. L. Bryce and J. Autschbach, *Can. J. Chem.*, 2009, **87**, 927.
161. J. Autschbach, *J. Chem. Phys.*, 2008, **129**, 094105.
162. J. Autschbach, *J. Chem. Phys.*, 2009, **130**, 209901.
163. W. Makulski and K. Jackowski, *J. Mol. Struct.*, 2004, **704**, 219.
164. R. E. Wasylishen, S. Mooibroek and J. B. Macdonald, *J. Chem. Phys.*, 1984, **81**, 1057.
165. N. F. Ramsey, *Nuclear Moments*, J. Wiley & Sons, Inc., New York, 1953.
166. J. S. Rigden, *Hydrogen: The Essential Element*, Harvard University Press, Cambridge, 2002.

167. *E.g.*, see I. Estermann, O. C. Simpson and O. Stern, *Phys. Rev.*, 1937, **52**, 535.
168. J. M. B. Kellogg, I. I. Rabi, N. F. Ramsey and J. R. Zacharias, *Phys. Rev.*, 1939, **56**, 728.
169. N. J. Stone, *At. Data Nucl. Data Tables*, 2005, **90**, 75.
170. R. K. Harris, E. D. Becker, S. M. C. de Menezes, R. Goodfellow and P. Granger, *Pure Appl. Chem.*, 2001, **73**, 1795.
171. A. Laaksonen and R. E. Wasylishen, *J. Am. Chem. Soc.*, 1995, **117**, 392.
172. M. G. H. Gustavsson and A.-M. Mårtensson-Pendrill, *Phys. Rev. A: At., Mol., Opt. Phys.*, 1998, **58**, 3611.
173. W. Makulski, *Magn. Reson. Chem.*, 2014, **52**, 430.
174. J. Vaara and P. Pyykkö, *J. Chem. Phys.*, 2003, **118**, 2973.
175. D. Brinkman, *Phys. Lett. A*, 1968, **27**, 466.
176. M. Jaszuński, A. Antušek, P. Garbacz, K. Jackowski, W. Makulski and M. Wilczek, *Prog. Nucl. Magn. Reson. Spectrosc.*, 2012, **67**, 49.
177. K. Jackowski and M. Jaszuński, *Concepts Magn. Reson., Part A*, 2007, **30**, 246.
178. ADF 2012, SCM, Theoretical Chemistry, Vrije Universiteit, Amsterdam, The Netherlands. http://www.scm.com (Accessed May 29, 2015).
179. G. te Velde, F. M. Bickelhaupt, S. J. A. van Gisbergen, C. F. Guerra, E. J. Baerends, J. G. Snijders and T. Ziegler, *J. Comput. Chem.*, 2001, **22**, 931.
180. C. F. Guerra, J. G. Snijders, G. te Velde and E. J. Baerends, *Theor. Chem. Acc.*, 1998, **99**, 391.
181. B. Fabricant and J. S. Muenter, *J. Chem. Phys.*, 1977, **66**, 5274.
182. S. A. Cooke and M. C. L. Gerry, *Phys. Chem. Chem. Phys.*, 2004, **6**, 4579.
183. S. A. Cooke and M. C. L. Gerry, *J. Mol. Spectrosc.*, 2005, **234**, 195.
184. S. A. Cooke, C. Krumrey and M. C. L. Gerry, *Phys. Chem. Chem. Phys.*, 2005, 7, 2570.
185. S. A. Cooke, C. Krumrey and M. C. L. Gerry, *J. Mol. Spectrosc.*, 2011, **267**, 108.
186. S. A. Cooke and M. C. L. Gerry, *J. Mol. Spectrosc.*, 2002, **216**, 122.
187. S. A. Cooke and M. C. L. Gerry, *J. Am. Chem. Soc.*, 2004, **126**, 17000.
188. C. J. Evans and M. C. L. Gerry, *J. Am. Chem. Soc.*, 2000, **122**, 1560.
189. S. A. Cooke, J. M. Michaud and M. C. L. Gerry, *J. Mol. Spectrosc.*, 2004, **695**, 13.
190. J. Cederberg, J. Nichol, E. Frodermann, H. Tollerud, G. Hilk, J. Buysman, W. Kleiber, M. Bongard, J. Ward, K. Huber, T. Khanna, J. Randolph and D. Nitz, *J. Chem. Phys.*, 2005, **123**, 134321.
191. J. Cederberg, L. Kang, C. Conklin and E. Berger, *J. Mol. Spectrosc.*, 2010, **263**, 142.
192. R. H. Hammerle, J. T. Dickinson, R. G. van Ausdal, D. A. Stephenson and J. C. Zorn, *J. Chem. Phys.*, 1969, **50**, 2086.
193. J. T. Dickinson, D. A. Stephenson and J. C. Zorn, *J. Chem. Phys.*, 1970, **53**, 1525.
194. D. A. Stephenson, J. T. Dickinson and J. C. Zorn, *J. Chem. Phys.*, 1970, **53**, 1529.

CHAPTER 3

Nuclear Magnetic Moments and NMR Measurements of Shielding

KAROL JACKOWSKI* AND PIOTR GARBACZ

Laboratory of NMR Spectroscopy, Faculty of Chemistry, University of Warsaw, Pasteura 1, 02-093 Warszawa, Poland
*Email: kjack@chem.uw.edu.pl

3.1 Introduction

NMR spectroscopy in the gas phase delivers unique information on molecules free from intermolecular interactions. Spectral parameters extrapolated to the zero-density limit are equivalent to their values obtained for an isolated molecule. It permits the experimental comparison of magnetic properties of two different nuclei at the stable external magnetic field when the two nuclei are in the same NMR sample or in the same molecule. Assuming that the magnetic moment of a proton is known with satisfactory accuracy and that the magnetic shielding of nuclei in small molecules can be precisely determined or calculated, it opens a way for the accurate determination of magnetic moments for many other nuclei. This method is limited by the number of gaseous compounds available for such a comparison but due to high sensitivity of modern spectrometers it can be easily extended also to liquid chemicals when binary gaseous solutions are applied in the experimental studies.

The improved values of nuclear magnetic moments are essential for nuclear physics and different molecular spectroscopies. NMR experiment

New Developments in NMR No. 6
Gas Phase NMR
Edited by Karol Jackowski and Michał Jaszuński

Published by the Royal Society of Chemistry, www.rsc.org

itself is based on the observation of nuclear magnetic moments and therefore the availability of nuclear magnetic moments may significantly change this method of spectroscopy. In this chapter it is shown that the straightforward measurement of nuclear magnetic shielding is possible for many nuclei and can be used for the alternative standardization of NMR spectra. It makes NMR spectroscopy more efficient, more universal and better connected with the quantum theory of shielding. This issue is discussed in detail with some noteworthy examples of general applications and, as concluded in this chapter, new methods for NMR studies are available due to the gas phase measurements.

Generally NMR techniques used for the observation of gaseous samples are not especially original and almost all the methods well known from liquid experiments can be also applied for gases. However, the preparation of gaseous samples requires unique laboratory equipment which must be suitable for the range of applied pressure and to the property of investigated chemicals. Firstly, it should be remembered that numerous gases and vapors are poisonous and may be extremely dangerous for a human being. Secondly, any high-pressure sample can be also a dangerous object for researchers if safety procedures are neglected. Our review will start with the description of experimental methods used in gas phase NMR investigations.

3.2 NMR Experimental Methods in the Gas Phase

3.2.1 Gas Samples

Gas samples are usually prepared by the condensation of pure gases or gaseous mixtures from the calibrated part of vacuum line to cylindrical glass ampoules and sealed with a propane-butane torch as shown in Figure 3.1. Long glass tubes with outside diameters (o.d.) of 5 mm and a reference capillary placed inside were formerly used for the measurements at room temperature.[1–3] The stabilization of temperature requires smaller samples and they are usually prepared in 4 mm o.d. and 5.5 cm long glass tubes. The volume of such ampoules and the calibrated part of a vacuum line are precisely measured using mercury. The sealed gas samples are fitted into standard 5 or 10 mm o.d. NMR tubes with liquid deuterated solvent for a lock system in the annular space.[4–6]

The above "classical method" of gaseous NMR studies is very good for the majority of observations if the final pressure of samples does not exceed 50 bar. It frequently correlates with the range of required observation as the linear dependence of shielding is often limited to the similar border line, approximately up to 40 bar.[7] Although the glass tubes of smaller internal diameter can be used at higher pressures, in each case it is highly recommended to carefully check each sample prior to use (*i.e.*, by heating at high temperature, $\sim$100 °C, for several hours). Using this technique Jameson *et al.*[8] studied xenon gas up to 200 bar, and Gordon and Dailey[9] examined methane and ethane up to 300 bar. Usage of high-quality ampoules allows to

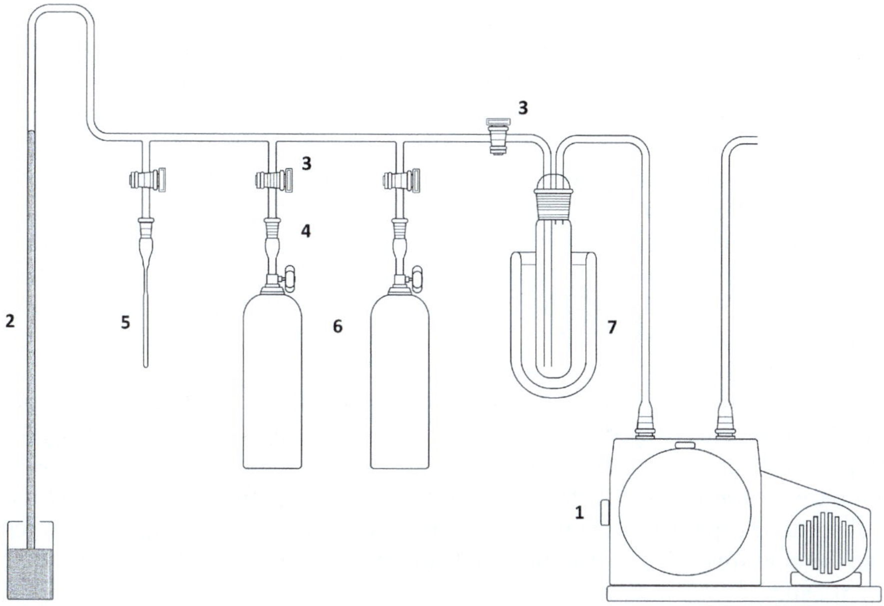

Figure 3.1 The scheme of a classical vacuum line used for the preparation of gaseous samples. The line consists of a vacuum pump (**1**), a mercury manometer (**2**), valves (**3**), and inlets/outlets with ground glass joints (**4**). A sample is prepared by freezing gases delivered from lecture bottles (**6**) to a glass ampoule (**5**) and then sealing the ampoule by a torch. The vacuum pump is separated from the line by a cold trap (**7**).

obtain narrow NMR lines (<1 Hz where it is possible), but this technique is difficult to handle in case of gases which solidify at very low temperatures, *e.g.*, for hydrogen, helium, and other small molecules. Moreover, the light gases can be observed up to very high pressure and for this purpose special equipment is required as described in detail in the next section.

3.2.2 High-pressure Techniques

The NMR measurements of pressurized samples require suitable experimental equipment. There are two commonly used approaches applied in the preparation of the high-pressure sample and sample handling during the measurements. In the first approach the NMR probe consists of a pressure bomb in which an rf coil and the sample are placed. The body of this kind of probe is built with a non-magnetic material of high mechanical strength. For this purpose beryllium copper alloy, titanium alloys and stainless steel are used.[10] Besides high safety, the advantage of this experimental setup is the possibility of obtaining good filling factor of the rf coil and consequently high signal-to-noise ratio. This approach was used in the first high-pressure experiments performed by Benedek and Purcell[11] in 1954 and then this experimental setup was further developed by several research groups.[12–16]

In the second approach one can use the high-pressure NMR cell which is usually made of the pressure-resistive material, *e.g.*, quartz, borosilicate glass, zirconia, and sapphire. The quartz cell permits measurements at pressure up to ~6 kbar,[17] glass and sapphire cells ~2 kbar,[18,19] and polyimide thick-walled tubes allow obtaining pressure up to ~1 kbar.[20] High-pressure cells are usually compatible with standard NMR probes and do not require a dedicated probe. Wall thickness of the high-pressure cell limits the filling factor of the coil. The highest available pressure in the high-pressure cell is determined by the material of the walls and the shape of the sample. For cylindrical samples of outer-to-inner diameter ratio 3.5–5.0 the strength can be estimated as 80%–90% of the maximum achievable value which corresponds to an infinitely long cylinder.[17]

Depending on the state of the sample and required pressures the sample can be pressurized by different external pressurizing fluids, *e.g.*, nitrogen, CS_2, or tetrachloroethylene.[21] In this case, the sample is placed in a closed container which is separated from the pressurizing fluid by a wall which transmits the pressure from the pressurizing fluid to the sample. Samples can be pressurized to extremely high pressure (~100 kbar) in a diamond anvil cell.[22–24]

Another possibility of obtaining the pressurized gaseous sample provides dedicated systems in which the gas is compressed in the high-pressure NMR cell. This approach can be exemplified on the system for studies of gaseous mixtures of hydrogen and noble gases at variable pressures up to 300 bar,[25] which is shown in Figure 3.2. In the first step the volume of the mechanical compressor is $V_c = 75$ mL and the mechanical compressor is filled with the first gas from the gas fill line. In the next step the second gas is transferred from a vacuum line to the high-pressure NMR tube (o.d. = 5 mm, i.d. = 3 mm). Then, the volume of the mechanical compressor is reduced, $V_c = 15$ mL, and the gas present in the mechanical compressor is transferred to a high-pressure NMR tube. The pressure of the mixed gases in the high-pressure NMR tube can be adjusted to the requested value by changing the volume of the mechanical compressor. Obtaining a highly homogenous magnetic field in high-pressure tubes of this type is usually more laborious than for glass ampoules.

3.3 Resonance Frequency in an Isolated Molecule

Modern NMR spectrometers are supplied with very stable superconducting magnets. The fixed magnetic field (B_0) enables us to measure the absolute resonance frequencies for selected nuclei as a function of gas density. In a gas of low density, the resonance frequency $\nu(\rho,T)$ can be written as an expansion in powers of the density ρ:[26]

$$\nu(\rho, T) = \nu_0(T) + \nu_1(T)\rho + \nu_2(T)\rho^2 + \cdots \tag{3.1}$$

where $\nu_0(T)$ is the frequency for an isolated molecule and $\nu_1(T)$ is a measure of the effects on this frequency due to binary collisions. The higher-order

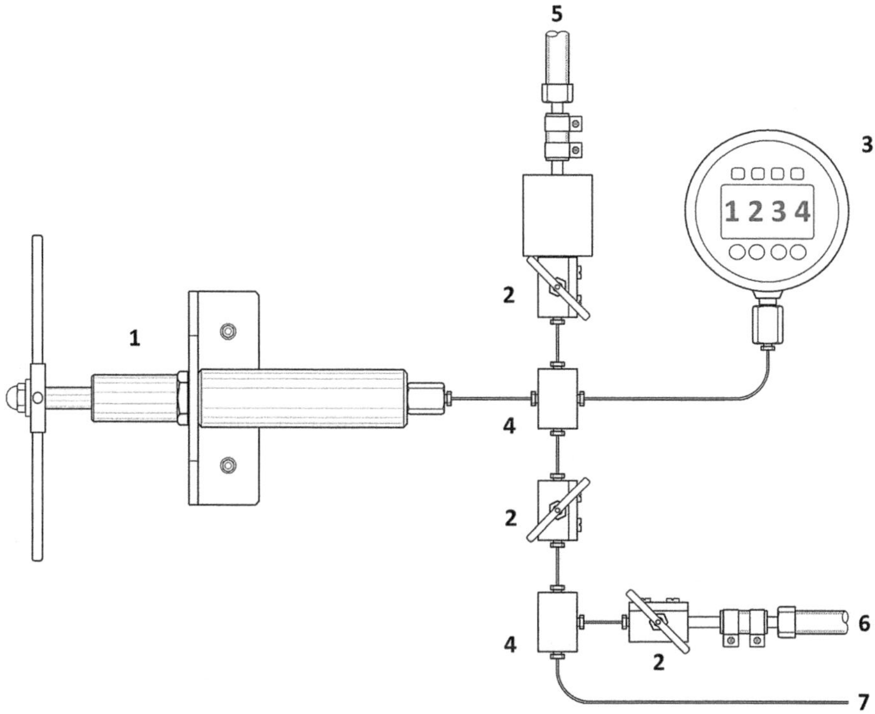

Figure 3.2 The high-pressure system which consists of a mechanical compressor (**1**), needle valves (**2**), a pressure gauge (**3**), and connectors (**4**). This system permits to mix two gases provided from inlets (**5**, **6**) and to transfer the mixture through the outlet (**7**) to a high-pressure cell, not shown in this picture.

terms, starting from $\nu_2(T)$, are negligible for low-density samples. Then the density dependence is linear and the two frequency parameters, $\nu_0(T)$ and $\nu_1(T)$, are easily available from eqn (3.1). All the frequency parameters in the latter equation are temperature-dependent and they are usually measured at constant temperature, 300 K. It is interesting that the $\nu_0(T)$ frequency parameter can be obtained also for a chemical compound with low vapor pressure at room temperature. Such an experiment is possible in a gas matrix when the NMR signal of a sample is strong enough to be observed at very low concentration. Molecules enriched in magnetic nuclei are recommended for these kinds of studies. Let us note that the gas matrix allows the separation of solute molecules by solvent molecules and in this way the strong solute–solute interactions are usually replaced by weaker solute–solvent molecular interactions. As a result more chemical compounds are available for NMR studies in the gas phase, including *e.g.* chemicals that contain heavy atoms or hydrogen bonding systems and for this reason they are liquid at standard ambient temperature and pressure. For a binary mixture of gas A, containing the nucleus X whose frequency ν(X) is observed, and gas B as the solvent, the corresponding equation becomes:

$$\nu^{A}(X) = \nu_0{}^{A}(X) + \nu_1{}^{AA}(X)\rho_A + \nu_1{}^{AB}(X)\rho_B + \cdots \tag{3.2}$$

where ρ_A and ρ_B are the densities of A and B components, respectively, and $\nu_0{}^{A}(X)$ is the frequency at the zero-density limit. For solvent density below 40 bar all higher terms in eqn (3.2) can be safely neglected, as the dependence of resonance frequency on density is linear. Then the coefficients $\nu_1{}^{AA}(X)$ and $\nu_1{}^{AB}(X)$ are the only terms responsible for medium effects, they contain the bulk susceptibility contributions ($\nu_{1b}{}^{A}$ and $\nu_{1b}{}^{B}$) and the terms taking account of intermolecular interactions during the binary collisions of A-A and A-B molecules: $\nu_1{}^{A\text{-}A}(X)$ and $\nu_1{}^{A\text{-}B}(X)$, respectively. In such experiments the density of A (ρ_A) is usually kept very low in order to eliminate completely the solute–solute molecular interactions and eqn (3.2) can be simplified to a short form:

$$\nu^{A}(X) = \nu_0{}^{A}(X) + \nu_1{}^{AB}(X)\rho_B \tag{3.3}$$

in which there is only one second virial term ($\nu_1{}^{AB}$) and it permits the determination of resonance frequency for an isolated A molecule ($\nu_0{}^{A}$) after linear extrapolation of the results to the zero-density point (Figure 3.3).

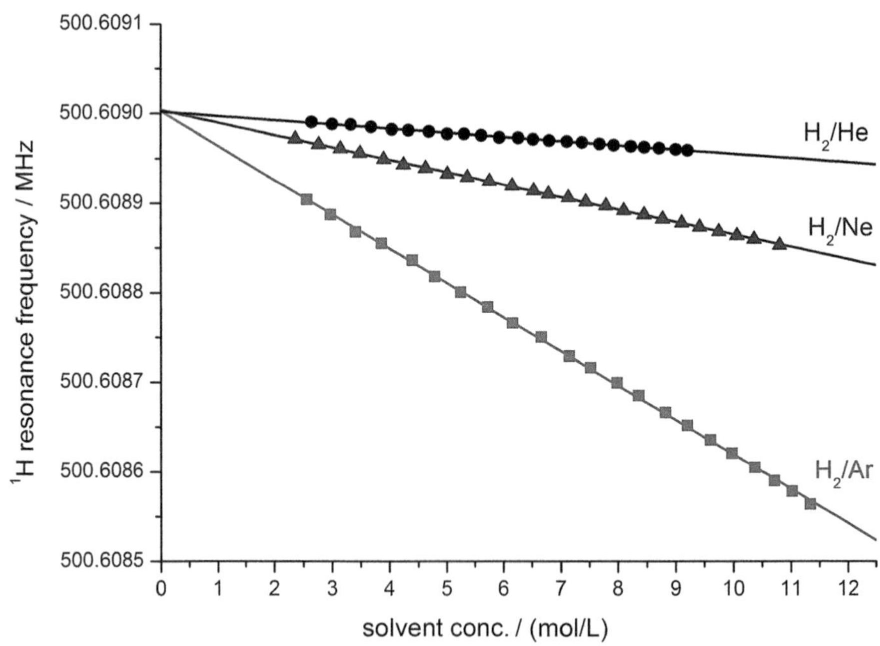

Figure 3.3 Density dependence of the ^{1}H NMR resonance frequency of hydrogen dissolved in noble gases recorded at temperature 300 K and magnetic field 11.75 T.

3.4 Nuclear Magnetic Moments

Most of the atomic nuclei have spin angular momenta I and their magnitude can be defined by:

$$I^{\text{length}} = \hbar\ [I(I+1)]^{1/2} \tag{3.4}$$

where $\hbar = h/2\pi$, h is Planck's constant, I is the maximum projection of the nuclear spin usually called the nuclear spin quantum number. For stable nuclei the spin quantum number I can have values from 0, 1/2, 1, 3/2, 2, ... up to 7 for different nuclides. If $I \neq 0$ the nucleus has also the nuclear magnetic moment (μ_I) which can be described as:

$$\mu_{I\text{length}} = \gamma_I \hbar\ [I(I+1)]^{1/2} \tag{3.5}$$

where γ_I is the magnetogyric ratio with dimensions of radians per tesla per second. Since the magnetic moment and angular momentum are parallel vectors, magnetic moments are also defined in the form of

$$\mu_I = \gamma_I \hbar I \tag{3.6}$$

where the magnitude of the magnetic moment depends on the maximum component of the nuclear angular momentum. Magnetic moments are usually described in units of nuclear magnetons, $\mu_N = 5.050\ 783\ 53 \times 10^{-27}$ J T^{-1},[29] then eqn (3.6) is simplified to

$$\mu_I = g_I \mu_N\ I \tag{3.7}$$

and the magnetic moment is finally represented by the nuclear g_I factor, a dimensionless value. For most nuclei $\boldsymbol{\mu_I}$ and $\boldsymbol{I}$ are parallel vectors and γ_I and g_I are positive, but some nuclei (*e.g.* ^{15}N, ^{17}O, or ^{29}Si) have $\boldsymbol{\mu_I}$ and $\boldsymbol{I}$ anti-parallel and γ_I, g_I are negative. Tables of all the nuclei with non-zero spins and magnetic moments are available in the literature.[30] However, there are a few simple rules based on the number of nucleons which permit the prediction of nuclear spins and magnetic moments. (1) An odd number of nucleons gives the half-integral values of I (1/2, 3/2, 5/2, ...) for different nuclei like ^1H, ^3He, ^{13}C, or ^{15}N (all with $I = 1/2$) and many others, ^{11}B ($I = 3/2$), ^{17}O ($I = 5/2$), ^{57}V ($I = 7/2$), or ^{83}Kr ($I = 9/2$). (2) An odd number of protons and odd number of neutrons gives nuclides with non-zero integral values of I, *e.g.* 1, 3, and 4 for ^2H, ^{10}B, and ^{40}K, respectively. (3) There are a few nuclei which have an even number of both protons and neutrons, and such nuclei have zero angular momentum ($I = 0$), as observed for ^4He, ^{12}C, ^{16}O, and several others.

It is important that the values of μ_I cannot be accurately predicted from theory and they have to be obtained experimentally. Different experimental techniques can be applied for the determination of nuclear magnetic moments but the most accurate data for stable nuclei were obtained from NMR spectra. However, the accuracy of nuclear magnetic dipole moments obtained from NMR spectra depends on the accuracy of the shielding constants, needed to extract from the experimental results the magnetic moments of bare nuclei. The literature values of magnetic moments of common stable nuclei, tabulated in the 1980s and still currently used,[27,28,30] are often of low accuracy (except for the proton magnetic moment), since very crude approximations have been used for absolute shielding constants and the experimental studies were performed for liquid and solid chemicals where the influence of strong intermolecular interactions is obvious. The need for re-measurements of nuclear magnetic moments has been previously noted by Laaksonen and Wasylishen[31] and by Gustavsson and Märtensson-Pendrill.[32] In the next sections it is shown how NMR measurements in the gas phase can deliver more accurate values of nuclear magnetic moments but first we need to present the most important magnetic moments of hydrogen isotope nuclei.

3.4.1 The Magnetic Moment of the Proton

Determination of the dipole magnetic moment of the proton played an important role in the development of nuclear physics. Measurements of the magnetic moment of the proton started in the 1930s after the pioneering experiment conducted by Otto Stern. That work was an extension of his previous experiments known today as the Stern–Gerlach experiment,[33,34] which were performed in 1922. At that time the importance of the experimental determination of the magnetic moment of the proton was not widely recognized, since one could expect, on the basis of the very successful theory of the electron developed by Dirac,[35] that the magnetic moment of the proton equals one nuclear magneton.

The measurement of the magnetic moment of the proton requires an experiment in which the effect of the magnetic field on the orientation of the proton spin is observed. In the very first experimental attempt the measured quantity was the deflection of a molecular beam passing throw a spatially inhomogeneous magnetic field. The experiment was based on the fact that particles, atoms, and molecules possessing a magnetic moment behave like magnetic dipoles, thus, their direction of flight is deflated in the inhomogeneous magnetic field. If the gradient of the magnetic field, velocity of the species, and geometry of the experimental setup are known then one can deduce the strength of interaction between given species and the magnetic field. A simplified sketch of this experiment is shown in Figure 3.4.

This method, applied by Stern and Estermann[36] in 1933 to a beam of molecular hydrogen, gave a value of 2.5 μ_N in clear contradiction to theoretical expectations. However, one can rationalize this discrepancy assuming that, unlike the electron, the proton is not an elementary particle[37] and according to the quark model consists of three quarks (two up and one down quark).[38,39]

Successive measurements of the magnetic moment of the heavier nuclide of hydrogen, deuteron, conducted independently by Stern[40] and Rabi[41] in 1934 gave values of 0.8 μ_N. Inferring from the data for the proton and deuteron the magnetic moment for the neutron one can obtain that, similar to

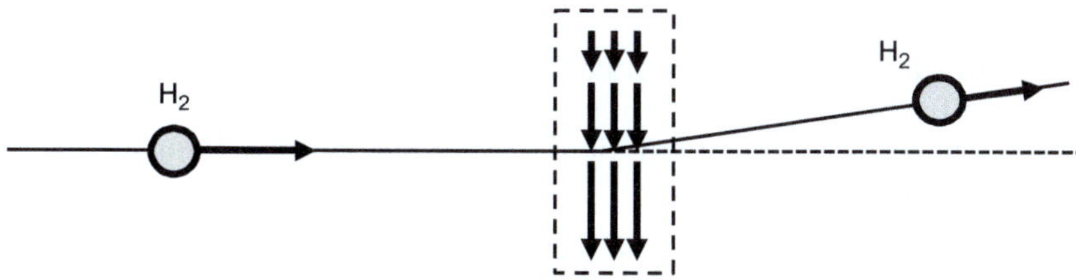

Figure 3.4 The collimated beam of hydrogen molecules passes through a region of an inhomogeneous magnetic field. For both spin isomers of hydrogen the beam is deflected due to the rotational magnetic moment of a hydrogen molecule. The proton magnetic moments contribute only to deflection for *ortho*-H_2.

the proton, the neutron is not an elementary particle, $\mu_n = -1.9\ \mu_N$. Indeed, the neutron consists of one up and two down quarks. The minus sign means that the spin angular momentum of the neutron rotates anticlockwise about the direction of the magnetic field. Moreover, the ratio of the magnetic moment of the proton to the magnetic moment of the neutron is close to −3/2, which supports the Standard Model of particle physics.[42,43] For a few of the lightest isotopes the nuclear magnetic moments follow predictions of the shell model of the nucleus.[44,45] For example, the magnetic moment of the next lightest NMR-active nucleus, helium-3, is approximately equal to the magnetic moment of the neutron, therefore, one can infer that contributions from two protons approximately cancel out.

The uncertainty of the measurement of the proton and the deuteron magnetic moments decreased with time from 10% in 1934 to 0.7% in 1939.[46,47] The method of the measurement gradually evolved to a molecular beam magnetic resonance method,[48] which substantially improved the accuracy of the determination of the magnetic moment of the proton. This improvement was due to application of an oscillating magnetic field instead of a static magnetic field.

The magnetic resonance method was used in the study conducted by Winkler *et al.*,[49] which provided the most accurate value of the proton for more than the last 40 years. In the experiment a molecular beam was not used, but frequencies of transitions in a hydrogen maser allowed one to determine the magnetic moment of the proton. This approach was based on the Larmor formula:

$$\nu = \frac{\mu_p}{2\pi I_p} B_0, \tag{3.8}$$

where ν is the proton spin precession frequency (Hz), μ_p is the magnetic moment of the proton (J T^{-1}), $I_p = 1/2\ \hbar$ is a component of the proton spin angular momentum along the quantization axis (J s), and B_0 is the strength of the magnetic field (T). This experiment takes advantage of the fact that the frequency of spin precession of the proton can be determined with much higher precision ($\sim 10^{-8}$) than the displacement of the molecular beam.

Thus, the key issue for the high-precision measurement of the magnetic moment of the proton is the simultaneous determination of the proton spin precession frequency and the strength of the magnetic field.

The strength of the magnetic field can be determined from the electron spin precession frequency:

$$\nu = \frac{\mu_e}{2\pi I_e} B_0, \tag{3.9}$$

where $\mu_e = -928.476\ 430(21) \times 10^{-26}$ J T^{-1} is the magnetic moment of the electron, $I_e = 1/2\ \hbar$ is a component of the electron spin angular momentum along the quantization axis. More specifically, one can observe the hyperfine spectrum of atomic hydrogen in the hydrogen atom maser. In this case several transitions determined by quantum states of the proton and the electron are observed in hydrogen atoms, see Figure 3.5. The frequencies of these transients depend on the strength of the magnetic field. It can be shown that for determination of the proton magnetic moment it is sufficient to measure the transition frequencies at one particular strength of the magnetic field. Using these frequencies, Winkler *et al.*[49] determined the electron-to-proton resonance frequency ratio in a hydrogen atom:

$$R = \frac{\nu_e(\mathrm{H})}{\nu_p(\mathrm{H})} = \frac{|\mu_e|}{\mu_p}. \tag{3.10}$$

In order to obtain a more accurate value one has to take into account that the field near the proton, $B_0' = (1 - \sigma_p(\mathrm{H}))B_0$, is slightly smaller than the magnetic field applied in the experiment, B_0. This is due to the fact that the electron shields the external magnetic field. In a similar way the magnetic field near the electron has to be corrected for the shielding of the proton, σ_p. The final result is $\mu_p = 2.792\ 847\ 356(23)\ \mu_N$. An analogous experiment was also performed for the atomic deuterium, *cf.* Table 3.1.

Recently a more accurate value was reported by Mooser *et al.*,[50] $\mu_p = 2.792\ 847\ 350(9)\ \mu_N$ from measurements of the single proton confined in the double Penning trap. In the Penning trap a charged particle is confined by application of the homogenous magnetic field and the spatially inhomogeneous electric field. The idea of determination of the proton magnetic moment using the Penning trap technique is based on the fact that these measurements provide the cyclotron frequency of the proton:

$$\nu_c = \frac{q_p}{2\pi m_p} B_0, \tag{3.11}$$

where q_p and m_p are the charge and the mass of the proton. Using the cyclotron frequency one can eliminate the B_0 field from eqn (3.8) and if the spin precession frequency is known find the magnetic moment of the single proton. The significance of this measurement method of the magnetic moment of the proton lies in the possibility of the measurement of the magnetic moments for single antiparticles for which contact with ordinary

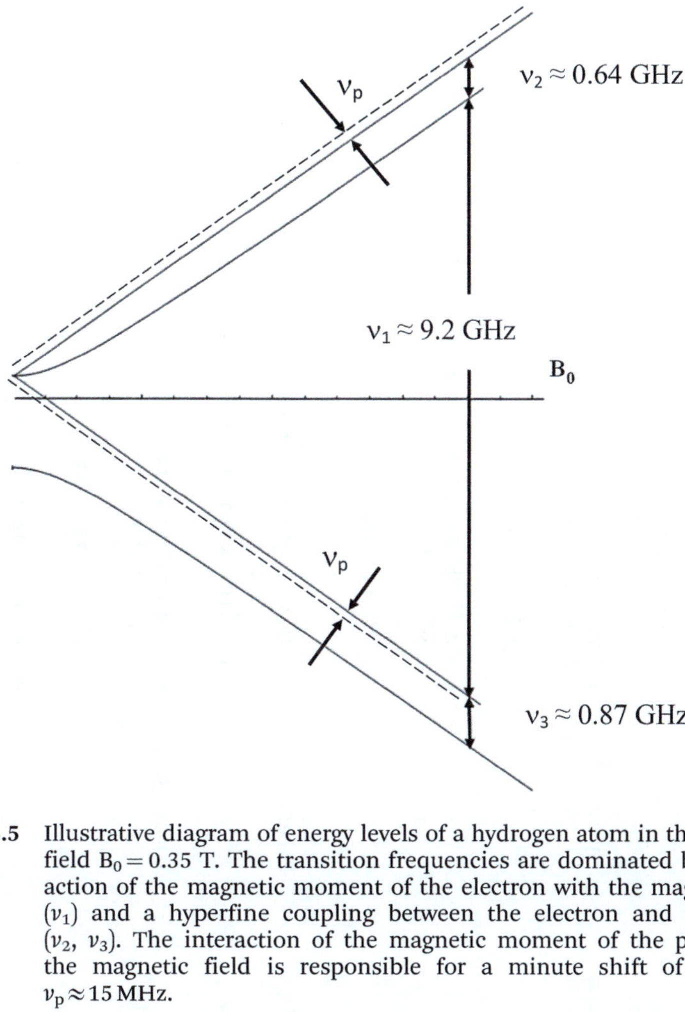

Figure 3.5 Illustrative diagram of energy levels of a hydrogen atom in the magnetic field $B_0 = 0.35$ T. The transition frequencies are dominated by an interaction of the magnetic moment of the electron with the magnetic field (ν_1) and a hyperfine coupling between the electron and the proton (ν_2, ν_3). The interaction of the magnetic moment of the proton with the magnetic field is responsible for a minute shift of the lines, $\nu_p \approx 15$ MHz.

matter has to be avoided. Measurement of the magnetic moment of the antiproton could provide a stringent test of matter/antimatter baryon symmetry,[51] because one can expect that the magnitude of the magnetic moment of the antiproton is the same as for the proton, but its sign is opposite. One can notice that all the experiments listed above provide the converged value of the proton magnetic moment, which is currently known with extremely small relative uncertainty $\approx 3 \times 10^{-9}$.

3.4.2 Nuclear Magnetic Moments from Gas Phase NMR Experiments

The magnetic moment of the proton can be used for accurate determination of the nuclear magnetic moments of other nuclei from the gas phase NMR

Table 3.1 The proton and deuteron magnetic moments determined from atomic maser experiments.[a,29,49]

Nucleus	Magnetic moment	Data source	Comment
Proton	$\mu_p = \left(\frac{\nu_e(\text{H})}{\nu_p(\text{H})}\right)^{-1}(1-\sigma_p(\text{H}))^{-1}(1-\sigma_e(\text{H}))\lvert\mu_e\rvert$	Experiment	The ratio of hyperfine structure transitions of atomic hydrogen:
	$\boldsymbol{\mu_p = 1.410\ 606\ 743(33)\times10^{-26}\ \text{J T}^{-1}}$		$\nu_e(\text{H})/\nu_p(\text{H}) = \mathbf{658.210\ 705\ 8(66)}$
	$\mathbf{2.792\ 847\ 356(23)}\ \boldsymbol{\mu_N}$	Theory (QED)	• The shielding of the proton by the electron in a hydrogen atom:
			$\sigma_p(\text{H}) = \mathbf{17.7354\times10^{-6}}$
			• The shielding of the electron by the proton in a hydrogen atom:
			$\sigma_e(\text{H}) = \mathbf{17.7054\times10^{-6}}$
Deuteron	$\mu_d = \frac{\nu_d(\text{D})}{\nu_e(\text{D})}(1-\sigma_d(\text{D}))^{-1}(1-\sigma_e(\text{D}))\lvert\mu_e\rvert$	Experiment	The ratio of hyperfine structure transitions of atomic deuterium:
	$\boldsymbol{\mu_d = 0.433\ 073\ 489(10)\times10^{-26}\ \text{J T}^{-1}}$		$\nu_d(\text{D})/\nu_e(\text{D}) = \mathbf{4.664\ 345\ 392(50)\times10^{-4}}$
	$\mathbf{0.857\ 438\ 230\ 8(72)}\ \boldsymbol{\mu_N}$	Theory (QED)	• The shielding of the deuteron by the electron in a hydrogen atom:
			$\sigma_d(\text{D}) = \mathbf{17.7461\times10^{-6}}$
			• The shielding of the electron by the deuteron in an hydrogen atom:
			$\sigma_e(\text{D}) = \mathbf{17.7126\times10^{-6}}$

[a]It was assumed the electron magnetic moment, μ_e, is equal to $\mathbf{-928.476\ 430(21)\times10^{-26}\ J\ T^{-1}}$.

experiment. In this case it is more convenient to measure the resonance frequencies at the same strength of the magnetic field for a molecule than for an atom. An advantage of this kind of measurement is the possibility of the extrapolation to the zero-density limit resonance frequencies, thus, the final result is free from influence of intermolecular interactions and bulk magnetic susceptibility effects. Several examples of nuclear magnetic moments of stable nuclei determined applying this method are given in Table 3.2.

If a volatile binary hydride which contains a given nucleus is available, then the magnetic moment of this nucleus can be determined from gas phase NMR measurements and computed (or evaluated using the experimental spin-rotation constant) nuclear magnetic shielding. Let us exemplify this method of determination of the nuclear magnetic moment on the studies of the magnetic moment of the phosphorus nucleus described by Lantto *et al.*[53] The shielding of the proton and the phosphorus nucleus in the phosphine molecule determined from quantum mechanical computations is respectively $\sigma^{H} = 29.305$ ppm and $\sigma^{P} = 614.758$ ppm. The resonance frequency extrapolated to the zero-density limit for the proton in this molecule is $\nu^{H} = 500.606\ 730\ 3$ MHz, while for the phosphorus nucleus it is $\nu^{P} = 202.595\ 027\ 8$ MHz. These two frequencies were determined at the same magnetic field strength, therefore, one can find the magnetic moment of the phosphorus nucleus from the formula:[56]

$$\mu_{P} = \frac{(1-\sigma^{H})\nu_{P}}{(1-\sigma^{P})\nu_{H}}\mu_{H}. \tag{3.12}$$

Using the value of the magnetic moment of the proton, 2.792 847 350 μ_N, one can find that the magnetic moment of the phosphorus nucleus is $\mu^{P} = 1.130\ 925\ \mu_N$. Several other binary hydrides were used in a similar manner for determination of the nuclear magnetic moments. For example, magnetic moments of ^{35}Cl and ^{37}Cl were determined from gas phase NMR measurements of hydrogen chloride,[54] magnetic moments of ^{29}Si and ^{73}Ge were obtained from data for SiH_4 and GeH_4,[55] magnetic moment of ^{17}O was determined from NMR spectra of $H_2{}^{17}O$,[56] and magnetic moment of ^{13}C was found using $^{13}CH_4$.[56]

An interesting case is the determination of the tritium magnetic moment from ^{1}H and ^{3}H NMR gas phase spectra of HT.[57] For this molecule adiabatic approximation predicts exactly the same shielding for the proton and the triton, therefore, only the ratio of the resonance frequencies of the proton and triton is necessary for determination of the magnetic moment of the triton. However, detailed investigation indicates that there is a small non-adiabatic contribution to shielding of the proton and the triton in HT (the difference is 24.20 ppb),[58] which has to be taken into account.

Many elements do not possess gaseous hydrides or their hydrides are unstable, therefore, in these cases it is advantageous to use an external reference nuclear magnetic moment, for instance gaseous ^{3}He. Similarly to

Table 3.2 Nuclear magnetic moments explored and improved on the basis of NMR measurements in the gas phase.[a]

Nucleus	Spin number, I	Natural abundance, %	Magnetic moment, μ_I/μ_N	Magnetogyric ratio, $\gamma_I = \mu_I/(\hbar I)$	g value, $g_I = (\mu_I/\mu_N I)$	Ref.
^{1}H	1/2	99.99	2.792 847 350 (9)	267.522 199 7 (9)	5.585 694 70 (2)	50
^{2}H	1	0.01	0.857 438 231 (7)	41.066 290 6 (3)	0.857 438 231 (7)	29
			0.857 438 235 (5)	41.066 291 8 (2)	0.857 438 235 (5)	58
^{3}H	1/2	0.00	2.978 962 45 (4)	285.349 855 (4)	5.957 924 90 (8)	29
			2.978 962 47 (1)	285.349 857 (1)	5.957 924 94 (2)	58
^{3}He	1/2	0.000137	− 2.127 625 31 (3)	− 203.801 688 (3)	− 4.255 250 62 (6)	60–62
^{10}B	3	19.90	1.800 463 6 (8)	28.743 90 (1)	0.600 154 5 (3)	64
^{11}B	3/2	80.10	2.688 378 (1)	85.838 42 (3)	1.792 252 0 (7)	64
^{13}C	1/2	1.07	0.702 369 4 (7)	67.278 80 (7)	1.404 739 (1)	56
^{14}N	1	99.63	0.403 572 3 (5)	19.328 76 (2)	0.403 572 3 (5)	56
^{15}N	1/2	0.37	− 0.283 057 (1)	− 27.113 6 (1)	− 0.566 141 (2)	56
^{17}O	5/2	0.04	− 1.893 547 (2)	− 36.275 94 (4)	− 0.757 418 8 (8)	56
^{19}F	1/2	100.00	2.628 34 (1)	251.764 (1)	5.256 68 (2)	56
^{21}Ne	3/2	0.27	− 0.661 776 2 (6)	− 21.130 15 (2)	− 0.441 184 1 (4)	52
^{29}Si	1/2	4.68	− 0.555 052 (3)	− 53.167 5 (2)	− 1.110 104 (6)	55
^{31}P	1/2	100.00	1.130 925 (5)	108.329 4 (5)	2.261 85 (1)	53
^{33}S	3/2	0.76	0.643 25 (2)	20.538 6 (6)	0.428 83 (1)	56
^{35}Cl	3/2	75.78	0.821 70 (1)	26.236 4 (3)	0.547 800 (7)	54
^{37}Cl	3/2	24.22	0.683 98 (1)	21.839 1 (3)	0.455 987 (7)	54
^{73}Ge	9/2	7.73	− 0.878 24 (5)	− 9.347 2 (5)	− 0.195 2 (1)	55
^{77}Se	1/2	7.63	0.533 56 (5)	51.109 (5)	1.067 1 (1)	52
^{83}Kr	9/2	11.49	− 0.970 730 (3)	− 10.331 62 (3)	− 0.215 717 8 (7)	65
^{119}Sn	1/2	8.59	− 1.045 1 (1)	− 100.11 (1)	− 2.090 2 (2)	55
^{129}Xe	1/2	26.44	− 0.777 96 (2)	− 74.519 (2)	− 1.555 92 (4)	66
^{131}Xe	3/2	21.18	0.691 845 (7)	22.0902 (2)	0.461 230 (5)	66

[a]Constants: $\mu_N = 5.050\ 783\ 53 \times 10^{-27}$ J T^{-1}, $\hbar = 1.054\ 571\ 726 \times 10^{-34}$ J s.

the magnetic moment of the proton, the magnetic moment of the helion-3 was found using as a primary reference the magnetic moment of the electron; however, this determination involved several intermediate measurements. More specifically, Phillips *et al.*[59] measured the ratio between the magnetic moment of the electron in a hydrogen atom and the magnetic moment of the proton in a water molecule at 34.7 °C. Next, the obtained value was referenced to temperature 25 °C by Petley and Donaldson.[60] The magnetic moment of shielded ^{3}He was determined from NMR resonance frequencies of water and ^{3}He recorded at the same magnetic field strength by Flowers *et al.*[61] Then, the magnetic moment of helion in a helium atom was corrected using shielding of ^{3}He by its two electrons computed by Rudziński *et al.*[62,63]

This method of measurement of the magnetic moment using ^{3}He as the reference was applied to the determination of magnetic moments of boron isotopes, ^{10}B and ^{11}B, by Jackowski *et al.*[64] Since BH_3 is unstable at room temperature, BF_3 was used in the experiment. Gaseous BF_3 was mixed with a small amount of ^{3}He and the resonance frequencies of ^{3}He, ^{10}B, and ^{11}B were measured. Applying an analogous formula to eqn (3.12) the magnetic moments of boron isotopes were found, $\mu^{B\text{-}10} = 1.800\ 463\ 6\ (8)\ \mu_N$ and $\mu^{B\text{-}11} = 2.688\ 378\ (1)\ \mu_N$. Following the same scheme magnetic moments for ^{83}Kr, ^{129}Xe, and ^{131}Xe were measured.[65,66]

One can remark that the determination of the magnetic moments is not limited to only stable nuclei. For radioactive nuclei which exhibit beta decay one can use an exotic form of nuclear magnetic resonance, β detected NMR method. In this method one can infer the nuclear magnetic moment from the anisotropy of the beta decay. For instance, using β-NMR nuclear magnetic moments of neon isotopes from ^{17}Ne to ^{25}Ne were determined.[67,68]

Several selected atomic nuclei for which magnetic moments were not studied by the gas phase NMR are collected in Table 3.3; other examples can be found in the comprehensive table published by Stone.[69] These data were mostly obtained from liquid-state NMR and they were not fully corrected for magnetic susceptibility of the sample and nuclear magnetic shielding. Therefore, error bars shown in the table reflect high precision of experimental determination of resonance frequencies rather than high accuracy of magnetic moments determination. One could improve these magnetic moments using gas phase NMR for volatile compounds of these nuclei. However, these studies require caution, since some of the listed compounds are highly toxic (*e.g.* H_2Se, AsH_3), unstable (BiH_3), and moisture-sensitive (*e.g.* WF_6, MoF_6).

3.5 Direct Measurements of Shielding

Having more accurate magnetic moments of atomic nuclei we can apply the new data back to the laboratory practice in NMR spectroscopy. First, let us note that it would be considerably better if the absolute values of shielding (σ_i) could be directly read from the NMR spectrum instead of chemical

Table 3.3 Selected magnetic moments which have not been determined from the gas phase NMR.

Nucleus	Spin	Gaseous compound which could be used for the gas phase NMR experiment	Boiling point/°C	Magnetic moment determined from liquid-state NMR, μ/μ_N	Ref.
^{57}Fe	1/2	$Fe(CO)_5$	103	0.090 623 00 (9)	70
^{75}As	3/2	AsH_3	−62.5	1.439 48 (7)	71, 72
^{77}Se	1/2	H_2Se	−41.3	0.535 074 3 (3)	73
^{81}Br	3/2	HBr	−66.4	2.270 562 (4)	74
^{95}Mo	5/2	MoF_6	34	−0.914 2 (1)	75
^{111}Cd	1/2	$Cd(CH_3)_2$	106	−0.594 886 1 (8)	76, 77
^{121}Sb	5/2	SbH_3	−17	3.363 4 (3)	75
^{125}Te	1/2	H_2Te	−2	−0.888 505 1 (4)	78
^{127}I	5/2	HI	−35.6	2.813 27 (8)	79
^{183}W	1/2	WF_6	17	0.117 784 76 (9)	70
^{199}Hg	1/2	$Hg(CH_3)_2$	93	0.505 885 489 (6)	30
^{209}Bi	9/2	BiH_3	17	4.110 6 (2)	75, 80

shifts. Such a procedure was outlined with the application of helium-3 gas as the universal reference standard of nuclear magnetic shielding.[63] It is based on the recent calculation of magnetic shielding in an isolated helium-3 atom ($\sigma_0(^3He) = 59.9674$ ppm)[62] and on our previous measurement of resonance frequency for the same atomic object.[52] As shown it permits the direct reading of shielding from ^{1}H, ^{13}C NMR spectra and can be extended on the other nuclei when their magnetic moments are accurately known. Here we will use only the scalar values of σ parameters as they are sufficient for the isotropic medium of gases.

Measuring two resonance frequencies (ν_X, ν_{He}) for molecule X and ^{3}He atom in the same external magnetic field we can determine any unknown shielding σ_X when the helium shielding parameter σ_{He} is known and the appropriate nuclear magnetic moments (μ_X, μ_{He}) are available with satisfactory accuracy:

$$\sigma_X = 1 - \frac{\nu_X}{\nu_{He}} \cdot \frac{|\mu_{He}|}{|\mu_X|} \cdot \frac{I_X}{I_{He}} (1 - \sigma_{He}) \tag{3.13}$$

where I_X, I_{He} are the spin numbers of X and He nuclei. Equation (3.13) can be rewritten using the magnetogyric ratios (γ_{He}, γ_X) as:

$$\sigma_X = 1 - \frac{\nu_X}{\nu_{He}} \cdot \frac{|\gamma_{He}|}{|\gamma_X|} \cdot (1 - \sigma_{He}) \tag{3.14}$$

or with the application of nuclear g_{He}, g_X factors, *cf.* eqn (3.6) and (3.7):

$$\sigma_X = 1 - \frac{\nu_X}{\nu_{He}} \cdot \frac{|g_{He}|}{|g_X|} \cdot (1 - \sigma_{He}) \tag{3.15}$$

Let us note that the moduli of μ, γ, and g can be replaced by their actual numerical values if one accepts the convention of positive and negative frequencies for magnetic nuclei in NMR experiments.[81] In such a case the resonance frequency of an isolated helium-3 atom is equal to −381.358 662 MHz while at the same magnetic field the proton frequency of liquid TMS is found to be + 500.607288 MHz.[52]

3.5.1 Referencing of Shielding Measurements

Equations (3.13)–(3.15) present the method of direct shielding measurement when an isolated helium-3 atom is used as the universal reference standard.[63] The isolated ^{3}He atom is "universal" as the reference standard because it can be applied to the shielding measurement of any magnetic nucleus. Its shielding constant is independent of temperature, well known from accurate calculations ($\sigma_0(^3\text{He}) = 59.9674$ ppm),[62] and therefore it is certainly the best choice for a primary reference of nuclear magnetic shielding. However, there are a few serious problems with the application of gaseous helium-3 in experimental NMR practice. First of all this gas is not easily available on the market and therefore it is also very expensive. Next, the resonance frequency of ^{3}He nucleus is in the region between ^{19}F and ^{31}P nuclei and this range of frequency is always omitted in standard NMR probes. Moreover, helium-3 gas can easily escape from glass containers (*e.g.* ordinary NMR tubes) due to its efficient diffusion. Several years ago Harris *et al.*[30] concluded that gaseous helium-3 could be a good primary reference of chemical shifts "... but this is not practicable". Fortunately all the difficulties with the application of helium-3 gas in NMR laboratory work could be overcome and an isolated atom of helium-3 was in fact successfully used as the primary reference of nuclear magnetic shielding (Figure 3.6).[63]

Equations (3.13)–(3.15) permit the measurement of observable shielding and it can be done also for liquid samples. Thus the value of absolute shielding can be transferred from helium-3 atom to pure liquid deuterated solvents which are usually present in routine NMR work for the stabilization of external magnetic field (deuterium lock). Table 3.4 presents the shielding of deuterons (σ_D*) found for pure liquid solvents which can be used as the secondary reference standards of shielding. The σ_D* parameters were precisely determined at 300 K relative to an isolated helium-3 atom. For the application of new reference standards eqn (3.13) can be modified as follows:

$$\sigma_X = 1 - \frac{\nu_X}{\nu_D} \cdot \frac{|\mu_D|}{|\mu_X|} \cdot \frac{I_X}{1}(1 - \sigma_D{}^*) \tag{3.16}$$

where ν_D is deuterium resonance frequency, μ_D is the magnetic moment of a deuteron, and the shielding reference value (σ_D*) is selected for the appropriate solvent from Table 3.4. Moreover, eqn (3.16) can be further simplified because the μ_X and I_X parameters are constant for the selected X nucleus so we write:[82]

$$\sigma_X = 1 - \frac{\nu_X}{\nu_D} \cdot C_X(1 - \sigma_D{}^*) \tag{3.17}$$

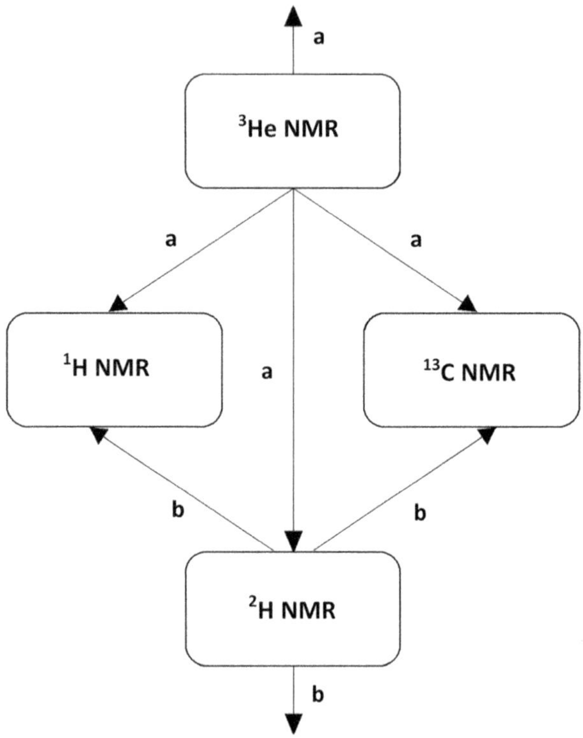

Figure 3.6 The measurements of shielding can be performed using gaseous helium-3 as the reference standard (option a) or after the transfer of shielding reference on deuterated liquid solvents (option b) and both these methods are exactly equivalent.

where C_X is a constant coefficient for the observed nucleus. For the most popular NMR nuclei the C_X values are established and the following relations can be used for the measurements of nuclear shielding:[63,82]

$$\sigma_H = 1 - \frac{\nu_H}{\nu_D} \cdot C_H(1 - \sigma_D{}^*) \tag{3.18}$$

$$\sigma_C = 1 - \frac{\nu_C}{\nu_D} \cdot C_C(1 - \sigma_D{}^*) \tag{3.19}$$

$$\sigma_N = 1 - \frac{\nu_N}{\nu_D} \cdot C_N(1 - \sigma_D{}^*) \tag{3.20}$$

where $C_H = 0.153\ 506\ 104$, $C_C = 0.610\ 389\ 782$, and $C_N = 1.514\ 602\ 904$. In eqn (3.13)–(3.20) the shieldings ($\sigma_D{}^*$, σ_{He}, σ_H, σ_C...*etc.*) are represented by their usual numerical values and must be multiplied by the factor of 10^6 in order to obtain the popular units in NMR – ppm (parts per million). The application of equations (3.18)–(3.20) permits the easy measurement of

Table 3.4 Absolute resonance frequencies [Hz] and shielding measurements [ppm] for selected liquid solvents. In this experiment the ^{2}H frequency was kept constant and equal to 76 846 401.5 Hz, the ^{1}H values refer to the residual protons in the deuterated solvents. The σ_D* shielding was determined using an isolated helium-3 atom as the universal reference standard, the values of σ_D* were applied for the measurements of ^{1}H and ^{13}C shielding (σ_H, σ_C) according to eqn (3.16). $^{0}\Delta H(^{2/1}H)$ [ppm] represent the primary isotope effects.[63]

No.	Solvent	Observed nuclei	Absolute frequencies [Hz]		σ_D*	σ_H	σ_C	$^{0}\Delta H(^{2/1}H)$
			^{1}H	^{13}C				
1	Cyclohexane-d_{12}	**$-CD_2-$**	500 608 190.4	125 881 059.3	31.834	31.687	160.522	−0.147
2	Acetone-d_6	**$-CD_3$**	500 608 183.3	125 881 408.9	30.570	30.439	156.479	−0.131
3	Methanol-d_4	**$-CD_3$**	500 608 184.7	125 883 599.3	29.593	29.464	138.108	−0.129
4	Water-d_2	**−OD**	500 608 170.3	–	28.837	28.730	–	−0.107
5	Benzene-d_6	**=CD-**	500 608 165.4	125 893 195.7	26.441	26.344	58.731	−0.097
6	Chloroform-d	**−CD**	500 608 191.6	125 886 807.5	26.389	26.239	109.421	−0.150
7	DMSO-d_6	**$-CD_3$**	500 608 177.7	125 882 747.6	30.574	30.454	145.852	−0.120
8	Toluene-d_8	**$-CD_3$**	500 608 187.6	125 880 282.7	31.525	31.387	166.390	−0.138
9	Acetonitrile-d_3	**$-CD_3$**	500 608 188.4	125 877 820.3	30.864	30.722	185.274	−0.142
10	Nitromethane-d_3	**$-CD_3$**	500 608 182.5	125 885 204.5	28.041	27.914	123.807	−0.127
11	Ethanol-d_6	**$-CD_3$**	500 608 183.3	125 879 927.6	32.020	31.891	169.692	−0.129

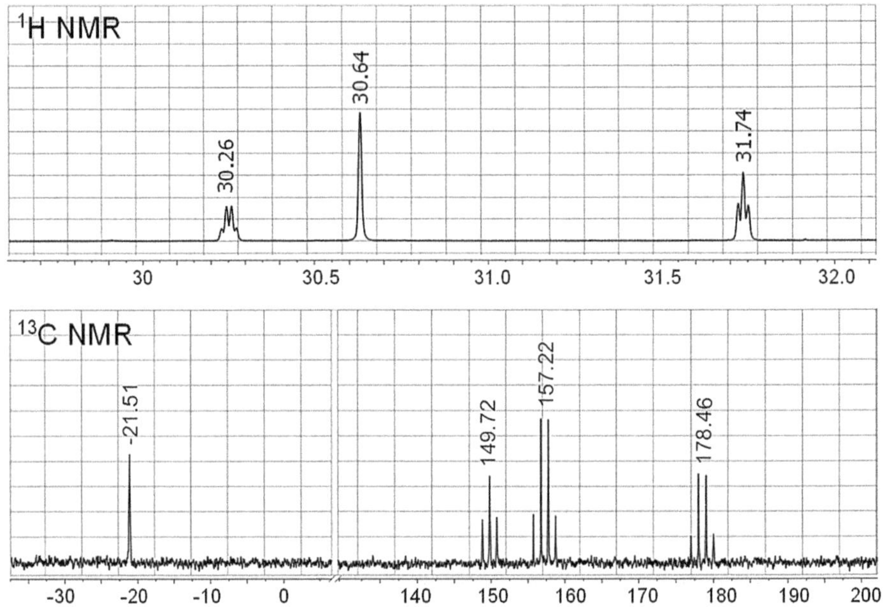

Figure 3.7 ^{1}H and ^{13}C NMR spectra of liquid methyl ethyl ketone obtained using benzene-d_6 as the external reference standard of shielding.

shielding for different nuclei from one sample as is illustrated in Figure 3.7, where ^{1}H and ^{13}C NMR spectra of methyl ethyl ketone are presented with the scale of absolute shielding. Here benzene-d_6 was applied as the external reference standard of shielding and the same solvent was used for the deuterium lock. A similar equation can be also written for fluorine-19:

$$\sigma_F = 1 - \frac{\nu_F}{\nu_D} \cdot C_F(1 - \sigma_D{}^*) \tag{3.21}$$

with $C_F = 0.379\ 087\ 277$ or other nuclei like boron-11 ($C_B = 0.478\ 413\ 866$), oxygen-17 ($C_O = 1.132\ 052\ 741$), silicon-29 ($C_{Si} = 0.772\ 395\ 075$), and phosphorus-31 ($C_P = 0.379\ 087\ 277$).[82] Let us note that all the C_X parameters are given with nine decimal digits as this is needed to obtain the shielding from appropriate equations with the precision of ± 0.01 ppm, which is usually required for NMR measurements. This precision of shielding measurements is obviously preserved as far as the C_X parameters are constant. Then the shielding measurements can be successfully used for the analysis of NMR spectra because all the shielding data will be also reproducible.

The problem of accuracy is a little different. The accuracy of the discussed C_X coefficients varies and it is usually noticeably lower. It depends on our knowledge of the appropriate nuclear magnetic moments. For example, the proton shielding can be determined using eqn (3.18) with the accuracy of ± 0.003 ppm[83] while the magnetic moment of carbon-13 nucleus was determined from the shielding of methane molecule,[56] which is known with lower accuracy (probably ± 0.9 ppm)[84] and eqn (3.19) cannot give higher accuracy than ± 0.9 ppm for carbon shielding. Similar accuracy of shielding measurements is available also for nitrogen-15 nucleus when eqn (3.20) is applied. Recalling that the range of shielding is a few hundred ppm for the latter nuclei the accuracy of ± 0.9 ppm is not so bad at all. For the other nuclei (*e.g.* fluorine-19, boron-11, oxygen-17, silicon-29, or phosphorous-31) the accuracy of shielding measurements is distinctly lower at this moment. However, we expect that availability of better values of magnetic dipole moments and shielding constants will significantly improve the accuracy of this approach to shielding measurements for many nuclei in the near future.

3.5.2 External and Internal Referencing of Shielding

Equations (3.18)–(3.20) can be used for routine measurements when any deuterated solvent selected from Table 3.4 is used as the external reference standard of shielding (*cf.* the ^{1}H and ^{13}C NMR spectra of ethyl methyl ketone in Figure 3.7). In such a case the shielding measurement will contain the bulk susceptibility correction (BSC). For measurements in the gas phase it does not matter because such results are usually extrapolated to the zero-density point where the BSC effect vanishes. In liquid NMR experiments the results will require the estimation of BSC when the external referencing is applied. Recently it was shown[85] that this problem can be simply overcome if investigated molecules are dissolved in the deuterated solvent which simultaneously serves as the internal reference standard and lock solvent. So long as the concentration of solute molecules is small this method should give good results because only a small fraction of solvent molecules is engaged in molecular interactions with solute molecules. The shielding measurements with the internal referencing are easy and free from the BSC problem.

On the other hand the joint application of external and internal referencing was recently explored for the studies of intermolecular interactions in monosubstituted benzenes.[82] ^{13}C magnetic shielding was observed for 11 C_6H_5X compounds when the solute molecules were dissolved in liquid benzene-d_6. As concluded the direct observation of shielding gives better results than the similar investigation of ^{13}C chemical shifts because there is no need to analyze the influence of intermolecular interactions on the molecules of reference standard.

3.6 Applications of Shielding Measurements

3.6.1 Standardization of NMR Spectra

At present the standardization of NMR spectra is exclusively based on the measurements of chemical shifts. Therefore the chemical shift, discovered in 1950,[87,88] became the most important parameter of experimental NMR spectroscopy. According to the IUPAC conventions[30,89–91] the chemical shift

(δ_S) is obtained from the difference of resonance frequencies of sample (ν^S) and reference standard (ν^R):

$$\delta_S/\text{ppm} = \frac{\nu^S - \nu^R}{\nu^R} \cdot 10^6 = \frac{\sigma^R - \sigma^S}{1 - \sigma^R} \cdot 10^6 \approx (\sigma^R - \sigma^S) \cdot 10^6 \tag{3.22}$$

where σ^R and σ^S are the shieldings of reference (R) and sample (S), respectively, when the external magnetic field (B_0) is constant. The chemical shifts and shielding parameters are small and therefore they are usually expressed in parts per million [ppm]. At this point let us note that the application of chemical shifts is limited:

- chemical shifts show only the relative values of shielding,
- there is no possibility for straightforward comparison of experimental and calculated results for shielding,
- usually each magnetic nucleus requires another reference standard, has its own scale of chemical shifts, and it splits NMR spectroscopy into many different methods,
- the opposite signs of shielding (σ^S) and appropriate chemical shift (δ_S) can be a source of numerous misunderstandings in the literature. Before 1972 the chemical shifts were usually defined in the same direction as the values of nuclear shielding.

The main problem with the use of chemical shifts is qualitatively presented in Figure 3.8 where the chemical shifts (δ_S) are shown for sample (S) and reference (R) molecules on the scale of nuclear magnetic shielding. The diagram has five levels for investigated molecules: (a) molecules are at the equilibrium geometry – available only for calculations, (b) as before plus vibrations at 0 K (zero-point vibration, ZPV) – also available only for calculations, (c) temperature increase to 300 K is included – this level is already available for experiments and calculations, (d) single measurements for gases and vapors – good for experiments but more difficult for calculations and (e) represents molecules in liquids – suitable for routine NMR measurements and the most difficult for calculations. As seen the shielding is modified first due to the increase of intramolecular motion with temperature (a–c) and even more as the result of intermolecular interactions (c–e). These changes of shielding are imposed unequally on sample and reference molecules and the chemical shift provides us with rather limited information on the shielding of sample molecules. As seen both shielding parameters (σ^S and σ^R) must be carefully analyzed for the accurate interpretation of each chemical shift.

The description of NMR results becomes a little more universal if the absolute resonance frequencies (Ξ_X)[92,93] are used for the standardization of spectra instead of chemical shifts.[30] On the Ξ scale the resonance frequency of protons in liquid TMS is exactly equal to 100.000 000 MHz under standard conditions. In general liquid TMS as the primary reference standard is not a

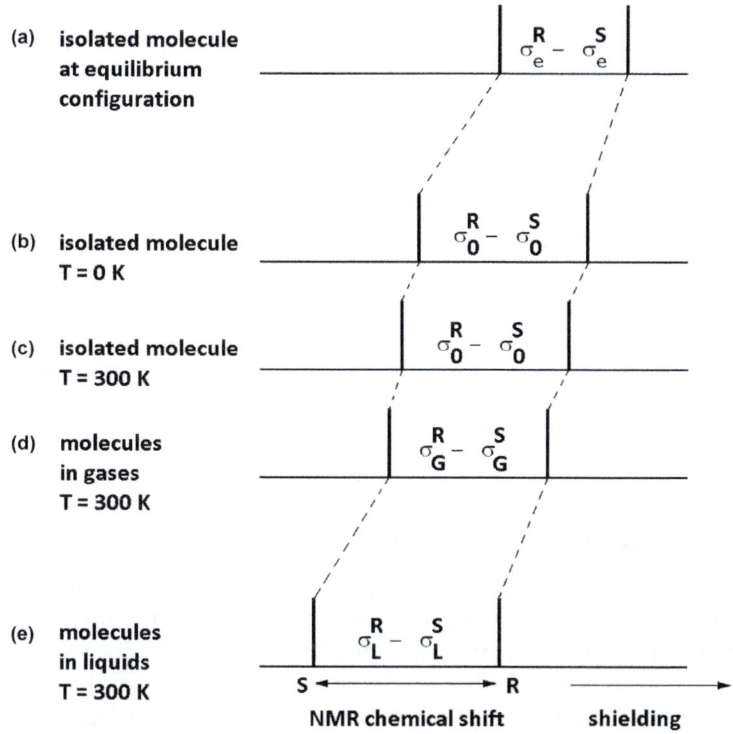

Figure 3.8 The NMR chemical shift is shown on the scale of nuclear magnetic shielding for the sample (S) and the reference molecules (R) in various conditions. Rovibrational movements usually decrease the shielding of molecules when temperature is increased (a–c) and intermolecular interactions enlarge the deshielding effects (c–e), which is well seen on the scale of shielding but not properly observed when only the chemical shift ($\sigma^R - \sigma^S$) is monitored.

good choice as its resonance frequency at the stable external magnetic field is still dependent on temperature and intermolecular interactions. Moreover, the Ξ scale requires from 8 to 10 digits for the precise description of each NMR signal, which is a bit difficult in everyday practice. Absolute resonance frequency is not easily connected to the shielding of nuclei. Altogether the Ξ scale is impractical in experimental work and used almost exclusively as the general parameter of magnetic nuclei.

The measurements of shielding performed according to eqn (3.13) can certainly be used for the standardization of NMR spectra. Let us note that the experimental shielding is always determined with the same precision as the appropriate chemical shift because it is based on the same reading of resonance frequencies, *cf.* eqn (3.13) and (3.22). It is simply enough to apply systematically the fixed, best accurate value of a nuclear magnetic moment (μ_X) for X nucleus and get the precise measurements of shielding suitable for the standardization of NMR spectra. Using for example the magnetic moments of Table 3.2 we can always get the same results of magnetic shielding in molecules containing the nuclei listed in this table and the precision of shielding measurements will be exactly the same as for the appropriate NMR chemical shifts.

The accuracy of shielding measurement is still dependent on the value of a given nuclear magnetic moment as previously described in Section 3.5.1, but new experiments and calculations will certainly improve the values of nuclear magnetic moments in the near future. Finally, the chemical shifts may become completely useless when the μ_X magnetic moments are known with satisfactory accuracy, *i.e.* the accuracy sufficient for the identification of chemical compounds on the basis of shielding reading in NMR spectrometers. In our opinion such a situation already exists for the shielding measurements of a proton and its isotopes[83] and to some extent also for carbon and nitrogen spectra.

3.6.2 Verification of Shielding Calculations

Verification of shielding calculations requires accurate measurements of shielding, which must be extracted from the appropriate chemical shifts. The situation is better when the shielding data are directly available from experiments as described in Section 3.5. This method was recently applied for the accurate measurements of ^{1}H isotropic shielding in 71 isolated molecules and the 115 benchmark values of ^{1}H shielding were established.[83] These data range over more than 20 ppm, from a maximum value of 43.92 ppm for HI to a minimum of 19.258 ppm for CF_3COOH. Because the errors in these 115 values are generally less than 0.01 ppm, they should prove invaluable to computational scientists interested in testing the general reliability of their computational methods. Similar results will soon also be obtained for the shielding of ^{13}C and ^{15}N nuclei and it will simplify the comparison of experimental and calculated shielding values.

However, theoretical chemical shifts are actually widely popular for comparison with experimental chemical shifts. Occasionally the correlation between the experimental and theoretical results is perfect both for very good[85] and very modest calculations,[86] but it does not prove the quality of shielding calculations. Sometimes the quality of the calculations can be fortuitously good due to the error compensation in eqn (3.22), where the calculated shielding values of investigated and reference molecules are used for the determination of calculated chemical shifts. It means that the comparison of experimental and theoretical chemical shifts has rather limited significance (for details see Figure 3.7). Only the verification of shielding values will reveal the real quality of performed calculations. For this reason the experimental data on nuclear magnetic shielding are undoubtedly valuable and the simplified method of their measurement is certainly important.

3.6.3 Primary Isotope Effects in Shielding

Equation (3.13) and its later modification (eqn (3.18)) allow us to study the primary isotope effects in nuclear shielding. It is something new for NMR measurements because chemical shifts can be used only for the observation of isotope effects from neighbor nuclei (the so-called secondary isotope effect; a detailed discussion of isotope effects is given by Jameson in Chapter 1 of this book). The primary isotope effect is observed when the shielding of a given nucleus and its isotope are measured and compared for the same atom in a molecule. In Table 3.4 the primary isotope effects, ${}^{0}\Delta H({}^{2/1}\mathrm{H}) = \sigma_D{}^* - \sigma_H$ are listed in the last column. They show that the residual protons in deuterated liquid solvents are less shielded than appropriate deuterons. These effects are obtained for liquids and their values can be modified by intermolecular interactions, especially by hydrogen bonds in liquids. Pure isotope effects in shielding can be observed only for isolated molecules.

Following the steps outlined in Section 3.5 it was possible to evaluate the shielding in isolated H_2, HD, and D_2 molecules. As summarized in Table 3.5, application of direct shielding measurements yields $\sigma_0(\mathbf{H}_2) = 26.293(5)$ ppm, $\sigma_0(\mathbf{H}\mathrm{D}) = 26.327(3)$ ppm, $\sigma_0(\mathrm{H}\mathbf{D}) = 26.339(3)$ ppm, and $\sigma_0(\mathbf{D}_2) = 26.388(3)$ ppm at 300 K. The primary isotope effects in hydrogen molecules are as follows: $\sigma_0(\mathrm{H}\mathbf{D}) - \sigma_0(\mathbf{H}_2) = +0.046(8)$ ppm and $\sigma_0(\mathbf{D}_2) - \sigma_0(\mathbf{H}\mathrm{D}) = +0.061(6)$ ppm. The secondary isotope effects are slightly smaller: $\sigma_0(\mathbf{H}\mathrm{D}) - \sigma_0(\mathbf{H}_2) = +0.034(8)$ ppm and $\sigma_0(\mathbf{D}_2) - \sigma_0(\mathrm{H}\mathbf{D}) = +0.049(6)$ ppm. The accuracy of the above results is fairly satisfactory for hydrogen NMR and the results themselves are consistent with the experimental values deduced from appropriate spin-rotation constants[94] and accurate calculations.[95,96]

3.6.4 ^{13}C Shielding Scale for NMR Measurements in Solids

Once the shielding parameter was transferred from an isolated helium-3 atom to pure liquids, it was also possible to establish the reference standards of shielding which can be applied for solids in ^{13}C NMR spectroscopy with the magic angle spinning (MAS) technique.[97] For this purpose the ^{13}C shielding of liquid TMS and solid fullerene C_{60} were determined with the use of a spherical sample. Helium-3 gas as the reference of shielding was applied

Table 3.5 Experimental and calculated shielding in isolated H_2, HD, and D_2 molecules.

	Gas-phase measurement[a]	Spin-rotation constant[b]	Computations[c]
$\sigma(\mathbf{H}_2)$	26.293(5)	26.2886(15)	26.2980
$\sigma(\mathbf{H}\mathrm{D})$	26.327(3)	26.3329(12)	26.3416
$\sigma(\mathrm{H}\mathbf{D})$	26.339(3)	26.3436(48)	26.3416
$\sigma(\mathbf{D}_2)$	26.388(3)	26.3884(20)	26.3930

[a]Measured using helium-3 nuclear magnetic shielding at 300K.[83]
[b]Paramagnetic shielding obtained from spin-rotation constant, diamagnetic shielding from computations.[94]
[c]*Ab initio* quantum mechanical computations.[95]

first, then the shielding reference was transferred to nitromethane-d_3 in the outer chamber of an NMR tube and finally the absolute ^{13}C shielding of TMS and fullerene were measured. Let us note that the measurements performed in a spherical cuvette are free from the BSC effect and therefore the reference data of shielding can be used in MAS experiments, which are also free from BSC.

3.6.5 Adsorbed Gases

Gas phase NMR can provide valuable information about properties of zeolites,[98] biopolymers,[99] and clathrates.[100] Among other methods, *e.g.*, mercury porosimetry and gas-absorption methods, ^{129}Xe NMR studies are especially useful for characterization of micropore materials, *i.e.*, materials with pore diameters less than 2 nm. From ^{129}Xe NMR studies one can infer about the size and symmetry of pores, the average number of atoms in a pore, and the rate of exchange between pores.[101] For studies of small pores it is advantageous to use ^{3}He, whose van der Waals radius, $r_{He} = 0.14$ nm, is approximately two times smaller than the radius of a xenon atom. For instance, introduction of ^{129}Xe into zeolite NaA takes several weeks,[101] while this material is penetrated by ^{3}He in a few seconds. ^{3}He has also higher sensitivity than ^{129}Xe, $\gamma(^3\text{He})/\gamma(^{129}\text{Xe}) \approx 3$, but it is less accessible.

Since nuclear magnetic shielding is sensitive to local electron density changes, studies of this property permit to characterize species on the surface of a microporous material. For example, shielding of ^{3}He introduced to 5 Å molecular sieve depends on interactions between ^{3}He and cations (Ca, Na), interaction between helium atoms, and the bulk magnetic susceptibility of the sample.[102–104] For pressure up to a few hundred bars contributions of ^{3}He–^{3}He interactions are neglectable.[25,105] Therefore, the shielding of ^{3}He introduced to 5 Å molecular sieve is mainly determined by a sum of a negative contribution from bulk magnetic susceptibility of the sieve and the positive contribution from paramagnetic ions in the sample. One can also notice that the density dependence for shielding of a nucleus in a gaseous molecule is usually linear, but this is not in general the case for a gas which is introduced into a solid host material. For example, Cheung and Wharry[106] reported that shielding depends on the gas density, ρ, as $1/\rho$ for a microporous material possessing two absorption sites.

3.7 Conclusions

As shown multinuclear magnetic resonance measurements in the gas phase and shielding calculations permit the determination of nuclear magnetic moments for many nuclei using the well-known magnetic moment of the proton and following the connection between resonance frequencies, nuclear magnetic moments, and shielding parameters.[107] The improved values of nuclear magnetic moments are important for nuclear physics and different kinds of molecular spectroscopy and especially for all NMR experiments. With the accurate nuclear magnetic moments it is possible to measure the magnetic shielding in molecules taking an isolated helium-3 atom as the primary reference standard for all magnetic nuclides. After the transfer of the shielding reference to the deuterons in liquid deuterated solvents the latter chemical compounds become the secondary reference standards of shielding and the simplified measurements of shielding can be performed in any laboratory on standard NMR spectrometers.[63] This new method of shielding measurements was launched on the basis of numerous studies in the gas phase[82] and it can be applied for investigation of various problems present in NMR spectroscopy. To summarize, it appears that the measurements of shielding may be alternatively used for the standardization of NMR spectra because this method has many advantages as compared with the description of spectra on the basis of chemical shifts:

(1) it unifies multinuclear methods into one NMR spectroscopy because the values of magnetic shielding have the same meaning,
(2) in contrast to chemical shifts the shielding of a nucleus in a molecule can be treated as the molecular parameter, which is also available from quantum chemical calculations,
(3) shielding can be measured with the same precision as the appropriate chemical shift because it is based on the same reading of resonance frequencies,
(4) the new method allows the determination of the primary isotope effects in molecules,
(5) it seems that the shielding measurements in liquids performed relatively to external or internal deuterated solvents in contrast to chemical shifts give practically the same results in the absence of strong intermolecular interactions like hydrogen bonding,
(6) the measurements of shielding can be easily extended on MAS NMR experiments in solids.[96]

As a new experimental method the measurement of magnetic shielding certainly requires many improvements, which must be added in future. First of all more accurate values of nuclear magnetic moments are needed, especially for heavier nuclei. It will improve the accuracy of shielding measurements for these nuclei. At present the magnetic shielding can be successfully measured with good accuracy in ^{1}H, ^{13}C, and ^{15}N NMR experiments. The measurements of shielding in liquids contain the BSC effects and it seems important to verify if the internal referencing of shielding measurements can be recommended as the general solution of this problem. Fortunately, in the experiments in gases the BSC effect can be easily removed by the extrapolation of the results to the zero-density limit.

Acknowledgements

This work was financially supported by the National Science Centre (Poland) grant, according to the decision No. DEC-2011/01/B/ST4/06588.

References

1. W. T. Raynes, A. D. Buckingham and H. J. Bernstein, *J. Chem. Phys.*, 1962, **36**, 3481.
2. L. Petrakis and H. J. Bernstein, *J. Chem. Phys.*, 1963, **38**, 1562.
3. F. H. A. Rummens, *Mol. Phys.*, 1971, **21**, 535.
4. K. Jackowski and W. T. Raynes, *Mol. Phys.*, 1977, **34**, 465.
5. C. J. Jameson, A. K. Jameson and S. M. Cohen, *J. Chem. Phys.*, 1973, **59**, 4540.
6. K. Jackowski, M. Jaszuński and W. Makulski, *J. Magn. Reson.*, 1997, **127**, 139.
7. C. J. Jameson, *Bull. Magn. Reson.*, 1980, **3**, 3.
8. A. K. Jameson, C. J. Jameson and H. S. Gutowsky, *J. Chem. Phys.*, 1970, **53**, 2310.
9. S. Gordon and B. P. Dailey, *J. Chem. Phys.*, 1961, **34**, 1084.
10. J. J. Koziol, X. Peng, C. Reiner and D. M. Campbell, *J. Magn. Reson., Ser. B*, 1993, **102**, 299.
11. G. B. Benedek and E. M. Purcell, *J. Chem. Phys.*, 1954, **22**, 2003.
12. J. Jones, *Rev. Sci. Instrum.*, 1972, **43**, 643.
13. J. G. Oldenziel and N. J. Trappeniers, *Physica A*, 1976, **82**, 565.
14. W. L. Earl, H. Vanni and A. E. Merbach, *J. Magn. Reson.*, 1978, **30**, 571.
15. L. Ballard, C. Reiner and J. Jonas, *J. Magn. Reson., Ser. A*, 1996, **123**, 81.
16. L. Ballard, A. Yu, C. Reiner and J. Jonas, *J. Magn. Reson.*, 1998, **133**, 190.
17. H. Yamada, K. Nishikawa, M. Honda, T. Shimura, K. Akasaka and K. Tabayashi, *Rev. Sci. Instrum.*, 2001, **72**, 1463.
18. H. Yamada, *Rev. Sci. Instrum.*, 1974, **45**, 640.
19. M. R. Arnold, H. R. Kalbitzer and W. Kremer, *J. Magn. Reson.*, 2003, **161**, 127.
20. H. Vanni, W. L. Earl and A. E. Merbach, *J. Magn. Reson.*, 1978, **29**, 11.
21. I. T. Horvath and J. M. Millar, *Chem. Rev.*, 1991, **91**, 1339.
22. S. Lee, K. Luszczynski, R. E. Norberg and M. S. Conradi, *Rev. Sci. Instrum.*, 1987, **58**, 415.
23. M. G. Pravica and I. F. Silvera, *Rev. Sci. Instrum.*, 1998, **69**, 479.
24. T. Meier, T. Herzig and J. Haase, *Rev. Sci. Instrum.*, 2014, **85**, 043903.
25. P. Garbacz, K. Piszczatowski, K. Jackowski, R. Moszyński and M. Jaszuński, *J. Chem. Phys.*, 2011, **135**, 084310.
26. A. D. Buckingham and J. A. Pople, *Discuss. Faraday Soc.*, 1956, **22**, 17.
27. P. Raghavan, *At. Data Nucl. Data Tables*, 1989, **42**, 189.
28. I. Mills, T. Cvitaš, K. Homann, N. Kallay and K. Kuchitsu, *IUPAC, Physical Chemistry Division: Quantities, Units and Symbols in Physical Chemistry*, Blackwell Science, Oxford, 1993.
29. P. J. Mohr, B. N. Taylor and D. B. Newell, *Rev. Mod. Phys.*, 2012, **84**, 1527.
30. R. K. Harris, E. D. Becker, S. M. C. de Menezes, R. Goodfellow and P. Granger, *Pure Appl. Chem.*, 2001, **73**, 1795, reprinted in *Magn. Reson. Chem.*, 2002, **40**, 489.

31. A. Laaksonen and R. E. Wasylishen, *J. Am. Chem. Soc.*, 1995, **117**, 392.
32. M. G. H. Gustavsson and M.-A. Märtensson-Pendrill, *Phys. Rev. A: At., Mol., Opt. Phys.*, 1998, **58**, 3611.
33. W. Gerlach and O. Stern, *Z. Phys.*, 1992, **9**, 349.
34. W. Gerlach and O. Stern, *Z. Phys.*, 1992, **9**, 353.
35. P. A. M. Dirac, *Proc. R. Soc. A*, 1928, **117**, 610.
36. I. Estermann and O. Stern, *Z. Phys.*, 1933, **85**, 17.
37. J. D. Bjorken and S. D. Drell, *Relativistic Quantum Mechanics*, McGraw-Hill, New York, 1964.
38. M. Gell-Mann, *Phys. Lett.*, 1964, **8**, 214.
39. G. Zweig, CERN Report no.8182/TH.401 and no.8419/TH.412, 1964.
40. I. Esterman and O. Stern, *Phys. Rev.*, 1934, **45**, 761.
41. I. I. Rabi, J. M. Kellogg and J. R. Zacharias, *Phys. Rev.*, 1934, **46**, 163.
42. M. A. B. Beg, B. W. Lee and A. Pais, *Phys. Rev. Lett.*, 1964, **13**, 514.
43. B. Sakita, *Phys. Rev. Lett.*, 1964, **13**, 643.
44. O. Haxel, J. H. D. Jensen and H. Suess, *Phys. Rev.*, 1949, **75**, 1766.
45. M. Goeppert-Mayer, *Phys. Rev.*, 1949, **75**, 1969.
46. I. I. Rabi, J. M. B. Kellogg and J. R. Zacharias, *Phys. Rev.*, 1934, **46**, 157.
47. J. M. B. Kellogg, I. I. Rabi, N. F. Ramsey, Jr. and J. R. Zacharias, *Phys. Rev.*, 1939, **56**, 728.
48. I. I. Rabi, S. Millman, P. Kusch and J. R. Zacharias, *Phys. Rev.*, 1939, **55**, 526.
49. P. F. Winkler, D. Kleppner, T. Myint and F. G. Walther, *Phys. Rev. A: At., Mol., Opt. Phys.*, 1972, **5**, 83.
50. A. Mooser, S. Ulmer, K. Blaum, K. Franke, H. Kracke, C. Leiteritz, W. Quint, C. C. Rodegheri, C. Smorra and J. Walz, *Nature*, 2014, **509**, 596.
51. R. Bluhm, V. A. Kostelecky and N. Russell, *Phys. Rev. D: Part. Fields*, 1998, **57**, 3932.
52. K. Jackowski, M. Jaszuński, B. Kamieński and M. Wilczek, *J. Magn. Reson.*, 2008, **193**, 147.
53. P. Lantto, K. Jackowski, W. Makulski, M. Olejniczak and M. Jaszuński, *J. Phys. Chem. A*, 2011, **115**, 10617.
54. M. Jaszuński, M. Repisky, T. B. Demissie, S. Komorovsky, E. Malkin, K. Ruud, P. Garbacz, K. Jackowski and W. Makulski, *J. Chem. Phys.*, 2013, **139**, 234402.
55. W. Makulski, K. Jackowski, A. Antušek and M. Jaszuński, *J. Phys. Chem. A*, 2006, **110**, 11462.
56. A. Antušek, K. Jackowski, M. Jaszuński, W. Makulski and M. Wilczek, *Chem. Phys. Lett.*, 2005, **411**, 111.
57. N. N. Aruev and Yu. I. Neronov, *Tech. Phys.*, 2012, **57**, 1579.
58. M. Puchalski, J. Komasa and K. Pachucki, *Phys. Rev. A: At., Mol., Opt. Phys.*, 2015, **92**, 020501(R).
59. W. D. Phillips, W. E. Cooke and D. Kleppner, *Metrologia*, 1977, **13**, 179.
60. B. W. Petley and R. W. Donaldson, *Metrologia*, 1984, **20**, 81.
61. J. L. Flowers, B. W. Petley and M. G. Richards, *Metrologia*, 1993, **30**, 75.

62. A. Rudziński, M. Puchalski and K. Pachucki, *J. Chem. Phys.*, 2009, **130**, 244102.
63. K. Jackowski, M. Jaszuński and M. Wilczek, *J. Phys. Chem. A*, 2010, **114**, 2471.
64. K. Jackowski, W. Makulski, A. Szyprowska, A. Antušek, M. Jaszuński and J. Jusélius, *J. Chem. Phys.*, 2009, **130**, 044309-5.
65. W. Makulski, *Magn. Reson. Chem.*, 2014, **52**, 430.
66. W. Makulski, *Magn. Reson. Chem.*, 2015, **53**, 273.
67. L. T. Baby, C. Bordeanu, M. Hass, H. Haas, L. Weissman and B. A. Brown, *J. Phys. G: Nucl. Part. Phys.*, 2004, **30**, 519.
68. W. Geithner, B. A. Brown, K. M. Hilligsøe, S. Kappertz, M. Keim, G. Kotrotsios, P. Lievens, K. Marinova, R. Neugart, H. Simon and S. Wilbert, *Phys. Rev. C: Nucl. Phys.*, 2005, **71**, 064319.
69. N. J. Stone, *At. Data Nucl. Data Tables*, 2005, **90**, 75.
70. W. Sahm and A. Schwenk, *Z. Naturforsch.*, 1974, **29a**, 1763.
71. C. D. Jeffries, H. Loeliger and H. H. Staub, *Phys. Rev.*, 1952, **85**, 478.
72. Y. Ting and D. Williams, *Phys. Rev.*, 1953, **89**, 595.
73. W. Koch, O. Lutz and A. Nolle, *Z. Naturforsch.*, 1978, **33a**, 1025.
74. J. Blaser, O. Lutz and W. Steinkilberg, *Z. Naturforsch.*, 1972, **27a**, 72.
75. W. G. Proctor and F. C. Yu, *Phys. Rev.*, 1951, **81**, 20.
76. W. G. Proctor, *Phys. Rev.*, 1950, **79**, 35.
77. P. W. Spence and M. N. McDermott, *Phys. Lett. A*, 1972, **42**, 273.
78. K. U. Buckler, J. Kronenbitter, O. Lutz and A. Nolle, *Z. Naturforsch.*, 1977, **32a**, 1263.
79. E. Yasaitis and B. Smaller, *Phys. Rev.*, 1951, **82**, 750.
80. Y. Ting and D. Williams, *Phys. Rev.*, 1953, **89**, 595.
81. M. H. Levitt, *J. Magn. Reson.*, 1997, **126**, 164.
82. M. Jaszuński, A. Antušek, P. Garbacz, K. Jackowski, W. Makulski and M. Wilczek, *Prog. Nucl. Magn. Reson. Spectrosc.*, 2012, **67**, 49.
83. P. Garbacz, K. Jackowski, W. Makulski and R. E. Wasylishen, *J. Phys. Chem. A*, 2012, **116**, 11896.
84. W. T. Raynes, R. McVay and S. J. J. Wright, *J. Chem. Soc., Faraday Trans. 2*, 1989, **85**, 759.
85. A. F. Maldonado and G. A. Aucar, *Chem. Phys.*, 2015, **459**, 125.
86. Ł. Jaremko, M. Jaremko, A. Buczek, M. A. Broda, T. Kupka and K. Jackowski, *Chem. Phys. Lett.*, 2015, **627**, 1.
87. W. G. Proctor and F. C. Yu, *Phys. Rev.*, 1950, 77, 717.
88. W. C. Dickinson, *Phys. Rev.*, 1950, 77, 736.
89. Commission on Molecular Structure and Spectroscopy, *Pure Appl. Chem.* 1972, 29, 627.
90. Commission on Molecular Structure and Spectroscopy, *Pure Appl. Chem.* 1976, 45, 217.
91. R. K. Harris, E. D. Becker, S. M. Cabral de Menezes, P. Granger, R. E. Hoffman and K. W. Zilm, *Pure Appl. Chem.*, 2008, **80**, 59, Reprinted in *Magn. Reson. Chem.*, 2008, **46**, 582.
92. W. McFarlane, *Proc. R. Soc. London, Ser. A*, 1968, **306**, 185.

93. W. McFarlane, *Annu. Rev. NMR Spectrosc.*, 1968, **1**, 135.
94. D. Sundholm and J. Gauss, *J. Mol. Phys.*, 1997, **92**, 1007.
95. M. Jaszuński, G. Łach and K. Strasburger, *Theor. Chem. Acc.*, 2011, **129**, 325.
96. K. Jackowski and W. Makulski, *Magn. Reson. Chem.*, 2011, **49**, 600.
97. N. S. Golubev and D. N. Shchepkin, *Chem. Phys. Lett.*, 2014, **591**, 292.
98. C. J. Jameson, A. K. Jameson, R. Gerald and A. C. de Dios, *J. Chem. Phys.*, 1992, **96**, 1676.
99. I. Moudrakovski, D. V. Soldatov, J. A. Ripmeester, D. N. Sears and C. J. Jameson, *Proc. Natl. Acad. Sci. U. S. A.*, 2004, **101**, 17924.
100. J. A. Ripmeester, C. I. Ratcliffe and J. S. Tse, *J. Chem. Soc., Faraday Trans.*, 1988, **84**, 3731.
101. J. Bonardet, J. Fraissard, A. Gédéon and M. Springuel-Huet, *Catal. Rev.: Sci. Eng.*, 1999, **41**, 115.
102. P. Garbacz and K. Jackowski, *Microporous Mesoporous Mater.*, 2015, **205**, 52.
103. S. Hayashi, *Microporous Mesoporous Mater.*, 2007, **101**, 3.
104. S. Hayashi, *Chem. Lett.*, 2006, **35**, 92.
105. A. Antušek, M. Jaszuński and A. Rizzo, *J. Chem. Phys.*, 2007, **126**, 074303.
106. T. T. P. Cheung, C. M. Fu and S. Wharry, *J. Phys. Chem.*, 1988, **92**, 5170.
107. K. Jackowski and M. Jaszuński, *Concepts Magn. Reson., Part A*, 2007, **30**, 246.

CHAPTER 4

Gas Phase NMR for the Study of Chemical Reactions: Kinetics and Product Identification

ALEXANDER A. MARCHIONE* AND BREANNA CONKLIN

Chemours Fluoroproducts Analytical, Chemours Co., Wilmington, DE, USA
*Email: alexander.marchione@chemours.com

4.1 Introduction

The value of nuclear magnetic resonance (NMR) spectroscopy as a tool for the study of reaction kinetics has long been appreciated. NMR is an inherently quantitative technique, assuming proper allowances for spin–lattice relaxation time are made. It is information rich, permitting facile identification and resolution of most organic molecules. Common ranges of reactant concentrations (M to mM) afford more than adequate sensitivity for experiments observing ^{1}H or ^{19}F (naturally abundant isotopes with high magnetogyric ratio γ), and with the advent of high-sensitivity cryosystems the use of lower γ, scarcer isotopes in kinetic studies is increasingly practical as well. NMR probes are compatible with common materials of construction of laboratory organic reactors (generally borosilicate glass), and often permit a relatively wide range of sample temperatures. Because of all of these advantages, the use of *in situ* NMR spectroscopy (*i.e.* performing reactions directly inside an NMR probe) for the study of reaction kinetics in solution is widespread.

Much less practiced is the use of NMR spectroscopy for the study of chemical reactions in the gas phase. The overwhelming majority of newly reported synthetic chemistry is in the condensed phase, so the rarity of

New Developments in NMR No. 6
Gas Phase NMR
Edited by Karol Jackowski and Michał Jaszuński

Published by the Royal Society of Chemistry, www.rsc.org

kinetic gas phase NMR experiments in academic research is understandable. However, a much greater fraction of industrial chemistry occurs in the gas phase. This includes reactions both desired (*e.g.* synthesis of monomers or volatile small molecules, telomerizations) and undesired (*e.g.* oxidation of organic solvents during a high-temperature drying process). It is primarily in the field of industrial chemistry that gas phase NMR has been, to date, most fruitfully applied to the study of reaction kinetics.

4.2 Experimental Considerations – Concentration, Pressure, Temperature, Vessel Design

4.2.1 Concentration and Pressure

Gas phase NMR experiments can be designed to accommodate a wide range of reaction conditions with relative ease. The concentration of an ideal gas at 298 K and 1 bar is 41 mM; for the sake of comparison, in 0.1 vol% ethyl benzene, which is a solution commonly used as a ^{1}H sensitivity standard, the title compound is present at 8.2 mM. Modern spectrometers yield a signal-to-noise ratio of at least 100 (often much higher) with a single transient on this standard; therefore concentrations in the range of tens of μM, corresponding to pressures in the mbar range near ambient temperature, are easily observed. With a few minutes devoted to signal averaging, gases at concentrations less than 1 μM can be routinely observed by ^{1}H or ^{19}F experiments, even under quantitative acquisition conditions. ^{13}C-observe experiments are intrinsically much less sensitive (having only 0.017% the receptivity of ^{1}H, barring isotopic enrichment), and therefore concentrations must approach tens of mM to permit an experiment of reasonable duration.

For NMR experiments in the gas phase, the discussion of concentration entails the consideration of pressure. The discussion in the preceding paragraph assumes an overall pressure in the range of *ca.* 0.1–10 bar. Outside of this range, two separate effects are observed. At higher pressure, both T_1 and T_2 relaxation times increase; the increase in T_1 renders signal-averaging a more time-consuming process, reducing effective sensitivity. At lower pressure, T_1 and T_2 diverge; T_1 increases while T_2 decreases, so the problem of slower signal-averaging is compounded by greater line widths and lower amplitudes. This effect was demonstrated on H_2 by Armstrong[1] (Figure 4.1), but is observed very generally. It is therefore much easier to detect a trace gaseous species in the presence of other gases (even if NMR-invisible) than to detect it alone. Let us note that the chemical shifts and spin–spin coupling constants of observed molecules may be slightly changed when the solvent gas is used in the NMR experiment. This issue is extensively discussed in Chapter 1 of this book.

The more conventional effect of pressure on experimental design is in the consideration of the integrity of the reaction ampule. Subambient pressure rarely poses any issues in this regard; conventional 5 mm "thin-walled" tubes, with wall thickness of <0.4 mm, can be sealed under high vacuum

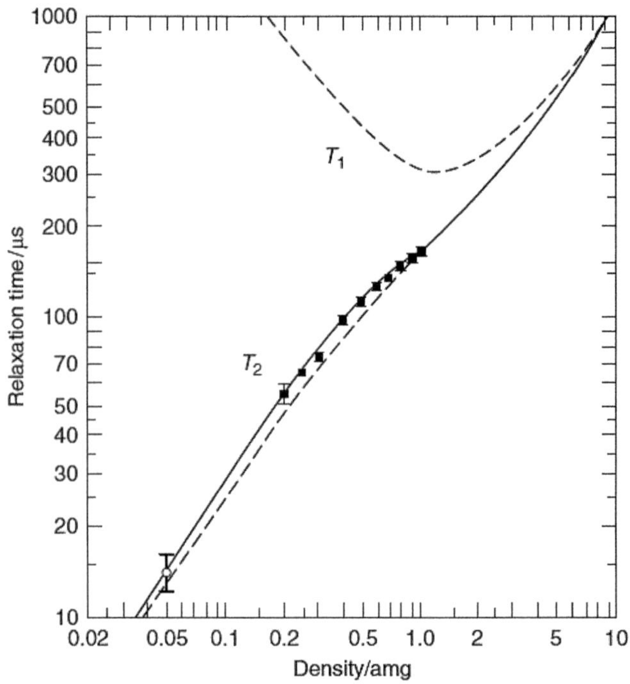

Figure 4.1 Density dependence of T_1 and T_2 in H_2 at 61 MHz. The solid curve for T_2 represents a theoretical prediction; the dotted curves for T_2 and T_1 represent best fits to experimental data.
(Reproduced from Ref. 1.)

Table 4.1 Approximate safe operating pressures of commercial NMR tubes.

Material	Outer diameter (mm)	Wall thickness (mm)	Maximum pressure (bar)[c]	Source
Borosilicate glass	5	0.38	5	a
Borosilicate glass	5	0.77	11	a
Borosilicate glass	10	0.92	6	a
Clear-fused quartz	10	0.92	16	a
Sapphire	10	1.5	300	b
Zirconia	5	1.0	3000	b

[a]http://www.wilmad-labglass.com/Support/NMR-and-EPR-Technical-Reports/NMR-003–Pressure-Performance-of-NMR—EPR-Sample-Tubes/
[b]http://daedalusinnovations.com/apparatus/high-pressure.html
[c]These values are estimates for informational purposes, and should not take the place of a thorough, process-specific safety review.

with no threat to their integrity. Positive internal pressure can pose more significant experimental challenges, and can require the use of stronger, more expensive materials in the construction of the reaction ampule. Table 4.1 estimates and compares the maximum safe operating pressure of some commercially available ampules or tubes of different diameters and materials of construction.

Appropriate safety precautions must accompany the charging of any of these ampules or tubes at significantly elevated pressure, or even at modest pressure near the upper limit of its tolerance. The best experimental designs protect the scientist by keeping the pressurized vessel in a protective container or portable barricade from the time the vessel is pressurized until it is inserted into the NMR probe, and later depressurized.

4.2.2 Temperature and Sample Temperature Calibration

The accessible range of temperature for an *in situ* reaction study is defined by the design of the NMR probe. Most commercial probes are equipped with relatively simple borosilicate glass or quartz Dewars, which permit thermal manipulation of the sample while minimizing heat flow toward or away from the probe circuitry. Such Dewars often permit operation both well below ambient temperature (173–193 K is a common lower operational temperature limit) and somewhat above (sometimes up to 423 K). Industrial chemical reactions of gases are often commonly performed above the upper temperature limits of common commercial probes. Two modifications in probe design have been applied to extend the useful temperature range of such probes. The first entails coating the interior of the Dewar with an IR-reflective material, applying the coating in a pattern such that the energy of the radio-frequency pulse is not prevented from reaching the sample. The second actively cools the portion of the probe outside of the Dewar with a recirculating fluid. This serves to protect not only the electronics of the probe, but the room-temperature shims in the bore of a vertical magnet, which can be damaged by excess heat leaking from the probe. Commercially available probes with such modifications operate safely for extended periods of time at nominal sample temperatures of 673 K or higher.

Even probes that are specially designed for high-temperature operation generally use a simple mechanism for sample heating: nitrogen gas flows across the surface of a resistive electric heater near the bottom of a probe, past a thermocouple near the bottom of the sample, past the sample from the bottom to the top, and eventually (mixed with cooler purging gas) out of the top of the upper stack of the magnet. This single-pass design invariably imparts some vertical temperature gradient to the sample, as well as the usual discrepancy between the effective sample temperature and the read temperature of the thermocouple some few millimeters below the sample. The precise calibration of sample temperature is thus somewhat challenging. Moreover, the simplest and most common sample temperature calibration methods, based on the dependence of the ^1H chemical shift of –OH resonances with temperature, have not been reported above 410 K.[2]

The authors have applied two methods for sample temperature calibration above 410 K. The first is simple thermocouple insertion. The tip of a long shielded thermocouple is placed in a tube or unsealed ampule, which can be filled either with air or with a liquid or solid medium (sand is convenient). The tube or ampule is then lowered into the probe like a standard sample, and the temperature read by the thermocouple in the sample vessel is used to calibrate the temperature read by the thermocouple embedded in the probe below the sample. This method has the advantage of simplicity, and by repositioning the thermocouple toward the top or bottom of the sample tube one can estimate the magnitude of the thermal gradient. The disadvantages of this method are that the tube or ampule must remain open, permitting heat loss through convective flow, and that the thermocouple wire itself may conduct heat from the tube. Insofar as these conditions are not experienced by a sealed vessel in a typical experiment, they create some uncertainty as to the accuracy of a temperature calibration by this method.

An alternative method, not heretofore published to the authors' knowledge, tracks the loss of signal intensity as the thermal spin polarization changes with increasing temperature. A spectrum of a gaseous standard is acquired at a low temperature, calibrated by a standard method, and the decrease in integrated signal intensity is used to calculate the effective sample temperature as the set temperature of the probe is raised. This method offers two advantages: a sealed ampule can be used, removing uncertainties about convective or conductive heat loss; and the effective mean temperature of the sample is obtained, which addresses uncertainties about the temperature gradient. The disadvantage of the method is that the Boltzmann distribution of spin states changes as T^{-1}; at higher temperatures, this change is quite modest relative to the intrinsic errors of measurement, and hence the precision of the method is limited. Figure 4.2 shows the results obtained from a tube containing 26 mM hexafluoropropene, studied by ^{19}F-observe experiments, in a commercial high-temperature probe. 373 K (as calibrated by a standard of ethylene glycol) was defined as the set point for these observations; the effective temperatures as a function of the probe set temperature were calculated by ratios of the signal intensity at the unknown T *vs.* that obtained at 373 K. The probe was carefully retuned at each temperature. The pulse width corresponding to the $\pi/2$ flip angle was invariant over this temperature series, a testament to the excellent thermal isolation afforded by the probe Dewar.

4.2.3 Vessel Design and Material-of-construction

Gaseous analytes pose unusual challenges in the selection and handling of sample vessels for an NMR experiment. The first requirement is that the tube or ampule be at least somewhat gas-tight. A standard NMR tube used for liquid-phase studies equipped with a standard polyethylene cap will generally retain any gas introduced to it for <1 h, possibly longer if the gas is particularly heavy and the tube is kept upright. Tighter-fitting caps (*e.g.* common polytetrafluoroethylene [PTFE] caps) are better, but will still leak over the course of hours, and of course are not compatible with pressures much above or below atmospheric. For any experiment at lower or higher pressure, any experiment requiring quantitation, or any experiment with toxic volatile analytes, a properly closed vessel is required.

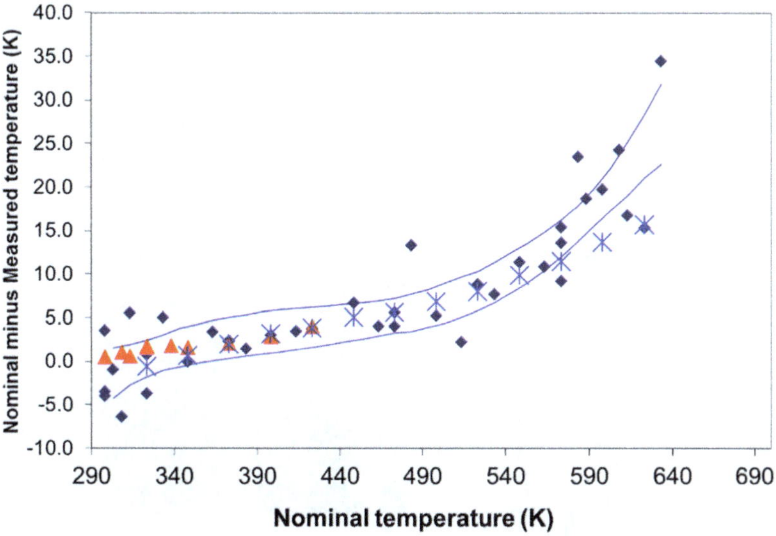

Figure 4.2 Comparison of calibration methods of an example high-temperature NMR probe. Triangles: ethylene glycol method. Asterisks: calibrated by thermocouple insertion in a mock ampule. Diamonds: calibrated by ratios of signal intensity. The solid lines represent 95% confidence intervals for a third-order polynomial fit of (nominal – measured) temperature *vs.* nominal temperature.

The simplest means of obtaining a gas-tight vessel constructed of common borosilicate glass or quartz is to seal the open end of such a tube or ampule with a torch. Borosilicate glass is conveniently sealed with common laboratory butane/air or propane/air torches; quartz requires a flame of higher temperature, most easily achieved by a propane/O_2 mixture. While common borosilicate glass or quartz tubes can be sealed in the absence of an applied vacuum, the presence of a negative pressure in the tube renders the seal stronger and easier to achieve. To this end, the use of vacuum manifold is very helpful. The analyte or reagent gas(es) can be transferred to an evacuated tube by condensation under *e.g.* liquid nitrogen or dry ice, a vacuum can be pulled on the chilled tube, and a torch applied to create the seal. In the authors' experience, a borosilicate tube properly sealed by this means is no more likely to fail at the seal than along the body.

An additional consideration in selecting reaction vessels for *in situ* NMR kinetic studies is that, in contrast to liquid-phase experiments, the temperature of the entire vessel must be regulated. Standard liquid-phase NMR tubes, usually 17–20 cm in length are poorly designed to meet this requirement; generally only the bottom 5–7 cm are thermostatted, and gaseous components can of course travel freely from the heated volume to the cooler volume above (and perhaps condense there, depending on their concentration, vapor pressure, *etc.*). It is therefore quite necessary, in most cases, to deny the gases access to an unthermostatted region of the vessel. The most

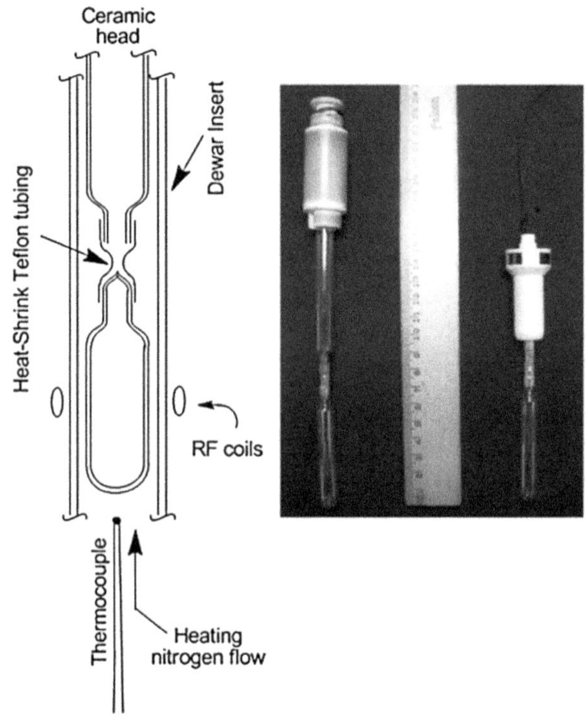

Figure 4.3 Schematic and photograph of a high-temperature NMR apparatus with a sealed ampule. Schematic based on image created by P. J. Krusic, DuPont Central Research and Development, retired.

direct approach in this regard is to keep the length of a sealed vessel within the thermostatted region of the NMR probe. An elegantly simple design to that end was introduced by Kating *et al.*[3] (Figure 4.3). The body of a 10 mm o.d. tube, borosilicate or quartz, is fused to a 5 mm o.d. neck, through which reagents are introduced, and at which the seal is created. The stub of the neck present after sealing is attached by fluoropolymer "shrink-wrap" tubing to an adapter, which is connected to a nylon string. The assembly is then manually loaded into an NMR probe. The resulting ampule presents a single surface to its contents, is as pressure-tolerant as the 10 mm tube from which it was formed (*vide supra*), and can withstand the entire temperature range of a high-temperature probe. (The only limitation in this regard is that the fluoropolymer of the shrink-wrap tubing deforms at $T > 573$ K, deflecting the sample and requiring a re-homogenization of the magnetic field.)

While the flame-sealing approach described above is generally convenient, it bears certain limitations. The first of these is that it requires subambient, or at worst ambient, pressure in the tube or ampule to be sealed. With most gases, this can be achieved by condensation under liquid nitrogen. With certain low-boiling gases (H_2 and O_2 are common examples in industrial chemical reactions), this approach either fails completely, or is complicated by the non-trivial vapor pressure of the gas at 77 K, the boiling point of N_2 at 1 bar. In such cases, a valved system is required.

The simplest and most common kind of valved system is that in which a J. Young-type valve is fused to the top of a borosilicate glass tube. In this case, a PTFE seal is pressed or released against a conical section of glass by screw action. When the seal is opened, gas can flow through a hole in the side of the PTFE plunger through the top. This kind of assembly is relatively inexpensive, forms a good long-term seal for systems at subambient pressure, and (in certain designs) is compatible with tubes at positive pressure as well. The biggest limitation to these tubes is the very different expansion coefficients of PTFE and borosilicate glass or quartz. If the area around the seal is heated, the PTFE expands and flows to some degree against the glass; upon cooling, the deformed PTFE will not create as effective a seal (if any at all).

Various designs exist for systems at higher pressures than can be tolerated by quartzware. Sapphire,[4] polyimide,[5,6] and zirconia[7] tubes can tolerate internal pressures two orders of magnitude higher than borosilicate glass or quartz. In all of these cases, access to the tube is controlled by a metallic valve, which is sealed to the tube. These seals can tolerate only modest temperatures, and cannot closely approach the thermostatted region of the probe when performing experiments at $T > ca.$ 423 K. Thus these assemblies naturally pose the problem of either limited range in an isothermal experiment, or a large temperature gradient over the course of the tube. Nevertheless they have found application in the study of numerous high pressure systems.[8–10] In particular, their temperature limitations are less important in the study of heterogeneous reactions between a gas and a solid reagent or catalyst, in which one can assume that the temperature of the solid at the bottom of the reaction vessel is well controlled.

In addition to temperature and pressure, the chemical compatibility of the surface of the reaction vessel with its contents is sometimes an important consideration. Certain strong fluorinating agents at elevated temperature, for example, are incompatible with SiO_2 of any kind, and may thus require a sapphire vessel for reasons other than pressure tolerance. The metallic portion of valved high-pressure vessels may be incompatible with a strong acid, or a particular adhesive in the seal may be incompatible with certain solvents.

4.3 Spectroscopic Considerations – Probe Design, Phase and Frequency Drift, Spectral Acquisition Schedule

4.3.1 Probe Design

The basic thermal aspects of a probe with extended upper temperature range are described above. The radiofrequency (RF) components of these probes are generally no different from their analogues with ordinary temperature ranges. High-temperature-capable probes have been constructed for both "narrow bore" (*ca.* 53 mm) and "wide bore" (89 mm) magnets, incorporating either one or two RF coils, one of which is often doubly tuned to permit a ^{2}H field-frequency lock. In the authors' experience, the special steps taken to achieve optimal thermal isolation of the sample do not meaningfully affect the applied power reaching the sample (and therefore the 90° pulse), nor the sensitivity of the probe.

High-temperature-capable probes do not generally accommodate the presence of a properly shielded pulsed-field gradient (PFG) coil. To the authors' knowledge, there are no commercially available probes with PFG coils and an operational temperature range above 453 K. Experiments that rely on gradient pulses have been found to be useful in gas-phase studies (*vide infra*), but to date they have been performed in probes with ordinary ranges of operational temperature.

4.3.2 Phase and Frequency Drift

Despite extraordinary precautions to achieve thermal isolation of the sample in many high-temperature-capable probes, thermal leakage to the RF electronics does occur over time. As the temperature of the electronics changes, the phase of the signal reaching the receiver changes, and with it the phases of the resonances in the frequency-domain spectrum. This is most troublesome in phase-sensitive two-dimensional (2D) acquisitions, especially as most processing software is not equipped to correct for a phase drift across the increments of a 2D spectrum. In magnitude-mode 2D experiments phase drift poses no problem, and in a kinetic series of one-dimensional (1D) experiments individual phase corrections can be applied to each spectrum, given the proper software. The problem of phase drift can be eliminated, or at least mitigated, by allowing the probe electronics to reach thermal equilibrium before commencing an experiment.

It is often difficult to obtain a field-frequency lock in a gas phase experiment. Operating at ordinary pressures, it is not possible to introduce a deuterated gas at sufficiently high concentration to produce a reliable lock signal. In some cases, one can insert a sealed capillary into a reaction tube or ampule containing a liquid phase lock solvent, or prepare a sealed tube containing the gas sample and insert it into a larger tube, with a liquid lock solvent in the annular space. However, the simpler approach is to acquire without a lock. Depending on the duration of the experiment and the drift rate of the magnet, a noticeable frequency drift can occur in the spectrum. This frequency drift can affect the accuracy of integrals obtained over defined regions in automated routines, as well as disturbing the transform in the indirect dimensions of 2D correlation and DOSY experiments.

4.3.3 Acquisition Parameters

The chief difference between gas phase and liquid phase NMR spectroscopy is the prevalence of the spin-rotation relaxation mechanism in the former.

The efficacy of spin-rotation relaxation depends, to some degree, on the geometry of the molecule in question (small, symmetric molecules like CH_4 or CF_4 undergo particularly fast longitudinal and spin–spin relaxation), but depends even more strongly on the observed nucleus. The rate of spin-rotation relaxation is given by

$$1/T_1\ (SR) = I_r^{\,2}C^2/9\hbar^2\tau_c \tag{4.1}$$

where I_r represents the molecular inertia moment, C is the spin-rotation constant, and τ_c is the molecular motion correlation time.[11] In general, the constant C is much greater for ^{19}F nuclei than for ^{13}C, and in turn much greater for ^{13}C than for ^{1}H. For the ^{1}H nuclei in many organic molecules at pressures near 1 bar, the spin-rotation relaxation mechanism is not necessarily dominant, and T_1 and T_2 relaxation times are similar to those commonly observed in liquid-phase analyses. In the case of ^{13}C, spin-rotation relaxation is dominant, and T_1 relaxation times in the range of 50–300 ms are common. In the case of ^{19}F, spin-rotation relaxation is even more efficient, and T_1 times are commonly 5–100 ms.

The prevalence of spin-rotation relaxation directly affects the choice of acquisition parameters, and even the time scale over which a kinetic study can be performed. In the case of ^{1}H, acquisition times and recycle delays similar to those used in liquid-phase analyses are appropriate (*e.g.* 1–3 s acquisition time, 15–60 s recycle delays for quantitative analyses). T_2 relaxation rates are commonly in the range of 0.1–0.4 s, permitting signal resolution comparable to the liquid phase. However, insofar as complete longitudinal relaxation is necessary for obtaining a series of kinetic spectra with quantitatively accurate signal intensities, a relatively long recycle delay between pulses is required, which may hinder the study of reactions occurring in the time scale of seconds to minutes. Likewise, spectral averaging to improve signal-to-noise ratio (S/N) comes with a heavy penalty of time. By contrast, in the case of ^{19}F, recycle delays of less than 1 s are generally adequate for quantitative analyses, permitting the study of faster reactions and much more efficient signal averaging. The penalty in spectral resolution in ^{19}F is generally mitigated by the superior spectral dispersion afforded by that nucleus. ^{13}C is an intermediate case, but recycle delays of 1–2 s are often sufficient for quantitative analyses, which permits the rapid signal averaging needed to yield sufficient S/N.

4.4 Survey of Published Studies

Gas phase NMR can in principle be applied to study the kinetics of any reaction involving volatile species with at least one NMR-active nucleus. In practice, all reported studies have pertained to organic chemistry. The body of literature on chemical reaction kinetics studied by gas phase NMR is not extensive, and is briefly summarized below.

The first reported use of NMR spectroscopy to follow a gas-phase reaction was by Haugh and Dalton[12] in 1975, in which the addition of hydrogen chloride to propene was studied over the course of days to months. In this study, the gaseous reagents were transferred into a borosilicate glass, quartz, or fluoropolymer-lined tube, each equipped with a fluoropolymer valve and held at ambient temperature or in an oven to effect reaction, with periodic ^{1}H spectral acquisition on a 2.3 T continuous-wave spectrometer. The reactions were performed at a total pressure of 6–30 bar. Incidental catalysis was observed both in the gas phase, from trace moisture and NO_x, and (to a lesser extent) by contact with the tube surface. The authors posited that irreversible gas-phase addition of HCl to propene occurs *via* the reaction of a transition-state adduct with HCl dimer.

The Dalton group employed the same methodology to study the addition of hydrogen chloride to 2-methylpropene in 1986.[13] This study was performed at much lower pressure, less than 1 bar, and in a typical 12 mm o.d. tube equipped with a valve. Here they discovered that surface catalysis provided the dominant reaction mechanism.

In the first of a series of papers on the technique, Krusic and co-workers at DuPont opened the field to ^{19}F NMR, and to more conventional heterogeneous catalysis. Their 1996 report[3] on the hydrogenation of perfluoro-2-butenes and perfluoro-2-pentenes demonstrated that the hydrogenation of such olefins was stereospecific, and that the reaction rates of the *trans* isomers were faster than those of the *cis*. Their study was performed on a 8.3 T spectrometer, and took advantage of the rapid relaxation rates of ^{19}F in the gas phase (57 ms acquisition time, 157 ms recycle delay). This work also introduced the use of the ampule design shown in Figure 4.3 above. Absolute quantitation was afforded by use of a known quantity of tetrafluoromethane as an inert internal standard.

In 1999,[14] Krusic *et al.* released a study on the thermal decomposition of 2,2,3-trifluoro-3-(trifluoromethyl)oxirane (trivially hexafluoropropylene oxide, or HFPO), according to the scheme below:

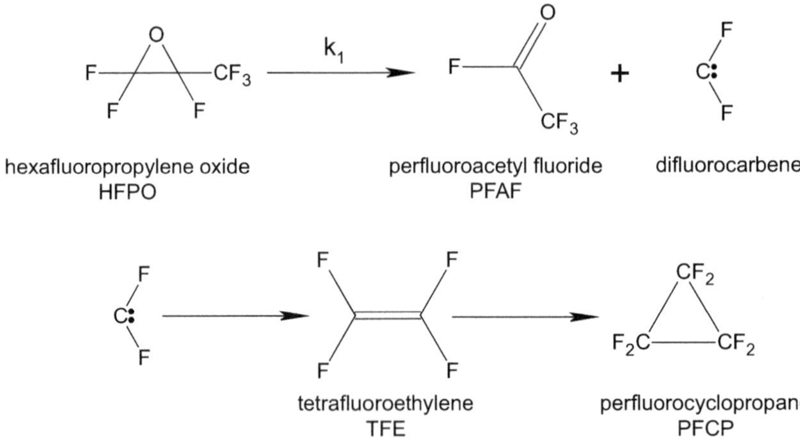

The reactions were studied in the temperature range 190–230 °C, at a total system pressure of *ca.* 1.6 bar. A linear fit of the plot of [HFPO]/[HFPO]$_{t=0}$ *vs.* time yielded first-order rate constants, consistent with a unimolecular decomposition (Figure 4.4), and an Arrhenius plot (Figure 4.5) and activation parameters for the decomposition of HFPO were reported.

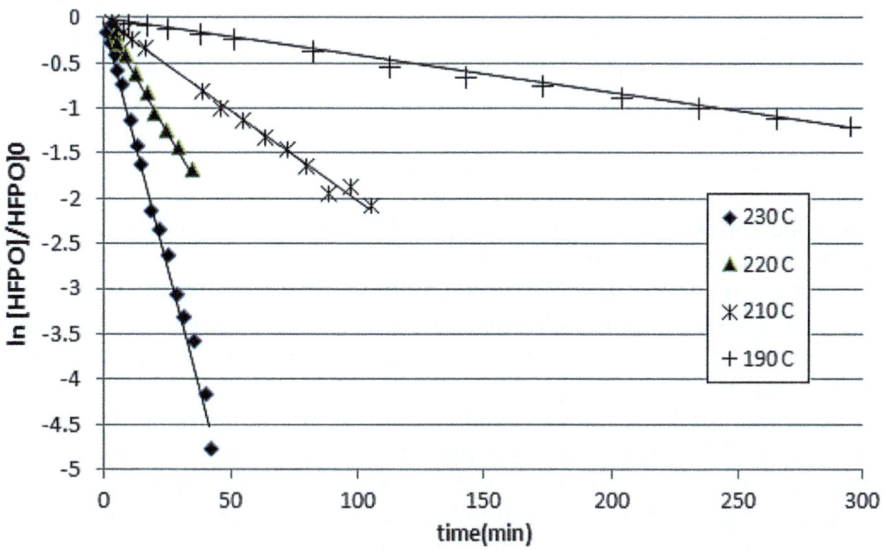

Figure 4.4 Plots of [HFPO]/[HFPO]$_{t=0}$ *vs.* time in the study of its pyrolysis. Figure reproduced from Ref. 14.

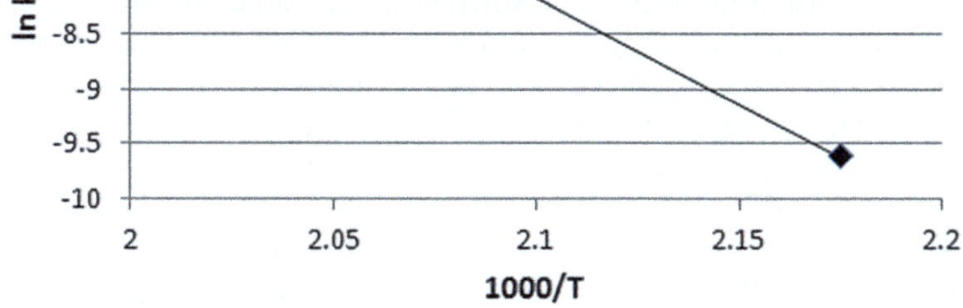

Figure 4.5 Arrhenius plot of pyrolysis of HFPO in the temperature range 463–503 K. Obtained Arrhenius parameters were $E_a = 38.7$ kcal mol^{-1}, $A = 10^{14.2}$ s^{-1}. Figure reproduced from Ref. 14.

In 2001, this group in collaboration with Dolbier reported[15] activation parameters for [2 + 2] and [2 + 4] cycloaddition reactions of tetrafluoroethene and chlorotrifluoroethene over the temperature range 132–210 °C by ^{19}F NMR. Two experimental aspects are of note. The first is the safe use of tetrafluoroethene, a highly explosive gas, the reactions of which are generally only studied behind protective barricades. This highlights an advantage of reaction studies by the NMR technique, in that the milligram quantities of reagents used in such studies are much safer to handle than an analogous reaction at the synthetic scale would be. The second experimental aspect of note is the use of short excitation pulses, corresponding to small flip angles, in the ^{19}F experiment. Given the wide spectral window attendant to ^{19}F fluorocarbon spectra (often >50 kHz even at magnetic fields no greater than 9.3 T), a short excitation pulse is often necessary to minimize inhomogeneities in excitation of the ^{19}F resonances. In this study, a 10–15° flip angle was employed.

In 2002, this group in collaboration with Smart[16] studied the isomerization of vinyl pentafluorocyclopropane by similar techniques in the temperature range 80–120 °C. A biradical mechanism was proposed for the rearrangement.

The same year,[17] this group reported on the rearrangement and dimerization of dichloromethylene cyclopropane by ^{1}H gas phase NMR. In this case, the superior resolution of ^{1}H NMR permitted accurate integration of distinct cyclopropyl resonances. Biradical transition states were proposed for both the rearrangement and dimerization reactions.

In 2004, Krusic and Roe[18] published kinetics of decomposition of the ammonium salt of a perfluorinated carboxylate by ^{19}F gas phase NMR. This was a heterogeneous study; the solid salt was present in the bottom of the reaction ampule until decomposition, whereupon it generated the volatile products n-$C_7F_{15}H$, CO_2 and NH_3. The evolution of the first of these was observed, and the kinetics of decomposition calculated from it.

In 2005, the previous study was extended to the parent carboxylic acid,[19] itself volatile at its decomposition temperature. The range of decomposition temperatures in the study extended beyond the operational temperature limit of the NMR probe employed (307 °C). The reaction ampule was therefore heated in a tube furnace and subjected to periodic spectral acquisition. Here catalysis by borosilicate glass and even clear-fused quartz was reported. This was the first study reported at a magnetic field of 9.3 T.

In 2008, Marchione *et al.*[20] issued the first report of a photochemical reaction studied by gas phase NMR, in the service of developing a new methodology for the estimation of atmospheric lifetimes of fluorinated gases. In this study, analyte gases were sealed in an ampule with Cl_2 and subjected to UV irradiation for timed increments, with ^{19}F spectra acquired between irradiation periods. The relative rates of decomposition of a gas under these conditions were found to correlate closely with the rates of decomposition in the troposphere, suggesting a new means of estimating the atmospheric lifetime for volatile species.

Finally, in 2010, Marchione *et al.*[21] reported the first application of ^{13}C NMR to gas phase kinetics, comparing the rate constants obtained for isomerization of quadricyclane to norbornadiene by ^{1}H and ^{13}C NMR, and demonstrating reasonable agreement. The analytes were present at a partial pressure of only 1 bar, and were not isotopically enriched. Efficient spin-rotation relaxation permitted a quantitative experiment with recycle delay of *ca.* 1 s, enabling significant signal-averaging for each spectrum in the space of 15 min. between data points.

4.5 Current Example

It may be illustrative to present an example in greater experimental detail. For this purpose, an experiment from a study of the decomposition kinetics of di-*t*-butyl peroxide (DTBP) is presented.[22] DTBP is a common organic peroxide initiator. The *t*-butyl moieties donate sufficient electron density to the oxygen atoms to render the molecule relatively stable, by the standards of dialkyl peroxides, and thermal decomposition is slow below *ca.* 100 °C. DTBP shows significant volatility around its decomposition temperature (b.p. 111 °C), and so its gas-phase decomposition is readily studied by ^{1}H gas phase NMR. Its kinetics of decomposition in the gas phase are of interest for its application as an initiator in that phase, and in fact have already been studied by other methods.[23]

In this study, a commercial sample of DTBP was washed with deionized water three times to remove acetone and *t*-butanol (some of its decomposition products), and moved into a nitrogen glove box with less than 5 ppm O_2 by volume. 8.2 μL DTBP were then transferred *via* microsyringe to an ampule of the design shown in Figure 4.3. The internal volume of the ampule, after sealing, was *ca.* 3.9 mL; the concentration of DTBP gas upon vaporization was therefore 10 mM. While still in the glove box, the reaction ampule was attached to a stopcock adapter, which was closed under nitrogen. The adapter was then moved to a vacuum manifold equipped with precision barometers and an oil-diffusion pump. Liquid nitrogen was applied to the ampule to freeze the DTBP within. The stopcock was then opened to evacuate the ampule, and closed again. 1,1,1-trifluoroethane is a ^{1}H-bearing gas that is highly resistant to free-radical attack, and it had previously been demonstrated to be inert toward the reactive *tert*-butoxy radicals generated upon DTBP decomposition. 1.57 mbar 1,1,1-trifluoroethane was introduced to the manifold (itself of known volume, 307 mL in this case), yielding 1.98×10^{-5} mol by the ideal gas law. The stopcock was opened to introduce this quantity of gas, and the ampule was then flame-sealed at the neck with a butane-air torch, with its bottom still immersed in liquid nitrogen. The sealed ampule was then allowed to thaw in air.

Prior to initiating the kinetic experiment, the NMR probe was allowed to equilibrate at the experimental temperature (423 K) for several hours. The set temperature for the probe was 431 K; the "true" temperature of the probe had previously been determined by calibration with a thermocouple inside a sand-filled tube as described above. The sample was attached to a ceramic head with a fluoropolymer sleeve, also as described above. The probe temperature was briefly lowered by 30 K, and the sample was lowered into the probe by nylon fishline tied to the top of the ceramic head. The sample was allowed to equilibrate in the probe for *ca.* 1 min., at which point a ^{1}H spectrum was acquired. The magnetic field was homogenized by optimization of the lineshapes in the ^{1}H spectrum in the frequency domain. The probe temperature was then returned to the experimental temperature; after the readings from the thermocouple stabilized (*ca.* 2 min.), an experiment was launched with a preset acquisition schedule, which called for more frequent spectral acquisition toward the beginning of the reaction and less frequent at the end. Spectral acquisition continued for 14 h, at the end of which time the ampule was lifted from the probe by the same nylon fishline.

Even allowing several hours for the probe to equilibrate at temperature before the experiment began, the kinetic series of spectra displayed some degree of phase drift, as well as frequency drift. The magnitude of this drift, and the resulting correction, are shown in Figure 4.6.

After the phase correction was applied, a simple spline baseline correction was applied to the series of spectra. Integration of the signals of interest followed. Because of the frequency drift in this long experiment without a field-frequency lock, the starting and ending points of the integrals (as defined by position within the spectral window) appropriate for the initial spectra were not appropriate for the final spectra, and *vice versa*. In less sophisticated software, this required manual integration and resetting of the starting and ending points with each spectrum. In the software used in

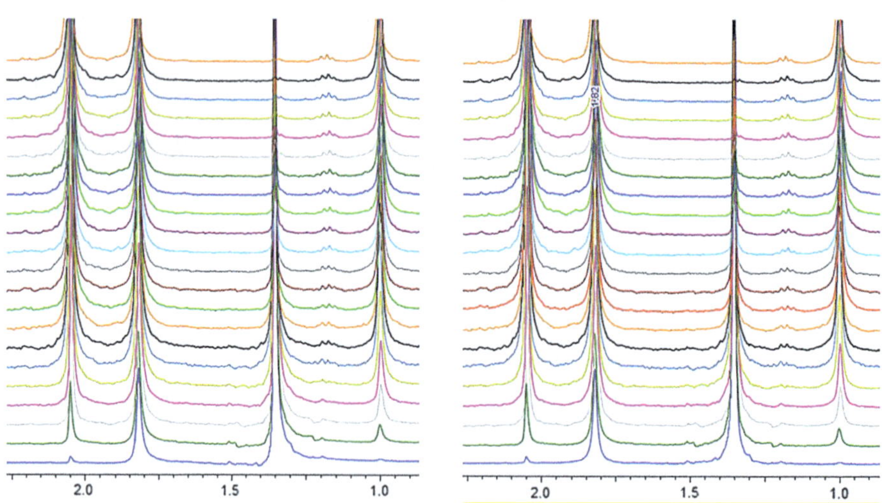

Figure 4.6 Series of kinetic ^{1}H NMR spectra (alkyl region) acquired during the pyrolysis of DTBP at 423 K. Left: Spectral series as acquired. Right: Phase-corrected series.

Figure 4.6 (Spectrus Processor© from Advanced Chemistry Development, Inc., Toronto, Canada), the integration regions can be flexibly defined over the series of spectra.

In the absence of other reactive species, DTBP undergoes decomposition according to the following mechanisms (top in the absence of collision with an H-atom donor, bottom in the case of such a collision):

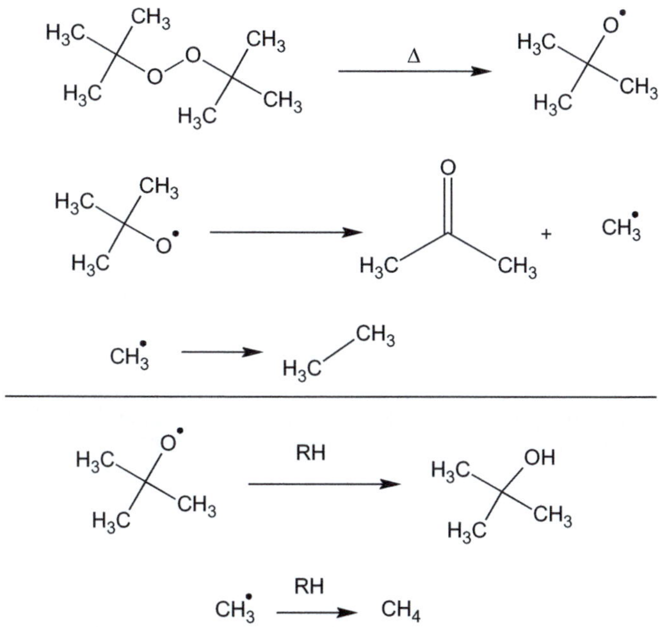

Under the reaction conditions presented above, the primary products of decomposition are acetone and ethane, with *t*-butanol and methane observed as minor by-products. Figure 4.7 shows the integral regions displayed in a selected spectrum at roughly the midpoint of the reaction. Acetone, ethane, and methane were clearly resolved (despite the broadness of the methane resonance) and were integrated in a straightforward fashion. The methyl resonance of *t*-butanol overlaps significantly with that of DTBP, and in fact can only be seen after more than 90% of the DTBP has decomposed. Deconvolution of the final spectrum revealed *ca.* 1% yield of *t*-butanol.

For the kinetic analysis, the integrals were appropriately normalized according to the number of equivalent ^{1}H nuclei associated with each. By comparison with the concentration of the internal standard, the normalized integral intensities were converted to molar concentrations, as shown in the concentration *vs.* time plot given in Figure 4.8.

The unimolecular decomposition of DTBP in this case was observed to obey simple first-order kinetics. The rate constant **k** for the reaction at 423 K can therefore be derived in a number of ways. The plot of [DTBP] *vs.* time can be fitted with an exponential decay of the form $Ae^{-\mathbf{k}t}$, or a plot of ln [DTBP]/$[DTBP]_{t=0}$ *vs.* time can be fitted with a line with the slope $-\mathbf{k}$, or

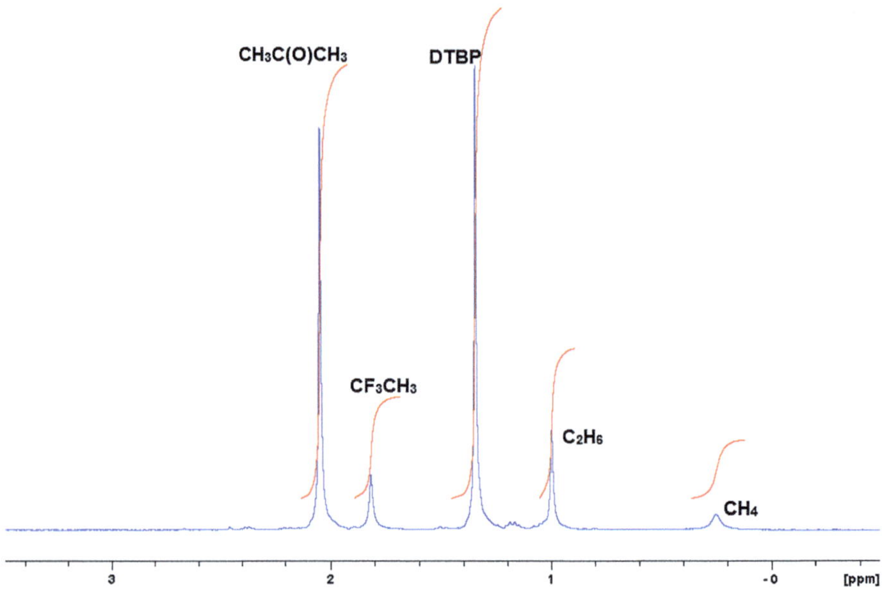

Figure 4.7 Integral regions of the ^{1}H spectrum used for analysis of the pyrolysis of DTBP, shown after 59% decomposition of DTBP.

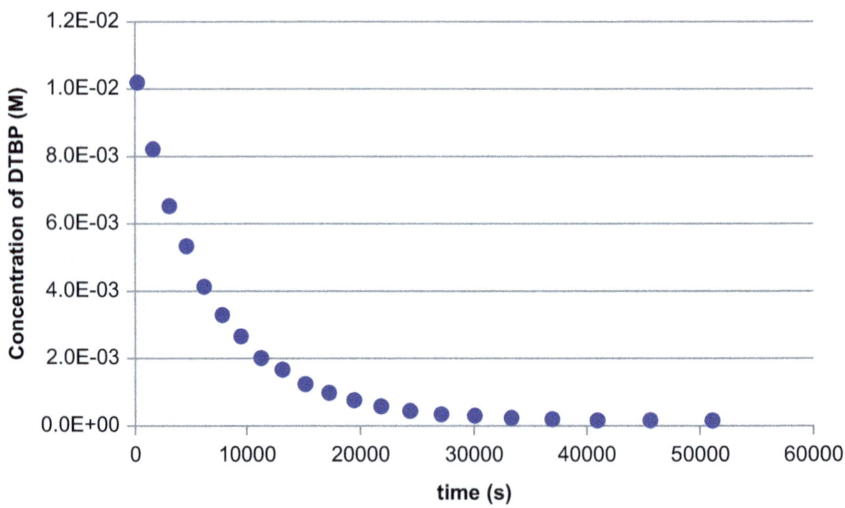

Figure 4.8 Concentration of DTBP *vs.* time upon heating at 150 °C.

the differential d [DTBP]/dt = −**k** [DBTP] can be integrated numerically over discrete units of time, and the residuals from the resulting model *vs.* the experimental data can be minimized to obtain the optimal value of **k**. In this simple case, all three methods were in good agreement (1.45×10^{-4} *vs.* 1.40×10^{-4} *vs.* 1.45×10^{-4} s^{-1}, respectively – well within experimental error).

Numerical integration is the most general approach, accommodating more complex reaction systems, and is usually the authors' preference. Given the complicating factor of spectral overlap between *t*-butanol and DTBP mentioned above, data collected after the point of 90% conversion of DTBP (in which the contribution of *t*-butanol to the DTBP integral was non-negligible) was excluded from these fits.

4.6 Characterization of Reaction Products – 2D Correlation Experiments and DOSY

A crucial component of the kinetic analysis of any reaction is the identification of all relevant products. In several of the reports cited above, the kinetic 1D spectra could not be assigned by inspection, and further experiments were required to enable that assignment. It is sometimes possible to employ conventional analytical techniques to aid in product identification. For example, a reaction ampule can be chilled and opened, and a suitable solvent injected into it. The resulting product solution can then be analyzed by *e.g.* gas chromatography or liquid-phase NMR.

While the approach of post-reaction solvation is often helpful, it is not without drawbacks. First, and most importantly, the identification of species by a separate technique does not, by itself, necessarily permit confident assignment of the gas phase NMR spectrum. Even an unambiguously assigned liquid phase NMR spectrum is often not definitive; the change in chemical shifts going from solution to gas phase (essentially the ultimate limit of the solvent effect) can be large and unpredictable, and what is often a significant difference between the temperature of reaction and the temperature of post-reaction solution-phase analysis can be a complicating factor as well. The second drawback is that in many cases the analytes of interest are fugitive, and do not permit a quantitative transfer from the closed reaction vessel to a solution in a different vessel. The third drawback is that the products of reaction in some of the reports given above are themselves very reactive, and are not necessarily compatible with most common solvents. These are all reasons to prefer product characterization in the reaction vessel, ideally under the same conditions used for the reaction itself.

For this reason, attention in our labs was turned to the exploration of two-dimensional NMR techniques for product characterization in the gas phase. Such experiments had previously been applied to the study of gas phase spin–spin coupling constants,[24–25] and multidimensional NMR experiments such as EXSY, or for the purpose of imaging, in the gas phase were known as well.[26–31] However, the use of two-dimensional NMR correlation experiments in the gas phase for structural identification of analytes, analogous to their most common application in the liquid phase, was reported for the first time only recently.[32]

4.6.1 Gas Phase Correlation Experiments

By far the most versatile experiment for the characterization of gaseous reaction products by NMR is the COSY.[33] This simple homonuclear scalar-coupling correlation experiment is sensitive and robust. It is generally successful even with inhomogeneous excitation of the spectral window (which is often unavoidable with ^{19}F). As discussed above, commercial high-temperature probes are not generally equipped with field gradient coils, and two-dimensional experiments utilizing those probes are therefore dependent on phase-cycling to remove undesired coherence pathways. The simple two-step phase cycle of the basic magnitude-mode COSY experiment tends to reduce spectral artifacts in comparison with the careful phase-cycling necessary to remove signals from ^{1}H bound to ^{12}C in a HMQC experiment, for example. For studies that can be performed in probes equipped with field gradient coils, the COSY offers a different advantage; the gradients applied for coherence selection can be placed immediately before and after the second pulse, minimizing the time during which diffusive loss can occur. By contrast, coherence selection in heteronuclear experiments requires a polarization transfer on the order of milliseconds, during which time substantial diffusive loss is observed in samples at ordinary pressures.

Figure 4.9 shows examples of ^{1}H (top) and ^{19}F (bottom) COSY spectra. The ^{1}H spectrum was acquired on a sample containing 1-pentene and tetramethylsilane, each at a partial pressure of *ca.* 1 bar, and the ^{19}F spectrum was acquired on a sample containing n-C_4F_9Cl and CF_3OCFCF_2, each at *ca.* 1.4 bar. The ^{19}F spectrum was acquired in only 6 s (*cf.* the ^{1}H spectrum acquired in *ca.* 1 h), highlighting the time-savings afforded by efficient spin–rotation relaxation in gas-phase ^{19}F spectroscopy.

Other homonuclear experiments have been successfully applied to gaseous analytes as well. Both the TOCSY[34] and J-resolved[35] experiments succeeded on the same 1-pentene/tetramethylsilane sample. These are likely to be applicable only to ^{1}H; ^{19}F spectral windows are generally too wide to effect safely a uniform and sufficient spin-lock, and the short spin–spin relaxation times in gas phase ^{19}F NMR preclude the use of a J-resolved pulse sequence, which generally incorporates rather lengthy delays for J evolution. As an instructive curiosity, the authors' labs obtained a ^{13}C INADEQUATE[36] spectrum of propene at natural isotopic abundance, at its own vapor pressure at 303 K. In contrast to the long recycle delays often required in the liquid phase for that experiment, a 2 s recycle delay sufficed.

Heteronuclear correlation experiments (generally ^{1}H-^{13}C or ^{19}F-^{13}C) can likewise be of great value in the identification of reaction products. For the reasons described above, sequences without gradient coherence selection are generally required. If the stability of the spectrometer permits effective cancellation of unwanted signals by phase cycling, the common sensitive-nucleus-observe sequences (HMQC, HSQC, HMBC)[37–39] can be usefully applied. Figure 4.10 shows HSQC spectra of the samples on which COSY spectra are shown in Figure 4.9. Each spectrum was acquired in *ca.* 1 h; in

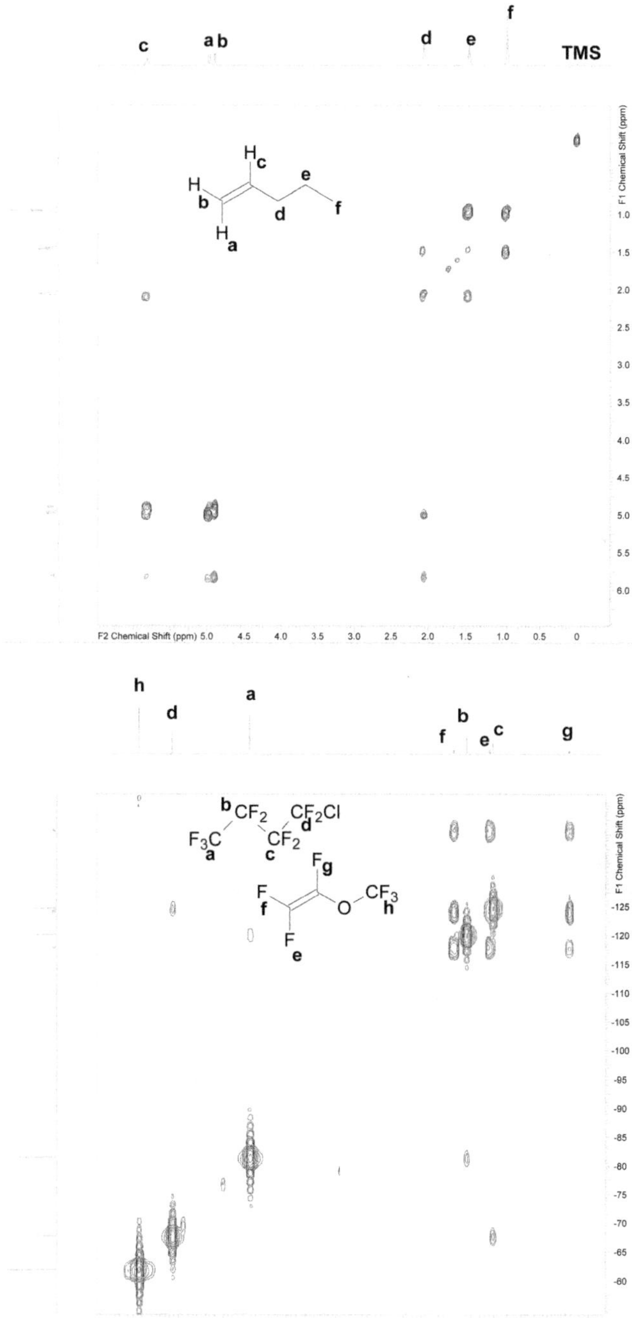

Figure 4.9 Gas phase COSY spectra. Top: ^{1}H COSY of 1-pentene + tetramethylsilane at 303 K. Bottom: ^{19}F COSY of n-C_4F_9Cl and CF_3OCFCF_2 acquired at 303 K.

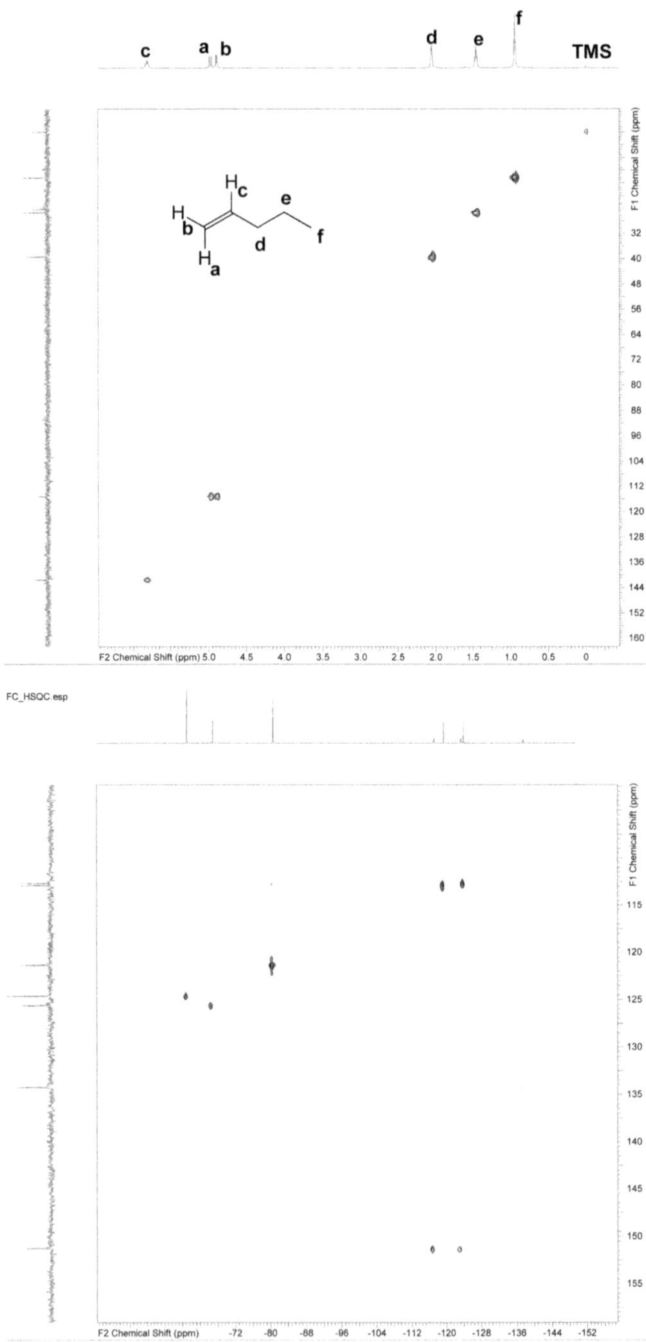

Figure 4.10 Gas phase HSQC spectra. Top: ^{1}H-^{13}C spectrum of 1-pentene and tetramethylsilane acquired at 303 K. Bottom: ^{19}F-^{13}C spectrum of n-C_4F_9Cl and CF_3OCFCF_2 acquired at 303 K.

this case, extensive signal averaging in each increment was required in the ^{19}F-observe HSQC to yield a 2D spectrum with an adequate signal-to-noise ratio. If suppression of ^{1}H-^{12}C (or ^{19}F-^{12}C) artifacts by phase cycling is not sufficient, or if higher resolution in the ^{13}C dimension is required, a simple HETCOR experiment[40] may be a useful alternative. In contrast to the liquid-phase, ^{13}C relaxation is generally faster in the gas phase than ^{1}H, and so the need to set recycle delays to accommodate ^{1}H longitudinal relaxation, a boon in liquid phase work, is actually a hindrance. Nevertheless, a useful spectrum of analytes present near ambient pressure can be obtained in an overnight acquisition.

The only major class of correlation experiments that have not been reported as successful on gaseous systems are those based on the nuclear Overhauser effect (*i.e.* NOESY, ROESY, or HOESY). The efficiency of the spin-rotation relaxation mechanism for ^{19}F and ^{13}C renders dipolar relaxation mechanisms irrelevant, and so it is no surprise that nOe-based experiments fail for such nuclei. For ^{1}H, by contrast, one might expect a detectable nOe, but a successful ^{1}H NOESY experiment has not been reported.

4.6.2 Gas Phase DOSY

The popularity of DOSY (diffusion-ordered spectroscopy) techniques has risen sharply since their introduction in 1992,[41] and they are now a ubiquitous tool in the characterization of mixtures in liquid-phase NMR spectroscopy. Likewise, NMR has been used for the determination of self-diffusion rates of gases for decades,[42] and the diffusion rates of gases have been studied by NMR as a probe of surface porosity of materials,[43–45] and in pulmonary magnetic resonance imaging.[46–47] Of interest to the characterization of chemical reaction products is high-resolution spectroscopy, as is commonly applied in the liquid phase. The efficacy of such experiments on sample gaseous mixture was demonstrated by the authors' labs in 2009.[48] It was found that, in general, spectral separation of gaseous species by translational diffusion rate was more effective in the gas phase than in solution. Figure 4.11 gives an illustrative example, in which the set of perfluoro-*n*-alkanes from $n = 1$–6 are easily separated at ambient temperature and 1.5 bar total pressure, using a basic gradient-compensated stimulated echo (GCSTE) sequence.[49] The same experiment achieved ready separation of perfluorobutane and perfluorocyclobutane, and partial separation of variously fluorinated ethanes (CFH_2CH_3 *vs.* CF_2HCH_3 *vs.* CF_3CH_3, *etc.*).

The most important experimental limitation is the need for a pulsed field gradient coil, with which most high-temperature probes are not equipped. However, because of the rapid diffusion rates observed in gaseous samples, only a very weak gradient is required, and in some spectrometers a sufficient gradient can be effected by the Z1 coil (often referred to as a "homospoil" pulse). The required gradient is so weak, in fact, that samples containing particularly small analytes at low total internal pressure can test the lower

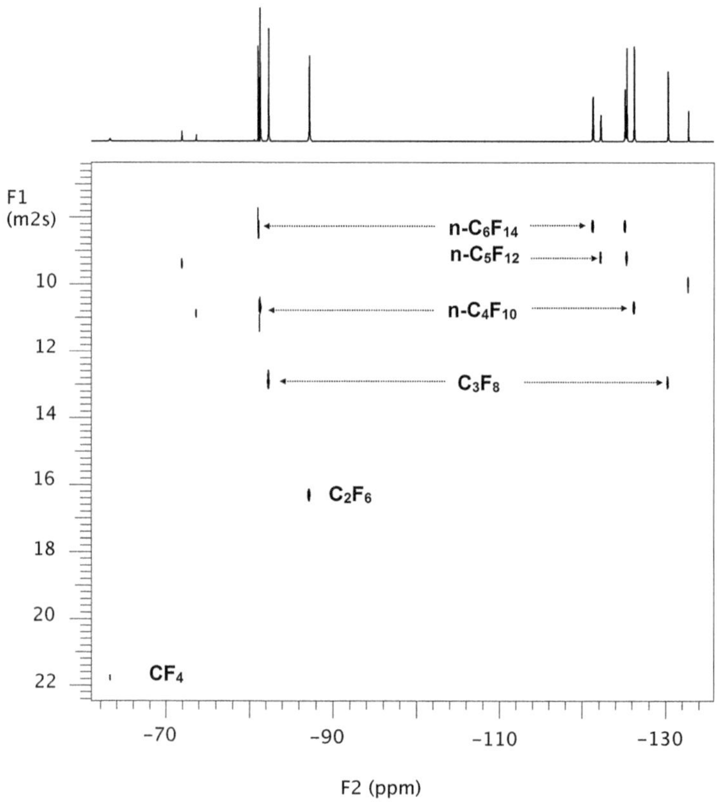

Figure 4.11 ^{19}F gas-phase gradient-compensated stimulated echo DOSY spectrum of $F(CF_2)_nF$, $n = 1–6$, at 298 K. Ordinate axis is in units of $10^{-7}\ m^2\ s^{-1}$.

limit of reproducibility in the pulsed field gradient controls (in both gradient strength and duration). For example, in a study of the self-diffusion rate of tetrafluoromethane at 298 K and 0.25 bar, a gradient strength of $9.0 \times 10^{-3}\ T\ m^{-1}$ was used for the first increment of the GCSTE; weaker gradients would have been desirable, but the data obtained from application of nominally weaker gradients suggested non-linearity of the gradient amplifier in that range. The absolute diffusion constants obtained from such experiments afforded good agreement with predicted values, supporting the value of the technique for physiochemical studies as well.

4.7 Conclusions and Outlook

Gas phase NMR spectroscopy is a sensitive, information-rich tool for the study of chemical reaction kinetics, particularly for *in situ* studies at elevated temperature. It has been applied to a number of industrially relevant problems involving both hydrocarbons and fluorocarbons, in both homogeneous and heterogeneous reactions. Two-dimensional correlation and diffusion experiments can be applied to aid in the identification of reaction products, permitting an unambiguous analysis of reaction kinetics.

Looking forward, the development of spectrometers of ever-greater sensitivity can only prove a boon for gas-phase experimentation. In particular, the advent of cryoprobe systems (in which the probe and preamplifier electronics are kept at cryogenic temperatures to minimize thermal noise) with extended ranges of sample temperatures may render ^{13}C-detected experiments far more practical. It is also interesting to speculate on the potential applicability of microcoil NMR circuitry to gaseous flow systems, analogous to those currently popular for laboratory-scale hydrogenation reactions. The combination of elevated pressure tolerance in capillary-type reactors with the per-spin boost in S/N for NMR coils of capillary-scale dimensions seems a happy convergence for the safe study of high-pressure reactions with good sensitivity.

Acknowledgements

The authors thank Ms. Rebecca Dooley for a careful review of the document, and many helpful suggestions for improvement.

References

1. R. L. Armstrong, *Magn. Reson. Rev.*, 1986, **12**, 91–135.
2. A. L. Van Geet, *Anal. Chem.*, 1968, **40**, 2227–2229.
3. P. M. Kating, P. J. Krusic, D. C. Roe and B. E. Smart, *J. Am. Chem. Soc.*, 1996, **118**, 10000–10001.
4. D. C. Roe, *J. Magn. Reson.*, 1985, **63**, 388–391.
5. H. Vanni, W. L. Earl and A. E. Merbach, *J. Magn. Reson.*, 1987, **29**, 11–19.
6. T. W. Swaddle, *Can. J. Phys.*, 1995, **73**, 258–266.
7. R. W. Peterson and J. A. Wand, *Rev. Sci. Instrum.*, 2005, **76**, 1-7.
8. D. C. Roe, P. M. Kating, P. J. Krusic and B. E. Smart, *Top. Catal.*, 1998, **5**, 133–147.
9. D. C. Roe, *Organometallics*, 1987, **6**, 942–946.
10. A. J. Wand, C. R. Babu, P. F. Flynn and M. J. Milton, *Biol. Magn. Reson.*, 2003, **20**, 121–160.
11. V. I. Bakhmutov, *Practical NMR Relaxation for Chemists*, Wiley, West Sussex, 2004.
12. M. J. Haugh and D. R. Dalton, *J. Am. Chem. Soc.*, 1975, **97**, 5674–5678.
13. F. Costello, D. R. Dalton and J. A. Poole, *J. Phys. Chem.*, 1986, **90**, 5352–5357.
14. P. J. Krusic, D. C. Roc and B. E. Smart, *Isr. J. Chem.*, 1999, **39**, 117–122.
15. A. B. Shtarov, P. J. Krusic, B. E. Smart and W. R. Dolbier, Jr., *J. Am. Chem. Soc.*, 2001, **123**, 9956–9962.
16. B. E. Smart, P. J. Krusic, D. C. Roe and Z.-Y. Yang, *J. Fluorine Chem.*, 2002, **117**, 199–205.

17. A. B. Shtarov, P. J. Krusic, B. E. Smart and W. R. Dolbier, Jr., *J. Org. Chem.*, 2002, **67**, 3464–3467.
18. P. J. Krusic and D. C. Roe, *Anal. Chem.*, 2004, **76**, 3800–3803.
19. P. J. Krusic, A. A. Marchione and D. C. Roe, *J. Fluorine Chem.*, 2005, **126**, 1510–1516.
20. A. A. Marchione, P. J. Fagan, E. J. Till, R. L. Waterland and C. LaMarca, *Anal. Chem.*, 2008, **80**, 6317–6322.
21. A. A. Marchione, D. C. Roe and P. J. Krusic, *Encyclopedia of Magnetic Resonance*, ed. R. K. Harris and R. E. Wasylishen, Wiley, Chichester, 2012, vol. 3, pp. 1726–1732.
22. A. A. Marchione, A. Moser and S. Golotvin, unpublished work.
23. E. T. Denisov, T. G. Denisova and T. S. Pokidova, *Handbook of Free Radical Initiators*, Wiley, Hoboken, 2003.
24. M. Wilczek, W. Kozminski and K. Jackowski, *Chem. Phys. Lett.*, 2002, **358**, 263–270.
25. E. Wielogorska, W. Makulski, W. Kozminski and K. Jackowski, *J. Mol. Struct.*, 2004, **704**, 305–309.
26. D. O. Kuethe, A. Caprihan, E. Fukushima and R. A. Waggoner, *Magn. Reson. Med.*, 1998, **39**, 85–88.
27. R. W. Mair, M. S. Rosen, R. Wang, D. G. Cory and R. L. Walsworth, *Magn. Reson. Chem.*, 2002, **40**, S29–S39.
28. M. S. Conradi, B. T. Saam, D. A. Yablonskiy and J. C. Woods, *Prog. Nucl. Magn. Reson. Spectrosc.*, 2006, **48**, 63–83.
29. M. Tomaselli, B. H. Meier, P. Robyr, U. W. Suter and R. R. Ernst, *Chem. Phys. Lett.*, 1993, **214**, 1–4.
30. F. Engelke, S. Bhatia, T. S. King and M. Pruski, *Phys. Rev. B: Condens. Matter Mater. Phys.*, 1994, **49**, 2730–2738.
31. C. Y. Cheng, J. Pfeilsticker and C. R. Bowers, *J. Am. Chem. Soc.*, 2008, **130**, 2390–2391.
32. A. A. Marchione, *J. Magn. Reson.*, 2011, **210**, 31–37.
33. W. P. Aue, E. Bartholdi and R. R. Ernst, *J. Chem. Phys.*, 1975, **64**, 2229–2246.
34. L. Baunschweiler and R. R. Ernst, *J. Magn. Reson.*, 1983, **53**, 521–528.
35. W. P. Aue, J. Karhan and R. R. Ernst, *J. Chem. Phys.*, 1976, **64**, 4226–4227.
36. A. Bax, R. Freeman and T. A. Frenkiel, *J. Am. Chem. Soc.*, 1981, **103**, 2102–2104.
37. L. Muller, *J. Am. Chem. Soc.*, 1979, **101**, 4481–4484.
38. G. Bodenhausen and D. J. Ruben, *Chem. Phys. Lett.*, 1980, **69**, 185–189.
39. A. Bax and M. F. Summers, *J. Am. Chem. Soc.*, 1986, **108**, 2093–2094.
40. R. Freeman and G. A. Morris, *J. Chem. Soc., Chem. Commun.*, 1978, 684–686.
41. K. F. Morris and C. S. Johnson, Jr., *J. Am. Chem. Soc.*, 1992, **114**, 3139.
42. P. E. Suetin and P. G. Zykov, *Diffuz. Gazakh Zhidk.*, 1972, 75–78.
43. D. Raftery, *Annu. Rep. NMR Spectrosc.*, 2006, **57**, 206–270.
44. R. W. Mair, D. G. Cory, S. Peled, C.-H. Tseng, S. Patz and R. L. Walsworth, *J. Magn. Reson.*, 1998, **135**, 478–486.

45. R. W. Mair, M. S. Rosen, R. Wang, D. G. Cory and R. L. Walsworth, *Magn. Reson. Chem.*, 2002, **40**, S29–S39.
46. D. O. Kuethe, A. Caprihan, E. Fukushima and R. A. Waggoner, *Magn. Reson. Med.*, 1998, **39**, 85–88.
47. M. S. Conradi, B. T. Saam, D. A. Yablonskiy and J. C. Woods, *Prog. Nucl. Magn. Reson. Spectrosc.*, 2006, **48**, 63–83.
48. A. A. Marchione and E. F. McCord, *J. Magn. Reson.*, 2009, **201**, 34–38.
49. M. D. Pelta, H. Barjat, G. A. Morris, A. L. Davis and S. J. Hammond, *Magn. Reson. Chem.*, 1998, **36**, 706–714.

CHAPTER 5

^{17}O and ^{33}S NMR Spectroscopy of Small Molecules in the Gas Phase

WŁODZIMIERZ MAKULSKI

Faculty of Chemistry, University of Warsaw, Pasteura 1, 02-093 Warszawa, Poland
Email: wmakul@chem.uw.edu.pl

5.1 Introduction

In his brief article entitled "Early Work on Gas-Phase Chemical Shifts", published in "NMR Encyclopedia" in 1996, W.T. Raynes remarked that "much is to be done" in this field of NMR spectroscopy.[1] Now, over a dozen years later, gas-phase studies are still scarce and restricted to a few laboratories. Regrettably so, as the properties of molecules in dilute gases more closely resemble those obtained from theoretical calculations; therefore, judgments regarding their accuracy and reliability must come from gas-phase experimental data. Also, intermolecular effects present in condensed phases can be strongly restricted or even fully eliminated in the gas phase. This allows a zero point to be established, whereby the effects of solvents on molecular properties and dynamics can be quantified.

Currently, gas-phase studies are focused on three major areas. Firstly, a considerable number of more recent experimental and theoretical studies have been aimed at determination of the nuclear magnetic shielding. The density dependence of chemical shifts can be used to obtain absolute values for the nuclear magnetic shielding σ. These results yield information about

New Developments in NMR No. 6
Gas Phase NMR
Edited by Karol Jackowski and Michał Jaszuński

Published by the Royal Society of Chemistry, www.rsc.org

the isotropic part of the intermolecular potential and provide tests for theoretical methods of calculating shielding constant. Several reviews by Jameson[2,3] discuss these studies. Fewer works were devoted to measurements and calculations of spin–spin couplings *J*. Secondly, experimental and theoretical studies of relaxation can yield information about collisional cross-sections and the anisotropic part of the intermolecular potential, which describes transitions between rotational states. These areas are a continuation of and improvement on the earlier studies. Finally, dynamic NMR spectroscopy in the gas phase can be used, as it is in liquids, to provide kinetic and thermodynamic data related to low-energy inter- and intramolecular reactions, such as conformational or isotopic exchange and molecular associations. Because gas-phase dynamic studies can use pressure as well as temperature as a variable, information not available from condensed-phase studies can be obtained in the gas phase. This includes the use of pressure-dependent rate constants to study intramolecular vibrational redistribution in molecules undergoing a low-energy dynamic process and to determine collisional efficiencies for dynamic processes in the bimolecular pressure region. True, Suarez and LeMaster have reviewed this area.[4–6]

This particular work is confined to the first part of the studied problems mentioned above related to ^{17}O and ^{33}S nuclei NMR experimental efforts. The establishing of absolute shielding scales for both nuclei is discussed in detail. The list of best experimental results of shielding and spin–spin couplings in many organic and inorganic simple molecules is given. These values were analyzed in the context of modern theoretical *ab initio* calculations.

Only in the modest range were the above-mentioned problems reviewed before.[7,8]

5.2 Background

5.2.1 Oxygen and Sulfur in Chemistry of Small Molecules

Oxygen (and sulfur to a certain degree) is a key constituent of many organic and inorganic compounds. Oxygen and sulfur belong to the *VIA* (or 16) group in the periodic table. The analogy in the properties of O- and S-containing compounds is well known. Both atoms can simply substitute each other in a variety of compound classes. Nevertheless, some distinctions are obvious. They come from very few circumstances: sulfur is much less electronegative than oxygen, sulfur can expand its valence shell to hold more than 8 electrons (10 or even 12) but oxygen cannot, covalent sulfur radius (0.104 nm) is twice that of oxygen (0.066 nm). As a consequence, O=O double bonds are much stronger than S=S double bonds while S–S single bonds are approximately twice as strong as O–O bonds. An exceptional class of chemical substances are sulfur oxides: simple- SO, SO_2, SO_3, S_2O, S_2O_2 and rings- S_5O, S_6O, S_7O, S_7O_2, and S_8O. Many of them are unstable in normal conditions and therefore cannot be analyzed in standard NMR experiments.

Table 5.1 Simple chemical compounds containing oxygen and sulfur atoms.[9]

Oxygen compounds	b.p.(°C)	Sulfur compounds	b.p.(°C)
H_2O (Water)	100	H_2S (Hydrogen sulfide)	−60.7
O_3 (Ozone)	−111.9	SO_2 (Sulfur dioxide)	−10
SO_3 (Sulfur trioxide)	44.8	SO_3 (Sulfur trioxide)	44.8
COS (Carbonyl sulfide)	−50	COS (Carbonyl sulfide)	−50
Cl_2O (Dichlorine monoxide)	3.8	SCl_2 (Sulfur dichloride)	59
$COCl_2$ (Fosgene)	8.3	$CSCl_2$ (Thiofosgene)	73.5
HOCN (Cyanic acid)	23.5	HSCN (Thiocyanic acid)[a]	
$OC(NH_2)_2$ (Urea)[a]		$SC(NH_2)_2$ (Thiourea)	150–160
CO_2 (Carbon dioxide)	-78.6^{sub}	CS_2 (Carbon disulfide)	46.3
H_2O_2 (Hydrogen peroxide)	150.2	H_2S_2 (Hydrogen disulfide)	70.7
CH_3OH (Methanol)	65.15	CH_3SH (Methanethiol)	6.2
$(CH_3)_2O$ (Dimethyl ether)	−24.8	$(CH_3)_2S$ (Dimethyl sulfide)	37.34
CH_3CHO (Acetaldehyde)	20.8	CH_3CHS (Thioacetaldehyde)[b]	
$(CH_3)_2CO$ (Acetone)	56.2	$(CH_3)_2SH$ (Thioacetone)[a]	
CH_3COOH (Acetic acid)	117.9	CH_3COSH (Thioacetic acid)	93
CF_3COOH (Trifluoroacetic acid)	72.4	CF_3COSH (Trifluorothiolacetic acid)[b]	

[a]Unstable.
[b]Unknown.

It is worth noting that most of the small organic and inorganic O- and S-containing compounds are normal gases and volatile liquids at room temperature (see Table 5.1). Some play an important role as pollutant gases in the Earth's atmosphere and they are observed as astrophysics objects. Most of them were analyzed with the use of NMR spectroscopy in the gas phase and it will be discussed here.

5.2.2 NMR Parameters of ^{17}O and ^{33}S Nuclei

There are three stable oxygen isotopes; because ^{16}O and ^{18}O both have $I=0$, ^{17}O is the only stable NMR-active nucleus. It is primarily the extremely low natural abundance of ^{17}O isotope, 0.037%, that has made oxygen a little-studied nucleus from the NMR point of view. Despite good sensitivity, its receptivity is still only 6.11×10^{-2} that of ^{13}C. Additionally, oxygen-17 is a quadrupolar nucleus (spin number $I=5/2$) with electric quadrupole moment $Q=-0.02578$ b ($1\text{b}=10^{-28}$ m^2).[10]

The common chemical shift reference is the naturally abundant ^{17}O water sample. It is not an ideal chemical shift reference because of its relatively large line width, but it is acceptable on account of the large line widths for most ^{17}O resonances. Recently the pure liquid D_2O is preferred. The relative peak positions of liquid H_2O/D_2O system in ^{17}O NMR spectrum were shown in Figure 5.1. The total range of ^{17}O shielding of diamagnetic compounds reaches about 1160 ppm. An extensive ^{17}O NMR spectroscopy review in two parts has appeared recently.[11–12]

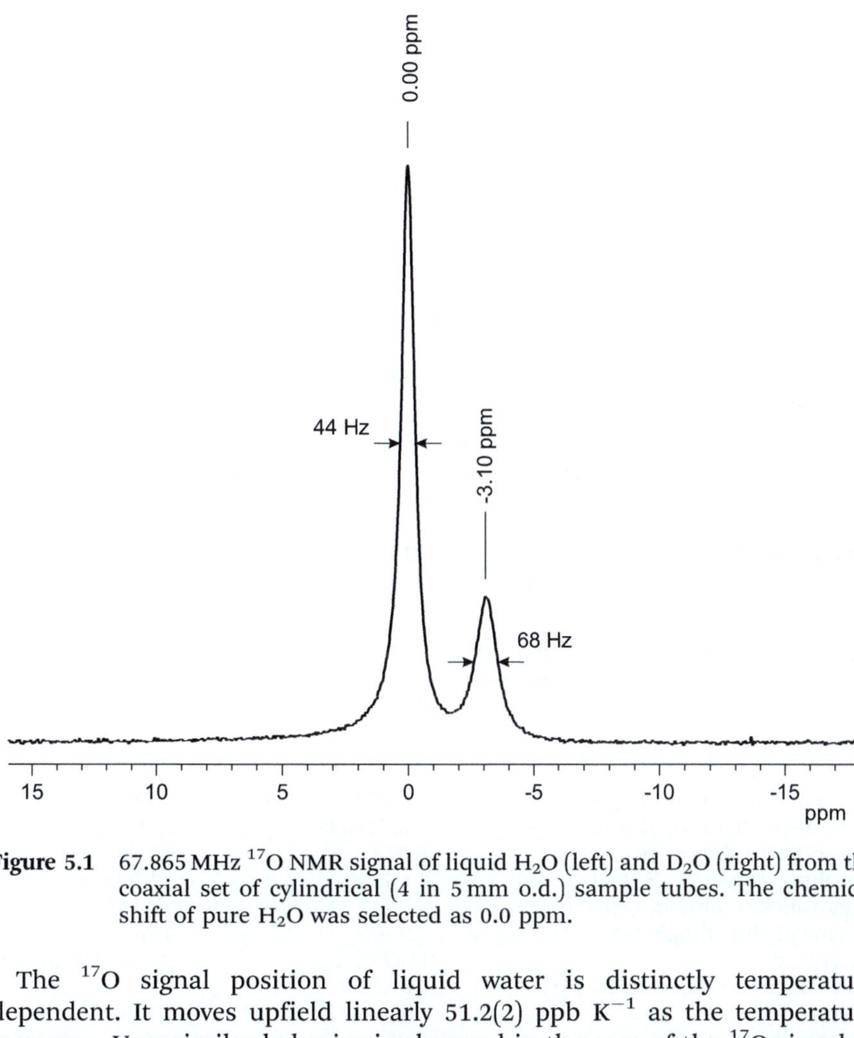

Figure 5.1 67.865 MHz ^{17}O NMR signal of liquid H_2O (left) and D_2O (right) from the coaxial set of cylindrical (4 in 5 mm o.d.) sample tubes. The chemical shift of pure H_2O was selected as 0.0 ppm.

The ^{17}O signal position of liquid water is distinctly temperature dependent. It moves upfield linearly 51.2(2) ppb K^{-1} as the temperature increases. Very similar behavior is observed in the case of the ^{17}O signal of heavy water (51.3(2) ppb K^{-1}, see Figure 5.2).

Sulfur is the only group VIA element with a poorly developed NMR because it has several unfavorable characteristics: low natural abundance (0.76% for the only active isotope ^{33}S), a low resonance frequency ($\Xi = 7.6760$ MHz), and a quadrupole moment $Q = -0.064(10)$ b.[10] It means that the observation of sulfur spectra is difficult. The ^{33}S NMR spectroscopy is even more poorly developed than that of oxygen. The well-known example is that of thiosulfate anion $S_2O_3{}^{-2}$ in the water solution, the ^{33}S NMR spectrum of which consists only of one signal which belongs to the internal sulfur atom. The thiosulfur resonance line is too broad to be observed.[13]

The IUPAC recommended reference of chemical shift is saturated $(NH_4)_2SO_4$ in D_2O as solvent ($\Delta_{1/2} \sim 10$ Hz). Alternatively, one can use 2M aqueous solution of $Cs_2(SO_4)_2$. The tetrahedral symmetry around the sulfur

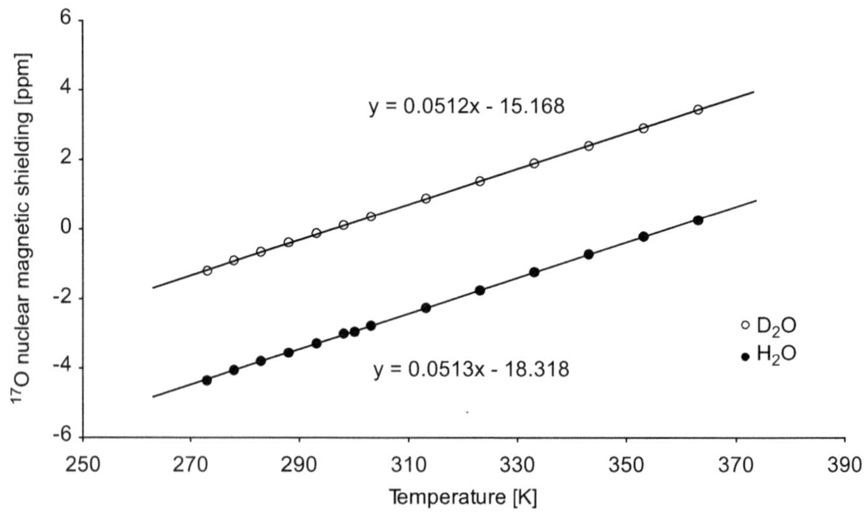

Figure 5.2 Temperature dependence of ^{17}O magnetic shielding in liquid H_2O and D_2O (bulk susceptibility corrections are included).

Table 5.2 NMR properties of the ^{17}O and ^{33}S nuclei.

Property	Nuclide ^{17}O	^{33}S
Nuclear spin	5/2	3/2
Nuclear magnetic moment, μ_N	−1.8935474(68)[29]	0.643251(16)[29]
Gyromagnetic ratio, γ_x [rad T^{-1} s^{-1}]	−3.62759521[29]	2.053864832[29]
Natural abundance, %	0.038	0.76
Absolute frequency, Ξ [MHz]	13.556457	7.676000
Quadrupole moment, Q [mb]	−0.02578	−0.678
Chemical shift range, ppm	1160	964
Reference	H_2O or D_2O	$(NH_4)_2SO_4$ in D_2O
Line width of reference, $\Delta_{1/2}$ [Hz]	44 or 68	10
Receptivity rel. to ^{1}H	1.11×10^{-5}	1.72×10^{-5}
Receptivity rel. to ^{13}C	0.065	0.101

atom in SO_4^{-2} leads to low asymmetry parameter and, as a consequence, long relaxation times and a narrow resonance signal. On this scale the secondary reference signal of CS_2 is −333 ppm. The spectral range of ^{33}S shieldings is about 1000 ppm. All the important NMR parameters of ^{17}O and ^{33}S are given in Table 5.2.[14]

5.2.3 ^{17}O and ^{33}S-labelled Compounds

Extremely low ^{17}O natural abundance could be overcome by utilizing oxygen-17 enriched material. Unfortunately, only limited labeled substances are commercially available: $C^{17}O$, $C^{17}O_2$, $^{17}O_2$, $H_2{}^{17}O$, $CH_3{}^{17}OH$, and $C_2H_5{}^{17}OH$.

Syntheses involving oxygen isotopes tend to apply rather straightforward organic reactions: substitution, addition-elimination, oxidation, and, moreover, photolytic processes. Obviously, any reaction involving oxygen-16 can be utilized to incorporate oxygen-17, oxygen-18, and, theoretically, oxygen-15.[15]

Up to now ^{33}S commercially labeled compounds are not accessible; ^{33}S- sulfur powder is available at high cost ∼15 000 €/1 g (Cambridge Isotope Laboratories Inc.). The substitution syntheses are troublesome and exceptional.

5.3 NMR Experiments in Gas Phase

5.3.1 Experimental Approach and Problems

It is well known that any molecular electromagnetic property of a gaseous substance can be described as virial expansion in powers of the density. Nuclear magnetic shielding (σ) defined by a virial expansion in molar volume (V_m) was first formulated by Buckingham and Pople.[16] Its dependence on density (ρ) is as follows:

$$\sigma(\rho, T) = \sigma_0(T) + \sigma_1(T)\rho + \sigma_2(T)\rho^2 + \cdots \quad (5.1)$$

where σ_0 is the shielding of an isolated molecule, (σ_1, σ_2) are due to intermolecular interactions, and all the shielding coefficients are temperature dependent (T). Usually eqn (5.1) becomes linear, then the σ_0 and σ_1 can be obtained directly from linear regression analyses and discussed from a phenomenological or theoretical point of view. These problems are discussed in detail by Jameson in Chapter 1 of this book.

Contributions to σ_1 include the influence of intermolecular interactions on shielding and the effect of bulk susceptibility. The bulk susceptibility correction can be generally removed from σ_1, so that only the true intermolecular effects are considered. Each σ_1 parameter is described by a complex function of the intermolecular separation and orientation between two interacting molecules (an intermolecular shielding surface). This function may be calculated by *ab initio* methods but the appropriate procedure requires enormous computational work and at present it has been completely obtained for only the atomic and simplest molecular systems.

Analogously to the shielding relationships, the nuclear spin–spin coupling is also modified by interactions of molecules in gases; the appropriate equation for the spin–spin coupling in a pure substance is similar to eqn (5.1):

$$J(T) = J_0(T) + J_1(T)\rho + J_2(T)\rho^2 + \cdots \quad (5.2)$$

where $J_0(\mathrm{T})$ is the spin–spin coupling for an isolated molecule and $J_1(T)$, $J_2(T)$, ... are due to intermolecular effects in the collisions of molecules.[8] This formula can be easily revised for mixtures of gases where $J_n(\mathrm{AB})$ coefficients are involved.

5.3.2 Gas Phase Experimental Characteristic

Except for some advantages mentioned above, the gas-phase experiments have self-evident limitations. These are sample volatility, sensitivity, and natural line widths which limit the range of densities and temperatures for the study. Relaxation times in gases are typically such that many intra- or intermolecular polarization transfer schemes which are routinely used in condensed phases cannot usually be used here.

Nevertheless, sometimes the accomplishment of such impulse sequences can be completed with success. A good example is the INEPT (Insensitive Nuclei Enhanced by Polarization Transfer) experiment performed with and without decoupling in H_2O mixed with different inert gases: Xe, Kr, CH_3F, and CHF_3. The sequence, optimized for ^1J(O,H) spin–spin coupling and $I = 5/2$, was used to measure spin–spin coupling and strengthen the weak oxygen signal at low concentrations.[17] The sequence preserves the acoustic ringing process and gives a flat basis spectrum line (see Figure 5.3).

Practically, to lengthen the relaxation processes one can suggest the use of $\pi/2$ pulses.

For the acquisition of 2D (two-dimensional) spectra, indispensable for the measurements of passive coupling constants, a modified PFG-HSQC sequence was proposed. The application of the commonly used standard HSQC technique with gradient echo-antiecho coherence selection was not successful due to signal attenuation caused by very effective diffusion in gases. We used instead the selection of doubly longitudinal two spin $2I_zS_z$ coherence by application of two opposite sign gradient pulses during both INEPT steps. This technique enables the cancellation of all unwanted transverse magnetization and is not sensitive to the effects of diffusion. In order to reduce signal losses owing to fast transverse relaxation, the refocusing period before t_2 data acquisition was omitted. Consequently, purely absorptive correlation signals appear with the active coupling in the antiphase along the F_2 domain.[18] In fact, the use of 2D spectra in gas phase precise measurements is scarce because of natural limitations of resolution in both transformed dimensions.

The easiest objects in gas phase NMR research are normal gases at ambient temperatures where relatively high oxygen or sulfur atom concentrations are available. The use of matrix solutions, *e.g.* mixtures of gases, makes the situation worse, but successful measurements are still possible. Sometimes trial experiments are performed in liquid cyclohexane solutions where intermolecular interactions are small and can simulate the gas phase.[19] Nevertheless it cannot replace the experiment in the gas phase with shielding extrapolation to zero density.

5.3.3 Absolute Shielding

Systematic computational studies of the specific magneto-electric parameters like electric field gradients, magnetic susceptibilities, polarizabilities,

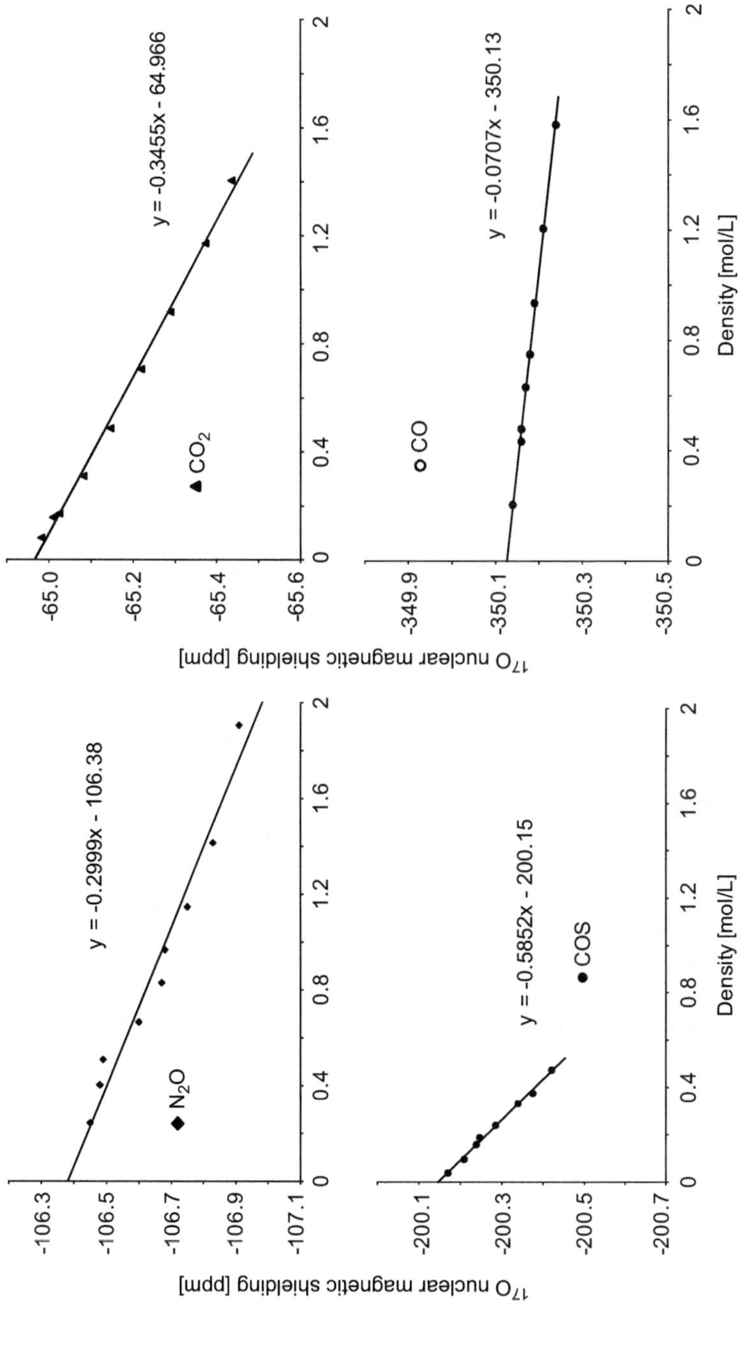

Figure 5.3 Density-dependent ^{17}O nuclear magnetic shielding of pure gaseous compounds: nitrous oxide (N_2O), carbon dioxide (CO_2), carbonyl sulfide (COS), and carbon monoxide (CO).

magnetic moments, chemical shifts, nuclear shieldings, and spin–spin couplings have become a routine procedure for chemical systems. The last three parameters are of prime importance in the context of this work. Quantum-chemical calculations of NMR properties are supplementary to experiments in the interpretation of NMR spectra. In recent years we have witnessed a spectacular rise in computation efficiency, especially for small chemical objects which are discussed in this work. Several review articles are noteworthy here.[20–24]

This chapter is not intended to present a theoretical approach to shielding. Nevertheless, some principal comments should be made. From a theoretical point of view, the nuclear magnetic shielding σ is a second-rank tensor with its symmetric part quantitatively expressed in terms of three principal components: σ_{11}, σ_{22}, and σ_{33}, which describe the change in the local magnetic fields at the nucleus position. In the isotropic matter like gaseous or liquid solutions rapid tumbling leads to an averaging of components and $\sigma_{iso} = 1/3\ (\sigma_{11} + \sigma_{22} + \sigma_{33})$ can be utilized. Nuclear magnetic shielding was discovered by physicists concerned with accurate measurements of nuclear magnetic moments. In the fundamental classic approach each shielding constant in a molecule can be divided into two terms $\sigma = \sigma^{d} + \sigma^{p}$, *i.e.* a diamagnetic and paramagnetic one. The field induced within an atom or molecule proportional to the applied field and opposite in sign is known as the diamagnetic contribution. Generally, σ^{d} derives from the second-order correction to the energy, involving only the ground state wavefunction, and its calculation seems to be straightforward.

The paramagnetic term arises from the mixing of certain excited states with the ground electronic state in the presence of a magnetic field. Calculations of this term are much more demanding (for details and references see Chapter 6).

In both calculations a gauge origin is chosen for the magnetic vector potential. The natural choice of gauge origin is the nucleus in question but the shielding values are observable properties, which have to be independent of the choice of the gauge origin. The situations for its parts, the diamagnetic and paramagnetic terms, are different, as they are both gauge dependent. This so-called gauge problem was successfully eliminated in the past.

It is known that external and internal magnetic fields discussed in each NMR experiment give rise to very small energetic effects. In this context the use of perturbation theory is the optimal method in the calculation of the NMR parameters. Several different *ab initio* or density-functional theory levels are used. Uncorrelated HF (Hartree–Fock), correlated MCSCF (Multi-configurational self-consistent field), second-order MP2 (Møller–Plesset perturbation theory), and CC (coupled cluster) methods prove to be useful. Particular quantum-chemical methods have their own advantages and specific limitations. For example, in practical DFT calculations a selection of exchange–correlation functional is always needed. A theoretical model approach entails not only the general method but also a basis set choice, a description of electron correlation effects, the gauge problem, and several specific details.

Because the basic considerations are devoted to molecules at their equilibrium geometry, when comparison with the experiment is necessary, the additional calculations of intermolecular and intramolecular effects should be made. When comparison with the gas phase experiment at room temperature is performed, rovibrational and temperature effects should at least be taken into account: $\Delta\sigma = \sigma_{vib} + \sigma_T$.

With the gauge origin at the nucleus in question, σ^p in Ramsey's expression (non-relativistic approach) is related to another molecular property, the nuclear spin-rotation constant C_I. This value arises from the coupling of the magnetic moment of a nucleus with the magnetic field generated by the molecular rotation at that nucleus. Ramsey[25] and Flygare[26] have shown that C_{gg} and $\sigma^p{}_{gg}$ are the related diagonal components of the spin-rotation tensor and the paramagnetic shielding tensor, respectively. The $\sigma^p{}_{gg}$ can be related to the components along the principal axes of the shielding tensor by a rotational transformation when the molecular geometry is known. Since σ^d can be calculated from the ground state wavefunction of the molecule, and sometimes can be estimated to within 0.1 to a few ppm of the known free neutral atom σ, the absolute shielding constant for a nucleus in a molecule can be determined from the nuclear spin-rotation constants (for details see Chapter 2).

The traditional quantum chemical methods have been exploited in the non-relativistic approximation, which are sufficient for the first and second row of the elements but recently fully relativistic calculations have become available. The theoretical calculations are crucial in establishing the absolute shielding ^{17}O and ^{33}S scales and will be discussed in subsequent sections.

Gradual progress in the determining of magnetic shielding constants has important consequences for recognizing the nuclear dipole moments of different nuclei. Originally, the measurements of spin numbers and nuclear magnetic moments were carried out in the molecular beam experiments (the molecular beam magnetic resonance method[27]) developed before classical NMR spectroscopy was discovered. Presently the experiments in external static strong magnetic field B_0 are recommended. For now, we can consider an equation[28,29] where the ratio of NMR frequencies for a pair of different nuclei can be used:

$$\Delta\mu_X^z = \frac{\nu_Y}{\nu_X} \cdot \frac{(1-\sigma_Y)}{(1-\sigma_X)} \cdot \Delta\mu_Y^z \tag{5.3}$$

allowing us to compute the magnetic moment μ_Y when all other quantities are known. The $\Delta\mu_Z^z$ is the transition-related change of the projection of magnetic moment on the field axis. It is obligatory to use the most accurate values of shielding constants (σ_X and σ_Y), which can be received as a result of the accomplished theoretical calculations (see Chapter 3 for details).

Accurate frequency ratios and chemical shifts, determined in this work from gas phase spectra, were taken for the above calculation. The appropriate shielding constants were recomputed from chemical shift values.

As a consequence, the nuclear magnetic dipole moments for ^{17}O and ^{33}S were determined with higher accuracy than ever before. H_2O, D_2O and CS_2, SF_6 molecules were taken as research objects. New results as μ_x and factors γ_x were included in Table 5.2.

The modified eqn (5.3) for extraction of one shielding value (see Chapter 3) has its own specific advantage. The parameter of absolute shielding can be transferred from helium-3 atoms to pure liquid deuterated solvents which are used in the common NMR measurements as lock systems.[30,31] The resonance frequencies of these compounds can serve as the secondary reference standard of nuclear magnetic shieldings. Simple equations utilized here are as follows:[29]

$$\sigma_O = 1 - 1.13205277(1 - \sigma_D^*) \tag{5.4}$$

$$\sigma_S = 1 - 1.9994642(1 - \sigma_D^*) \tag{5.5}$$

The deuterium shielding values of several organic and inorganic solvents were carefully evaluated.[31] Strictly speaking these relationships can be successfully used for the gaseous samples only. The liquid samples need an analysis of the macroscopic magnetic susceptibility corrections, which is possible, but for concentrated solutions may be inconvenient.

5.3.4 Spin–Spin Coupling

The scalar coupling constants can be defined as the mixed second derivative of the total electronic energy with respect to the magnetic moments of the two nuclei involved.[22] The indirect nuclear spin–spin interaction can be expressed as nuclear spin–spin coupling tensor J_{KL} or in terms of the reduced coupling tensor K_{KL}. The relationship between these values is

$$J_{KL} = h\frac{\gamma_K}{2\pi} \cdot \frac{\gamma_L}{2\pi} \cdot K_{KL} \tag{5.6}$$

where γ_K and γ_L are the gyromagnetic ratios of the two nuclei. Again, in the gas or liquid phases, only the isotropic part of the indirect coupling tensor is observed. The reduced values are better in the comparison of the coupling between different nuclei because of the elimination of the effect of gyromagnetic factors of both nuclei.

The J_{KL} value in the non-relativistic approximation can be composed of four components: FC (Fermi contact), PSO (paramagnetic spin–orbit), DSO (diamagnetic spin–orbit), and SD (spin–dipole):[32]

$$J_{KL} = J_{KL}{}^{DSO} + J_{KL}{}^{PSO} + J_{KL}{}^{FC} + J_{KL}{}^{SD} \tag{5.7}$$

A number of standard second-order perturbation quantum chemical methods, within a relativistic and non-relativistic framework, have been developed for the calculation of these values. Among the different mechanisms contributing to J_{KL}, the FC term is usually the most important in particular for one-bond coupling.

It is worth noting that relativistic effects are not so important for spin–spin couplings involving the ^{17}O nucleus and probably for those involving ^{33}S nuclei.

5.4 ^{17}O and ^{33}S Shielding from Gas Phase Measurements

5.4.1 $C^{17}O$ Molecule as Reference of Oxygen Shielding

This simple diatomic molecule was taken to establish the absolute shielding scales in ^{13}C and ^{17}O resonances. The suitable calculations are based on ^{13}C and ^{17}O nuclear spin-rotation constants measured for $^{13}C^{16}O$ and $^{12}C^{17}O$ isotopomers. While the ^{13}C nuclear spin-rotation constant, $C_I(^{13}C)$, is known with high precision, the corresponding value for ^{17}O, $C_I(^{17}O)$ has been less accurately established. The first fully experimental ^{17}O absolute shielding scale was provided on the basis of $C_I(^{17}O) = -30.4(12)$ kHz.[33]

This spin-rotation constant was determined from the $J = 1 \leftarrow 0$ rotational transition for $^{12}C^{17}O$ in interstellar space, namely the Bok globule B335. Using expressions for dia- and paramagnetic terms in the isotropic value of the oxygen magnetic shielding tensor, the $\sigma_{v=0}(^{12}C^{17}O) = -42.3(172)$ ppm value corresponding to the ground vibrational state was evaluated. This result was rather poor, as the uncertainty for the measured absolute chemical shielding reached over a dozen or so ppm. On the other hand, the same shielding constant at 300 K has been determined just on the basis of quantum chemical calculations and equals $-59.34(200)$ ppm.[34] The slightly different value $-62.3(15)$ ppm was suggested on the basis of extensive multiconfiguration self-consistent field (MCSCF) calculations for water molecules.[35] Taking the difference between the so-called experimental and the fully computed oxygen shielding, an error estimate of about ± 2 ppm seems plausible for the calculated values.

Later on, the very precise value of oxygen spin-rotation constant for $^{12}C^{17}O$ molecule $C_I(^{17}O) = -31.609(41)$ kHz was measured by the rotational Lamb-dip technique.[36] This allows for the establishment of a revised experimental oxygen magnetic shielding scale.[37] Using the equilibrium value of $C_I(^{12}C^{17}O) = -31.561$ kHz, one gets $\sigma^p = -752.74$ ppm while the diamagnetic contribution $\sigma_{\perp}{}^d = 462.453$ ppm and $\sigma_{||}{}^d = 410.200$ ppm were calculated. As a consequence, the σ_e equal to -56.79 ppm will be employed. Rotational-vibrational corrections to σ_e (-5.945 ppm) allow for the establishment of a new absolute shielding scale for oxygen at 300 K ($-62.735(590)$ ppm). Based on this revised scale and on experimentally known oxygen chemical shifts, $\sigma_{300K}(H_2O)_{liquid}$ is 287.5(6) ppm. The last value was further slightly upgraded to 287.4(6) ppm and was next used to recalculate all established absolute shielding of ^{17}O nuclei in small molecules measured from gas phase experiments in our laboratory.[38]

5.4.2 The "Isolated" Water Molecule

Without doubt water is the most known substance in the world. As a starting point we can set about studying the single water monomer molecule. Very few experimental studies have been carried out to determine the gas phase NMR parameters of the water molecule. Earlier observation of the resonance lines in water molecules is confined to the systems in which the rapid proton-deuteron exchange connected with the presence of hydrogen bonds can be effectively stopped. This condition was achieved in our study[17] by resorting to examination of the system in the gas phase. Fluoromethane gaseous matrices were taken for the study, as they have considerable dipole moments and are capable of interacting strongly with water even in the gas phase, and the experiments could be carried out at room temperature (300 K). The final results comprise the shielding value of monomolecular water $\sigma_0(^1H) = -2.724(10)$ ppm and $\sigma_0(^{17}O) = 35.220(70)$ ppm relative to the liquid water sample.

On the other hand, the absolute oxygen shielding was calculated on the basis of the spin-rotation constant $C_0 = -25.12(12)$ kHz measured by the Lamb-dip technique with high accuracy. Calculations, besides the paramagnetic part of shielding, contain the computed diamagnetic part of the shielding as well as all vibrational and temperature corrections, which finally leads to $\sigma_0(^{17}O) = 325.3(3)$ ppm at 300 K.[39] This result was confirmed by the purely theoretical result of 325.6 ppm composed of equilibrium value 337.7 ppm, next corrected by vibrational terms −11.7 ppm and the temperature term of −0.4 ppm. The experimental value was recommended as the absolute shielding scale reference for ^{17}O NMR spectroscopy ($\sigma_0 = 287.1$ ppm for liquid H_2O, according to our ^{17}O measurements).[17] On this scale the nuclear magnetic shielding of CO molecule is $\sigma_0(^{17}O) = 63.0$ ppm in excellent agreement with the value used in previous works on the oxygen shielding scale ($\sigma_0(^{17}O) = 62.7$ ppm).[39] This proposal against the older one utilizing carbon monoxide as a reference will be discussed later.

5.4.3 ^{17}O Magnetic Shielding of Small Molecules

The measurements of density-dependent ^{17}O shielding for gaseous samples of CO_2, N_2O, OCS, and CO were the first results of this kind.[40] In each case an increase in density diminishes the shielding (see Figure 5.4); similar data have previously been found for many other nuclei. The density dependence is linear for all substances, which means that $\sigma_2(T)$ and higher coefficients in eqn (5.1) can be neglected here. $\sigma_0(T)$ is obtained by extrapolation to the zero-density limit. Measurements of $\sigma_1(300\ K)$ terms prove that intermolecular interactions can efficiently modify the magnetic shielding of oxygen nuclei in the gas phase. The second virial coefficient reflects the range in chemical shifts of the nucleus being used as a probe. Similar results were collected for other small-molecule compounds *i.e.* SO_2, SO_3,[41] and a

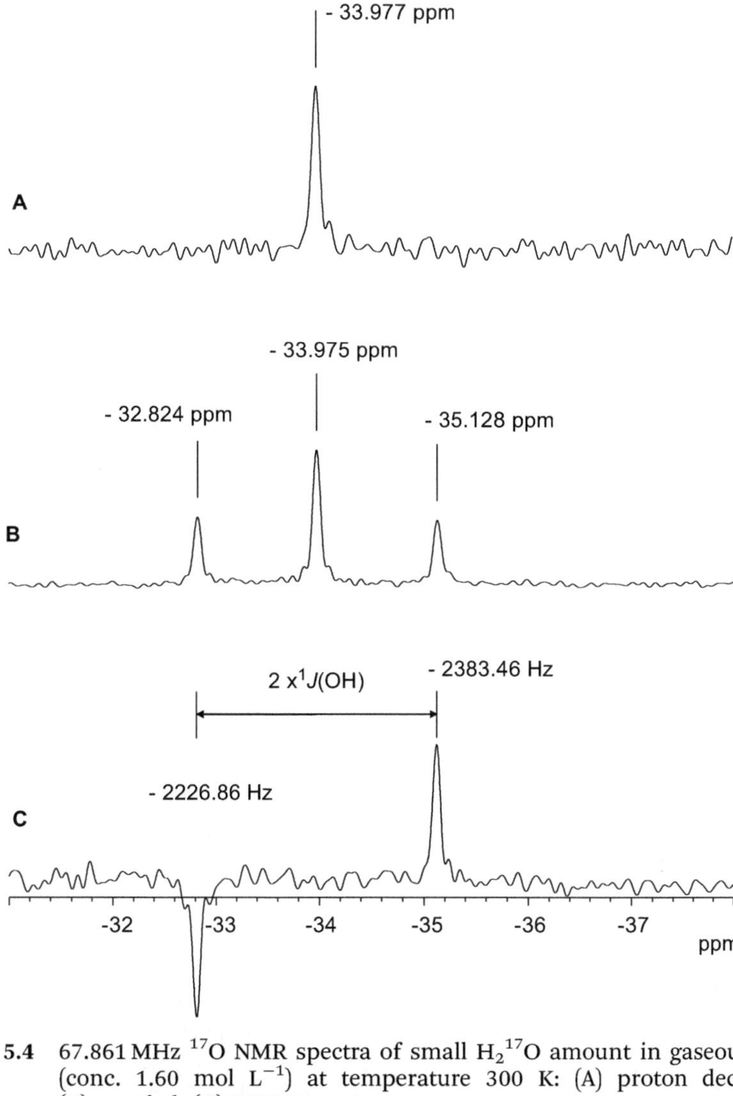

Figure 5.4 67.861 MHz ^{17}O NMR spectra of small $H_2{}^{17}O$ amount in gaseous CH_3F (conc. 1.60 mol L^{-1}) at temperature 300 K: (A) proton decoupled, (B) coupled, (C) INEPT spectrum.

series of dimethyl ethers $(CH_3)_2O$, CH_3OCD_3, and $(CD_3)_2O$.[42] This kind of experimental procedure cannot be carried out for low volatility compounds. In such a case, one can add a buffer gas to the system and ^{17}O enriched compounds are needed for this purpose. The experimental spectra of $H_2{}^{17}O$ are shown in Figure 5.4. The graph of density functions is analogous to that in Figure 5.3.[17] So, the tabulation of the first adequate absolute shielding scale could be now established. Experimental ^{17}O magnetic shielding constants resulting from extrapolations to zero-pressure limit of some polyatomic molecules are collected in Table 5.3.

Table 5.3 Experimental ^{17}O chemical shifts and magnetic shielding parameters from gas phase experiments at 300 K.

Molecule	Nucleus	δ_0, ppm	σ_0, ppm	Solvent	$\sigma_1-\sigma_{1b}$, ppm×ml×mol^{-1}	Ref.
Water (gas)	$H_2\underline{O}$	−35.22	325.3(9)			17
				Xe	−1148(40)	17
				Kr	−592(40)	17
				CH_3F	−818(20)	17
				CHF_3	−1305(40)	17
Water (liquid)	$H_2\underline{O}$	0	287.4(6)			38
Deuterium oxide (liquid)	$D_2\underline{O}$	3.1	290.2(7)			
Carbon monoxide	$C\underline{O}$	350.13(1)	−63.05	CO	−119(12)	40
Carbon dioxide	$C\underline{O}_2$	64.97(10)	222.08(70)	CO_2	−432(12)	40
Carbonyl sulfide	$C\underline{O}S$	200.15(10)	86.93(70)	COS	−720(12)	40
Nitrous oxide	$N_2\underline{O}$	106.44(20)	180.64(80)	N_2O	−335(25)	40
Sulfur dioxide	$S\underline{O}_2$	518.41(2)	−231.3	SO_2	+869(50)	41
	$S\underline{O}_2$ (333 K)	518.52(2)	−231.4		+709(50)	41
Sulfur trioxide	$S\underline{O}_3$ (333 K)	232.07(2)	55.0(6)	SO_3	−948(150)	41
Methyl ether	$(CH_3)_2\underline{O}$	−48.98	336.06(60)	$(CH_3)_2O$	−1400(120)	42
	$(CH_3)_2\underline{O}$ (333 K)	−49.65	336.73(60)		−1575(50)	42
	$(CD_3)_2\underline{O}$	−51.54	338.62(60)		−1394(120)	42
	$(CD_3)_2\underline{O}$(333 K)	−52.08	339.16(60)		−1550(50)	42
Methanol	$CH_3\underline{O}H$	−42.51	329.61	CH_3F	−106.5(20)	43
				CHF_3	−980.1(300)	43
	$CH_3\underline{O}D$	−44.36	331.46	CH_3F	−152.4(30)	43
				CHF_3	−1023.0(400)	43
Ethanol	$C_2H_5\underline{O}H$	0.31(5)	287.4(6)	Xe	−390(67)	44
				SF_6	−968(48)	44

The proper theoretical evaluation of shielding constants for gaseous molecules needs several specific calculations. The choice of method suggests a few quantitative steps. Undoubtedly, the geometry optimization, basis sets, electron correlation, and rovibration effects are most important. The experimental conditions are fulfilled only when ZPV (zero-point vibration) and temperature effects are calculated. These kinds of calculations for 24 nuclei involved in 21 simple compounds were performed by Auer.[45] They show that ZPV corrections fluctuate between −2.8 ppm in the acetone molecule up to −15.7 ppm in oxetane C_3H_8O. Temperature effects are much smaller, of the order of 1 ppm, and can be of both signs.

In order to compare the experimental and those purely theoretical shieldings of ^{17}O nuclei, an isolated water molecule was chosen as practical reference and all experimental shieldings relative to it were recalculated from past measurements.[17,40–44] These values are shown in Figure 5.5 against the best theoretical results. The linear correlation is quite good (correlation coefficient 1.004) but significant overestimations of theoretical results is striking. The source of this discrepancy can be seen in systematic errors of theoretical calculations or experimental overestimation of reference shielding values. The last updated calculations of shielding scale were based on accurate rotational microwave data for $H_2{}^{17}O$ and $C^{17}O$ molecules. The best estimate shielding constants with relativistic corrections are: 328.4(3) and −59.05(59) ppm, respectively;[46] ~3 ppm higher than that presented in Table 5.3. This difference can partly explain the controversies mentioned above but it is clear that further experiments are needed.

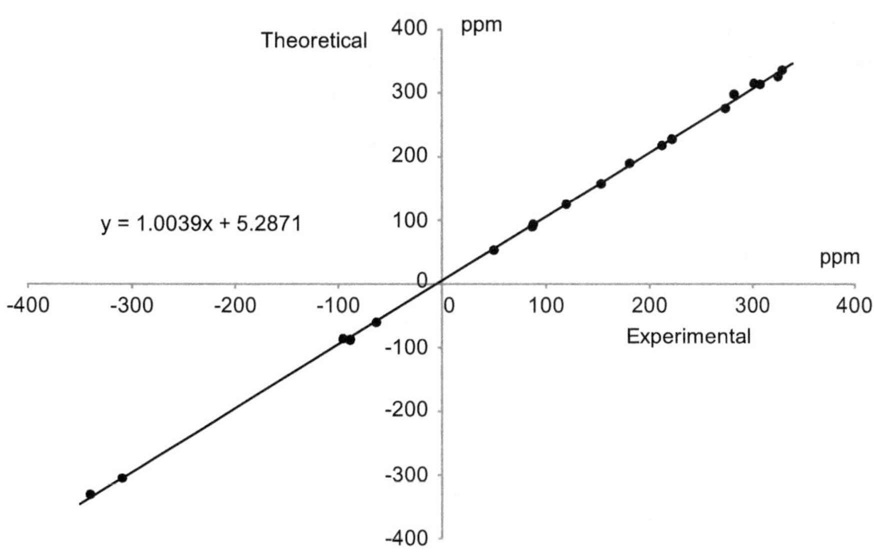

Figure 5.5 The correlation between experimental and theoretical ^{17}O shielding constants for organic and inorganic molecules.

5.4.4 $CO^{33}S$ as the Reference of Sulfur Shielding

Special attention should be paid to the carbonyl sulfide (COS) molecule, which can be a bridge between three kinds of resonances: ^{13}C, ^{17}O, and ^{33}S. As a CO molecule for ^{17}O NMR shielding scale, a COS molecule for ^{33}S plays a crucial role. Firstly Wasylishen[47] proposed the use of the value of 843(12) ppm as the ^{33}S shielding of an isolated COS molecule. This result was obtained using the spin-rotation constant for COS, measured by a molecular beam experiment, Flygare's procedure,[26] and the CHF estimation (Coupled Hartree–Fock) of the diamagnetic shielding in the free sulfur atom. Afterwards the diamagnetic contribution of a COS was calculated using the CCSD(T) method and a more accurate value 817(12) ppm was obtained.[48] A sulfur atom is terminal in COS molecule and the electric field gradient at the ^{33}S nucleus is rather large (0.1877). It gives a wide ^{33}S NMR signal of COS even for the pure liquid ($\Delta\nu_{1/2} \sim 440$ Hz) and leads to extremely bad conditions for studying this molecule in the gas phase. Nevertheless the use of an improved RIDE technique enabled good results to be obtained – the ^{33}S NMR signal was observed as a function of density.[18] It was the first observation of the ^{33}S signal in the gas phase of a substance at different pressures. The shieldings of a few secondary liquid references (liquid SF_6, CS_2, and SO_2) were measured relative to primary isolated molecule reference (COS) and the absolute shielding constants of sulfur nuclei for the liquid references were determined.[18]

Extensive knowledge of nuclear shieldings in a COS molecule needs determination of the ZPV and temperature effects. It should be done on a theoretical and an experimental basis if at all possible. The total temperature dependence for $^{12}C^{17}O^{32}S$, $^{13}C^{16}O^{32}S$, and $^{12}C^{16}O^{33}S$ isotopomers of the shielding constants were calculated using a few correlated wavefunctions.[48] The contribution of the zero-point-vibration motion is large in comparison to the part arising from vibrational excitations at 170–420 K temperature range. Different contributions due to bending and stretching modes are discussed here in detail. In the range of investigated temperatures (278–373 K) the ^{13}C shielding of COS is diminished by -0.104 ppm, -2.75, and -3.0 ppm for ^{13}C, ^{33}S, and ^{17}O, respectively. In this work, the equilibrium and rovibrationally averaged ^{17}O and ^{33}S nuclear quadrupole coupling constants were evaluated. They are $\chi(^{17}O) = -1.084$ MHz and $\chi(^{33}S) = -31.46$ MHz. Both values correspond well with the calculated and experimental results.

5.4.5 Uniqueness of the SF_6 Molecule

Among many sulfur compounds SF_6 makes an exception due to its electronic structure. In this highly symmetrical structure (O_h point group, octahedral molecular geometry), the equilibrium bond length was determined as $r_e = 1.5560(1)$ Å.[49] The consequence of this simple, symmetric structure is chemically inert and extraordinarily stable in the presence of most materials up to high temperatures ~200 °C. Its symmetry reduces the electric field gradient at the sulfur nucleus practically to zero, and its ^{33}S line width at half maximum height is only ~1 Hz. It permits the observation of a very precise sulfur resonance line of SF_6 in liquid and even gaseous state.

We have used this opportunity to observe density-dependent ^{33}S shielding in SF_6 ($\sigma_0 = 379.9(20)$ ppm).[50] The sulfur shielding was measured with high accuracy. It is in better agreement with the relativistic shielding result ($\sigma_0 = 392.6$ ppm) than with the non-relativistic one ($\sigma_0 = 350.0$ ppm).[51] The $^1J(^{33}S,^{19}F)$ was measured very precisely by INEPT sequence (see Section 5.7 of this chapter). Generally, the intermolecular effects on sulfur shielding in SF_6 are modest and arising from the central position of the S atom in this stable molecule.

5.4.6 Other Sulfur Containing Compounds

Only very scarce results from gas phase are known up to now. Besides SF_6 these are: COS ($\sigma_0 = 817(2)$ ppm) and SO_2 ($\sigma_0 = -152.5(20)$ ppm). This means that the ^{33}S screening parameters are established mainly for liquid substances and discussed in terms of chemical shifts (δ). As we see from the gas-to-liquid shifts, which are significant, this fact strongly limits the possibility of comparing experimental results with theory.

The special class of the species interesting from our point of view is sulfur-oxygen compounds. Sulfur can form many simple oxides like: SO, SO_2, SO_3, SO_4, S_2O, S_2O_2. Only two of them are stable in bulk condition, namely SO_2 and SO_3. The remaining substances are very reactive and unstable gases, hardly maintained as macroscopic samples. Only in the case of SO_2 could sulfur shielding be measured as a function of density.[41] The chemical shifts of liquid SO_3 were also presented. All ^{17}O and ^{33}S shielding constants for simple sulfur oxides were calculated at the DFT theory level by B3LYP, B971, and HCTH functionals used and displayed in Table 5.4.[52] The experimental results are also attached. The semiexperimental $\sigma_0(SO^{17}O) = -309(26)$ ppm

Table 5.4 ^{17}O and ^{33}S magnetic shielding constants calculated at the DFT theory levels[52] with respect to the experimental data.

		B3LYP	B971	HCTH	Experiment
SO_3	^{17}O	14.3	22.3	6.7	55.3[39]
	^{33}S	46.7	70.4	84.9	68.4[39]
SO_2	^{17}O	−299.9	−289.2	−290.1	−231.0,[39] −309.0[49]
	^{33}S	−286.3	−258.1	−235.0	−152.5,[39] −255.0[49]
SO	^{17}O	42.2	48.0	77.7	
	^{33}S	61.9	84.8	114.7	
S_2O	^{17}O	−545.7	−539.0	−501.6	
	^{33}S	−656.8	−634.8	−507.8	
		−772.4	−754.5	−616.3	
S_2O_2	^{17}O	−243.9	−238.9	−194.6	
	^{33}S	−563.7	−538.1	−409.1	

and $\sigma_0(^{33}SO_2) = -255(23)$ ppm are presented as received from calculations of diamagnetic shielding, while the paramagnetic term comes from the nuclear magnetic spin-rotation coupling constant measured by high-resolution microwave Fourier transform spectra.[53] The quality of DFT calculations can be estimated by comparing with experimental results for SO_2 and SO_3 molecules. The dependence of final results on the different class of functionals is visible here.

5.4.7 Intermolecular Interactions

A single experiment can never describe the complete gas phase behavior. Only the series of samples filled by gas at different pressures and densities can extract the σ_1 parameter, which is responsible for intermolecular interactions during bimolecular collisions (*cf.* eqn (5.1)). The coefficient σ_1 contains the bulk susceptibility correction σ_{1b} depending on the volume susceptibility of the medium substance χ_v and for a cylindrical tube parallel to the direction of external magnetic field $\sigma_{1b} = -(4\pi/3)\chi_v$. If this correction is subtracted from σ_1 factor, $\sigma_1 - \sigma_{1b}$ parameter is obtained, which is a sum of different kinds of intermolecular interactions. The σ_1 and $\sigma_1 - \sigma_{1b}$ parameters established for ^{17}O nuclei in simple molecules are presented in Table 5.3. They refer to the interactions of pure gaseous substances and buffer gases taken in excess in mixtures. Several simple observations can be made here. Firstly, it can be seen that almost all the coefficients are negative, which is observed as the universal behavior. The only exception is the SO_2 molecule. Characteristic changes in $\sigma_1 - \sigma_{1b}$ coefficients are observed in the series of smallest molecules CO, N_2O, CO_2, and COS (−119, −335, −432, and −720 ppm mL mol^{-1}). The increase of terms is connected not only with dipole moments of these species (0.112 D, 0.166 D, 0.000 D, and 0.715 D) but also with their polarizabilities and geometrical structures. The molecular interactions in dimethyl ether (dipole moment 1.30 D) change the oxygen shielding much more efficiently than in the previous molecules.

The density dependence of ^{33}S NMR shielding was measured for SF_6 molecule in pure gas and its binary mixtures with xenon, carbon dioxide, and ammonia.[50] The observed intermolecular effects ($\sigma_1 - \sigma_{1b}$) are fairly modest, *i.e.* −132(7), −225(20), −117(22), and −233(9) ppm mL mol^{-1}, respectively. It is worth mentioning that the same coefficient for fluorine nuclei in a pure SF_6 substance is more significant..[54,55] The second result of $\sigma_1 - \sigma_{1b}$ (^{33}S) was reported for pure sulfur dioxide gas and is equal to −11 650(700) ppm mL mol^{-1} at 333 K,[41] displaying strong sensitivity on intermolecular interactions. The interaction parameters are more substantial in peripheral atoms than for those located inside each molecule. The calculations of intermolecular shielding surfaces are needed to provide adequate insights into these effects. Because oxygen and sulfur compounds are large systems containing so many electrons these calculations were not feasible up to now.

5.5 Effects of Condensation

One of the best indicators of intermolecular interactions during condensation is the gas-to-solution shift or gas-to-solid state shift. The gas-to-liquid (solution) $\Delta\sigma_{GL}$ parameter has been evaluated according to the formula:

$$\Delta\sigma_{GL} = \delta_{LIQ} - \delta_{GAS} - (4\pi/3)\chi_v \tag{5.8}$$

where χ_v is the bulk susceptibility of a liquid compound.

The gas-to-liquid shifts reveal the influence of intermolecular effects when coming from gas to pure liquid state. In NMR spectroscopy mostly the diamagnetic substances are used and their volume susceptibility χ_v is negative but the measured $-(4\pi/3)\chi_v$ contribution is positive. The molar susceptibility values (χ_M) are determined by Gouy or Evan's balance[56] and the results are characterized by rather small accuracy. Accordingly, the errors of calculated $\Delta\sigma_{GL}$ parameters are rather substantial. The ^{17}O NMR gas-to-liquid shifts for a number of simple compounds are presented in Table 5.5. From the numerous values it is obvious that the intermolecular effects are significant and cannot be ignored; they vary from −38.22 ppm for water up to +24.3 ppm for acetaldehyde. Moreover, the intermolecular effects change

Table 5.5 ^{17}O NMR gas-to-liquid shifts in the series of simple chemical compounds (in ppm).

Molecule/Nucleus	$\Delta\sigma_{GL}$	Ref.
Water, $H_2\underline{O}$	−36.2	21
	−38.22	17
Fluorine monoxide, $F_2\underline{O}$	−18	33
Methyl acetate, $CH_3CO\underline{O}CH_3$	−8.6	38
Methyl formate, $HCO\underline{O}CH_3$	−8.6	38
Methanol, $CH_3\underline{O}H$	−8.3	42
Ethanol, $C_2H_5\underline{O}H$	−8.7	43
Methyl ether, $(CH_3)_2\underline{O}$	−5.19	40
(at 333 K)	−4.86	40
$(CD_3)_2\underline{O}$	−4.96	40
(at 333 K)	−4.67	40
Ethyl ether, $(C_2H_5)_2\underline{O}$	−2.5	38
Furan, $C_4H_4\underline{O}$	−2.4	38
Trimethylene oxide, $(CH_2)_3\underline{O}$	0.9	38
1,3-Dichlorotetrafluoroacetone, $(CF_2Cl)_2C\underline{O}$	2.1	38
Trifluoroacetic anhydride, $(CF_3C\underline{O})_2CO$	3.3	38
Methyl acetate, $CH_3C(\underline{O})OCH_3$	5.4	38
Methyl trifluoroacetate, $CF_3C(\underline{O})OCH_3$	6.3	38
Vinyl formate, $HC(\underline{O})OC_2H_3$	8.0	38
Ethyl formate, $HC(\underline{O})OC_2H_5$	8.1	38
Methyl formate, $HC(\underline{O})OCH_3$	9.2	38
Acetyl chloride, $CH_3C\underline{O}Cl$	11.0	38
Propionaldehyde, $C_2H_5CH\underline{O}$	16.1	38
Acetone, $(CH_3)_2C\underline{O}$	19.7	38
$(CD_3)_2C\underline{O}$	20.1	38
Acetaldehyde, $CH_3CH\underline{O}$	24.3	38

^{17}O shielding constants of different oxygen positions in the opposite directions. Generally, for compounds possessing carbonyl atoms an increase of shielding is observed whereas the hydroxyl group in water and alcohols is subjected to deshielding effects. The well-known deshielding effect in alcohols (for both the ^{1}H and ^{17}O resonances) is obviously connected with hydrogen-bonded liquid structure. Paradoxically, the strong positive shift in acetaldehyde can also be interpreted in terms of weak hydrogen bonds during C–H—O interactions.[57] The etherate bonded oxygen atoms (furan, methyl ether, ethyl ether) show relatively small negative effects, which means that these are weakly associated liquids.

The gas-to-liquid shift for water is of prime importance when the structure of liquid water is discussed in terms of NMR parameters. Additionally, this difference can be a source of the absolute shielding value taken for the reference standard, that is a liquid water sample. The gas-to-liquid shift for ^{17}O nucleus was measured by Florin and Alei[61] in the higher temperature sample. The very new results were obtained using the shielding value of an isolated water molecule, σ_0's instead of σ_{GL}'s: $\Delta\sigma_{GL}(^{17}O) = -38.22(5)$ ppm and $\Delta\sigma_{GL}(^{1}H) = -4.63(3)$ ppm.[17] These new experimental results constitute the best reference point for several theoretical calculations. They are presented in Table 5.6. There are different approaches used to model the gas-to-liquid shifts. It is known that the continuum model, which describes the electrostatic effects in dielectric solvent, is inadequate in this case.[62] Other approaches which involve molecular motions and using molecular dynamics or the supermolecule calculations are better suited for water simulation. The procedures involved consist in the optimization of the geometry for small aggregates and calculations of oxygen shielding changes upon increasing the number of molecules in clusters. Benchmark calculations of the shielding constants of the water dimer show significant deshielding effects on both oxygen nuclei involved (−1.32 ppm and −5.61 ppm at CCSD level) compared with a single molecule.[63]

The study[64] extended to the water clusters $(H_2O)_n$, $n = 2$–6, 12, and 17, gave the average environment-induced changes in the isotropic ^{17}O shielding from −15.5 ppm when $n = 3$ up to −24 ppm when $n = 6$. In the $(H_2O)_{17}$ size

Table 5.6 ^{1}H and ^{17}O gas-to-liquid shifts ($\Delta\sigma_{GL}$) of water in experiments and calculations (ppm).

Method	$\Delta\sigma(^{17}O)_{GL}$	$\Delta\sigma(^{1}H)_{GL}$	Ref.
Experimental			
	−36.1	−4.26	21
	−38.22	−4.338	17
Theory			
IGLO/HF/molecular dynamics	−24.3	−5.14	58
IGLO/DFT/MD	−41.2	−5.27	
MD/DFT/GGA (liquid)	−36.6	−5.83	59
$(H_2O)_{13}$	−37.6	−3.22	
QM/MM	−38.1	−2.91	60

cluster, modelled on the two coordination shells system, the largest shift is related to the adding of the first shell consisting of four molecules (−47.3 ppm), though the next shell contributes to a smaller extent (−6.8 ppm). This cluster can serve as a simple liquid water model. It is interesting that in the ^{17}O shielding for the central water monomer the deformation-induced shift is also observed. The appropriate ^{1}H and ^{17}O gas-to-liquid shift results from theory and experiment are collected in Table 5.6.

The second object investigated by measuring sulfur spectra in gaseous and liquid states was the SF_6 molecule. The appropriate spectra were registered in 13 assorted solvents and in the gas, extrapolated to the zero pressure limit. The $\Delta\sigma_{GL}(^{33}S)$ shifts are negative and vary from −0.57 ppm in hexafluorobenzene up to −2.49 ppm in diiodomethane solutions, *i.e.* the sulfur nucleus is deshielded in liquids.[65] The solvent effects in this case probably come from two sources: magnetic anisotropy of solvent molecules and dispersion interactions. They are approximately four times weaker than the ^{19}F NMR gas-to-solution shifts of the same solute. It is interesting that good correlation between both $\Delta\sigma_{GL}(^{33}S)$ and $\Delta\sigma_{GL}(^{19}F)$ results has been found.[65]

5.6 Isotope Effects on Chemical Shifts and Spin–Spin Coupling

When an isotope label is present in a molecule, every neighboring resonant nucleus may experience a slight modification of some NMR parameters. Several past reviews have described isotope effects on NMR parameters, *i.e.* chemical shifts (IECS) and spin–spin coupling constants (IECC).[66,67] The origin and general rules of both phenomena were given by Jameson.[68] IECS can be written in terms of the chemical shifts or nuclear magnetic shielding differences:

$$^{n}\Delta X(^{m/M}Y) = \delta_x(^{M}Y) - \delta_x(^{m}Y) = \sigma_x(^{m}Y) - \sigma_x(^{M}Y) \tag{5.9}$$

where X is the nucleus observed and ^{m}Y, ^{M}Y are the neighboring nuclei. In this convention the isotope effects are usually negative *i.e.* a heavier isotope produces a shift to the lower frequencies or to the higher field. They are called secondary isotope effects because the substitution sets in another place than the nucleus being measured.

5.6.1 Isotope Effects Observed on ^{17}O and ^{33}S Nuclei

As mentioned before, the very efficient relaxation of both nuclei in most cases prevents observation of precise NMR frequencies and then the subtle isotope effects, especially in the gas phase. The one exception is deuter/proton substitution in the hydroxyl group of methanol molecules where $^{1}\Delta^{17}O(H, D) = -1.85$ ppm is observed in the zero pressure limit.[43] This effect is strongly dependent on the density of buffer gases CHF_3 and CH_3F. The

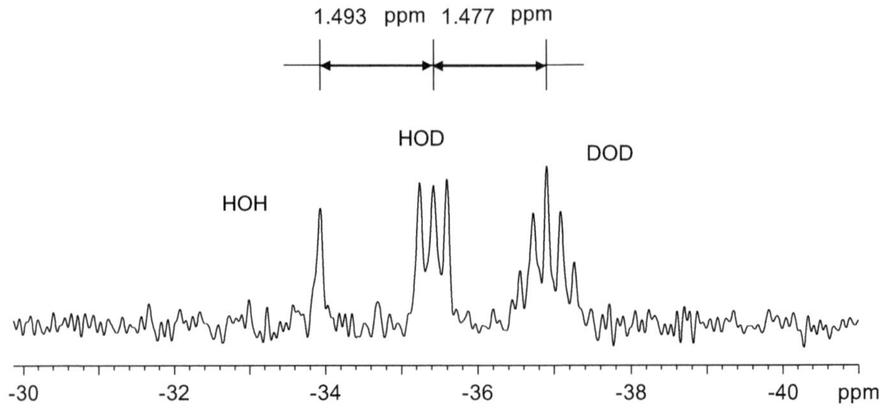

Figure 5.6 67.861 MHz ^{17}O spectrum of $H_2O/HDO/D_2O$ mixture in gaseous CH_3F at pressure ~34 atm (conc. 1.4 mol L^{-1}), and temperature 300 K.

same substitution in liquid methanol gives $^1\Delta^{17}O(H, D) = -1.60$ ppm. In the case of all substituted protons the isotope shift is even larger $^{1,2}\Delta^{17}O(CH_3OH, CD_3OD) = -2.81$ ppm. Assuming the additivity of isotope effects one can suppose that $^2\Delta^{17}O(CH_3OH, CD_3OH)$ and $^2\Delta^{17}O(CH_3OD, CD_3OD)$ are ~1.2 ppm (~0.4 ppm for one H/D substitution). These experimental results were later confronted with theoretical predictions of the rovibrational effects on chemical shifts.[69] Likewise in the series of liquid methyl ethers $(CH_3)_2O$, CH_3OCD_3, and $(CD_3)_2O$ the oxygen spectra show a distinct isotope effect $^2\Delta^{17}O(^{2/1}H) = -1.30(2)$ ppm and $-1.29(2)$ ppm, ^{17}O NMR signal shifts towards lower frequency-higher shielding for heavier isotopomers. Again the 0.43 ppm for one H/D atom exchange occurs.

It was for a long time of interest to determine isotope effects in water molecules in the gaseous state. The relative natural abundances of all isotopic water molecules are: $H_2{}^{16}O$ (99.78%), $H_2{}^{18}O$ (0.20%), $H_2{}^{17}O$ (0.038%), $HD^{16}O$ (0.0149%), $D_2{}^{16}O$ (0.022%), and $HT^{16}O$ (trace). $H_2{}^{17}O/D_2O$ mixed with an excess of fluorinated methane molecules: CH_4, CH_3F, CH_2F_2, CHF_3, and CF_4 was explored by ^1H NMR spectroscopy. Unfortunately, the low concentration of ^{17}O-enriched water in gaseous mixtures prevents the measurements of the $^1\Delta^{17}O$ effect in the full range of pressure used. The exception was 1.4 mol L^{-1} solution in gaseous CH_3F buffer (see Figure 5.6). The appropriate isotope effects found from this spectrum are $^1\Delta^{17}O(H_2O, HOD) = -1.493(10)$ ppm and $^1\Delta^{17}O(HOD, D_2O) = -1.477(10)$ ppm. So, small non-additivity is observed.

5.6.2 Isotope Effects Observed on Other Nuclei

Sometimes it is interesting to observe the presence of different isotopes of oxygen (^{16}O, ^{17}O, and ^{18}O) or sulfur (^{32}S, ^{33}S, ^{34}S, and ^{36}S) on other sorts of nucleus, for example ^1H or ^{19}F. The fluorine atom is especially well suited to measure some subtle frequency differences because of its large chemical shift range (1000 ppm), high sensitivity, and half-integer spin. The spectacular measurements of the isotope effects were presented for SF_6 molecule in the liquid phase when the very weak ^{19}F NMR signal of $^{36}SF_6$ (0.02%) isotopomer was observed for the first time.[70] All ^{19}F shieldings systematically increase from $^{32}SF_6$ through $^{33}SF_6$, $^{34}SF_6$ to $^{36}SF_6$. In this case the results for gaseous and liquid samples are almost the same, so the condensation process has negligible effect on the isotope effects measured.

5.7 Spin–Spin Coupling Involving ^{17}O or ^{33}S Nuclei

^{17}O and ^{33}S NMR signals are usually broad and the observation of J coupling constants appears rare. A new summary of J constants was given recently by Gerothanassis.[12] Older values can be found in Kintzinger's book.[71]

Due to the low natural abundance of ^{17}O, it is preferable to observe ^{17}O and not X in order to detect a $^{n}J(O, X)$ coupling. In addition to sensitivity considerations (which could be removed by isotopic enrichment), one also has to take into account the quadrupolar "washing out" of the coupling which occurs if the ^{17}O relaxation time is too short. For the coupling between ^{17}O and exchangeable hydrogen atoms, the rate of change must be minimized through pH and temperature control or by addition of electrolytes. Even in an optimum case, the ^{17}O quadrupolar broadening could be greater than the coupling constants. The coupling could then be extracted from a lineshape analysis. Different methods have been used depending on the nature of the problem: coupling with several protons, coupling with one nucleus of spin $\frac{1}{2}$, coupling with quadrupolar nuclei. The coupling constant $^{n}J(^{17}O, X)$ in simple molecules measured in different conditions, but preferably gaseous environments, together with some reduced values $^{n}K(^{17}O, X)$ are given in Table 5.7.

Many attempts have been made to avoid the proton exchange in liquids and precisely establish the $^{1}J(^{17}O, ^{1}H)$ spin–spin coupling. Usually, the small and controlled amount of water in different solvents was used to having a system free from hydrogen bonding. The potential source of errors is proton exchange processes present at higher concentrations. For the first time this coupling was observed in the water vapor in equilibrium with liquid at elevated temperatures 175 and 215 °C.[72] The vapor ^{17}O signal was a well-resolved 1 : 2 : 1 triplet due to proton coupling but of very limited signal to noise ratio (see Table 5.7). The best results were so far achieved for pure gaseous water vapor (20% ^{17}O enrichment) mixed with CH_3F and CF_3H.[17] The splitting signals were observed in ^{1}H (sextet) and ^{17}O (triplet) NMR spectra; these last spectra were used to extract spin–spin coupling. The density dependence coefficient in the range 100–200 Hz mL mol^{-1} was observed and average $^{1}J(^{17}O, ^{1}H)$ free from intermolecular interactions was measured as 78.22(10) Hz at 300 K. This result significantly differed from experimental values observed in ammonia[72] and in neat liquid[73] and only slightly from that observed in organic solutions.[74–76] The sign of $^{1}J(^{17}O, ^{1}H)$

Table 5.7 $^{n}J(^{17}O, X)$ spin–spin coupling constants of small molecules in experiment.

Compound	Conditions	$^{n}J(^{17}O, X)$, Hz	$^{n}K(^{17}O, X) \times 10^{19} J^{-1} T^{-1}$	Ref.
HOH	Liquid vapor equilibrium, 488 K	−79		72
	Gas phase, extrapolated	−78.22	48.02	17
	0.1% in cyclokaxane-d_{12}, 293 K	−78.7		74
	0.5 mol% in nitromethane-d_3	−80.6		75
	1.6 mol% in nitromethane-d_3, 300 K	−81.07		76
	343 K	−80.36		
	Liquid	−89.8	55.1	73
	10 mol% in ammonia	−80		72
CH_3OH	Gas phase, extrapolated	−79.45	48.77	43
	Liquid	−85.5	52.5	77
C_2H_5OH	Gas phase, extrapolated	−78.48	48.18	44
	Liquid	−83.6	51.3	77
$(CH_3)_2CO$	Liquid	22		78
$HCOOCH_3$	Liquid	38.1	23.6	79
$HCOOCH_3$		7.5	4.6	
$HCOOCH_3$		10.5	6.5	
^{13}CO	In chloroform-d	16.4	40	80
	In cyclohexane	16.44		54
	Gas at 35 atm	11.95		
$O^{13}CO$	In chloroform-d	16.1	39.3	80
$^{13}CH_3{}^{13}CHO$	Liquid	28.9	70.5	54
$^{15}N^{15}NO$	Gas at 50 atm	51.47	311.6	54
$^{14}N^{14}NO$	Supercritical state	37	314.2	81
	In CD_3CN	35.5		82
FOF	Liquid at 133 K	300(30)	196	47
FOOF		424	277	83
OOO		101.5	459	84

is known to be negative. Just recently the $^{1}J(^{17}O, ^{1}H)$ in methanol[43] and ethanol[44] were also extrapolated to zero-point of density of buffer gases CH_3F, CHF_3, and Xe, SF_6, respectively. The resonance lines were measured in the normal coupled mode ^{17}O NMR spectra with small relaxation delay time (0.01 s). The signal-to-noise ratio was more than 25 in each spectrum. The convolution procedure was necessary to receive proper spin–spin coupling values (see Figure 5.7). The influence of geminal $^{2}J(^{17}O, ^{1}H)$ spin–spin coupling was neglected here. A small diminishing effect with reducing density can be seen in the case of the three mentioned couplings. The $^{1}J(O, H)$ measured and convoluted in gaseous sample of methanol is shown in Figure 5.7. It is significantly lower than measured in liquids before.

When spin–spin couplings are calculated, the four terms mentioned above should be considered (FC, PSO, SO, and DSO). The new theoretical results for $^{1}J(O, H)$ in water molecule are presented in Table 5.8.

In the case of water, $^{1}J(OH)_{ZPV}$ correction is $^{1}J_{rovib.} = -4.58$ Hz[85] or -4.34 Hz,[86] whereas in the CO molecule $^{1}J(OC)_{rovib.}$ is 0.58 Hz.[87] The best calculation of $^{1}J(OH) = -78.52$ Hz in water,[88] which is corrected for

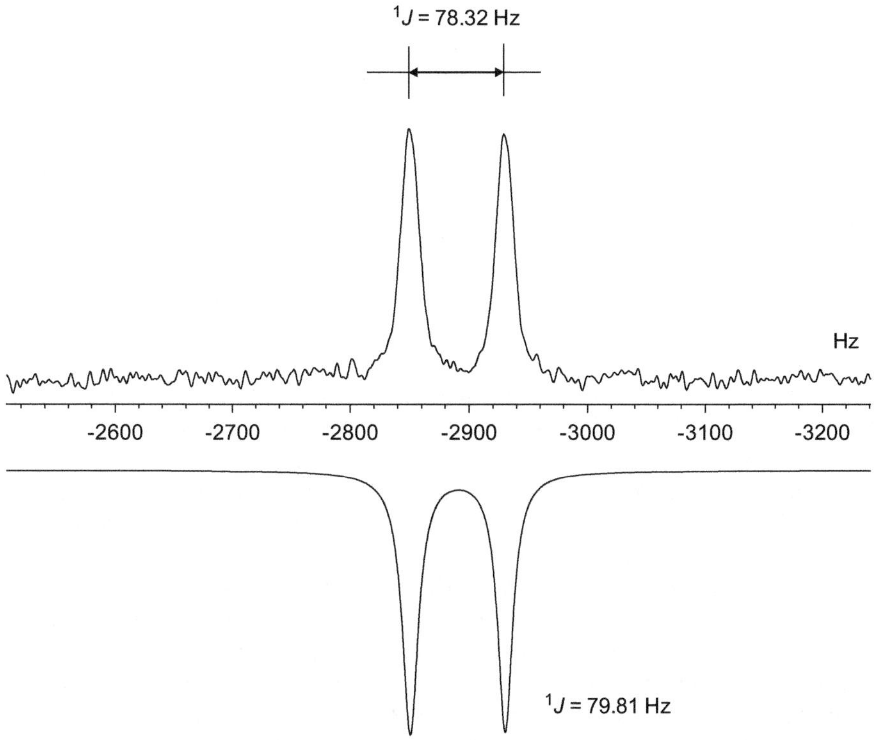

Figure 5.7 Experimental and convoluted ^{17}O NMR spectrum of $CH_3{}^{17}OH$ dissolved in CH_3F at high pressure (30.41 atm).

Table 5.8 Calculated isotropic values of different contributions to the $^1J(^{17}O, X)$ in oxygen-containing molecules.

Molecule/Method	J(FC)	J(OP)	J(SD)	J(OD)	J(O, X)	Ref.
H_2O						
CC3	−66.12	−11.79	−0.57	−0.04	−78.52	88
SOPPA/CCSD	−69.092	−11.943	−0.485	−0.035	−81.55	86
MCRPA	−72.083	−11.451	−0.411	−0.034	−83.934	
SOPPA	−70.34	−11.56	−0.47	−0.05	−82.42	89
SOPPA/CCSD	−68.56	−11.51	−0.47	−0.05	−80.6	
DFT	−64.2	−12.8	−0.7	−0.2	−77.7	
	−65.5	−9.1	−0.2	−0.1	−74.9	
CO						
CC3	6.92	13.1	−4.82	0.1	15.3	88
SOPPA	10.2	14.42	−4.31	0.1	20.41	89
SOPPA/CCSD	8.76	14.11	−4.37	0.1	18.6	
	9.0	12.8	−5.7		17.2	90
CO_2						
SOPPA/CCSD	14.8	5.0	−2.7		17.1	90

rovibrational contributions, agrees very well with our experimental value; the difference is less than 0.5%. As usual for one-bond spin–spin couplings the Fermi-contact term makes the main part. The PSO term is also important. Interestingly, for ^{1}J(CO) in carbon oxide, the situation is quite different. The paramagnetic spin–orbit term is dominating because of multiple C–O bonding. For comparison, the results of calculations of individual terms to one bond spin–spin couplings in carbon oxides CO and CO_2 are included in Table 5.8.

^{2}J(O, X) spin–spin couplings are small and were obtained from ^{17}O line shape analysis.[79] The exceptionally high value of J coupling of methyl formate is in agreement with the general theory of geminal coupling constants between nuclei adjacent to a carbonyl group.

Great difficulty encountered in measuring the spin–spin couplings arises in particular when both coupled nuclei are quadrupolar. A good example is the nitrous oxide molecule in the main natural abundance isotopomer where both nitrogen atoms are ^{14}N isotopes with spin number $I=1$. The ^{14}N–^{17}O spin–spin coupling $^{1}J(^{14}\text{N}, ^{17}\text{O})=37$ Hz was measured for the first time in the supercritical state[81] where a significant drop in viscosity was achieved and relatively narrow resonance lines in ^{17}O spectrum with $\Delta_{1/2} \sim 11$ Hz were observed. Meticulous inspection of this spectrum shows that $^{2}J(^{14}\text{N}, ^{17}\text{O})$ should be less than 5 Hz. The supercritical phases did not become a general experimental method in NMR spectroscopy on account of the drastic pressure and temperatures required. Afterwards the one-bond ^{14}N–^{17}O spin–spin coupling in nitrous oxide $^{1}J(^{14}\text{N}, ^{17}\text{O})=35.8(3)$ Hz value was measured in CD_3CN solution by fitting the badly resolved triplet with the aid of the QUADR program taking into consideration the fixed spin–lattice relaxation time.[91]

Sometimes the isotope labelling of a particular molecule can solve the problem of spin–spin coupling measurements. The $^{1}J(^{15}\text{N}, ^{17}\text{O})$ coupling may be approximately estimated on the basis of the gyromagnetic ratios $\gamma(^{14}\text{N})/\gamma(^{15}\text{N})=-1.40276$. This gives the approximate value $^{1}J(^{15}\text{N}, ^{17}\text{O})=50.2$ Hz, which is in good agreement with our recent measurement of ^{17}O NMR doublet peak in a 50 atm sample of $^{15}\text{N}^{15}\text{NO}$ isotopomeric form of nitrous oxide $^{1}J(^{15}\text{N}, ^{17}\text{O})=51.47(50)$ Hz (see Figure 5.8).

Up to now only one ^{n}J(S, X) spin–spin coupling was measured in the gas phase, the one-bond ^{1}J(S, F) in SF_6 molecule.[47,50] It has a negative sign. This value is almost constant across the pressure domain in gaseous pure sulfur hexafluoride at 7–25 atm as shown in Figure 5.9. The first virial coefficient, when linear approximation of function eqn (5.2) is taken into account, is small but unexpectedly positive $^{1}J(\text{SF})=+58.1$ Hz ml^{-1} mol^{-1}. Similar behavior was observed in the case of the $^{1}J_1(^{11}\text{B}, ^{19}\text{F})=+154$ Hz ml^{-1} mol^{-1} term for ^{1}J(B, F) spin–spin coupling measured in the gaseous boron trifluoride BF_3 at 300 K.[92] This value is due to the intermolecular interactions of SF_6 molecules during bimolecular collisions in gas.

At the same time this coupling was explored in several different chemical environments, *i.e.* in liquids, in TLC (thermotropic liquid crystals), or in supercritical state.[49] Normal behavior is observed; the spin–spin coupling is slightly greater (as absolute value) in liquids than in the gaseous state and

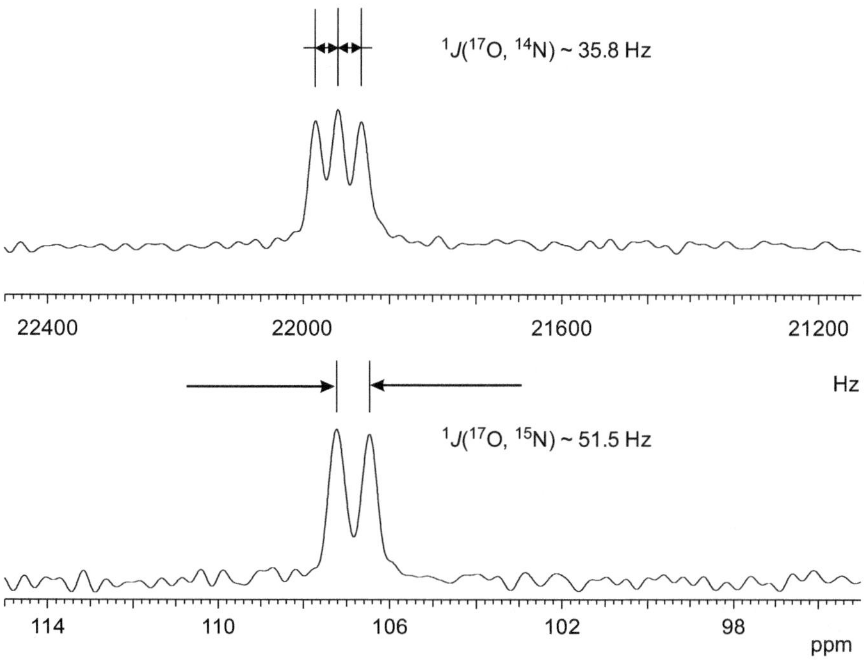

Figure 5.8 67.861 MHz ^{17}O NMR signals of $^{14}N_2O$ and $^{15}N_2O$ nitrous oxide molecules obtained from high-pressure samples ($\sim$50 atm). Spin–spin splitting due to neighbor ^{14}N ($I=1$) and ^{15}N ($I=1/2$) nuclei are well seen.

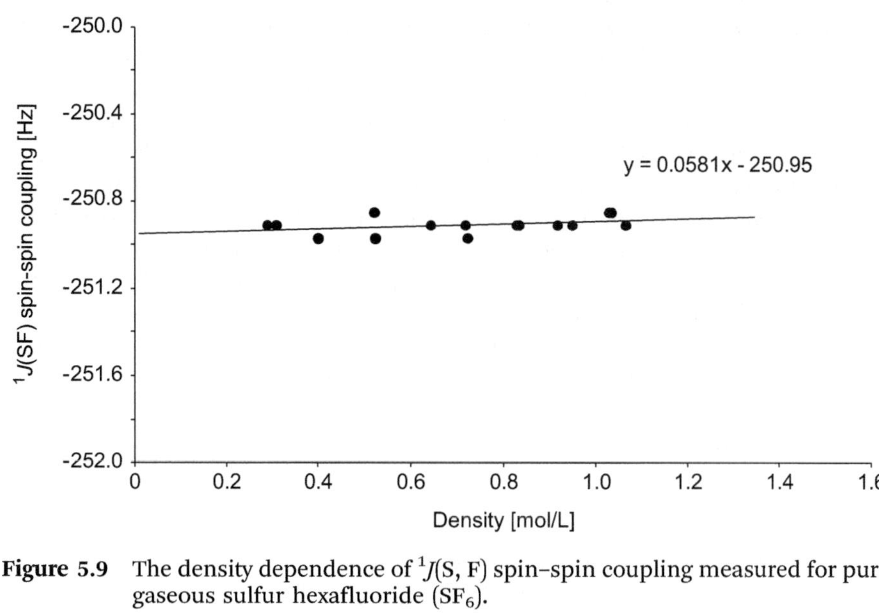

Figure 5.9 The density dependence of ^1J(S, F) spin–spin coupling measured for pure gaseous sulfur hexafluoride (SF_6).

Table 5.9 Theoretical and experimental ^1J(S, F) spin–spin coupling in sulfur hexafluoride (SF_6) molecule.

Experiment[49]		Theory[87]	
Gas phase (isolated molecule)	−250.95	DFT/UGBS2P	−320.46
Liquid (298 K)	−251.6	DFT/contracted TZ	−272.49
Liquid (223 K)	252.0(2), 251.8(2)[46]	DFT/uTZ	−306.91
Supercritical fluid	−251.4	DFT/uTZ-w	−316.49
In carbon disulfide	−253.08	DFT/uTZ-wd2	−316.00
In ethanol	−253.48	DFT/uTZ-wd4	−315.86
In *n*-pentane	−253.37	DFT/contracted G	−304.67
In *n*-hexane	−253.38	DFT/uG	−294.82
In hexafluorobenzene	−253.38	DFT/uG-w	−308.60
In acetonitrile	−253.38	DFT/uG-wd2	−308.95
In hexamethylodisiloxane	−253.38		
In acetone	−253.78		
In tetrachloromethane	−253.98		
In dimethylformamide	−254.00		
In cyclohexane	−254.08		
In benzene	−255.18		
In dimethylsulfoxide	−255.19		
In methyl iodide	−255.74		
In ZLI 2806 and HAB	−253.3[93]		
In ZLI 3125	−253.9[93]		
In Phase 4	−255.5[93]		

the coupling for supercritical state places itself just between these two results. The ^1J(S, F) values established for the TLC[93] are very similar to those measured in different isotropic liquids. Also, a small non-linear effect was observed when the temperature of the liquid SF_6 sample was reduced up to −252 Hz at 245 K. Finally, it is clear that the indirect spin–spin coupling between the ^{33}S and ^{19}F nuclei in SF_6 is insensitive to its chemical environment and changes not more than 2–3% of its gaseous value.

The experimental and theoretical results for ^1J(S, F) were collected in Table 5.9. The DFT/B3LYP calculations using the different modifications of basis sets were performed.[87] All the results seem to be underestimated; the total spin–spin coupling originates mainly from FC and PSO contributions. Both are of negative sign and also give the ^1J(SF) negative. A proper comparison with the experiment cannot be fully made because of the unknown rovibrational effects. Nevertheless, it is clear that they are far from experimental expectations and need more advanced calculations. The basic results show promise for sulfur hexafluoride use in the interpretation of the new experimental technique that involved inclusions into solid state and inhalation in MRI medical treatments.[93]

5.8 Summary

^{17}O and ^{33}S NMR are less popular than many other similar methods of NMR spectroscopy. It is because of the low receptivity of resonance signals from

^{17}O and ^{33}S nuclei and the appearance of nuclear electric quadrupole moments. Nonetheless oxygen and sulfur are such important elements in chemistry that the application of ^{17}O and ^{33}S NMR is systematically growing due to the availability of modern high-field spectrometers. The development of ^{17}O and ^{33}S spectroscopies requires basic information on appropriate spectral parameters and such data are available from gas phase NMR measurements.

An effort has been made to summarize recent progress in experimental and theoretical data of ^{17}O and ^{33}S NMR spectroscopy achieved in the gas phase within the last two decades. ^{17}O density-dependent magnetic shielding was observed for numerous chemical compounds and oxygen shielding parameters were collected for important molecules like water, methyl alcohol, and many others. It includes the shielding of isolated molecules (σ_0) and the second virial coefficient (σ_1), which is observed as the result of intermolecular interactions in the gas phase. The measurements were often extended on the deuterium isotopomers of chemical compounds, which allowed the determination of isotope effects in oxygen shielding. Similar measurements for ^{33}S nucleus were done for a few available chemical compounds. The most complete results were obtained for sulfur hexafluoride where the incredible narrow signal permitted the precise observation of sulfur shielding and spin–spin splitting. The spin–spin couplings involving oxygen and sulfur were measured in many available compounds. New original results from gas phase NMR measurements are presented, *e.g.* the density-dependent $^1J(S, F)$ in gaseous sulfur hexafluoride and $^1J(C, O)$ in carbon monoxide. Several theoretical computations were accomplished to predict the above experimental data. As the result of joint experimental and theoretical studies the absolute shielding scales and the nuclear magnetic moments $\mu(^{17}O)$ and $\mu(^{33}S)$ were established with good accuracy.

The further development of ^{17}O and ^{33}S NMR experimental methods is inevitable and more valuable results are also expected from the gas phase. Upcoming progress can be achieved by ultrafast NMR spectroscopy with cold probe technology, which makes the experiment 4 to 16 times shorter than in the conventional procedure. The higher accessible magnetic fields up to $\sim$20 T (1 GHz NMR systems) and the new pulse sequences developing polarization methods will be important. The new sequences known from solid state, like spin-lock in quadrupole nuclei, can be implemented to the high resolution spectroscopy.

Acknowledgements

This work was financially supported by the National Science Centre (Poland) grant, according to the decision No. DEC-2011/01/B/ST4/06588.

References

1. W. T. Raynes, Early Work on Gas-Phase Chemical Shifts, in *Encyclopedia of NMR*, ed. D. M. Grant and R. K. Harris, John Wiley, Chichester, 1996, p. 571.

2. C. J. Jameson, *Chem. Rev.*, 1991, **91**, 1375–1395.
3. C. J. Jameson and A. C. de Dios, Theoretical and Physical Aspects of Nuclear Shielding, in *Specialist Periodical Reports-Nuclear Magnetic Resonance*, RSC Press, 2013, vol. 42, pp. 45–77.
4. N. S. True, Gas Phase Applications of NMR Spectroscopy, in *Encyclopedia of Spectroscopy and Spectrometry*, ed. J. C. Lindon, Academic Press, 2nd edn, 2010, pp. 749–756.
5. C. Suarez, *Chem. Educ.*, 1998, **3**, 1–18.
6. C. B. LeMaster, *Prog. Nucl. Magn. Reson. Spectrosc.*, 1997, **31**, 119–154.
7. K. Jackowski, *J. Mol. Struct.*, 2001, **563–4**, 159–162.
8. K. Jackowski, *Int. J. Mol. Sci.*, 2003, **4**, 135–142.
9. *CRC Handbook of Chemistry and Physics*, ed. D. R. Lide, CRC Press, Boca Raton, 2000–2001, vol. 81.
10. N. J. Stone, *At. Data Nucl. Data Tables*, 2005, **90**, 75–176.
11. J. P. Gerothanassis, *Prog. Nucl. Magn. Reson. Spectrosc.*, 2010, **56**, 95–197.
12. J. P. Gerothanasis, *Prog. Nucl. Magn. Reson. Spectrosc.*, 2010, **57**, 1–110.
13. J. F. Hinton and D. Buster, *J. Magn. Reson.*, 1984, **58**, 324–327.
14. R. K. Harris, E. D. Becker, S. M. Cabral de Menezes, P. Granger, R. E. Hoffman and K. W. Zilm, *Pure Appl. Chem.*, 2008, **80**, 59–84.
15. *^{17}O NMR Spectroscopy in Organic Chemistry*, ed. D. W. Boykin, CRC Press, Boca Raton, 1991.
16. A. D. Buckingham and J. A. Pople, *Discuss. Faraday Soc.*, 1956, **22**, 17–21.
17. W. Makulski, M. Wilczek and K. Jackowski, to be published.
18. K. Jackowski, W. Makulski and W. Koźmiński, *Magn. Reson. Chem.*, 2002, **40**, 563–565.
19. K. Jackowski, M. Jaszuński and W. Makulski, *J. Magn. Reson.*, 1997, **27**, 139–143.
20. *Calculation of NMR and EPR Parameters. Theory and Applications*, ed. M. Kaupp, M. Bühl and V. G. Malkin, Wiley-VCH, Weinheim, 2004.
21. T. Helgaker, M. Jaszuński and K. Ruud, *Chem. Rev.*, 1999, **99**, 293–352.
22. J. Vaara, *Phys. Chem. Chem. Phys.*, 2007, **9**, 5399–5418.
23. A. Bagno and G. Saielli, *Theor. Chem. Acc.*, 2007, **117**, 603–619.
24. J. C. Facelli, *Concepts Magn. Reson., Part A*, 2004, **20**, 42–69.
25. N. F. Ramsey, *Phys. Rev.*, 1950, **78**, 699–703.
26. W. H. Flygare, *Chem. Rev.*, 1974, **74**, 653–687.
27. J. B. M. Kellogg and S. Millman, *Rev. Mod. Phys.*, 1946, **18**, 323–352.
28. A. Antušek, K. Jackowski, M. Jaszuński, W. Makulski and M. Wilczek, *Chem. Phys. Lett.*, 2005, **411**, 111–116.
29. M. Jaszuński, A. Antušek, P. Garbacz, K. Jackowski, W. Makulski and M. Wilczek, *Prog. Nucl. Magn. Reson. Spectrosc.*, 2012, **67**, 49–63.
30. K. Jackowski, M. Jaszuński, B. Kamieński and M. Wilczek, *J. Magn. Reson.*, 2008, **193**, 147–149.
31. K. Jackowski, M. Jaszuński and M. Wilczek, *J. Phys. Chem. A*, 2010, **114**, 2471–2475.

32. T. Helgaker, M. Jaszuński and M. Pecul, *Prog. Nucl. Magn. Reson. Spectrosc.*, 2008, **53**, 249–268.
33. R. E. Wasylishen, S. Mooibroek and J. B. Macdonald, *J. Chem. Phys.*, 1984, **81**, 1057–1059.
34. D. Sundholm, J. Gauss and A. Schäfer, *J. Chem. Phys.*, 1996, **105**, 11051–11059.
35. J. Vaara, J. Lounila, K. Ruud and T. Helgaker, *J. Chem. Phys.*, 1998, **109**, 8388–8397.
36. G. Cazzoli, L. Dore, C. Puzzarini and S. Beninati, *Phys. Chem. Chem. Phys.*, 2002, **4**, 3575–3577.
37. R. E. Wasylishen and D. L. Bryce, *J. Chem. Phys.*, 2002, **117**, 10061–10066.
38. W. Makulski and K. Jackowski, *J. Mol. Struct.*, 2003, **651–653**, 265–269.
39. C. Puzzarini, G. Cazzoli, M. E. Harding, J. Vázquez and J. Gauss, *J. Chem. Phys.*, 2009, **131**, 234304.
40. W. Makulski and K. Jackowski, *Chem. Phys. Lett.*, 2001, **341**, 369–372.
41. W. Makulski and K. Jackowski, *J. Mol. Struct.*, 2004, **704**, 219–222.
42. W. Makulski, *J. Mol. Struct.*, 2005, **744–747**, 439–446.
43. W. Makulski, *J. Mol. Struct.*, 2008, **872**, 81–86.
44. W. Makulski, M. Godlewski and K. Jackowski, to be published.
45. A. A. Auer, *J. Chem. Phys.*, 2009, **131**, 024116.
46. S. Komorovsky, M. Repisky, E. Malkin, K. Ruud and J. Gauss, *J. Chem. Phys.*, 2015, **142**, 091102.
47. R. E. Wasylishen, C. Connor and J. O. Friedrich, *Can. J. Chem.*, 1984, **62**, 981–985.
48. K. Jackowski, M. Jaszuński, W. Makulski and J. Vaara, *J. Magn. Reson.*, 1998, **135**, 444–453.
49. V. Boudon, J. L. Doménech, D. Bermejo and H. Willner, *J. Mol. Spectrosc.*, 2004, **228**, 392–400.
50. K. Jackowski, M. Wilczek, W. Makulski and W. Koźmiński, *J. Phys. Chem. A*, 2002, **106**, 2829–2832.
51. K. Ruud, T. B. Demissie and M. Jaszuński, *J. Chem. Phys.*, 2014, **140**, 194308.
52. A. Tulewicz, *Theoretical investigations of sulfur trioxide dimers and trimmers structures*, Master's thesis, Faculty of Chemistry, University of Warsaw, 2010.
53. H. S. P. Müller, J. Farhoomand, E. A. Cohen, B. Brupbacher-Gatehouse, M. Schäfer, A. Bauder and G. Winnewisser, *J. Mol. Spectrosc.*, 2000, **201**, 1–8.
54. W. Makulski, unpublished.
55. C. J. Jameson, A. K. Jameson and S. M. Cohen, *J. Chem. Phys.*, 1977, **67**, 2771–2778.
56. *Magnetic Susceptibility Measurements*, ed. C. J. O'Connor and S. J. Lippard, Progress in Inorganic Chemistry, Wiley, 1982, vol. 29, p. 203.
57. M. A. Blatchford, P. Raveendran and S. L. Wallen, *J. Am. Chem. Soc.*, 2002, **124**, 14818–14819.

58. T. S. Pennanen, J. Vaara, P. Lantto, A. J. Sillanpää, K. Laasonen and J. Jokisaari, *J. Am. Chem. Soc.*, 2004, **126**, 11093–11102.
59. B. G. Pfrommer, F. Mauri and S. G. Louie, *J. Am. Chem. Soc.*, 2000, **122**, 123–129.
60. E. E. Fileti, H. C. Georg, K. Coutinho and S. Canuto, *J. Braz. Chem. Soc.*, 2007, **18**, 74–84.
61. A. E. Florin and M. Alei, *J. Chem. Phys.*, 1967, **47**, 4268–4269.
62. M. Cossi and O. Crescenzi, *J. Chem. Phys.*, 2003, **118**, 8863–8872.
63. M. Pecul, J. Lewandowski and J. Sadlej, *Chem. Phys. Lett.*, 2001, **333**, 139–145.
64. H. Cybulski and J. Sadlej, *Chem. Phys.*, 2006, **323**, 218–230.
65. M. Wilczek, W. Makulski and K. Jackowski, *Mol. Phys. Rep.*, 2000, **29**, 183–185.
66. P. E. Hansen, *Prog. Nucl. Magn. Reson. Spectrosc.*, 1988, **20**, 207–255.
67. D. B. Davies, J. C. Christofides and R. E. Hoffman, Isotope Shifts in NMR Spectroscopy – Measurements and Applications, in *Isotopes: Essential Chemistry and Applications*, ed. J. R. Jones, RSC Press, 1988.
68. C. J. Jameson, Gas Phase Studies of Intermolecular Interactions and relaxation, in *Encyclopedia of NMR*, ed. R. K. Harris and R. E. Wasylishen, Wiley, Chichester, 2nd edn, 2007, vol. 4, p. 2197.
69. A. A. Auer, *Chem. Phys. Lett.*, 2009, **467**, 230–232.
70. W. Makulski, *Mol. Phys. Rep.*, 2001, **33**, 82–86.
71. J. P. Kintzinger, Oxygen NMR, Characteristic Parameters and Applications, in *NMR Basic Principles and Progress*, ed. P. Diehl, E. Fluck and R. Kosfeld, Springer, Berlin, 1981, vol. 17, pp. 1–64.
72. M. Alei and A. E. Florin, *J. Phys. Chem.*, 1969, **73**, 863–864.
73. R. E. Wasylishen and J. O. Friedrich, *Can. J. Chem.*, 1987, **65**, 2238–2243.
74. N. M. Sergeyev, N. D. Sergeyeva, Yu. A. Strelenko and W. T. Raynes, *Chem. Phys. Lett.*, 1997, **277**, 142–146.
75. N. M. Sergeyev, N. D. Sergeyeva and W. T. Raynes, *J. Magn. Reson.*, 1999, **137**, 311–315.
76. L. J. Burnett and A. H. Zeltmann, *J. Chem. Phys.*, 1974, **60**, 4636–4637.
77. H. Versmold and C. Yoon, *Bunsen-Ges. Phys. Chem., Ber.*, 1972, **76**, 1164–1168.
78. M. Broze and Z. Luzz, *J. Phys. Chem.*, 1969, **73**, 1600–1602.
79. C. Delseth, J. P. Kintzinger, T. T. T. Nguyen and W. Niederberger, *Org. Magn. Reson.*, 1978, **11**, 38.
80. R. E. Wasylishen, J. O. Friedrich, S. Mooibroek and J. B. Macdonald, *J. Chem. Phys.*, 1985, **83**, 548–551.
81. J. M. Robert and R. F. Evilia, *J. Am. Chem. Soc.*, 1985, **107**, 3733–3735.
82. Yu. A. Strelenko and N. M. Sergeyev, *J. Mol. Struct.*, 1996, **378**, 61–65.
83. I. J. Solomon, J. N. Keith, A. J. Kacmarek and J. K. Raney, *J. Am. Chem. Soc.*, 1968, **90**, 5408–5411.
84. A. Dimitrov, K. Seppelt, D. Scheffler and H. Willner, *J. Am. Chem. Soc.*, 1998, **120**, 8711–8714.

85. J. Casanueva, J. San Fabián, E. Díez and A. L. Esteban, *J. Mol. Struct.*, 2001, **565–6**, 449–454.
86. R. D. Wigglesworth, W. T. Raynes, S. P. A. Sauer and J. Oddershede, *Mol. Phys.*, 1998, **94**, 851–862.
87. W. Deng, J. R. Cheeseman and M. J. Frisch, *J. Chem. Theory Comput.*, 2006, **4**, 1028–1037.
88. A. A. Auer and J. Gauss, *J. Chem. Phys.*, 2001, **115**, 1619–1622.
89. T. Enevoldsen, J. Oddershede and S. P. A. Sauer, *Theor. Chem. Acc.*, 1998, **100**, 275–284.
90. B. Wrackmeyer, *Z. Naturforsch.*, 2004, **59b**, 286–290.
91. Yu. A. Strelenko, V. N. Torocheshnikov and N. M. Sergeyev, *J. Magn. Reson.*, 1990, **89**, 123–128.
92. K. Jackowski, W. Makulski, A. Szyprowska, A. Antušek and M. Jaszuński, *Magn. Reson. Chem.*, 2009, **47**, 857–861.
93. H. Tervonen, J. Saunavaara, L. P. Ingman and J. Jokisaari, *J. Phys. Chem. B*, 2006, **110**, 16232–16238.

CHAPTER 6

Accurate Non-relativistic Calculations of NMR Shielding Constants

ANDREJ ANTUŠEK[a] AND MICHAŁ JASZUŃSKI*[b]

[a] Slovak University of Technology in Bratislava, ATRI, Faculty of Materials Science and Technology in Trnava, Paulinska 16, 917 24 Trnava, Slovak Republic; [b] Institute of Organic Chemistry, Polish Academy of Sciences, Kasprzaka 44/52, 01-224 Warsaw, Poland
*Email: michal.jaszunski@icho.edu.pl

6.1 Introduction

In 1950, in a study of ammonium nitrate undertaken to determine the nuclear magnetic dipole moment of ^{14}N, Proctor and Yu[1] observed that there is a "first resonance followed by a second one of equal amplitude". It was attributed[2] to "some nasty chemical phenomenon ... which could terribly impede our progress in trying to measure the magnitude of nuclear magnetic moments". Another article published in the same issue of Physical Review by Dickinson[3] was entitled "Dependence of the F^{19} Nuclear Resonance Position on Chemical Compound". As stated by Proctor and Yu "...this effect is almost twice as large as the total diamagnetic correction calculated for the atom. These calculations, however, do not hold for the polyatomic molecules which we have studied...". Today this effect – shielding of the nuclei by the electrons, dependent on the molecular structure – would not be called *nasty*, although indeed it complicates the determination of nuclear magnetic moments. In particular, to obtain accurate values of these

New Developments in NMR No. 6
Gas Phase NMR
Edited by Karol Jackowski and Michał Jaszuński

Published by the Royal Society of Chemistry, www.rsc.org

moments from NMR spectra quantum chemical calculations of the shielding constants in the sample polyatomic molecules are required.

Ab initio methods of quantum chemistry are nowadays widely used to determine spectroscopic parameters such as NMR shielding constants. In this chapter, we give a short account of the underlying theory, and briefly overview the problems arising in the implementation of different methods. In the following discussion of the applications the emphasis is on the accuracy of the calculated shielding constants. We present several examples demonstrating that for small molecules state-of-the art theoretical methods may be used in spite of their relatively high computational cost and, moreover, one can estimate the error bars of the results.

We shall focus on the theoretical studies of the shielding constants. However, one should keep in mind that the molecular property of interest is in fact a 3×3 tensor. In the standard NMR spectrum in an isotropic medium for a freely rotating molecule only the average value is measured, so the individual tensor components are rarely discussed, even though in the calculations first all the components are determined. Secondly, in a standard experiment the observed property is the chemical shift, that is the difference between the shielding constant in the reference molecule and in the sample. Moreover, these properties depend on the environment (intermolecular forces, reduced but relevant also in the gas phase) and temperature,[4] so strictly speaking the observed parameter is not really a constant. All these factors have to be taken into account in the comparison of computed and measured values. They become significant when accurate *ab initio* methods, which provide not only a reliable value of the shielding constant but also an estimate of the error bars, are used in the calculations and the results are applied to predict or interpret the experimental NMR spectrum.

6.2 Non-relativistic Theory of NMR Parameters

In NMR, the interactions of nuclear magnetic moments in a molecule with the external magnetic field **B** and the mutual interactions of these moments are described by an effective Hamiltonian[5]

$$H^{\mathrm{NMR}} = -\sum_{K} \mathbf{B}^{\mathrm{T}}(\mathbf{1} - \boldsymbol{\sigma}_K)\mathbf{m}_K + \frac{1}{2}\sum_{K \neq L} \mathbf{m}_K^{\mathrm{T}}(\mathbf{D}_{KL} + \mathbf{K}_{KL})\mathbf{m}_L \tag{6.1}$$

where we used bold face to denote vectors and tensors, $\boldsymbol{\sigma}_K$ is the shielding tensor, $\mathbf{D}_{KL}$ describes the direct spin–spin coupling and $\mathbf{K}_{KL}$ is the reduced indirect spin–spin coupling tensor. The nuclear magnetic dipole moment is proportional to the nuclear spin $\mathbf{I}_K$, $\mathbf{m}_K = \gamma_K \mathbf{I}_K$, where γ_K is the gyromagnetic ratio for nucleus K. Both parameters in the NMR effective Hamiltonian given by eqn (6.1), the shielding and the spin–spin coupling, describe the dependence of the NMR spectrum on the molecular electronic structure, which is not explicitly considered – H^{NMR} acts within the space of the accessible nuclear spin states. In particular, the shielding constant of the nucleus K

represents the term bilinear in the external magnetic field induction **B** and the nuclear magnetic moment $\mathbf{m}_K$.

Soon after the development of NMR, early in the 1950s, Ramsey presented the theoretical methods of quantum chemistry needed to evaluate the main parameters of the NMR spectrum – the shielding constants[6] and the indirect spin–spin coupling constants.[7] It should be kept in mind that in a standard quantum chemical calculation, for instance performed to determine the molecular structure, the nuclei are described as point charges; in the calculation of the shielding constants the nuclear magnetic moments can be treated as small additional perturbations. Consequently, when we add to the molecular Hamiltonian the operators describing the external magnetic field and the nuclear magnetic moments, the energy $E(\mathbf{B},\mathbf{m}_K)$ may be determined as a function of these perturbation parameters.

The shielding has been defined by Ramsey as the derivative of the energy with respect to the external magnetic field and the magnetic moment of the nucleus of interest, and as such it can be computed applying perturbation theory. We shall focus here on the results for electronic ground states of closed-shell systems; these are of practical interest for comparison with experimental gas phase NMR data. Extracting the relevant bilinear term, corresponding to eqn (6.1), from the Taylor expansion about the unperturbed energy one finds that the shielding tensor $\boldsymbol{\sigma}_K$ is given by the mixed second derivative of the total energy with respect to the perturbations **B** and $\mathbf{m}_K$, computed in the absence of magnetic fields

$$\boldsymbol{\sigma}_K = \mathbf{1} + \left.\frac{\partial^2 E(\mathbf{B}, \mathbf{m}_K)}{\partial \mathbf{B} \partial \mathbf{m}_K}\right|_{\mathbf{B}=0, \mathbf{m}_K=0} \tag{6.2}$$

where we included the direct interaction of the magnetic moment with the external field.

For a closed-shell molecule in the ground electronic state, using standard time-independent perturbation theory yields for the shielding tensor the following sum-over-states expression

$$\boldsymbol{\sigma}_K = \langle 0|\mathbf{h}_{BK}^{\text{dia}}|0\rangle - 2\sum_{n\neq 0} \frac{\langle 0|\mathbf{h}_K^{\text{pso}}|n\rangle\langle n|(\mathbf{h}_B^{\text{orb}})^{\text{T}}|0\rangle}{E_n - E_0} \tag{6.3}$$

where the summation runs over singlet states and $E_n - E_0$ is the difference between the energy of an excited state, $|n\rangle$, and the energy of the ground state, $|0\rangle$. The operators corresponding to the perturbations linear in **B**, linear in $\mathbf{m}_K$, and bilinear in **B** and $\mathbf{m}_K$ are, respectively (in atomic units)

$$\mathbf{h}_B^{\text{orb}} = \frac{1}{2}\sum_i \mathbf{l}_{iO} = -\frac{1}{2}\sum_i \mathrm{i}\mathbf{r}_{iO} \times \nabla_i \tag{6.4}$$

where **l** is the angular momentum operator, the vector $\mathbf{r}_{iO}$ defines the position of the electron with respect to the chosen gauge origin O,

$$\mathbf{h}_K^{\text{pso}} = \alpha^2 \sum_i \frac{\mathbf{l}_{iK}}{r_{iK}^3} \tag{6.5}$$

and

$$\mathbf{h}_{BK}^{\text{dia}} = \frac{\alpha^2}{2}\sum_i \frac{(\mathbf{r}_{iO}\cdot\mathbf{r}_{iK})\mathbf{1} - \mathbf{r}_{iK}\mathbf{r}_{iO}^{\text{T}}}{r_{iK}^3} \tag{6.6}$$

where α is the fine-structure constant.

For a freely rotating molecule in an isotropic medium, with the external magnetic field directed along the Z axis, the NMR spin Hamiltonian may be written as

$$H_{\text{iso}}^{\text{NMR}} = -\sum_K B(1-\sigma_K)m_{K,Z} + \frac{1}{2}\sum_{K\neq L} K_{KL}\mathbf{m}_K^{\text{T}}\cdot\mathbf{m}_L \tag{6.7}$$

where the shielding constant σ_K is the trace of the tensor

$$\sigma_K = \frac{1}{3}\text{Tr}(\boldsymbol{\sigma}_K) \tag{6.8}$$

The direct spin–spin coupling representing the classical interaction of two magnetic dipoles vanishes for freely rotating molecules in gas phase NMR spectroscopy. Similarly, the individual components of the shielding tensor σ_K are not observed, therefore we shall concentrate in what follows on the analysis of the shielding constants. Nevertheless, it should be repeated that in the calculations all the tensor components are systematically determined.

The first term in eqn (6.3), an expectation value of the $\mathbf{h}_{BK}^{\text{dia}}$ operator in the ground state of the molecule, is known as the diamagnetic shielding. The second term, corresponding to the sum over states contribution, can be evaluated as a response property[8] or applying analytic second derivative methods,[9] and is known as the paramagnetic shielding (note that the excited states and the excitation energies are not determined in any practically used efficient computational method). We recall here that both terms depend on $\mathbf{r}_{iO}$, hence on the chosen gauge origin, thus the partition into dia- and paramagnetic contributions also depends on this choice.

When the gauge origin is placed at the nucleus of interest, the paramagnetic contribution is proportional to $\mathbf{C}_K^{\text{el}}$, the electronic part of the spin-rotation constant of this nucleus in the same molecule

$$\boldsymbol{\sigma}_K^{\text{para}} = \frac{1}{2\gamma_K}\mathbf{C}_K^{\text{el}}\mathbf{I} \tag{6.9}$$

where $\mathbf{I}$ is the moment-of-inertia tensor. This equation, known as Ramsey's[6] (for diatomics) and Flygare's[10,11] (for polyatomic molecules) formula, is valid only in the non-relativistic approach. It reflects the similarity of non-relativistic perturbing operators needed to compute the spin-rotation constant and the paramagnetic contribution to the shielding.

We refer to modern textbooks for a description of state-of-the-art methods of electronic structure theory.[12,13] In particular, Sauer[13] describes in detail the methods applied to determine numerous properties characterizing interactions of a molecule with electromagnetic fields as well as the underlying theory; we have only briefly summarized above the results leading to the calculation of the shielding constants. As a consequence of the advances in the *ab initio* methods of quantum chemistry, the progress in their implementation and in the performance of the computer hardware and software in the last 60 years, increasingly sophisticated methods are used to determine the NMR parameters, but the theory in all the non-relativistic methods is essentially related to Ramsey's equations. A discussion of different perturbation theory approaches, with the description of how they have been developed, implemented and applied for increasingly accurate approximations to the reference, unperturbed wavefunctions can be found in many reviews[14–18] and in several chapters of the books devoted to theoretical studies of NMR and EPR parameters.[19,20]

In addition, let us mention that significant progress has been achieved in the application of relativistic methods to the study of shielding constants (see *e.g.* the review by Autschbach[21] and Chapter 8 of this volume). They are undoubtedly needed in the treatment of molecules with heavy nuclei, and they are increasingly useful when highly accurate shielding constants are required – the error bars of state-of-the-art non-relativistic results for small molecules are often smaller than the relativistic effects.

To simplify the discussion of the theoretical results we shall present the values for isolated molecules, unless otherwise stated. In the calculations performed for molecular electronic ground states the standard procedure is to begin with a fixed molecular geometry. In accurate *ab initio* studies for small molecules this is the equilibrium geometry and the dependence on the nuclear motion (often significant) is next analyzed to describe the rovibrational effects. In this manner the temperature dependence of the shielding constants can be determined, and thus values suitable for comparison with experiment are obtained from theory.[22–24] We note that although it is a cumbersome procedure, there is practically no other way to make a proper comparison of theory and experiment; one cannot obtain experimentally the spectrum of a molecule with clamped nuclei. In approximate calculations one can use a single experimental geometry, for instance corresponding to the molecular structure at 300 K.

6.3 Analysis of the Shielding Constants within *Ab Initio* Electronic Structure Methods

We shall briefly summarize in this section some problems arising in the calculation of the shielding constants, and concentrate on the methods that have been developed to solve these problems and/or are implemented in the presently used computational programs.

We begin with the choice of the basis set, recalling that for a magnetic field perturbation it should ensure gauge-invariance of the results. Next, we discuss the treatment of the electron correlation effects, with the focus on the hierarchy of correlated methods, which allows for systematic improvement of the results beginning from the Hartree–Fock (HF) model towards the Full Configuration Interaction (FCI) approach.

Other issues that have to be considered are the relativistic effects and, for the comparison with experiment, the zero-point vibrational (ZPV) and temperature effects, as well as the influence of the intermolecular interactions on the measured shielding constants. The theory and methods applied to evaluate these effects are described in detail in other chapters of this volume (see in particular Chapters 3, 7, and 8), therefore we provide only a short description of these subjects.

6.3.1 Basis Sets in the Calculation of NMR Shielding Constants

In the calculation of properties characterizing the interaction of a molecule with an external magnetic field this field is represented by a vector potential, which is not uniquely defined – the gauge origin may be chosen arbitrarily. The ensuing dependence of the results on the gauge origin is unphysical; it complicated and delayed accurate studies of the shielding constants (in comparison to the calculation of other molecular properties). In practice, this problem nowadays can be eliminated by using perturbation-dependent basis sets; the most common approach of this type is based on the application of gauge-including atomic orbitals (GIAOs).[25,26] Another approach – continuous transformation of the origin of the current density leading to formal annihilation of its diamagnetic contribution (CTOCD-DZ)[27,28] – has been recently applied in a hierarchy of coupled cluster methods by Garcia Cuesta *et al.*[29] However, the implementation of such approaches requires an additional programming effort and even now some generally available computer packages do not enable a gauge-invariant calculation of the shielding constants for accurate reference wavefunctions (which, within the same package, can be used for other purposes, *e.g.* to determine the molecular structure).

The first problem to be considered is the construction of the basis set. The standard basis sets are optimized for calculations of molecular energies. Even though the application of GIAOs improves the convergence of the computed shielding constants with the extension of the basis set, to obtain accurate results for polyatomic molecules large basis sets have to be used. One of the reasons is that NMR probes the electron density close to the nucleus, whereas standard basis sets are not sufficiently flexible to describe this region. Although a proper description of the wavefunction near the nucleus may not be important for typically valence properties like molecular geometries or electric polarizabilities, it is essential for the analysis of the NMR parameters. In practice, a significant improvement of the results may be achieved for instance by adding (or at least by uncontracting) the tight p-type functions.[30]

Another more efficient procedure is to apply smaller basis sets optimized by Jensen[30,31] specifically for the study of NMR shielding constants (similarly, there are basis sets optimized for the calculation of the spin–spin coupling constants;[32–34] in this case the flexibility in the nuclear region is even more important). A method that allows for systematic optimization of the basis set in a controlled manner has been described and applied to study NMR parameters by Manninen and Vaara.[35]

A different solution which has been proposed is to use locally dense basis sets, larger for the atoms of interest, smaller for the other atoms in the molecule (for recent applications, see Reid *et al.*[36] and Rusakov *et al.*[37]).

6.3.2 Electron Correlation Effects

The shielding constants computed in the Hartree–Fock approximation are often not sufficiently accurate and a variety of methods that account for electron correlation effects has been developed.[14,15,18,19] We shall not discuss here density functional theory (DFT) calculations. Although for large molecules sometimes only the DFT-based approaches can be practically used, for small molecules, such as usually studied in the gas phase, more reliable results are obtained by applying *ab initio* methods. Moreover, within the *ab initio* methods there is a well-defined hierarchy of approximations (see for instance J. A. Pople's Nobel Lecture[38]). In other words, in contrast to DFT values, *ab initio* results may be refined in a systematic manner, thus their reliability and accuracy can be estimated *a priori* without any reference to experimental data.

In the analysis of shielding constants in closed-shell molecules at (or close to) their equilibrium geometries one can begin with the Hartree–Fock approach. The Møller–Plesset second-order perturbation theory (MP2) constitutes the first step in the hierarchy of correlated methods.[39] It was shown recently that scaled MP2 approaches, based on the application of different scaling factors for same spin and opposite spin perturbation contributions, give improved shielding constants.[40] However, state-of-the-art results are nowadays obtained applying the coupled cluster analytic second derivative methods. These methods, introduced by Gauss and co-workers,[41–44] enable calculations of the shielding constants at increasingly accurate levels of approximation. The CCSD – coupled cluster singles-and-doubles, CCSD(T) – CCSD with a perturbative triples correction, and CCSDT (singles–doubles–triples) methods are implemented within the CFOUR program package.[45] Similar calculations at the CCSDTQ (-quadruples) and higher levels, up to the full configuration interaction level, can be performed applying the MRCC program suite.[46]

Unfortunately, the coupled cluster methods are computationally expensive due to their unfavorable time scaling: approximately as $o^2\nu^4$ for CCSD and $o^3\nu^4$ for CCSD(T), where o and ν are the number of occupied and virtual orbitals, respectively. Therefore, for instance a CCSD(T) calculation for C_2H_4 is $\approx$100 times longer than for H_2O (assuming basis sets of the same size for C and O atoms). Applications of the CCSD(T) method for benchmark purposes thus have been limited to small molecules; with increasing number of atoms these calculations become unfeasible in reasonable time.

We note that although the coupled cluster methods are not variational, once an asymmetric form is used to calculate the shielding constants, the first-order perturbation corrections to the wavefunction with respect to the nuclear magnetic moments are not needed.[47] In the efficient implementation of the coupled cluster analytic second-derivative methods the shielding constants are expressed in terms of one-electron density matrices D_{pq} and matrix elements of the one-electron Hamiltonian h_{pq}. In a somewhat simplified notation, with two superscripts denoting the orders in the double perturbation expansions of D_{pq} and h_{pq} with respect to the external magnetic field and nuclear magnetic moment, the asymmetric expression used to evaluate a component of the shielding tensor can be written as[42,47]

$$\sigma = \sum_{pq} D_{pq}^{(00)} h_{pq}^{(11)} + \sum_{pq} D_{pq}^{(10)} h_{pq}^{(01)} \tag{6.10}$$

where we omitted the subscript K and the tensor indexes; $h_{pq}^{(11)}$ and $h_{pq}^{(01)}$ represent the matrix elements of the appropriate tensor components of $\mathbf{h}_{BK}^{\mathrm{dia}}$ (eqn (6.6)) and vector components of $\mathbf{h}_{K}^{\mathrm{pso}}$ (eqn (6.5)) operators, respectively. Although the final expression which gives the shielding constants, eqn (6.10), contains only contributions from the one-electron density matrix, two-electron contributions are needed in the calculation of $D_{pq}^{(10)}$, the first derivatives of the densities. However, to calculate all the shielding constants one has to solve the set of coupled cluster perturbation theory equations only for three vector components of the external magnetic field, independently of the size of the molecule.

Presently, the coupled cluster method gives undoubtedly the most accurate and reliable non-relativistic results. At a given level of approximation, extending systematically the basis set one can analyze the convergence in the one-electron space and thus estimate the error bars with respect to the basis set limit. Including systematically in a sequence of coupled cluster calculations higher and higher connected excitations one can analyze for the chosen basis set the convergence in the N-electron space, that is convergence to the corresponding FCI limit. In other words, one can obtain an estimate of the error bars in the description of the electron correlation effects. Combined together, these estimates give an approximate value of the error bars with respect to the FCI result in a complete basis set – that is, with respect to the solution of the Schrödinger equation.

6.3.3 Relativistic Effects

This chapter is essentially dedicated to non-relativistic calculations; relativistic methods are treated in Chapter 8. They are important mainly when a molecule includes heavy nuclei; however, in accurate studies of NMR properties the relativistic effects have to be evaluated even for compounds of light elements. Therefore, we mention here a few ways to describe these effects and we shall discuss some results in Section 6.4, Applications.

Different methods have been developed to determine the shielding constants at the relativistic level. One can start with a non-relativistic scheme and treat the relativistic effects as a correction, using Breit–Pauli perturbation theory (BPPT).[48] In this approach we begin with the standard Schrödinger equation and relativity is treated as another perturbation (in addition to external field and nuclear magnetic moment perturbations). Corrections to the non-relativistic value can also be determined in the so-called linear response within the elimination of the small component approach.[49] In other approaches which have been developed and successfully implemented the relativistic shielding constants are obtained directly, using either two-component or four-component methods.[50,51] In these latter methods, whenever one is interested in the magnitude of the relativistic contributions, analogous (*e.g.* using the same DFT functional) non-relativistic calculations are separately performed. The relativistic contribution determined in this manner may be treated next as a correction to be added to the best available non-relativistic shielding constant. However, it should be kept in mind that such approximations rely on the additivity of different contributions to the computed properties.

6.3.4 Zero-point Vibrational and Temperature Effects

The value of the shielding constant computed at the molecular equilibrium geometry does not correspond to that experimentally measured; one has to take into account the zero-point vibrations and the temperature dependence of the properties. Proceeding in accordance with the Born–Oppenheimer approximation, one should compute the potential energy surface and the property surface, determine the vibrational wavefunctions and finally use them to compute the rovibrational corrections to the shielding constant evaluated at the equilibrium geometry. The theory and different methods applied in practice are discussed in detail in Chapter 7, therefore here we only briefly outline the basic concepts of some approaches used to evaluate these corrections.

The rovibrational corrections may be efficiently calculated using thermally averaged values of the normal coordinates.[22–24] For small polyatomic molecules practically useful expressions can be determined considering the truncated Taylor expansion[24]

$$\sigma = \sigma_{\text{eq}} + \sum_i \frac{\partial \sigma}{\partial Q_i} \langle Q_i \rangle + \frac{1}{2} \sum_{ij} \frac{\partial^2 \sigma}{\partial Q_i \partial Q_j} \langle Q_i Q_j \rangle \tag{6.11}$$

where σ_{eq} denotes the shielding at the equilibrium geometry, $\langle Q_i \rangle$ is the average value of the vibrational normal coordinate Q_i, and similarly $\langle Q_i Q_j \rangle$ is the average of the product $Q_i Q_j$. These averages may be computed for particular rovibrational states; in the comparison with experimental shielding constants the corresponding temperature averages determined by applying Boltzmann averaging over rovibrational states are used. They are given by

$$\langle Q_i \rangle = -\frac{1}{4\omega_i^2} \sum_j \frac{k_{ijj}}{\omega_j} \coth\left(\frac{\omega_j}{2k_B T}\right) \tag{6.12}$$

and

$$\langle Q_i Q_j \rangle = \frac{1}{2\omega_i} \delta_{ij} \coth\left(\frac{\omega_i}{2k_B T}\right) \tag{6.13}$$

where ω_i is the harmonic vibrational frequency, k_{ijj} the cubic force constant, δ_{ij} denotes the Kronecker delta, and k_B is the Boltzmann constant (we did not include here the terms due to centrifugal distortion; see Chapter 7).

In practice, the levels of theory and basis sets used for the calculation of potential energy surface, required to determine the vibrational wavefunctions, and for the calculation of shielding constant derivatives, may differ. This increases often the efficiency of the calculations; for instance, in the analysis of the potential energy surface one can use smaller basis sets.

6.3.5 Intermolecular Interactions

The shielding constants determined from *ab initio* calculations are most accurate and reliable when computed for an isolated, single molecule. Theoretical description of the effects of the environment, in particular in the condensed phase, is much more complicated and requires numerous additional approximations. Therefore, for the comparison of theory and experiment, one should attempt to extract an isolated molecule value from the available experimental data (we note that such a procedure is also in agreement with the definition of a *molecular* property). In the liquid phase, the measured shielding constants are affected by the intermolecular forces, as shown for instance by the solvent effects. In the gas phase the dependence of the shielding on gas density is weaker, nevertheless it is also observable and it should be examined whenever possible.

Let us suppose that we measure the shielding $\sigma^A(X)$ in a binary mixture of the gas A, containing the nucleus X, and another gas B as the solvent. It has been observed[52] that at low pressures $\sigma^A(X)$ depends linearly on ρ_A and ρ_B – the densities of A and B, respectively. In the case of very low density of A only the A–B intermolecular interactions have to be considered and $\sigma^A(X)$ can be expressed as

$$\sigma^A(X) = \sigma_0^A(X) + \sigma_1^{AB}(X)\rho_B = \sigma_0^A(X) + \sigma_{1\text{bulk}}^B \rho_B + \sigma_1^{A-B}(X)\rho_B \tag{6.14}$$

where the first term, $\sigma_0^A(X)$, is the shielding in the zero-density limit, which we can compare with the theoretical value obtained for an isolated molecule. The other terms describe the solvent effects with $\sigma_{1\text{bulk}}^B$ representing the bulk susceptibility correction and the last term describing the effects due to A–B binary collisions.

It has been repeatedly demonstrated that these approximations work successfully, for instance by the fact that $\sigma_0^A(X)$ values obtained for the same solute and different gaseous solvents B are identical.[53] In principle, all the terms appearing in eqn (6.14) can be estimated from theory: the bulk susceptibility is approximately proportional to the magnetizability of the solvent gas B, and $\sigma_1^{A-B}(\mathrm{X})$ can be estimated analyzing the shielding constants in the A–B dimer. In practice, the latter contribution can be evaluated precisely only for very small systems (A and/or B being atoms), since all the geometries of the A–B dimer should be considered.

6.4 Applications

In this section, we discuss the computational strategies leading to accurate prediction of the shielding constants. To enable a comparison of the results with the available isolated molecule values extracted from gas phase NMR spectra we consider the role of the intermolecular interactions, but we do not present any results obtained applying continuum solvation models. We focus here on wavefunction-based methods, although also some density functional theory calculations will be mentioned. The selected examples of shielding constant calculations which we analyze in detail are mostly taken from the literature.

6.4.1 Approaching Accurate NMR Shielding Constants: Two Examples

We begin with two examples illustrating the magnitude of different effects contributing to the computed shielding constants. To obtain an accurate value of the shielding constant and estimate its error bars one has to take into account the basis set and electron correlation effects, vibrational and temperature effects, and relativistic corrections. All the computed contributions to the total NMR shielding constants of ^{11}B and ^{19}F in BF_3 molecule[54] and ^{1}H and ^{33}S in H_2S[55] are summarized in Table 6.1. The convergence of the results with the extension of the basis set has been satisfying for both molecules, so we tabulate only the values obtained with the largest basis sets.

First, one needs to estimate the accuracy of the results obtained for molecular equilibrium geometry. In our two examples the correlation effects are calculated at CCSD and CCSD(T) levels of theory. The CCSD contribution recovers the dominant part of the correlation effects, the additional contribution from non-iterative triple excitations is much smaller, as usual. The correlation effects constitute a few percent of the total shielding constants; clearly they have to be accounted for when accurate results are required for these molecules. At the same time, small magnitude of the CCSD(T) contribution indicates that in this case higher-level coupled cluster calculations, which very rarely can be performed in practice, are not really required.

Table 6.1 Contributions to the shielding constants in BF_3 and H_2S (in ppm).[a]

	^{11}B		^{19}F		^{33}S		^{1}H	
HF[b]	103.92	(106.2)	342.73	(103.3)	710.51	(96.6)	30.57	(101.7)
CCSD-HF[b]	−5.67	(−5.8)	−8.35	(−2.5)	23.74	(3.2)	0.00	(0.0)
CCSD(T)-CCSD[b]	−0.97	(−1.0)	−3.13	(−0.9)	3.67	(0.5)	−0.01	(0.0)
ZPV[c]	−0.21	(−0.2)	−2.59	(−0.8)	−20.86	(−2.8)	−0.40	(−1.3)
300 K[c]	−0.03	(−0.0)	−0.25	(0.1)	−0.89	(−0.1)	−0.03	(−0.1)
Δ_{rel}[d]	0.81	(0.8)	3.36	(1.0)	19.55	(2.7)	−0.09	(−0.3)
Total	97.85	–	331.77	–	735.72	–	30.05	–

[a]Percentage of the total shielding constant is shown in the parentheses.
[b]cc-pCVQZ basis set.
[c]cc-pVTZ basis set.
[d]Uncontracted cc-pCV5Z basis set.

NMR shielding constants calculated at the equilibrium geometry do not account for the vibrational effects. As shown in Table 6.1, the ZPV corrections, which describe the effect of the nuclear motion on the shielding constant at the lowest vibrational level, are very significant; they cannot be neglected when we aim for highly accurate results. The temperature-related effects are much smaller for all four tabulated shielding constants, but, as discussed in Section 6.3.4, evaluation of these contributions on top of ZPV corrections is often fairly simple and for comparison with experimental data they should also be considered.

The CCSD(T) equilibrium geometry value of the shielding, combined with ZPV and temperature corrections, provides typically an accurate non-relativistic approximation to the total shielding constant. However, when comparing with experiment, one has to describe also the relativistic effects which are inevitably present in all the experimental data. The tabulated relativistic corrections Δ_{rel} were evaluated for both molecules at the uncorrelated level, as the difference between relativistic Dirac–Hartree–Fock (DHF) and non-relativistic Hartree–Fock values obtained using the same basis sets. Although BF_3 includes only light atoms, the relativistic effects contribute about 1% to the total shielding constants. The relativistic contribution to σ(S) in H_2S is significantly larger, constituting 2.7% of the total shielding constant. In both molecules, the relativistic corrections are similar in magnitude to the corresponding correlation and ZPV contributions, indicating that to obtain very accurate values of the shielding constants one needs to take into account the relativistic corrections even for molecules consisting of light elements. These examples illustrate also the general trend, that the importance of the relativistic effects increases with the Z number (nuclear charge) of the atoms in the molecule.

Finally, an estimate of the error bars and an improved estimate of the total shielding constants may be determined following an analysis of the convergence patterns of different contributions. For instance, the recommended value of σ(S) in H_2S, 740.3(3.0) ppm,[55] was slightly larger than the calculated one, and it is in agreement with the new semi-experimental result, 742.9(4.6) ppm.[56]

6.4.2 Basis Sets Effects

The simplest way to estimate the basis set effects for small molecules is to study the changes of the computed shielding constants with the extension of the basis set. An example of the convergence of the results (taken from the work of Auer *et al.*[23]) is given in Table 6.2, which shows ^{13}C shielding constants in selected molecules. All these results have been obtained at the CCSD(T) level, using the same geometries for each basis set (and using GIAOs). As shown in this table, even at such a high level of approximation it was possible in practice to reach convergence to the basis set limit; the differences between the values determined using the largest two basis sets are negligible. For larger molecules such a "brute force" approach is not possible, in particular since one should describe simultaneously electron correlation effects at an appropriate level. The construction of specific basis sets, suitable for shielding constant calculations, has been discussed in Section 6.3.1; another option is to use extrapolation techniques.

Different methods of extrapolation relying on large basis set results for DFT and/or MP2 methods have been successfully applied to estimate the basis set error in smaller basis set CCSD(T) calculations.[57,58] The CCSD(T) results obtained with the aug-cc-pCVnZ, $n = 3$ and 4 basis sets were systematically extrapolated to determine the basis set limit by Teale *et al.*[59] We note that separate extrapolation methods, using different formulas, were used for the HF contributions and for the correlation contributions to the shielding. For 20 molecules the basis set limits at the HF and MP2 levels were successfully estimated by extrapolation in the work of de Oliveira and Jorge.[60] Various schemes of extrapolation to the complete basis set limit of shielding constants in small water clusters, computed using different correlated and DFT methods, have been examined by Armangue *et al.*[61] In this case the authors concluded that two-point extrapolation methods cannot be used with confidence.

6.4.3 Electron Correlation Effects

6.4.3.1 Accurate Calculations for Two-electron Systems

Not surprisingly, the most detailed analysis of the shielding constants has been performed for two-electron systems. The best results have been

Table 6.2 Basis set dependence of ^{13}C shielding constants (in ppm).[a]

	tzp	qz2p	pz3d2f	13s9p4d3f	15s11p4d3f
CH_4	201.2	199.0	199.0	198.8	198.8
C_2H_2	128.2	123.5	123.3	122.7	122.6
CO	9.6	6.3	4.1	3.0	3.0
CO_2	66.2	61.1	61.5	60.4	60.2
HCN	92.2	86.9	85.9	85.1	85.0
CH_2O	17.7	8.1	4.4	2.9	2.8

[a]CCSD(T) results taken from Auer *et al.*;[23] the basis set contractions are: tzp – 9s5p1d/5s3p1d, qz2p – 11s7p2d/6s4p2d, pz3d2f – 13s8p3d2f/8s5p3d2f.

obtained for atomic ^{3}He; in this study not only the relativistic effects but also quantum-electrodynamic and nuclear mass effects have been taken into account, leading finally to $\sigma(^3\text{He}) = 59.96743(10)$ ppm.[62] It has been stated in IUPAC recommendations[63] that $\sigma(^3\text{He})$ in gaseous helium would be a good universal standard of chemical shifts, but unfortunately it is not a practical one.

There are numerous *ab initio* and experimental studies of the shielding constants in hydrogen molecule (see Sundholm and Gauss,[64] Jaszuński *et al.*,[65] and references therein). In this case, the experimental spin-rotation constants can be used to determine the paramagnetic contribution to the shielding, and the effects of nuclear motion lead to observable differences between isotopomers. In the recent studies σ(H) in H_2 was found to be 26.2886 ppm when the experimental spin-rotation constant was used,[64] and slightly larger values were obtained from the calculations: 26.2983 ppm[64] and 26.29498 ppm.[65] Similar accuracy has been reached for the other isotopomers, and there is fair agreement with experimental data of the computed shielding differences between isotopomers. It appears that an analysis of the non-adiabatic effects is needed to improve the accuracy of these results, in particular to determine the difference between the H and D shielding in HD. Very recently, such calculations for HD and HT were utilized to obtain deuteron and triton magnetic moments.[66]

6.4.3.2 Benchmark Studies

There are in the literature several studies which assess the accuracy of various *ab initio* methods in the prediction of NMR shielding constants. In particular, for small molecules large basis sets may be presently applied in CCSD(T) calculations, thus providing accurate benchmark values, often used to investigate the performance of different DFT functionals. Calculations of this type have been performed for many nuclei commonly used in NMR, such as ^{13}C,[23] ^{17}O,[67] ^{19}F,[22] ^{15}N and ^{31}P[68] (altogether, more than 70 molecules were examined in these studies). In other works the shielding constants of ^{17}O in H_2O[69,70] and ^{33}S in H_2S and SO_2[71] were examined in detail.

Shielding constants determined applying theoretical methods can be compared with corresponding values based on experimental data. These experimental shielding constants are often derived using the theoretical prediction for one molecule, for which the most accurate theoretical value can be established and experimental chemical shifts with respect to that reference molecule. As an example, ^{19}F shielding constants in selected molecules computed at the Hartree–Fock level, correlated levels and applying two DFT functionals are shown in Table 6.3 (see Harding *et al.*[22] for the results for many other molecules and more details). The conclusion of this study was that the shielding constants calculated at the CCSD(T) level with large basis sets and including vibrational corrections are in very good agreement with the experimental values – the differences usually do not exceed a few ppm. Although very high accuracy can be reached for most of

Table 6.3 Comparison of calculated and experimental ^{19}F shielding constants (in ppm).[a]

	HF	MP2	CCSD	CCSD(T)	DFT BP86	DFT B3LYP	Theory[b]	Exp.
FH	413.7	423.8	418.1	418.4	409.9	410.4	410.3	409.6
CH_3F	486.6	486.6	481.2	479.7	460.9	465.2	472.9	470.6
CH_2F_2	367.2	353.9	355.9	351.4	317.4	326.3	340.7	338.7
CHF_3	303.4	287.7	291.8	286.9	249.0	258.9	275.2	273.7
CF_4	281.6	270.8	273.6	269.8	234.0	242.3	259.5	258.6
F_2	−177.3	−171.3	−175.2	−189.0	−269.4	−258.7	−214.9	−233.2

[a]Data taken from Harding *et al.*[22]
[b]The best theoretical prediction: large basis set CCSD(T) results, with the vibrational and temperature corrections.

the molecules in the test set, there are some cases where the theoretical prediction differs noticeably from experiment (this difference was largest for F_2 molecule, almost 20 ppm even for the best applied method and basis set).

In a recent extensive benchmark study[59] of more than 70 shielding constants, in addition to the CCSD and CCSD(T) methods also the performance of many DFT functionals was assessed. It is important to estimate the performance of various functionals in such analysis, because DFT is the method chosen for most of the routine applications. A large series of molecules, containing elements important for organic chemistry such as ^{1}H, ^{13}C, ^{15}N, ^{17}O, ^{19}F, ^{31}P, and ^{33}S, was considered and for comparison with experimental data the zero-point vibration effects were included. The mean absolute errors of CCSD and CCSD(T) shielding constants in comparison with experiment are 5.5 ppm and 2.9 ppm, respectively. DFT functionals included in this study systematically underestimate absolute shielding constants, with the deviation from CCSD(T) results exceeding 20 ppm for functionals like BLYP, B3LYP, and PBE (somewhat better results are obtained for the KT2 functional,[72] developed specially for NMR properties). Moreover, adding the vibrational corrections to the CCSD(T) results leads to further improvement of the agreement with experiment, whereas for DFT it increases the deviation from experimental data. More recently, in a similar study the CCSD(T)/cc-pVQZ values of H and C shielding constants in a series of somewhat larger molecules were used as a benchmark to analyze a variety of simpler approximations.[73] In this work the HF, DFT – for a number of functionals – and MP2 shielding constants have been determined using a variety of basis sets, the accuracy of different method/basis set combinations was analyzed and a new scaled MP2 approach was proposed.

For small molecules, the contributions of higher-order excitations in the coupled cluster approach can also be evaluated. The magnitude of these contributions is illustrated in Table 6.4 by the values of the oxygen and proton shielding constants in the water molecule, computed at the CCSD, CCSD(T), CCSDT, and CCSDTQ levels of theory.[74] The largest calculations were carried out using cc-pCVTZ basis set for oxygen (cc-pVTZ for hydrogen), in this context a relatively large correlation consistent basis set – in such

Table 6.4 Shielding constants in H_2O calculated using different methods (in ppm).

	σ(O)			σ(H)		
Basis set[a]	DZ	TZ	CTZ	DZ	TZ	CTZ
HF	347.957	335.625	336.047	31.380	30.836	30.834
CCSD	355.238	343.710	343.852	31.822	31.200	31.175
CCSD(T)[b]	355.115	344.482	344.681	31.874	31.236	31.210
CCSDT	354.994	344.269	344.449	31.879	31.240	31.215
CCSDTQ	355.177	344.362	344.545	31.877	31.244	31.218

[a]The total number of functions in DZ, TZ, CTZ basis set is 24, 58, and 71, respectively.
[b]For other water molecule results, see the text.

calculations there are $\approx 2\times10^8$ excitation amplitudes to be determined. The correlation contributions to σ(O) are Δ(CCSD-HF) = 7.805 ppm, Δ(CCSDT-CCSD) = 0.597 ppm, and Δ(CCSDTQ-CCSDT) = 0.096 ppm, decreasing fast with the level of the coupled cluster approximation. Similarly, the CCSD(T) and CCSDTQ values in the case of σ(O) in water differ by only 0.136 ppm. All the contributions, up to the FCI result, have been evaluated in the DZ basis by Kállay and Gauss.[44] The differences between their FCI results, equal to σ(O) = 355.18 and σ(H) = 31.877 ppm, and the tabulated CCSDTQ values are negligible.

A CCSD(T) calculation with a large aug-cc-pCV6Z basis set gives σ(O) = 337.6 ppm,[69] in perfect agreement with the recent semi-experimental value, 337.5(3) ppm.[56] Other correlated methods, giving fairly accurate results with less severe scaling than CCSD(T), have also been used in water molecule calculations. For instance, the second-order polarization propagator approximation (SOPPA) and the second-order polarization propagator approximation with coupled cluster singles and doubles amplitudes SOPPA(CCSD)[75] give for σ(O) in H_2O 335.18 ppm and 333.29 ppm, respectively. Different coupled cluster approximations – CC2 and CC3 – give for σ(O) in H_2O 340.82 ppm and 336.70 ppm.[29] All these values differ from the quoted CCSD(T) result by less than 5 ppm. We note, moreover, that this CCSD(T) result was obtained applying GIAOs, whereas the other values have been determined with an alternative method of gauge-dependence treatment, CTOCD.

To conclude this part, although DFT is frequently the inexpensive method of first choice, coupled cluster methods should be recommended for the calculation of NMR shielding constants. They provide usually excellent results, and their application is not limited to benchmark studies of small molecules. When the CCSD(T) approach becomes too expensive, shielding constants can be successfully computed using CCSD (see below for some examples). We note, however, that sometimes the convergence within the hierarchy of coupled cluster approximations is exceptionally slow, and in such case even the CCSD(T) results may not be sufficient. Accurate prediction of the shielding constants is notoriously difficult for electron rich molecules like F_2, F_2O or O_3; similarly, one may expect lower than average accuracy for molecules with triple bonds (see Section 6.4.3.3). Larger contributions to the shielding due to improved description of triple coupled cluster excitations (and/or due to higher-order excitations) may be expected for these molecules. For example, in a study of six small molecules at the CCSDT level Gauss[43] has shown that, in contrast to the other molecules, a significant improvement with respect to CCSD(T) is obtained for ozone. Unfortunately, such calculations – in particular applying large basis sets – become very expensive due to the scaling of the coupled cluster methods.

6.4.3.3 *Estimating Electron Correlation Effects*

As discussed in the previous section, in most cases CCSD(T) values obtained applying large basis sets can be treated as benchmark results. When the computed results converge with the level of approximation and with the extension of the basis set (as it happens for most of the small molecule calculations discussed earlier), the results are reliable and their error bars may be estimated by analyzing the convergence patterns. In this section we present a few examples illustrating the analysis of the shielding constants when large basis set CCSD(T) calculations appear to be insufficient, are not feasible or not really needed for practical purposes.

There are practically two possibilities to determine absolute shielding constants in a selected compound, as described above. The first is a straightforward *ab initio* calculation, the second is to combine the paramagnetic contribution derived from the experimental measurement of the spin-rotation constant and the diamagnetic shielding from *ab initio* calculations. The latter treatment, based on Flygare's formula, eqn (6.9), was recently used to compare the obtained semi-experimental absolute shielding scales with those computed for ^{17}O and ^{33}S.[69,71] We recall that *ab initio* calculation of the paramagnetic contribution is the computationally demanding part – as a response property, it is very sensitive to the choice of the basis set and requires a proper description of electron correlation effects. On the other hand, the diamagnetic term is an expectation value and often may be accurately predicted at a lower level of theory. However, when the relativistic effects are important the non-relativistic Flygare's formula becomes unreliable, thus one has to be careful in the analysis of the results for molecules including heavy nuclei. A known example is tin; it was shown[76] that the application of this formula leads to an error of $\approx$1000 ppm in the Sn absolute shielding scale.

Although the correlation effects on NMR shielding constants are usually relatively small, in particular for molecular systems with single bonds, they may be very significant even for diatomic molecules at their equilibrium geometries. An interesting example of the systems where electron correlation plays a substantial role is the series of triply bonded molecules AsN, AsP, and As_2.[77] The correlation contributions to σ(As) have magnitudes of hundreds of ppm and change its sign, from deshielding at the HF level to shielding at the correlated level (see Table 6.5). The large difference between CCSD(T) and CCSD makes it clear that full treatment of triple excitations may change the results, and higher coupled cluster excitations may contribute

Table 6.5 Arsenic shielding constants in AsN, AsP, and As_2 molecules (in ppm).

	HF	CCSD	CCSD(T)
As in AsN			
cc-pVTZ	123.65	722.71	824.96
cc-pVQZ	188.12	733.51	835.17
cc-pV5Z	49.04	606.04	714.52
cc-pVTZs,p-unc	−30.21	585.03	695.41
cc-pVQZs,p-unc	−23.74	552.60	664.28
cc-pV5Zs,p-unc	−25.15	538.87	651.27
As in AsP			
cc-pVTZ	−170.69	295.62	384.01
cc-pVQZ	−101.36	323.54	407.76
cc-pV5Z	−250.17	174.23	268.52
cc-pVTZs,p-unc	−339.51	132.71	230.04
cc-pVQZs,p-unc	−331.63	109.14	204.86
cc-pV5Zs,p-unc	−330.65	95.72	193.99
As in As_2			
cc-pVTZ	−520.62	96.66	214.35
cc-pVQZ	−442.13	119.03	235.36
cc-pV5Z	−616.02	−48.98	81.81
cc-pVTZs,p-unc	−709.04	−80.20	49.14
cc-pVQZs,p-unc	−703.44	−118.15	13.96
cc-pV5Zs,p-unc	−706.23	−136.02	0.16

non-negligibly, but for these systems such calculations are too demanding. Moreover, in these AsX calculations the standard cc-pVnZ (n = T, Q, 5) basis sets are obviously not saturated; they are insufficient even at the HF level. As shown in Table 6.5, uncontracting the s- and p-functions brings the Hartree–Fock results closer to the corresponding basis set limit. The basis set convergence of the coupled cluster shielding constants is also improved within the cc-pVnZs,p-unc series, and apparently when these basis sets are applied the treatment of the correlation effects becomes the main issue. This is not surprising; slow convergence in the coupled cluster hierarchy of approximations reflects the multi-reference character of the triply bonded AsX molecules (similarly to the ozone example mentioned above). Although a calculation of the shielding constants within the Multiconfiguration SCF (MCSCF) linear response approach is in principle feasible, and should provide a reasonable description of the dominant correlation effects, showing that the results converge with the extension of the active space might be a problem.

In DFT, using for comparison the KT2 functional and the cc-pVQZ-unc basis set, for the As shielding constant in AsN, AsP, and As_2 one obtains 491.06, −35.13, and −273.70 ppm, respectively. Thus, approximately 90%, 67%, and 60% of the CCSD(T) correlation correction to As shielding is recovered in DFT, somewhat less than in CCSD calculations. Although in the studies of the shielding constants for most diatomic molecules one can now systematically reach high accuracy and show that the error bars do not exceed several ppm, it appears that for molecules of this AsX series in practice it is not yet possible.

Finally, let us present some examples of *ab initio* studies of organic compounds. In many cases the absolute shielding constants are not needed; the chemical shifts are necessary or sufficient to interpret the NMR spectrum, identify the molecule, or determine its structure. The standard procedure is to compute the shielding constants in the molecule of interest and in the reference molecule at the same level of approximation, or – if there are a few nuclei of the same type in one molecule – to consider the relative chemical shifts.

A spectacular illustration of the role of electron correlation was provided by a study of the ^{13}C spectrum of 1-cyclopropylcyclopropylidenemethyl cation.[78] The calculations at the HF level do not allow even to assign correctly the peaks in the experimental spectrum to specific carbon nuclei, whereas all the ^{13}C chemical shifts computed at the CCSD(T) level agree with the experimental data to within 2.2 ppm. In this study, absolute shielding constants have been converted to chemical shifts using the corresponding data for CH_4 and tetramethylsilane (TMS).

In CCSD(T) calculations for molecules with more than 15–20 second-row atoms it is difficult to apply a sufficiently large basis set. A reasonable method of choice, which in practice combines reliability with acceptable computational cost, is CCSD. As an example illustrating the application of CCSD let us mention a study of the shielding constants in *o*-benzyne.[79] This molecule contains two equivalent triply bonded carbon atoms, and tensor properties of the shielding of these atoms are of interest. Calculations of the shielding in molecular systems with triple bonds are sensitive to the used correlation methods (see the AsX example above); nonetheless, CCSD results for *o*-benzyne show good qualitative agreement with the experimental observation that the shielding of triply bonded carbons is significantly lower than for other carbon atoms in the molecule. Among the tested DFT functionals, the Keal–Tozer functional KT1[80] performed satisfactorily for the triply bonded carbon shielding. For instance, the CCSD, KT1 and experimental values of these shielding constants are 1.3, -3.3, and 3.7 ppm, and for the neighboring carbon atoms 67.6, 61.4, and 59.5 ppm, respectively (we note that in the experiment *o*-benzyne was inside a molecular container). In general, the quality of DFT shielding constants determined with various functionals was improved when the calculations were carried out at the geometry optimized applying the same functional.

A useful approximation is obtained assuming additivity of higher-order correlation and basis set effects. For instance, one can compute CCSD shielding constants (in this case the lower level of approximation) using a large basis set and add as a correction the difference between CCSD(T) and CCSD values determined in a smaller basis, which allows for CCSD(T) calculations. A similar approach has been used successfully in a study of $C_{10}H_{10}$ annulene.[81] The total shielding constants were evaluated for three hypothetical isomers, combining larger basis set MP2 values with the corrections given by the differences of smaller basis set CCSD(T) and MP2 results. A comparison of computed in this approach ^{13}C NMR spectra with the experimental spectrum enabled the determination of the actual molecular structure.

6.4.4 Relativistic Effects

The relativistic methods applied in the analysis of shielding constants are described in detail in Chapter 8, thus here we present only a few examples.

As mentioned in Section 6.3.3, the first possibility to describe relativistic effects on the NMR shielding is to treat them as corrections, defined by perturbations of the non-relativistic Hamiltonian within Breit–Pauli perturbation theory.[48] This approach can be implemented on top of uncorrelated, correlated, or DFT approximations. For instance, by applying BPPT on top of HF and MCSCF methods in a study of group XIV hydrides,[82] the total relativistic correction to the shielding of Si in SiH_4 was found to be 13.65 and 13.61 ppm, respectively. This is in good agreement with 13.11 ppm, obtained as the difference between DHF and HF shielding constants.[55]

Presently, the shielding constants also can be computed directly at the relativistic DFT level. The total relativistic correction to phosphorus shielding in PH_3 computed within BPPT, using KT2 density functional and Hartree–Fock methods, respectively, is 18.78 ppm and 18.31 ppm.[83] The corresponding correction obtained as the difference between Dirac-KT2 and non-relativistic KT2 results is 18.42 ppm, and between DHF and HF shielding constants it is 18.97 ppm. In a similar previous calculation with a fixed gauge-origin the DHF *vs.* HF difference was smaller, 15.36 ppm,[55] possibly due to the basis set incompleteness. In the more recent study, four-component GIAO approach (as implemented within the DIRAC code[84]) was applied, so the new estimate of the relativistic contribution to the shielding, given by the difference with respect to non-relativistic values, is presumably more precise.

As another example, we present two very different approaches to the calculation of xenon atom shielding constants in xenon fluorides. In the first approach, the shielding constants were calculated at DHF level and next corrected for electron correlation comparing the non-relativistic MP2 and HF values (Δ_{MP2}), assuming additivity of relativistic and correlation effects.[85] In the other approach, scalar-relativistic effects were included in a spin-free exact-two-component approach (SFX2C-1e) with a coupled cluster reference wavefunction.[86] These scalar-relativistic results do not account for the spin–orbit (SO) effects, therefore literature values of the appropriate correction (Δ_{SO}) have been added. The results obtained by these two approaches are shown in Table 6.6 (experimental shielding constants were determined using the chemical shifts given by Cheng *et al.*[86] and $\sigma(\text{Xe}) = 6965$ ppm[87]). The agreement with the experimental data is good for XeF_2 and XeF_4 molecules. For XeF_6, both approaches predict 1089 ppm (a coincidence); a value

Table 6.6 Shielding constants of Xe in xenon fluorides (in ppm).

	DHF	Δ_{MP2}	Total	SFX2C-1e CCSD(T)	Δ_{SO}	Total	Exp.[a]
XeF_2	3393	216	3609	3845	−222	3623	3579
XeF_4	1605	−243	1362	1446	−108	1338	1342
XeF_6	1645	−556	1089	1089	N/A	1089	1540

[a]For conversion of chemical shifts to shielding constants, see the text.

differing from the experimental one by 450 ppm. The use of an unsaturated basis set, smaller than for XeF_2 and XeF_4, is probably the source of this discrepancy in the first approach, while in the second approach the discrepancy is probably due to the missing spin–orbit correction.

In general, the additivity of the correlation effects and relativistic corrections calculated at the uncorrelated level cannot be guaranteed. To illustrate a success of this approximation we present the results for $^1J(XH)$ spin–spin coupling constants in the CH_4, SiH_4, GeH_4, and SnH_4 series of molecules.[88] The spin–spin coupling constants, in contrast to absolute shielding constants, are observed in the standard NMR spectrum and a direct and reliable comparison of calculated values with experimental data from gas phase NMR is possible. Table 6.7 shows that for CH_4, SiH_4, and GeH_4 treating the relativistic contributions as additive corrections leads to excellent agreement with experiment, and even for SnH_4, where the relativistic contribution constitutes about 50% of the non-relativistic value, the difference between the total theoretical and experimental value is only $\approx$9%. The rule of thumb is that when the correlation effects or relativistic effects are small then assuming their additivity leads to satisfactory results.

Finally, it should be kept in mind that relativistic effects may be significant even for light nuclei. If there is a heavy atom in the molecule, it affects the shielding of the neighboring nuclei (the so-called HALA, heavy atom on light atom effect[19]). On the other hand, the role of the relativistic effects may be diminished if we are interested only in the chemical shifts – the effects often largely cancel out and thus an approximate analysis may be sufficient when we do not need the shielding constants.

We have not discussed here the practical aspects of relativistic four- and two-component calculations. The results, as shown for instance by the

Table 6.7 Comparison of the theoretical and experimental $^1J(XH)$ values (in Hz).[a]

	CH_4	SiH_4	GeH_4	SnH_4
Equilibrium, CCSD	120.13	−188.82	−83.77	−1396
Vibrational correction	5.4	−7.59	−1.33	−21.55
Relativistic correction	0.23	−4.13	−13.01	−680
Total	125.75	−200.78	−98.11	−2098
Experiment	125.304(10)	(−)201.01(2)	(−)96.973(15)	−1933

[a]See Antušek *et al.*[88] and references therein for details.

comparison of shielding constants in mercury compounds determined using three different methods,[89] may strongly depend on the basis set, DFT functional used, nuclear model *etc.* We refer again to Chapter 8 for a more detailed description of these issues.

6.4.5 Zero-point Vibrational and Temperature Effects

Numerous examples illustrating the calculation of ZPV and temperature effects, as well as their magnitude for different nuclei, are discussed in Chapter 7. Thus, here we mention only some practical aspects of these calculations.

The parameters needed to determine the ZPV and temperature corrections to the shielding, such as the cubic force constants and the derivatives of the shielding constants (appearing in eqn (6.11)–(6.13)), are in practice often evaluated numerically.[9,90] For this purpose shielding constants at several geometries have to be calculated, in addition to equilibrium geometry. The symmetry of the molecule at most of these geometries may be lower than at the equilibrium, making the analysis of the vibrational corrections computationally demanding. Therefore, these corrections are often calculated using a less demanding approach – for instance, in the BF_3 and H_2S examples discussed above (see Table 6.1), ZPV was evaluated at the CCSD level using the cc-pVTZ basis set, instead of the CCSD(T)/cc-pCVQZ approach applied at the equilibrium geometry.

The contribution to the shielding constants describing the temperature effects is also evaluated applying the vibrational wavefunctions; it can be computed taking into account Boltzmann distribution over the vibrational levels. These effects are usually much smaller than the ZPV contribution, often below the accuracy of the calculated equilibrium shielding constants. However, once the ZPV correction has been computed, the temperature dependence of the shielding constant may be easily determined, and moreover it may be directly compared with experimental data.

We have focused on the role of the nuclear motion in the comparison of accurate *ab initio* results with experimental data. There is another interesting and important aspect of the nuclear motion effects. Namely, the shielding constants computed for a specific molecular geometry do not depend on the nuclear masses, therefore to determine the dependence of the shielding on isotopic substitution the rovibrational effects have to be taken into account. This dependence may be observed in experiment, and also in gas phase NMR spectra. As an example of the calculations for a polyatomic molecule let us mention the study by Auer[91] of the secondary isotope effects of methanol. The calculated values were in very good agreement with the gas-phase experimental data. The larger effects were fairly well reproduced at the HF level and considering only the ZPV corrections, but noticeably improved results were obtained at the CCSD(T) level, including in addition the temperature contributions.

6.4.6 Intermolecular Interactions

Until now, we have discussed calculations of the shielding constants of isolated molecules. The theoretical value of the shielding constant then corresponds to the experimental gas phase NMR value extrapolated to zero pressure (zero density). When the density is non-zero the intermolecular interactions have to be taken into consideration.

Interaction-induced chemical shifts in the gas phase can be described within theory by the virial expansion of the shielding constants. The experimental evidence indicates that in the gas phase the shielding depends linearly on the density, meaning that it is only necessary to consider binary interactions within the theory. Consequently, calculations of the shielding surfaces, which can be performed in the "supermolecule" approach for the dimers of interest, should be sufficient, and the interaction-induced chemical shifts can be theoretically described using only the second virial coefficient (in the expansion of the shielding constant).

Full quantum mechanical description of the virial coefficients is feasible for instance for atomic gases; for light rare-gas dimers at ambient temperature and pressure the estimated interaction-induced chemical shifts were very small.[92] Nevertheless, at the pressure of 10 bars (conditions which can be reached in gas phase NMR spectroscopy) such a shift may be observable, because for ^{21}Ne in Ne–Ar mixture it has a magnitude of -0.1 ppm (there is no stable NMR-active isotope of argon). A detailed theoretical study of the second virial coefficient of Xe nuclear shielding was performed by Hanni *et al.*[93] The interaction-induced shifts in Xe dimer were computed at the HF, MCSCF, MP2, and CCSD levels, with counterpoise correction for the basis set superposition error, and the temperature dependence of the shifts was evaluated. In this dimer, the effects are $\approx$30 ppm, they have been observed, and the CCSD results are in good agreement with the experimental data of Jameson *et al.*[94]

The interaction-induced chemical shifts of ^{2}H, ^{3}He, and ^{13}C were analyzed in a theoretical and experimental study of gaseous He–H_2, Ne–H_2, Ar–H_2, and He–CO_2 binary systems.[95] The averaged chemical shifts due to the interaction were calculated using the shielding surfaces of the interacting systems and taking into consideration Boltzmann statistics. For weakly interacting systems, the interaction-induced chemical shifts are masked in experiment by the more significant effect of bulk susceptibility. Thus, an important prerequisite for the comparison of theoretical and experimental induced shifts is a proper estimate of the bulk susceptibility corrections, based on accurate *ab initio* calculations of the solvent gas magnetizability. The contributions of the intermolecular interactions to the total σ_1^{AB} (see eqn (6.14)) extracted from the experiment were estimated in this manner. These σ_1^{A-B} contributions, describing the binary interactions, constitute for the studied dimers only a few percent of the total σ_1^{AB} (up to $\approx$20% for the strongest interacting system). The agreement between experimental and theoretical interaction-induced chemical shifts was generally satisfying.

A similar "supermolecule" approach may also be applied to estimate the role of the environment in the condensed phase. For example, the shielding constants of solvated alkali and alkali earth metal cations in water were evaluated using clusters with increasing number of water molecules as models of these ions.[96,97] For light metal cations the calculations of the shielding are feasible at CCSD level up to filled first solvation shell (see Table 6.8 for some examples). The convergence of the metal shielding constant with the number of water molecules in these clusters is very regular and the molecules in the second solvation shell practically do not affect the shielding of the central metal cation. Thus, in principle, ion shielding constants computed for the isolated structures which model the metal-water clusters with the first solvation shell filled provide an accurate approximation to the shielding constants of these ions in water.

In the studies of the heavier cations relativistic corrections were calculated and found to be practically independent of the number of water molecules in the cluster. In addition, an estimate of the nuclear motion corrections is needed for this type of complex. The results obtained from Car–Parrinello molecular dynamics for Mg^{2+}–water cluster have shown that the dynamical effects are smaller than the uncertainties due to basis set incompleteness.[97] Finally, as shown in Table 6.8, DFT overestimates the electron correlation contribution to the shielding of these solvated cations.

Table 6.8 Shielding constants of selected metal ions in clusters with increasing number of water molecules (in ppm).[a]

# of water molecules	0	1	2	3	4	5	6
Be^{2+}							
HF	130.91	117.17	111.90	111.36	114.23	114.01	113.93
CCSD	130.91	115.31	109.30	108.53	111.64	111.39	111.29
B3LYP	130.77	111.13	104.70	103.61	107.00	106.80	106.75
KT2	134.21	116.37	108.76	107.06	109.90	109.70	109.68
Na^{+}							
HF	623.81	606.26	590.36	580.27	579.01	580.57	583.53
CCSD	623.44	603.80	585.74	574.22	572.81	573.65	–
B3LYP	623.68	601.00	579.96	566.67	565.10	–	–
KT2	628.35	606.37	586.11	572.03	570.42	567.95	570.33
Mg^{2+}							
HF	695.15	651.26	610.81	592.75	587.33	597.33	604.55
CCSD	695.15	647.06	601.83	581.57	575.32	585.62	593.28
B3LYP	695.02	641.41	588.50	566.10	559.08	569.32	577.26
KT2	699.91	648.16	596.89	572.15	562.58	570.75	577.07

[a]Benchmark CCSD results for clusters with the first solvation layer filled are underlined. ANO-RCC basis was used set for the ions and cc-pVDZ for water molecules; for details see Antušek *et al.*[96,97]

6.4.7 Determining the Nuclear Magnetic Dipole Moments

The main obstacle in NMR measurements of nuclear magnetic dipole moments, performed for many stable nuclei in 1950–1970, was the large inaccuracy of the shielding constants, leading to small but systematic errors in the determination of these moments. High accuracy of presently computed *ab initio* shielding constants opens the possibility for new measurements of nuclear magnetic moments, with the precision significantly increased in comparison to the older literature data. An unknown magnetic moment of interest $m_{X,Z}$ is related to a known reference magnetic moment $m_{R,Z}$ by the equation

$$m_{X,Z} = \frac{\nu_X}{\nu_R} \frac{(1 - \sigma_R)}{(1 - \sigma_X)} \frac{I_X}{I_R} m_{R,Z} \tag{6.15}$$

where ν_X/ν_R is the ratio of experimental NMR resonance frequencies, measured in the same external magnetic field, for nuclei X and R. Using computed *ab initio* accurate shielding constants σ_R and σ_X one can obtain the magnetic moment of nucleus X; see for instance Antušek *et al.*[98] A nucleus for which the magnetic dipole moment has been established with high accuracy, preferably a proton, should be chosen as the reference R. The measured frequencies and the corresponding shielding constants in eqn (6.15) should be for precisely the same species, so the best results are obtained when isolated molecules and gas-phase data are considered, but once the effects of the environment are accounted for liquid phase frequency ratios can also be used (as, for instance, for the alkali and alkali-earth metals discussed above). The method and the set of nuclear magnetic moments which have been updated were recently reviewed[74] (see also Chapter 3). An interesting application utilizing the corrected nuclear magnetic moments is a new standardization of NMR spectroscopy and the possibility of direct measurement of shielding constants, as demonstrated for a series of hydrocarbons.[74] We note that there is also a demand for more precise nuclear magnetic moments, especially for heavy nuclei, coming from the community of atomic physicists (see *e.g.* Sunnergren *et al.*[99]).

6.4.8 Available Software Packages

There are numerous quantum chemistry programs which enable calculations of NMR shielding constants; many of these software packages can be obtained free of charge. Most of the results presented here were obtained applying the CFOUR[45] program, which enables the use of HF, MP2, and a sequence of coupled cluster methods including CCSD, CCSD(T), and CCSDT. In the CFOUR package all these methods are implemented with GIAO orbitals; also the spin-rotation constants can be calculated at the coupled cluster level of theory.[100] Within Kállay's MRCC[46] program, interfaced to CFOUR, one can apply higher coupled cluster approximations, which are suitable for benchmark NMR shielding calculations of small molecules.

In addition to the HF method many correlated methods are available in the Dalton[101] package. One can compute the shielding constants using the MCSCF method, SOPPA, and SOPPA(CCSD) methods (see *e.g.* Ligabue *et al.*[75] for an application) and CC2 and CC3 methods (see *e.g.* Garcia Cuesta *et al.*[29]). Besides, shielding constants may be determined applying density functional theory, with a variety of DFT functionals implemented.

There are also several programs which yield the relativistic values of the shielding constants, such as ADF,[102] DIRAC,[103] and ReSpect[104] (with GIAO implementation ensuring gauge invariance of the results in each of these packages).

A more complete list of quantum chemistry packages suitable for NMR shielding calculations can be found in Facelli's recent review.[105] This list includes many programs which enable simple calculations for large molecules (such as a DFT study of a molecule with a thousand atoms), programs that provide estimates of solvent effects and/or can be used for solid state NMR. We have mentioned above mainly the programs which incorporate the most advanced methods. It should be pointed out that most of these programs are regularly updated, the new releases are user-friendly, and the implemented black-box methods become more efficient. Presumably, for small molecules application of these methods yields the most reliable results, best suited for comparison with experimental gas phase values.

6.5 Conclusions

The on-going development of *ab initio* methods of quantum chemistry is reflected by the increasing accuracy of theoretically determined NMR shielding constants. For small molecules consisting of first- and second-row atoms the accuracy of the computed shielding constants can now be established analyzing the convergence of the results with the extension of the basis set and within the hierarchy of methods describing electron correlation effects. The estimated error bars, for instance for the carbon atoms, are $\approx$5–10 ppm. To increase the accuracy of the results one should take into account the rovibrational corrections, which are often of the same order of magnitude, and which may also be computed applying *ab initio* methods. In general, for this type of molecule the relativistic effects are small.

For molecules including also third-row atoms, similar calculations become significantly more expensive. Demonstrating the convergence of the computed values is at best tedious, and a safer estimate of the error bars might be 15–20 ppm. Moreover, the relativistic effects may also contribute $\approx$20 ppm. Nonetheless, considering the much larger span of the shielding constants, this accuracy may be satisfying.

It is not so easy to obtain accurate results for molecules including heavier elements. Non-relativistic calculations are much more complicated, because there are more electrons (hence more occupied orbitals) and large basis sets are needed. At the same time, it becomes essential to describe the relativistic effects, which affect not only the shielding constant of the heavy element, but also the shielding of the other nuclei in the molecule. These effects may be accounted for, but not yet with simultaneous reliable description of the electron correlation applying a large basis set. Although accurate results for molecules including heavy elements may be determined, proving that they are reliable and estimating their error bars is a challenge.

We have focused here on theoretical studies of small isolated molecules; we did not discuss the significant progress in the calculation of shielding constants of large molecules (including hundreds of atoms[106]), because these are not easily studied in the gas phase. Moreover, there is a variety of new methods which enable a description of the environment effects, affecting the comparison with experimental condensed phase NMR measurements.[107–110] One should also note systematic progress in *ab initio* studies of open-shell, paramagnetic systems.[111]

The foremost application of state-of-the-art calculations for isolated, small molecules is to provide the absolute shielding scales for the nuclei of interest. Once a single precise value for a single molecule is established, one can proceed using the chemical shifts – measured or computed – to determine the shielding constant of the nucleus of interest for the standard chemical reference. For instance, once the calculations are performed for PH_3 molecule, one can determine in this manner the phosphorus shielding constant in 85% water solution of H_3PO_4 (which obviously cannot be calculated with the same accuracy). Another dividend of accurate *ab initio* calculations is that they supply benchmark values needed to estimate the quality of approximate theoretical methods, such as DFT, which can be applied next to study large molecules and/or the effects of the environment.

Moreover, although in experiment the chemical shifts are in everyday use, the absolute shielding constants may be easily and successfully applied to interpret the spectra (see Chapter 3). In particular, when accurate nuclear magnetic moments are known, in this approach there is no need to have the same nucleus in the sample and in the reference compound. Finally, we recall that in some applications, for instance in the determination of nuclear magnetic dipole moments, the absolute shielding constants are needed and their accuracy determines the accuracy of the results.

Acknowledgements

We are indebted to Magdalena Pecul, Joanna Sadlej, and Stephan P. A. Sauer for helpful comments. We acknowledge financial support by the National Science Centre (Poland) grant, according to the decision No. DEC-2011/01/B/ST4/06588.

References

1. W. G. Proctor and F. C. Yu, *Phys. Rev.*, 1950, 77, 717.
2. See http://www.ebyte.it/library/hist/ProctorWG_Reminiscences.html.
3. W. C. Dickinson, *Phys. Rev.*, 1950, 77, 736.

4. C. J. Jameson, *Chem. Rev.*, 1991, **91**, 1375.
5. A. Abragam, *The Principles of Nuclear Magnetic Resonance*, Oxford University Press, Oxford, 1961.
6. N. F. Ramsey, *Phys. Rev.*, 1950, **78**, 699.
7. N. F. Ramsey, *Phys. Rev.*, 1953, **91**, 303.
8. J. Olsen and P. Jørgensen, *J. Chem. Phys.*, 1985, **82**, 3235.
9. J. F. Stanton and J. Gauss, *Int. Rev. Phys. Chem.*, 2000, **19**, 61.
10. W. H. Flygare, *J. Chem. Phys.*, 1964, **41**, 793.
11. W. H. Flygare, *Chem. Rev.*, 1974, **74**, 653.
12. T. Helgaker, P. Jørgensen and J. Olsen, *Molecular Electronic-Structure Theory*, J. Wiley, Chichester, 2000.
13. S. P. A. Sauer, *Molecular Electromagnetism, A Computational Chemistry Approach*, Oxford University Press, Oxford, 2011.
14. T. Helgaker, M. Jaszuński and K. Ruud, *Chem. Rev.*, 1999, **99**, 293.
15. J. Gauss and J. F. Stanton, *Adv. Chem. Phys.*, 2002, **123**, 355.
16. P. Lazzeretti, Electric and Magnetic Properties of Molecules, in *Handbook of Molecular Physics and Quantum Chemistry*, ed. S. Wilson, Molecules in the Physico-chemical Environment: Spectroscopy, Dynamics and Bulk Properties, John Wiley and Sons, London, 2003, vol. 3, p. 53.
17. G. A. Aucar, R. H. Romero and A. F. Maldonado, *Int. Rev. Phys. Chem.*, 2010, **29**, 1.
18. T. Helgaker, S. Coriani, P. Jørgensen, K. Kristensen, J. Olsen and K. Ruud, *Chem. Rev.*, 2012, **112**, 543.
19. *Calculation of NMR and EPR parameters. Theory and applications*, ed. M. Kaupp, M. Bühl and V. G. Malkin, Wiley-VCH, Weinheim, 2004.
20. *High Resolution NMR Spectroscopy: Understanding Molecules and their Electronic Structures*, ed. R. H. Contreras, Elsevier, Amsterdam, 2013.
21. J. Autschbach, *J. Phil. Trans. A*, 2014, **372**, 20120489.
22. M. E. Harding, M. Lenhart, A. A. Auer and J. Gauss, *J. Chem. Phys.*, 2008, **128**, 244111.
23. A. A. Auer, J. Gauss and J. F. Stanton, *J. Chem. Phys.*, 2003, **118**, 10407.
24. T. A. Ruden and K. Ruud, Ro-Vibrational Corrections to NMR Parameters, in *Calculation of NMR and EPR Parameters. Theory and Applications*, ed. M. Kaupp, M. Bühl and V. G. Malkin, Wiley-VCH, Weinheim, 2004, p. 153.
25. F. London, *J. Phys. Radium*, 1937, **8**, 397.
26. K. Wolinski, J. F. Hinton and P. Pulay, *J. Am. Chem. Soc.*, 1990, **112**, 8251.
27. P. Lazzeretti, M. Malagoli and R. Zanasi, *Chem. Phys. Lett.*, 1994, **220**, 299.
28. A. Ligabue, S. P. A. Sauer and P. Lazzeretti, *J. Chem. Phys.*, 2007, **126**, 154111.
29. I. Garcia Cuesta, J. Sánchez Marín, A. Sánchez de Merás, F. Pawlowski and P. Lazzeretti, *Phys. Chem. Chem. Phys.*, 2010, **12**, 6163.
30. F. Jensen, *J. Chem. Theory Comput.*, 2008, **4**, 719.

31. F. Jensen, *J. Chem. Theory Comput.*, 2015, **11**, 132.
32. H. Kjær and S. P. A. Sauer, *J. Chem. Theory Comput.*, 2011, **7**, 4070.
33. F. Jensen, *J. Chem. Theory Comput.*, 2006, **2**, 1360.
34. U. Benedict, A. A. Auer and F. Jensen, *J. Chem. Phys.*, 2008, **129**, 064111.
35. P. Manninen and J. Vaara, *J. Comput. Chem.*, 2006, **27**, 434.
36. D. M. Reid, R. Kobayashi and M. A. Collins, *J. Chem. Theory Comput.*, 2014, **10**, 146.
37. Y. Y. Rusakov, I. L. Rusakova and L. B. Krivdin, *Magn. Reson. Chem.*, 2015, **53**, 485.
38. J. A. Pople, *Rev. Mod. Phys.*, 1999, **71**, 1267.
39. J. Gauss, *J. Chem. Phys.*, 1993, **99**, 3629.
40. M. Maurer and C. Ochsenfeld, *J. Chem. Theory Comput.*, 2015, **11**, 37.
41. J. Gauss and J. F. Stanton, *J. Chem. Phys.*, 1995, **102**, 251.
42. J. Gauss and J. F. Stanton, *J. Chem. Phys.*, 1996, **104**, 2574.
43. J. Gauss, *J. Chem. Phys.*, 2002, **116**, 4773.
44. M. Kállay and J. Gauss, *J. Chem. Phys.*, 2004, **120**, 6841.
45. CFOUR, a quantum chemical program package written by J. F. Stanton, J. Gauss, M. E. Harding, P. G. Szalay with contributions from A. A. Auer, R. J. Bartlett, U. Benedikt, C. Berger, D. E. Bernholdt, J. Bomble, L. Cheng, O. Christiansen, M. Heckert, O. Heun, C. Huber, T.-C. Jagau, D. Jonsson, J. Jusélius, K. Klein, W. J. Lauderdale, D. A. Matthews, T. Metzroth, L. A. Mück, D. P. O'Neill, D. R. Price, E. Prochnow, C. Puzzarini, K. Ruud, F. Schiffmann, W. Schwalbach, C. Simmons, S. Stopkowicz, A. Tajti, J. Vázquez, F. Wang, J. D. Watts and the integral packages MOLECULE (J. Almlöf and P. R. Taylor), PROPS (P. R. Taylor), ABACUS (T. Helgaker, H. J. Aa. Jensen, P. Jørgensen, and J. Olsen), and ECP routines by A. V. Mitin and C. van Wüllen For the current version, see http://www.cfour.de.
46. MRCC, a quantum chemical program suite written by M. Kállay, Z. Rolik, I. Ladjánszki, L. Szegedy, B. Ladóczki, J. Csontos and B. Kornis. See also Z. Rolik, L. Szegedy, I. Ladjánszki, B. Ladóczki and M. Kállay, *J. Chem. Phys.* **139**, 094105 (2013), as well as: www.mrcc.hu.
47. J. Gauss and J. F. Stanton, *J. Chem. Phys.*, 1995, **103**, 3561.
48. J. Vaara, P. Manninen and P. Lantto, Perturbational and ECP Calculation of Relativistic Effects in NMR Shielding and Spin–Spin Coupling, in *Calculation of NMR and EPR parameters. Theory and applications*, ed. M. Kaupp, M. Bühl and V. G. Malkin, Weinheim, 2004, Wiley-VCH, p. 209.
49. D. G. Zaccari, M. C. Ruiz de Azúa, J. I. Melo and C. G. Giribet, *J. Chem. Phys.*, 2006, **124**, 054103.
50. J. Autschbach, in *High Resolution NMR Spectroscopy: Understanding Molecules and their Electronic Structures*, ed. R. H. Contreras, Elsevier, 2013, vol. 3, p. 69.
51. M. Reiher and A. Wolf, *Relativistic Quantum Chemistry*, Wiley-VCH, 2009.
52. W. T. Raynes, A. D. Buckingham and H. J. Bernstein, *J. Chem. Phys.*, 1962, **36**, 3481.
53. K. Jackowski, M. Wilczek, W. Makulski and W. Koźmiński, *J. Phys. Chem. A*, 2002, **106**, 2829.

54. K. Jackowski, W. Makulski, A. Szyprowska, A. Antušek, M. Jaszuński and J. Jusélius, *J. Chem. Phys.*, 2009, **130**, 044309.
55. A. Antušek and M. Jaszuński, *Mol. Phys.*, 2006, **104**, 1463.
56. S. Komorovsky, M. Repisky, E. Malkin, K. Ruud and J. Gauss, *J. Chem. Phys.*, 2015, **142**, 091102.
57. T. Kupka, M. Stachów, J. Kaminsky and S. P. A. Sauer, *Magn. Reson. Chem.*, 2013, **51**, 482.
58. M. Sun, I. Y. Zhang, A. Wu and X. Xu, *J. Chem. Phys.*, 2013, **138**, 124113.
59. A. M. Teale, O. B. Lutnæs, T. Helgaker, D. J. Tozer and J. Gauss, *J. Chem. Phys.*, 2013, **138**, 024111.
60. P. J. P. de Oliveira and F. E. Jorge, *J. Phys. B: At. Mol. Phys.*, 2008, **41**, 145101.
61. L. Armangue, M. Sola and M. Swart, *Mol. Phys.*, 2013, **111**, 1332.
62. A. Rudziński, M. Puchalski and K. Pachucki, *J. Chem. Phys.*, 2009, **130**, 244102.
63. R. K. Harris, E. D. Becker, S. M. C. de Menezes, R. Goodfellow and P. Granger, *Pure Appl. Chem.*, 2001, **73**, 1795, reprinted in *Magn. Reson. Chem.*, **40**, 489, 2002.
64. D. Sundholm and J. Gauss, *Mol. Phys.*, 1997, **92**, 1007.
65. M. Jaszuński, G. Łach and K. Strasburger, *Theor. Chem. Acc.*, 2011, **129**, 325.
66. M. Puchalski, J. Komasa and K. Pachucki, *Phys. Rev. A*, 2015, **92**, 020501(R).
67. A. A. Auer, *J. Chem. Phys.*, 2009, **131**, 024116.
68. E. Prochnow and A. A. Auer, *J. Chem. Phys.*, 2010, **132**, 064109.
69. C. Puzzarini, G. Cazzoli, M. E. Harding, J. Vázquez and J. Gauss, *J. Chem. Phys.*, 2009, **131**, 234304.
70. C. Puzzarini, G. Cazzoli, M. E. Harding, J. Vázquez and J. Gauss, *J. Chem. Phys.*, 2015, **142**, 124308.
71. T. Helgaker, J. Gauss, G. Cazzoli and C. Puzzarini, *J. Chem. Phys.*, 2013, **139**, 244308.
72. T. W. Keal and D. J. Tozer, *J. Chem. Phys.*, 2003, **119**, 3015.
73. D. Flaig, M. Maurer, M. Hanni, K. Braunger, L. Kick, M. Thubauville and C. Ochsenfeld, *J. Chem. Theory Comput.*, 2014, **10**, 572.
74. M. Jaszuński, A. Antušek, P. Garbacz, K. Jackowski, W. Makulski and M. Wilczek, *Progr. Nucl. Magn. Reson. Spectrosc.*, 2012, **67**, 49.
75. A. Ligabue, S. P. A. Sauer and P. Lazzeretti, *J. Chem. Phys.*, 2003, **118**, 6830.
76. E. Malkin, S. Komorovsky, M. Repisky, T. B. Demissie and K. Ruud, *J. Phys. Chem. Lett.*, 2013, **4**, 459.
77. A. Antušek, M. Jaszuński and M. Olejniczak, *Comput. Theor. Chem.*, 2011, **970**, 54.
78. J. F. Stanton, J. Gauss and H.-U. Siehl, *Chem. Phys. Lett.*, 1996, **262**, 183.
79. T. Helgaker, O. B. Lutnæs and M. Jaszuński, *J. Chem. Theory Comput.*, 2007, **3**, 86.

80. T. W. Keal, D. J. Tozer and T. Helgaker, *Chem. Phys. Lett.*, 2004, **391**, 374.
81. D. R. Price and J. F. Stanton, *Org. Lett.*, 2002, **4**, 2809.
82. M. Jaszuński and K. Ruud, *Mol. Phys.*, 2006, **104**, 2139.
83. P. Lantto, K. Jackowski, W. Makulski, M. Olejniczak and M. Jaszuński, *J. Phys. Chem. A*, 2011, **115**, 10617.
84. M. Iliaš, T. Saue, T. Enevoldsen and H. J. Aa. Jensen, *J. Chem. Phys.*, 2009, **131**, 124119.
85. A. Antušek, M. Pecul and J. Sadlej, *Chem. Phys. Lett.*, 2006, **427**, 281.
86. L. Cheng, J. Gauss and J. F. Stanton, *J. Chem. Phys.*, 2013, **139**, 054105.
87. P. Lantto and J. Vaara, *J. Chem. Phys.*, 2007, **127**, 084312.
88. A. Antušek, D. Kędziera, K. Jackowski, M. Jaszuński and W. Makulski, *Chem. Phys.*, 2008, **352**, 320.
89. V. Arcisauskaite, J. I. Melo, L. Hemmingsen and S. P. A. Sauer, *J. Chem. Phys.*, 2011, **135**, 044306.
90. J. Gauss and J. F. Stanton, *Chem. Phys. Lett.*, 1997, **276**, 70.
91. A. A. Auer, *Chem. Phys. Lett.*, 2009, **467**, 230.
92. A. Antušek, M. Jaszuński and A. Rizzo, *J. Chem. Phys.*, 2007, **126**, 074303.
93. M. Hanni, P. Lantto, N. Runeberg, J. Jokisaari and J. Vaara, *J. Chem. Phys.*, 2004, **121**, 5908.
94. C. J. Jameson, A. K. Jameson and S. M. Cohen, *J. Chem. Phys.*, 1975, **62**, 4224.
95. P. Garbacz, K. Piszczatowski, K. Jackowski, R. Moszynski and M. Jaszuński, *J. Chem. Phys.*, 2011, **135**, 084310.
96. A. Antušek, D. Kędziera, A. Kaczmarek-Kędziera and M. Jaszuński, *Chem. Phys. Lett.*, 2012, **532**, 1.
97. A. Antušek, P. Rodziewicz, D. Kędziera, A. Kaczmarek-Kędziera and M. Jaszuński, *Chem. Phys. Lett.*, 2013, **588**, 57.
98. A. Antušek, K. Jackowski, M. Jaszuński, W. Makulski and M. Wilczek, *Chem. Phys. Lett.*, 2005, **411**, 111.
99. P. Sunnergren, H. Persson, S. Salomonson, S. M. Schneider, I. Lindgren and G. Soff, *Phys. Rev. A*, 1998, **58**, 1055.
100. J. Gauss and D. Sundholm, *Mol. Phys.*, 1997, **91**, 449.
101. K. Aidas, C. Angeli, K. L. Bak, V. Bakken, R. Bast, L. Boman, O. Christiansen, R. Cimiraglia, S. Coriani, P. Dahle, E. K. Dalskov, U. Ekström, T. Enevoldsen, J. J. Eriksen, P. Ettenhuber, B. Fernandez, L. Ferrighi, H. Fliegl, L. Frediani, K. Hald, A. Halkier, C. Hättig, H. Heiberg, T. Helgaker, A. C. Hennum, H. Hettema, E. Hjertenæs, S. Høst, I.-M. Høyvik, M. F. Iozzi, B. Jansik, H. J. Aa. Jensen, D. Jonsson, P. Jørgensen, J. Kauczor, S. Kirpekar, T. Kjærgaard, W. Klopper, S. Knecht, R. Kobayashi, H. Koch, J. Kongsted, A. Krapp, K. Kristensen, A. Ligabue, O. B. Lutnæs, J. I. Melo, K. V. Mikkelsen, R. H. Myhre, C. Neiss, C. B. Nielsen, P. Norman, J. Olsen, J. M. H. Olsen, A. Osted, M. J. Packer, F. Pawlowski, T. B. Pedersen, P. F. Provasi, S. Reine, Z. Rinkevicius, T. A. Ruden, K. Ruud, V. Rybkin, P. Salek,

C. C. M. Samson, A. Sanchez de Meras, T. Saue, S. P. A. Sauer, B. Schimmelpfennig, K. Sneskov, A. H. Steindal, K. O. Sylvester-Hvid, P. R. Taylor, A. M. Teale, E. I. Tellgren, D. P. Tew, A. J. Thorvaldsen, L. Thøgersen, O. Vahtras, M. A. Watson, D. J. D. Wilson, M. Ziolkowski and H. Ågren, *Wiley Interdisciplinary Reviews: Computational Molecular Science*, 2014, **4**, 269, see also: Dalton, a Molecular Electronic Structure Program, Release DALTON2013.0 (2013), see http://daltonprogram.org.

102. E. J. Baerends, J. Autschbach, A. Berces, F. M. Bickelhaupt, C. Bo, P. M. Boerrigter, L. Cavallo, D. P. Chong, L. Deng, R. M. Dickson, D. E. Ellis, M. van Faassen, L. Fischer, T. H. Fan, C. Fonseca Guerra, S. J. A. van Gisbergen, J. A. Groeneveld, O. V. Gritsenko, M. Gruning, F. E. Harris, P. van den Hoek, C. R. Jacob, H. Jacobsen, L. Jensen, G. van Kessel, F. Kootstra, E. van Lenthe, D. A. McCormack, A. Michalak, J. Neugebauer, V. P. Osinga, S. Patchkovskii, P. H. T. Philipsen, D. Post, C. C. Pye, W. Ravenek, P. Ros, P. R. T. Schipper, G. Schreckenbach, J. G. Snijders, M. Sola, M. Swart, D. Swerhone, G. teVelde, P. Vernooijs, L. Versluis, L. Visscher, O. Visser, F. Wang, T. A. Wesolowski, E. van Wezenbeek, G. Wiesenekker, S. Wolff, T. Woo, A. Yakovlev, T. Ziegler; ADF2014.01, SCM, Theoretical Chemistry, Vrije Universiteit, Amsterdam, The Netherlands; http://www.scm.com, 2014.
103. DIRAC, a relativistic ab initio electronic structure program, Release DIRAC13 (2013), written by L. Visscher, H. J. Aa. Jensen, R. Bast and T. Saue, with contributions from V. Bakken, K. G. Dyall, S. Dubillard, U. Ekström, E. Eliav, T. Enevoldsen, E. Faßhauer, T. Fleig, O. Fossgaard, A. S. P. Gomes, T. Helgaker, J. K. Lærdahl, Y. S. Lee, J. Henriksson, M. Iliaš, Ch. R. Jacob, S. Knecht, S. Komorovský, O. Kullie, C. V. Larsen, H. S. Nataraj, P. Norman, G. Olejniczak, J. Olsen, Y. C. Park, J. K. Pedersen, M. Pernpointner, K. Ruud, P. Sałek, B. Schimmelpfennig, J. Sikkema, A. J. Thorvaldsen, J. Thyssen, J. van Stralen, S. Villaume, O. Visser, T. Winther and S. Yamamoto (see http://www.diracprogram.org).
104. ReSpect, version 3.2, 2013; Relativistic Spectroscopy DFT program of authors M. Repisky, S. Komorovsky, V. G. Malkin, O. L. Malkina, M. Kaupp, K. Ruud, with contributions from R. Bast, U. Ekstrom, S. Knecht, I. Malkin Ondik, E. Malkin.
105. J. C. Facelli, *Prog. NMR Spectrosc.*, 2011, **58**, 176.
106. M. Maurer and C. Ochsenfeld, *J. Chem. Phys.*, 2013, **138**, 174104.
107. J. J. Eriksen, J. M. H. Olsen, K. Aidas, H. Ågren, K. V. Mikkelsen and J. Kongsted, *J. Comput. Chem.*, 2011, **32**, 2853.
108. C. Steinmann, J. M. H. Olsen and J. Kongsted, *J. Chem. Theory Comput.*, 2014, **10**, 981.
109. M. Bühl and T. van Mourik, *WIREs Comput. Mol. Sci.*, 2011, **1**, 634.
110. M. Dračinský, H. M. Möller and T. E. Exner, *J. Chem. Theory Comput.*, 2013, **9**, 3806.
111. J. Vaara, Chemical Shift in Paramagnetic Systems, in *High Resolution NMR Spectroscopy: Understanding Molecules and their Electronic Structures*, ed. R. H. Contreras, Elsevier, 2013, ch. 3, pp. 41–67.

CHAPTER 7

Rovibrational and Temperature Effects in Theoretical Studies of NMR Parameters

RASMUS FABER,[a] JAKUB KAMINSKY[b] AND STEPHAN P. A. SAUER*[a]

[a] Department of Chemistry, University of Copenhagen, Universitetsparken 5, DK-2100 Copenhagen Ø, Denmark; [b] Department of Molecular Spectroscopy, Institute of Organic Chemistry and Biochemistry, 166 10 Prague, Czech Republic
*Email: sauer@kiku.dk

7.1 Methods for Calculation of Rovibrational Corrections

In the Born–Oppenheimer approximation the wavefunction of a molecule is factorized into a nuclear and an electronic part and the electronic energy as a function of the nuclear geometry forms a potential energy surface (PES) for the motion of the nuclei. This PES can be evaluated using any of the various electronic structure methods available.

Molecular properties such as NMR parameters can also be calculated at a given nuclear geometry using a wide range of methods.[1] Since NMR properties do not depend on nuclear momentum, one can similarly calculate the NMR property in the Born–Oppenheimer approximation for any nuclear geometry and think of this as constituting a property surface $P(\mathbf{q})$. However, in the Born–Oppenheimer approximation the correct value of a property is

New Developments in NMR No. 6
Gas Phase NMR
Edited by Karol Jackowski and Michał Jaszuński

Published by the Royal Society of Chemistry, www.rsc.org

not simply the value P_{eq} calculated with an electronic structure method at the geometry corresponding to the minimum of the potential energy surface, the so-called equilibrium geometry. Rather, one has to calculate the expectation value of the property as a function of the nuclear coordinates, *i.e.* the property surface $P(\mathbf{q})$, with the nuclear wavefunction Ψ of the appropriate vibrational state

$$\frac{\langle\Psi|P(\mathbf{q})|\Psi\rangle}{\langle\Psi\mid\Psi\rangle} \tag{7.1}$$

However, traditionally NMR properties are calculated at the equilibrium geometry, and it is therefore convenient to define a vibrational correction ΔP as the difference between the expectation value of the property surface and the value at the equilibrium geometry

$$\Delta P = \frac{\langle\Psi|P(\mathbf{q})|\Psi\rangle}{\langle\Psi\mid\Psi\rangle} - P_{eq} \tag{7.2}$$

Calculations of vibrational corrections require thus a reasonable description of both the nuclear wavefunction and the property surface. Many complex, iterative methods for calculating the vibrational wavefunction exist, usually tailored for the calculation of vibrational transitions (see for instance the recent review article[2] and references therein), which have also been used for the calculation of vibrational corrections to NMR parameters.[3,4] However, the use of these iterative methods requires detailed information about the potential energy surface, which can be costly to calculate for all but the smallest molecules. In addition a detailed description of the property surface, probably beyond quadratic terms, is required in order to truly benefit from the very accurate wavefunction. Due to the high computational cost of calculating NMR properties, and in particular spin–spin coupling constants, a description beyond quadratic terms in the property surface has rarely been considered for NMR properties.[5]

An attractive approach is to use perturbation theory in order to obtain an expression which depends on only a limited number of parameters. While such an approach is less accurate than iterative methods, this will often not be a problem since vibrational corrections are usually one or two orders of magnitude smaller than the equilibrium values and the accuracy of these is limited by the correlation method and basis set used in the calculation. A discussion of perturbation theory should also be helpful in order to determine which terms are the most important to include in variational approaches.

7.1.1 Perturbation Theory Approach

The vibrational wavefunction of a molecule cannot be calculated analytically due to the anharmonicity of the PES. However, one can employ perturbation theory with the anharmonic terms in the PES as perturbation, expand the wavefunction in orders of this perturbation (using λ as an order parameter), and truncate the expansion at an appropriate low order

$$\Psi = \Psi^{(0)} + \lambda\Psi^{(1)} + \lambda^2\Psi^{(2)} + \cdots \tag{7.3}$$

First of all one must find a functional form for the property surface. The simplest choice is a power series expansion in nuclear coordinates around the equilibrium geometry. Assuming that the nuclear displacements due to vibrational motion are small, one may expect the terms of low power in the displacements to be most important, such that only these need to be averaged using the fully perturbed vibrational wavefunction, while a lower order approximation suffices for the terms in the property surface of higher power. One way to ensure such a gradual truncation of the expansion of the property surface is to employ a double perturbation theory approach and explicitly include the perturbation order parameters in the expansion,

$$P(\mathbf{q}) = P_{\mathrm{eq}} + \lambda \sum_i \left.\frac{\partial P}{\partial q_i}\right|_{\mathbf{q}=\mathbf{0}} q_i + \frac{1}{2}\lambda^2 \sum_{i,j} \left.\frac{\partial^2 P}{\partial q_i \partial q_j}\right|_{\mathbf{q}=\mathbf{0}} q_i q_j + \cdots \tag{7.4}$$

where the coordinates $\mathbf{q}$ could be any displacement coordinates ($\mathbf{q}=\mathbf{0}$ at equilibrium), but in the following we assume them to be the reduced normal coordinates of the molecule.

The vibrational expectation value can then be expanded as (note the importance of including the normalization term)

$$\begin{aligned}\frac{\langle\Psi|P(\mathbf{q})|\Psi\rangle}{\langle\Psi\mid\Psi\rangle} = \Bigg[P_{\mathrm{eq}} + \lambda^2\Bigg(&\left\langle\Psi^{(1)}\mid\Psi^{(1)}\right\rangle P_{\mathrm{eq}} + 2\sum_i \left\langle\Psi^{(1)}|q_i|\Psi^{(0)}\right\rangle \left.\frac{\partial P}{\partial q_i}\right|_{\mathbf{q}=\mathbf{0}} \\ &+ \frac{1}{2}\sum_{i,j}\left\langle\Psi^{(0)}|q_iq_j|\Psi^{(0)}\right\rangle \left.\frac{\partial^2 P}{\partial q_i\partial q_j}\right|_{\mathbf{q}=\mathbf{0}}\Bigg) + \cdots\Bigg] \\ &\cdot\left(1 - \lambda^2\left\langle\Psi^{(1)}\mid\Psi^{(1)}\right\rangle + \cdots\right)\end{aligned} \tag{7.5}$$

where terms like $\langle\Psi^{(0)}|q_i|\Psi^{(0)}\rangle$ have been neglected on the assumption that $|\Psi^{(0)}\rangle$ is based on a harmonic approximation.

This way of expanding the expectation value has the nice feature that the equilibrium value is obtained as the zeroth order value, whereas collecting the terms quadratic in λ gives the lowest order vibrational correction terms, *i.e.* vibrational corrections at the vibrational second-order level of perturbation theory (VPT2):

$$\Delta^{\mathrm{VPT2}}P = 2\sum_i \left\langle\Psi^{(1)}|q_i|\Psi^{(0)}\right\rangle \left.\frac{\partial P}{\partial q_i}\right|_{\mathbf{q}=\mathbf{0}} + \frac{1}{2}\sum_{i,j}\left\langle\Psi^{(0)}|q_iq_j|\Psi^{(0)}\right\rangle \left.\frac{\partial^2 P}{\partial q_i\partial q_j}\right|_{\mathbf{q}=\mathbf{0}} \tag{7.6}$$

7.1.2 Vibrational Corrections

In this section only purely vibrational effects are considered. This is usually the starting point for rovibrational contributions, since at 0 K all molecules are in the rotational ground state and the zero-point correction includes only vibrational terms.

For the purpose of a perturbation expansion of the vibrational Hamiltonian, it is usually convenient to express the Hamiltonian in terms of normal coordinates, **Q**.

These are defined by a linear transformation

$$\sqrt{m_n}\Delta R_{n\alpha} = \sum_i l_{n\alpha,i} Q_i \tag{7.7}$$

of the mass weighted displacements $\Delta R_{n\alpha} = R_{n\alpha} - R_{n\alpha}^{\text{eq}}$ of the α (= x, y, or z) coordinates of nucleus n with mass m_n, where $R_{n\alpha}^{\text{eq}}$ is the α coordinate in the equilibrium geometry. The elements $l_{n\alpha,i}$ of the transformation matrix are chosen in such a way that the matrix of second derivatives of the potential energy surface becomes diagonal

$$\left.\frac{\partial^2 V}{\partial Q_i \partial Q_j}\right|_{\mathbf{Q}=\mathbf{0}} = \delta_{ij}(2\pi c\omega_i)^2 \tag{7.8}$$

Here ω_i is a harmonic frequency (in cm^{-1}). Sometimes the expression is simplified further by introducing the dimensionless *reduced normal coordinates*, **q**,

$$q_i = \sqrt{\frac{2\pi c\omega_i}{\hbar}} Q_i \tag{7.9}$$

An appropriate choice of zeroth-order Hamiltonian will be the kinetic energy and the quadratic term from a normal coordinate expansion of the PES. Higher order terms will be included as successively higher order perturbations

$$\hat{H}^{\text{vib}(0)} = \frac{hc}{2} \sum_i \omega_i \left(\frac{p_i^2}{\hbar^2} + q_i^2 \right) \tag{7.10}$$

$$\hat{H}^{\text{vib}(1)} = \frac{hc}{6} \sum_{ijk} k_{ijk} q_i q_j q_k \tag{7.11}$$

$$\hat{H}^{\text{vib}(2)} = \frac{hc}{24} \sum_{ijkl} k_{ijkl} q_i q_j q_k q_l \tag{7.12}$$

where p_i is the conjugated momentum operator of the reduced normal coordinate q_i and k_{ijk} and k_{ijkl} are third and fourth partial derivatives of the potential energy surface with respect to reduced normal coordinates.

Since $\hat{H}^{\text{vib}(0)}$ is the sum of harmonic oscillator Hamiltonians, the corresponding eigenfunctions are products of harmonic oscillator wavefunctions, $\chi_{\upsilon_i}(q_i)$,

$$X_{\boldsymbol{\upsilon}}(\mathbf{q}) = \prod_i \chi_{\upsilon_i}(q_i) \tag{7.13}$$

$$E_{\boldsymbol{\upsilon}}^{(0)} = hc\sum_i \left(\upsilon_i + \frac{1}{2}\right)\omega_i \tag{7.14}$$

with υ_i (collected in vector $\boldsymbol{\upsilon}$) being the quantum number for the oscillator along mode i, and the zeroth-order wavefunction, $\Psi^{(0)}$, is the configuration with the desired set of harmonic quantum numbers $\boldsymbol{\upsilon}^r$, *i.e.*

$$\Psi^{(0)} = X_{\boldsymbol{\upsilon}^r} \tag{7.15}$$

Furthermore, in this basis matrix elements over displacement operators are given as

$$\left\langle X_{\boldsymbol{\upsilon}'} \middle| \prod_i q_i^{n_i} \middle| X_{\boldsymbol{\upsilon}} \right\rangle = \prod_i \langle \upsilon_i' | q_i^{n_i} | \upsilon_i \rangle \tag{7.16}$$

where the matrix elements on the right are zero for any i, where $n_i - |\upsilon_i' - \upsilon_i|$ is negative or odd.

In order to evaluate eqn (7.6) only the part of the first-order wavefunction that can contribute to the $\langle \Psi^{(1)} | q_i | \Psi^{(0)} \rangle$ type of matrix elements is needed. By comparison with eqn (7.16), this can be written as

$$\Psi^{(1')}(\mathbf{q}) = \sum_i \left(a_i^{+(1)}\chi_{\upsilon_i^r+1}(q_i) + a_i^{-(1)}\chi_{\upsilon_i^r-1}(q_i)\right)\prod_{j\neq i}\chi_{\upsilon_j^r}(q_j) \tag{7.17}$$

with the coefficients given by first-order perturbation theory

$$a_i^{\pm(1)} = \mp\frac{1}{6\omega_i}\left(k_{iii}\langle \upsilon_i^r \pm 1 | q^3 | \upsilon_i^r \rangle + 3\sum_{j\neq i} k_{ijj}\langle \upsilon_i^r \pm 1 | q | \upsilon_i^r \rangle \langle \upsilon_j^r | q^2 | \upsilon_j^r \rangle\right) \tag{7.18}$$

Inserting these results along with the relevant one-oscillator matrix elements into eqn (7.6), one obtains

$$\Delta^{\text{VPT2}}P = -\frac{1}{2}\sum_i \frac{1}{\omega_i}\frac{\partial P}{\partial q_i}\bigg|_{\mathbf{q}=\mathbf{0}} \sum_j k_{ijj}\left(\upsilon_j^r + \frac{1}{2}\right) + \frac{1}{2}\sum_i \left(\upsilon_i^r + \frac{1}{2}\right)\frac{\partial^2 P}{\partial q_i^2}\bigg|_{\mathbf{q}=\mathbf{0}} \tag{7.19}$$

Inserting $\boldsymbol{\upsilon}^r = \mathbf{0}$ yields the expression commonly used for calculation of zero-point vibrational corrections.[6,7]

A very nice feature of eqn (7.19) is that the off-diagonal second derivatives of the property surface do not contribute to the vibrational correction at this level. This greatly reduces the cost of calculating the property surface by an order of the number of vibrational degrees of freedom.

7.1.3 The Effective Geometry Approach

While expressions based on an expansion around the equilibrium geometry like the one above are the most common, other approaches are possible. Ruud, Åstrand, and Taylor[8] proposed to use an expansion point found by minimizing the sum of the potential energy and the harmonic vibrational correction. This *effective geometry* will then satisfy the condition

$$k_i^{\text{eff}} + \sum_j \frac{k_{ijj}^{\text{eff}}}{2}\left(\upsilon_j + \frac{1}{2}\right) = 0 \tag{7.20}$$

for all vibrational modes i, where the "eff" superscript denotes the expansion around the effective geometry.

Since the expansion is then no longer with respect to the equilibrium geometry, the Hamiltonian will contain an additional term due to the gradient of the PES at the expansion point. Ruud *et al.* treat this as a first-order perturbation,

$$\hat{H}^{\text{eff}(1)} = \frac{hc}{6}\sum_{ijk} k_{ijk}^{\text{eff}} q_i^{\text{eff}} q_j^{\text{eff}} q_k^{\text{eff}} + hc\sum_i k_i^{\text{eff}} q_i^{\text{eff}} \tag{7.21}$$

By repeating the calculation of the previous section with this first-order Hamiltonian, it can be seen that eqn (7.20) is equivalent to requiring that the average displacement from the effective geometry vanishes to first order†

$$\langle \Psi^{(1,\text{eff})} | q_i^{\text{eff}} | \Psi^{(0,\text{eff})} \rangle = 0 \tag{7.22}$$

Optimizing the geometry to satisfy eqn (7.20) explicitly is not practical, since this would require the calculation of the cubic force field in each optimization step. Instead an approximation consistent to second order can be calculated from an expansion around the equilibrium geometry:

$$\Delta^{\text{eff}} q_i = 2\left\langle \Psi^{(1)} | q_i | \Psi^{(0)} \right\rangle = -\frac{1}{2\omega_i}\sum_j k_{ijj}\left(\upsilon_j + \frac{1}{2}\right) \tag{7.23}$$

Having thus found the effective geometry, the normal coordinates with respect to the effective geometry, q_i^{eff}, can be found.‡ The vibrational correction when expanded around the effective geometry is

$$\Delta^{\text{VPT2eff}} P = P_{\text{eff}} - P_{\text{eq}} + \frac{1}{2}\sum_i \left(\upsilon_i + \frac{1}{2}\right) \left.\frac{\partial^2 P}{(\partial q_i^{\text{eff}})^2}\right|_{\mathbf{q}^{\text{eff}}=0} \tag{7.24}$$

†Furthermore the second-order term is zero by symmetry.

‡Assuming the translational and rotational degrees of freedom are projected out.

Whereas eqn (7.19) explicitly references the cubic force constants, in eqn (7.24) they are hidden in the definition of the effective geometry. Therefore, the computational effort required is about the same for the two expressions. When considering effects beyond the pure zero-point vibrational correction, it should, however, be noted that the effective geometry depends on the harmonic quantum numbers, which means the effective geometry will depend, for instance, on temperature.

7.1.4 Rotational Contributions

So far we have only considered motion within the internal coordinate system of the molecule. But in a gas at non-zero temperature many of the molecules will also have a non-zero rotational energy. This rotational motion is of course much faster than the NMR time scale and is the reason for using only the isotropic values of NMR parameters in gas and liquid phase spectroscopy. What we will consider in this section is therefore not this primary effect of molecular rotation, but rather the influence of the rotational motion on the vibrational motion of the molecule.

In the previous sections only the purely vibrational kinetic and potential energy in the molecular coordinates were included. By comparison with the Watson Hamiltonian,[9] one can identify the additional terms for coupling to the rotational motion

$$\hat{H}^{\text{Watson}} - \hat{H}^{\text{Vib}} = \frac{1}{2}(\hat{\mathbf{J}} - \hat{\boldsymbol{\pi}})\boldsymbol{\mu}(\hat{\mathbf{J}} - \hat{\boldsymbol{\pi}}) - \frac{1}{8}\hbar^2 \text{Tr}\boldsymbol{\mu} \tag{7.25}$$

where $\boldsymbol{\mu}$ is the inverse effective moment of inertia tensor,

$$\boldsymbol{\mu} = \mathbf{I}'^{-1} \tag{7.26}$$

$$I'_{\alpha,\beta} = I_{\alpha,\beta} - \frac{\hbar}{2\pi c} \sum_{i,j,k} \frac{q_j q_k}{\omega_j \omega_k} \zeta_{ji}^{\alpha} \zeta_{ki}^{\beta} \tag{7.27}$$

$\hat{\boldsymbol{\pi}}$ is the vibrational angular momentum,

$$\hat{\pi}_\alpha = \sum_{i,j} q_i p_j \sqrt{\frac{\omega_j}{\omega_i}} \zeta_{ij}^{\alpha} \tag{7.28}$$

and ζ_{ij}^{α} is the Coriolis coupling constant,

$$\zeta_{ij}^{\alpha} = \sum_n (l_{n\beta,i} l_{n\gamma,j} - l_{n\gamma,i} l_{n\beta,j}) \tag{7.29}$$

where α, β, γ is a cyclic permutation of x, y, z and $l_{n\beta,i}$ are again components of the transformation matrix to normal coordinates, eqn (7.7).

This term should then be expanded in order of perturbation theory in a suitable manner, for instance by equating the perturbation order with the power of the normal coordinate. Note that we have, as is customary, ignored the last term of eqn (7.25).

The expansion becomes then[10]

$$H^{\text{rovib}(0)} = \frac{1}{2}\sum_{\alpha}\frac{\hat{J}_{\alpha}^{2}}{I_{\alpha\alpha}^{e}} + H^{\text{vib}(0)} \tag{7.30}$$

$$H^{\text{rovib}(1)} = -\sum_{\alpha}\frac{\hat{\pi}_{\alpha}\hat{J}_{\alpha}}{I_{\alpha\alpha}^{e}} - \frac{1}{2}\sum_{i}\sqrt{\frac{\hbar}{2\pi c\omega_i}}q_i\sum_{\alpha,\beta}a_i^{\alpha\beta}\frac{\hat{J}_{\alpha}\hat{J}_{\beta}}{I_{\alpha\alpha}^{e}I_{\beta\beta}^{e}} + H^{\text{vib}(1)} \tag{7.31}$$

where $a_i^{\alpha\beta}$ is the linear expansion coefficient of the moment of inertia in the normal coordinates

$$a_i^{\alpha\alpha} = 2\sum_{n}\sqrt{m_n}\left(R_{n\beta}^{\text{eq}}l_{n\beta,i} + R_{n\gamma,i}^{\text{eq}}l_{n\gamma,i}\right) \tag{7.32}$$

$$a_i^{\alpha\beta} = -2\sum_{n}\sqrt{m_n}\left(R_{n\alpha}^{\text{eq}}l_{n\beta,i}\right) \qquad \alpha \neq \beta \tag{7.33}$$

The zeroth-order Hamiltonian is that of a rigid rotor with no coupling to the vibrational motion. Therefore the eigenfunctions will be the ones of the purely vibrational problem times a rigid rotor wavefunction, $|R\rangle$. The integrals in eqn (7.6) factorize therefore in integrals of the vibrational wavefunctions over the normal coordinates times an overlap integral of the rotational wavefunctions, which implies that the first and zeroth-order rovibrational wavefunctions must have the same rigid rotor wavefunction $|R\rangle$.

The first term of $H^{\text{rovib}(1)}$ is quadratic in the normal coordinates, and thus will not contribute to the first-order linear displacement. The second term gives the first-order centrifugal distortion term. Adding this term to the Hamiltonian modifies the coefficient for the first-order wavefunction by

$$a_i^{\pm(1,\text{cent})} = \pm\frac{\langle \upsilon_i \pm 1|q_i|\upsilon_i\rangle}{\omega_i hc}\left(\frac{1}{4\pi}\sqrt{\frac{h}{c\omega_i}}\sum_{\alpha}\frac{a_i^{\alpha\alpha}}{(I_{\alpha\alpha}^{e})^2}\langle R|\hat{J}_{\alpha}^{2}|R\rangle\right) \tag{7.34}$$

and the first-order contribution to the displacement is augmented by[11]

$$2\left\langle \Psi^{(1,\text{cent})}|q_i|\Psi^{(0)}\right\rangle = \frac{1}{2\omega_i}\frac{1}{2\pi c}\sqrt{\frac{1}{hc\omega_i}}\sum_{\alpha}\frac{a_i^{\alpha\alpha}}{(I_{\alpha\alpha}^{e})^2}\langle R|\hat{J}_{\alpha}^{2}|R\rangle \tag{7.35}$$

7.1.5 Temperature Averaging

At non-zero temperatures not all molecules will be in the state of lowest energy. While electronic excitation energies are usually so large compared to common thermal energies that thermal electronic excitation can be neglected, this is not the case for rotational and vibrational excitation energies.

Temperature can be taken into account by calculating the thermal averages of the expressions given in the previous sections.

The expression in eqn (7.19) is linear in terms of the type $\upsilon_i + \frac{1}{2}$, and the thermal average can therefore be found by simply taking the Boltzmann average of this term. If it is furthermore assumed that the energy of a vibrational state is adequately described by the harmonic approximation, the expression

$$\left\langle \upsilon_i + \frac{1}{2} \right\rangle^T = \frac{\sum_\upsilon \left(\upsilon + \frac{1}{2}\right) \mathrm{e}^{-hc\omega_i\upsilon/kT}}{\sum_\upsilon \mathrm{e}^{-hc\omega_i\upsilon/kT}} = \frac{\mathrm{e}^{-hc\omega_i/kT}}{1 - \mathrm{e}^{-hc\omega_i/kT}} + \frac{1}{2} = \frac{1}{2}\coth\left(\frac{hc\omega_i}{2kT}\right) \tag{7.36}$$

is obtained.

For the rotational motion, no general solution even to the zeroth-order problem can be found. Since the spacing between the rotational energy levels is small for all but the smallest of molecules, the thermal average may be approximated by treating it classically and using the equipartition theorem. Since $\langle R|J_\alpha^2|R\rangle/2I_{\alpha\alpha}^e$ is a quadratic energy contribution, one obtains

$$\frac{1}{2}\langle R|J_\alpha^2|R\rangle = \frac{1}{2}kTI_{\alpha\alpha}^e \tag{7.37}$$

Adding eqn (7.19) and (7.35) and inserting the above expression yields[11]

$$\Delta^{\mathrm{VPT2}}P = -\frac{1}{2}\sum_i \frac{1}{\omega_i}\frac{\partial P}{\partial q_i}\bigg|_{\mathbf{q}=\mathbf{0}} \left(\frac{1}{2}\sum_j k_{ijj}\coth\left(\frac{hc\omega_j}{2kT}\right) - \frac{kT}{2\pi c}\sqrt{\frac{1}{hc\omega_i}}\sum_\alpha \frac{a_i^{\alpha\alpha}}{I_{\alpha\alpha}^e}\right) + \frac{1}{4}\sum_i \coth\left(\frac{hc\omega_i}{2kT}\right)\frac{\partial^2 P}{\partial q_i^2}\bigg|_{\mathbf{q}=\mathbf{0}} \tag{7.38}$$

as an approximation to the rovibrational correction at a finite temperature.

7.1.6 Secondary Isotope Effects

In NMR spectroscopy isotopic substitution will give a large shift of the NMR parameters for an atom since different isotopes have different spins and g-factors. For the calculation of chemical shifts this is of little interest to us, because the change in Larmor frequency between isotopes is usually orders of magnitude larger than the range observed during an experiment, and spectra will thus show only resonances for one isotope of the given atom. It is possible to observe spin–spin coupling to a different isotope of a neighboring atom, but the pattern thus obtained will be further complicated by the secondary isotope effect on the chemical shift.

In this section we will instead discuss a secondary effect of isotopic substitution, which is caused by the fact that the differing masses of different isotopes will also change the vibrations of the molecule. This will mean that the vibrational corrections to a given NMR parameter will also depend on the masses of all atoms in the molecule, and this shift upon isotopic substitution is indeed observable.[12,13] Furthermore, non-standard isotopic compositions are necessary in order to observe coupling constants in highly symmetric compounds, where certain nuclei are otherwise equivalent.

For the purpose of studying isotope effects, it is apparent that the normal mode picture discussed above and the simple expressions they yield have certain shortcomings. Since the normal modes are transformed mass weighted coordinates, they will necessarily change when the masses of the nuclei change. While expressions like eqn (7.19) and (7.38) are attractive in that they depend only on diagonal second derivatives of the property surface, these derivatives are along a set of isotope-dependent coordinates. The derivatives can of course be transformed from the normal coordinates of one isotopic composition to another, but in order to do this also the mixed second derivatives must be known. Assuming that the number of different isotopic compositions of interest is lower than the number of vibrational degrees of freedom, it may therefore be advantageous to simply recalculate the property derivatives from scratch. A similar problem appears for the semi-diagonal cubic force field if calculated from analytical gradients, but if it is calculated from analytical Hessians it is possible to construct the full cubic force field and thus transform to an arbitrary set of coordinates.

For symmetric molecules it is possible to reduce the cost of calculating the full set of second derivatives. Mixed second derivatives along two modes belonging to different irreducible representations will necessarily be zero and these can be left out of the calculation. Calculations for the full set of possible isotropic compositions can then be carried out at a reasonable cost for certain molecules by establishing the property surface along a set of symmetry adapted coordinates as *e.g.* done in the work of Raynes and coworkers.[14–23]

7.1.7 Alternative Perturbation Expansions

It should be noted that most authors do not include perturbation order parameters explicitly in the expansion of the property surface. However, if this is not done, there is no natural limit at which the contribution from higher order terms will vanish completely even with a low-order expansion of the wavefunction. So while eqn (7.19) is the most common expression used for the calculation of vibrational corrections, it is usually only obtained after an *ad hoc* truncation of the property surface, since the remaining terms which should in this case appear to first order, $\langle\Psi^{(1)}|q_i^2|\Psi^{(0)}\rangle$, vanish due to the symmetry of harmonic oscillator functions.

With no ordering of terms built into the basic ansatz, it becomes unclear why the truncation of the property surface should be done after the quadratic terms, and arbitrarily high-order derivatives will indeed give non-zero contributions, even when using only the first-order perturbed wavefunction. Indeed a number of different truncations have been considered, for instance keeping the full third-order property surface or by keeping the diagonal third and fourth derivatives.[24] Still the high computational cost of calculating higher derivatives and the limited accuracy of higher order numerical differentiation make inclusion of higher derivatives unsuitable for routine applications.

Other approaches include treating also the quartic part of the potential as first-order perturbation.[25,26] If this is done consistently, the first-order wavefunction will include terms which differ from the reference wavefunction by an even number of harmonic quantum numbers and thus the $\langle\Psi^{(1)}|q_iq_j|\Psi^{(0)}\rangle$ terms will be non-zero, requiring the calculation of the off-diagonal property derivatives.[26]

Finally an intrinsic problem arising with any expansion using harmonic oscillator functions to zeroth order is the treatment of modes with multiple minima such as internal rotations or inversion modes such as the umbrella motion in ammonia. The harmonic oscillator functions will necessarily be localized in one potential well and no level of perturbation theory will be able to repair that. One simpler solution than abandoning perturbation theory completely[5] is to treat only the offending mode numerically and use the simple perturbation treatment for the rest.[27,28]

7.1.8 Calculation of the Required Parameters

While eqn (7.19) and (7.38) relate the vibrational corrections to quantities, which can be calculated by electronic structure theory, exactly how this is done deserves some additional consideration. The basic problem is that the quantities required are defined as the third (cubic force field and property gradient) and fourth (property Hessian) derivatives of the electronic energy.[§] However, whereas implementations of analytic first and second derivatives are fairly common, only very few implementations of up to fourth derivatives of the electronic energy have been reported so far.[29,30] Though these implementations allow calculations of the purely geometric derivatives, it seems that they have yet to be extended to allow for calculation of geometric derivatives of NMR properties. So while the development of high-order analytic derivative techniques is an ongoing process, in the following it will be assumed that analytic derivative approaches are only available up to second order.

For calculation of the cubic force fields there are, besides the rarely available fully analytical approach, two feasible schemes; numerical first derivatives of analytic Hessians and numerical second derivatives of analytic gradients.[¶,31–33] Since only the semi-diagonal parts of the cubic force field

§Assuming that the property itself is defined as the second derivatives, as is the case for the paramagnetic spin-orbit contribution to shielding constant and the Fermi-contact, spin-dipolar, and paramagnetic spin-orbit contribution to coupling constants.

¶Due to the efficiency of analytic gradient calculations and the numerical error associated with such an approach, fully numerical third derivatives of the energy should be avoided whenever possible.

are needed in eqn (7.19), the last approach is actually the cheapest in terms of computation time, as only gradient calculations on one geometry displaced in either direction along each normal mode, in total $6N$ gradient calculations, are required. In comparison the approach using analytic Hessians will usually require one Hessian calculation in either direction along each normal mode, that is $6N$ Hessian calculations, which will be about $3N$ times as expensive! The Hessian approach as described here allows determination of the full cubic and parts of the quartic force field, but if our interest is only eqn (7.19) this information is wasted. Still the Hessian approach is in general favored, in particular for black box approaches, since the error from numerical differentiation is much lower for first than for second derivatives.

Similarly the evaluation of the property surface derivatives can be done from $6N+1$ property calculations. In this case, however, there is no way to avoid calculation of second derivatives by numerical methods, and therefore the lack of numerical stability of the derivatives of the property surface puts a limit on the accuracy of computed vibrational corrections. It is important to keep in mind that the step length used for numerical differentiation is a compromise; too small and numerical errors will dominate, too large and the derivatives will be contaminated by higher order terms. The influence of higher order terms can be suppressed by using more points in the numerical differentiation, which allows the use of larger step lengths in order to reduce the numerical error. But this increases the number of property calculations to be carried out and thus the total computational cost of the vibrational averages. Because the numerical error of the differentiation step really depends on the numerical accuracy with which the points of the property surface have been determined, it is important to stress that the tightest possible convergence thresholds should be used in the electronic structure calculations.

To make matters even worse, calculations of spin–spin coupling constants on non-equilibrium geometries face an increased risk of running into triplet instabilities.[34,35] Unrelaxed coupled cluster approaches have been shown to avoid such problems,[36] but these methods will often be too costly for routine applications. DFT approaches are often used instead and while the pure LDA and GGA functionals should be free from instability problems,[‖] there may be some reason for concern regarding popular hybrid functionals such as B3LYP.[34,35]

7.2 Examples of Vibrational Corrections to Shieldings

In this chapter we want to review briefly some important contributions to this field. First we focus on zero-point vibrational corrections (ZPVCs) for

‖At least at geometries likely to be used for numerical differentiation of property surfaces.

different atom types and discuss differences between different classes of molecules. In the second part we address some methodological aspects of ZPVC calculations (basis set effect, electron correlation, relativistic effects, *etc.*). In the end we summarize general trends of vibrational corrections to magnetic shieldings and present some practical hints for their calculations.

7.2.1 Vibrational Corrections to Hydrogen Shieldings

Zero-point vibrational corrections to hydrogen NMR shieldings, σ_H, have been studied extensively during the last 20 years on many more systems than corrections to any other type of atom. In this chapter we will discuss trends in these corrections separately for hydrogens in hydrocarbons, hydrogens attached to atoms different from carbon, and aromatic hydrogens.

7.2.1.1 Aliphatic Hydrogens

Ruud *et al.*[37] reported zero-point vibrational corrections to aliphatic hydrogen shieldings in 34 different molecules comprising hydrocarbons, alcohols, aldehydes, ethers, organic acids, and amines. They observed that the ZPVCs to σ_H within the same or very similar functional groups are independent of the rest of the molecule and exhibit thus a high transferability as will be discussed in detail in Section 7.2.7. In Table 7.1 the zero-point vibrational corrections to alkyl hydrogens in methane derivatives (methane, methanol, formaldehyde, and formic acid) are shown as examples. It is apparent that the corrections vary only within a range of 0.05 ppm, which is remarkable taking into account the very different electronic structure in these molecules. Considering all 34 molecules in this study, it can be concluded that the ZPVCs to alkyl hydrogen shieldings are always negative and approximately in the range from −0.4 to −0.9 ppm. Although this represents a non-negligible contribution, the ZPVCs amount commonly to less than 3% of the electronic contribution. However, the calculated ZPVCs to aliphatic hydrogen shieldings bring the theoretical chemical shifts, calculated at the Hartree–Fock level, closer to experiment.

Table 7.1 Zero-point vibrational corrections (in ppm) to hydrogen shieldings in methane derivatives, calculated at the Hartree–Fock level, taken from ref. 37.

	ZPVC
CH_4	−0.59
CH_3OH	−0.60
HCHO	−0.59
HCOOH	−0.55

Table 7.2 ZPVC (in ppm) to hydrogen shieldings in -OH and -NH group calculated at the Hartree–Fock level.[37]

	ZPVC (-OH)		ZPVC (-NH)
Methanol	0.26		
Ethanol	0.51	Ethanamine	−0.19
1-Propanol	0.52	1-Propanamine	−0.17
2-Propanol	0.41	2-Propanamine	−0.19
Propen-2-ol	0.60		

7.2.1.2 Hydrogens Bound to Heteroatoms

It is obvious that the difference in character of hydrogens in -XH groups (X = O, N, S, *etc.*) and of alkyl hydrogens leads to different ZPVCs to their shieldings. From Table 7.2 it is seen that hydroxylic hydrogens have ZPV corrections of similar size but with the opposite sign to aliphatic hydrogens. One can observe a slight dependence of the ZPV correction on the position of hydroxyl group and that the presence of a multiple bond in the vicinity of the -OH group increases the ZPVC of the hydroxylic hydrogen. Contrary to -OH group, corrections to hydrogens in -NH groups are again negative but significantly smaller than corrections for aliphatic hydrogens. Furthermore, the changes due to different positions of the amino group are negligible.

7.2.1.3 Aromatic Molecules

Zero-point vibrational corrections to hydrogen shieldings of aromatic hydrogens (investigated for benzene, toluene, aniline, phenol, and benzoic acid) are approximately −0.4 ppm. Analyzing the data it becomes apparent that the ZPV corrections are almost independent of the position of the hydrogens relative to a functional group in the aromatic ring, which is in contrast to the equilibrium geometry values of these hydrogen shieldings. In general the functional groups do not significantly influence the ZPVC to the shieldings of aromatic hydrogens.

7.2.2 Vibrational Corrections to Carbon Shieldings

In this section we will discuss the effect of vibrational averaging on carbon shieldings. We focus again mostly on organic molecules, as studies of their ZPV corrections prevail in the literature, but we include also carbon mono- and dioxide, which are popular test systems in many studies. Furthermore we distinguish again between aliphatic and aromatic molecules. Note that the fullerene C_{70} is to the best of our knowledge the largest system for which the ZPVCs to ^{13}C NMR shieldings have been calculated so far.[38]

7.2.2.1 Aliphatic Molecules

Table 7.3 summarizes zero-point vibrational corrections to ^{13}C NMR shieldings obtained for 20 carbon atoms in 17 different molecules.[39,40]

Table 7.3 Vibrational corrections (ppm) to ^{13}C NMR shieldings.[39,40] Corrections calculated at several levels as indicated.

	HF	MP2	DFT	CCSD(T)
CH_4	−3.01	−2.56, −3.10	−3.20	−2.96
C_2H_6	−4.11	−3.82, −4.20	−4.70	
C_2H_4	−4.50	−3.53, −5.50	−5.00	
C_2H_2	−4.31	−3.09, −4.90	−3.80	−3.84
CH_3F	−3.55	−3.70		
CH_3OH	−3.90	−3.98		
CH_3NH_2	−4.10	−4.06		
CH_3CN	−2.67	−2.40		
CH_3CHO	−3.63	−3.47		
CH_3COCH_3	−3.38	−3.25		
CO	−2.51	−1.02, −3.50	−2.40	−1.78
CO_2	−1.71	−0.89, −2.00	−1.50	−1.34
CH_3CHO	−2.24	−1.34		
CH_3COCH_3	−1.99	−0.80		
HCN	−2.11	−1.15	−3.15	−1.62
CH_3CN	−1.45	−1.52		
CIH_2CCH_2	−2.02	−1.58		
CH_2CCH_2	−2.79	−1.74		
CF_4	−1.01	−1.36		
CH_2O	−3.14	−2.46	−8.57	

It appears that the ZPV corrections can lie in the range between −0.8 and −5.5 ppm and amount to approximately 2–3% of the equilibrium geometry values. For special cases, however, they can reach up to 33% (carbonyl in acetaldehyde; ZPVC −1.34 ppm, $\sigma_{eq.}$ −2.60 ppm, $\sigma_{exp.}$ −6.8 ppm). The ZPV corrections seem to be highly systematic, *i.e.* they are negative in all the cases presented here, which implies that the ZPV corrections on the chemical shifts are smaller. Note that for all molecules in the test set inclusion of vibrational corrections improves the agreement with the experimental values.

It appears that the ZPV corrections are quite similar for carbons of $-CH_3$ groups in different molecules. For example, for the smaller set of molecules with a methyl-group, there is only a slight correlation between the decreasing inductive effect of the attached group and an increasing ZPV correction in the sequence $CH_3CN < CH_3CHO = CH_3COCH_3 < CH_3F < CH_3OH < CH_3NH_2$. However, the difference between the maximal and minimal ZPV correction in this small set is only 1.7 ppm and methane does not fit in this sequence, as it ought to have the highest correction but has a ZPVC comparable to CH_3CN. Vibrational contributions seem to be also higher in absolute values for methyl carbons than for any other type of carbon atoms.

7.2.2.2 Aromatic Molecules

If one takes the ^{13}C shieldings in benzene, toluene, phenol, aniline, and benzoic acid as representatives of carbon shieldings in aromatic molecules, one finds ZPV corrections of approximately 3–6 ppm in absolute values.[37] The highest ZPVC (−6.19 ppm) is found, however, for the non-aromatic carbon in the $-CH_3$ group of toluene. The exception is benzoic acid, where all aromatic carbons have zero-point vibrational corrections lower than 1 ppm, while the carboxylic carbon has large negative correction (−12.34 ppm). One should also note that the ZPV corrections to the aromatic ring carbons in benzoic acid are both negative and positive depending on the position in the ring. A strong dependence of the ZPV correction on the position of the carbon relative to the functional group was observed also for other systems. Depending on the nature of the substituent the magnitude of the correction to the ^{13}C shielding can either increase or decrease with the distance from the *ipso*-carbon. Thus, for example in aniline, the ZPV correction decreases from the *ortho*- > *meta*- > *para*-positions, while in toluene it goes the other way around.

7.2.3 Vibrational Corrections to Nitrogen Shieldings

A deeper analysis of the effects of zero-point vibrational corrections to ^{15}N NMR shieldings is hampered by two facts. Firstly, ZPVCs in nitrogen-containing compounds are investigated in only a few theoretical studies and no systematic trial set of representative molecules has been created. Secondly, there are even fewer gas phase NMR data available. Thus, most studies concentrate mainly on the N_2 or NH_3 molecules. Nevertheless, we will gather available data for different molecules in order to get an idea of the influence of typical N-bonding environments (*i.e.* N-containing molecules with single, double, or triple bonds) on the ZPVC to ^{15}N shieldings.

Table 7.4 summarizes available ZPV corrections to nitrogen shielding in several N-containing molecules calculated at different levels of theory.[8,40–44] The corrections are in the order of −3 to −9 ppm, which is far from negligible considering the scale of ^{15}N shieldings ranging from −60 to 250 ppm for this set. However, the ZPV corrections are rather uniform over this set, which implies that their effect on the ^{15}N chemical shifts will be small. It is difficult to detect a relationship between the ZPVC and the character of the bonds to the studied nitrogen from such a small test set. Nevertheless, it appears according to the CCSD(T) results that ZPV corrections are a bit larger for molecules, where nitrogen is attached by a single bond. One should also

Table 7.4 Zero-point vibrational corrections (in ppm) to ^{15}N shieldings.[8,40–44]

	HF	DFT	MP2	CCSD(T)
N_2	−4.7		−0.8	−3.3, −4.0
NH_3	−6.7, −7.0	−10.7, −5.9	−5.5, −6.8, −7.5	−6.8
HCN	−9.3	−13.25	−5.2	−8.2
NNO	−8.5		−3.0	−6.8
N**N**O	−4.7		0.0	−3.1
CH_3NH_2	−8.2		−7.5	−8.7
CH_3NO_2		−11.8		

notice that the effect of electron correlation is in general high in systems with nitrogen in multiple bonds. The effect of electron correlation will be discussed in more detail in Section 7.2.8.1.

7.2.4 Vibrational Corrections to Oxygen Shieldings

The inclusion of ZPV corrections is also vital for an accurate description of ^{17}O NMR shieldings.[8,20,43–47] As seen from Table 7.5, ZPV corrections range from a few ppm up to −30 ppm for F_2O or even higher as shown by Kupka *et al.*,[43] who observed a ZPV correction of −45 ppm for ^{17}O in H_2CO. The inclusion of vibrational corrections calculated even at the Hartree–Fock level improves the agreement with experiment and reduces the mean average deviation by about 10 ppm.[45] Auer also concluded that, with the exception of some special molecules, Hartree–Fock-based ZPVC represent a cost-effective alternative at least for an estimate, since they differ from higher-level results by only a few ppm (see Section 7.2.8.1 for details). Thus, by including vibrational corrections it is possible to reproduce shieldings with deviations from experiment smaller than 10 ppm (only for F_2O, $CH_2(CH_3O)_2$ or oxetane the deviations are around 15 ppm). Considering that the scale of ^{17}O shieldings spans almost 800 ppm, these deviations amount to only a few percent.

Table 7.5 Zero-point vibrational corrections (ppm) to ^{17}O NMR shieldings[8,20,43–47] calculated at several levels as indicated.

	HF	MP2	DFT	CCSD(T), MCSCF
CO	−5.60	−3.50, −7.8	−5.5	
CO_2	−6.20	−5.00, −6.7	−5.2	
H_2O	−10.30, −11.05	−9.30	−10.19	−11.68, −11.62, −10.93, −9.86
N_2O	−15.90	−7.70		
OCS	−6.80	−5.30		
F_2O	−23.80	−29.90		
CH_3OH	−10.60	−10.40		− 11.97
$(CH_3)_2CO$	−1.80	−2.80	−7.80	
CH_3CHO	−11.10	−10.10		
CH(**O**)OCH$_3$	−12.60	−11.20		
CH(O)**O**CH$_3$	−8.50	−9.00		
$CH_2(OCH_3)_2$	−10.80	−11.20		
HC(**O**)OC$_2$H$_3$	−14.40	−13.30		
HC(O)**O**C$_2$H$_3$	−10.10	−11.20		
$(C_2H_5)_2O$	−12.80	−13.90		
HC(**O**)OC$_2$H$_5$	−13.10	−12.10		
HC(O)**O**C$_2$H$_5$	−10.00	−11.00		
C_4H_4O	−10.90	−11.20		
ppo-C_3H_6O	−13.10	−15.20		
oxt-C_3H_6O	−16.20	−15.70		
C_4H_8O	−9.20	−9.40		
C_4H_6O	−10.30	−11.50		

Interestingly, the fluctuation of ZPVCs to ^{17}O shieldings is rather small, while the equilibrium geometry values vary significantly among different molecules. For example, the vibrational correction to the shielding of the oxygen atom in hydroxyl groups is usually of minor importance compared to the equilibrium geometry value. On the other hand, the correction is almost as large as the equilibrium geometry value for the shielding of the carbonyl oxygen in benzoic acid.[37] Based on the set of small carbonyl-containing molecules it was concluded that this could be a general trend for oxygen atoms in carbonyl groups.[37] However, these findings are in contrast with earlier observations made for water and CO,[20,44,45] where quite a substantial ZPVC was found for the shielding of the single-bonded oxygen in water and only a modest ZPVC for the double-bonded oxygen in CO. Analyzing the results of Auer[45] one can conclude that it is mostly true for the carbonyl oxygens in esters, where the vibrational corrections can reach 15% of equilibrium geometry value, while for alcoxy-oxygens in esters it is at most 9%. For the rest of the molecules and even CH_3CHO the ZPVC to the oxygen shieldings represented only about 3% of the equilibrium geometry value.

7.2.5 Vibrational Corrections to Fluorine Shieldings

Zero-point vibrationally averaged nuclear shielding constants have been thoroughly studied for ^{19}F. Åstrand and Ruud[48] showed for a set of 24 fluorine-containing organic compounds that these corrections are negative for all molecules in the set and range from −0.64 ppm for *m*-difluorobenzene to −34.78 ppm for 2-fluorobutane. Harding *et al.*[49] added moreover some inorganic compounds to the set (*e.g.* OF_2, F_2, HOF, *etc.*) and found their ZPV corrections to be −15 ppm and larger. Considering that the experimental (or equilibrium geometry) fluorine shieldings are around 250 ppm and larger, one notes that the ZPVC amount to only about 2% of the equilibrium geometry value.

Table 7.6 shows ZPVCs to ^{19}F shieldings in fluoromethanes calculated at the Hartree–Fock level.[48] Only for fluoromethane the averaged value of 475.85 ppm is relatively close to experiment (most likely fortuitous), while for the other compounds the error is between 20 and 25 ppm. This is partly due to the lack of electron correlation in the calculations, as all corrections were calculated at the Hartree–Fock level. This was confirmed by the MP2

Table 7.6 Zero-point vibrational corrections (ppm) to ^{19}F NMR shieldings in fluoromethanes.

	Hartree–Fock[48]		MP2[49]		Exp.[50]
	ZPVC	$\sigma_{calc.}$	ZPVC	$\sigma_{calc.}{}^{a}$	$\sigma_{exp.}$
CH_3F	−7.32	475.85	−8.81	472.9	470.6
CH_2F_2	−8.57	358.40	−9.67	340.7	338.7
CHF_3	−8.00	298.03	−8.91	275.2	273.7
CF_4	−6.29	281.22	−6.83	259.5	258.6

[a]Equilibrium values of σ_F were calculated at the CCSD(T)/13s9p4d3f level.

calculations of ^{19}F ZPVC performed by Harding *et al.*[49] In particular, it has been pointed out that the accuracy of the equilibrium geometry crucially influences the accuracy of ZPV averages and electron correlation is essential in order to get reasonable equilibrium structures of F-containing molecules.

Interestingly, the ZPVC is similar for fluoroethane, 1,1-, and 1,2-difluoroethane (*ca.* −11 ppm), while the corrections to the shieldings in 1,1,1-trifluoroethane and hexafluoroethane are almost half of this value. Thus, the $-CF_3$ moiety seems to behave differently from $-CF_2H$ and $-CFH_2$ groups. Similarly, corrections for inner fluorine atoms in 2-fluoropropane (−20.19 ppm) and 2-fluorobutane (−34.78 ppm) are almost twice the size of the corrections for terminal fluorines in 1-fluoropropane (−11.97 ppm) and 1-fluorobutane (−15.78 ppm). Contributions for fluorobenzenes are relatively small (*ca.* −1.5 ppm); only *p*-difluorobenzene has a considerably larger ZPVC of −16.98 ppm. For most of these molecules the correction from the geometry shift expressed by the effective geometry was almost uniform, but the anharmonic contribution from the shielding derivatives varied substantially and dominated for most ZPVCs. Note that only for *m*-difluorobenzene the contribution from the shielding anharmonicity was positive, which, however, could be due to the missing electron correlation.

When comparing to experimental fluorine shifts, the absolute shieldings are traditionally converted to the δ_F scale using the formula $\delta_F = 188.7\ \text{ppm} - \sigma_F$, where 188.7 ppm is the shielding of liquid $CFCl_3$. Calculated ZPVCs improve the agreement of the calculated chemical shifts with the experimental values, but the differences are still about 30 ppm due to the missing treatment of electron correlation and solvent effects.

7.2.6 Vibrational Corrections to Phosphorus and Transition Metal Shieldings

Studies dealing with vibrational corrections to NMR shieldings of phosphorus are far less frequent. Besides the more widely studied PH_3, corrections have been described only for small related systems like PF_3, $P(CH_3)_3$, CH_3–PH_2, and even more seldom are studies with phosphorus bound by multiple bonds like in HCP or CH_3–CP. The ZPV corrections were negative in all these systems and reach values of up to −50 ppm for HCP. They represent thus a minor correction for molecules with single-bonded phosphorous atoms, where they amount to about 2% of the equilibrium geometry values, but can reach more than 10% of the equilibrium value for molecules with multiple-bonded phosphorous atoms like the 17% in HCP.[41,51,52]

Bühl and co-workers[53–55] published rovibrational corrections of magnetic shielding constants of transition metals calculated at the DFT level. The predicted corrections were relatively small for vanadium in $VOCl_3$ and manganese in MnO_4^- but slightly larger for iron in $Fe(Co)_5$. They also confirmed that DFT-based classical simulations can provide qualitatively similar corrections to the standard quantum-mechanical zero-point vibrational correction approaches. Obviously, quantitative differences (−98 ppm *vs.* −288 ppm) between these two approaches were observed in particular for the $Fe(CO)_5$ species. Bühl *et al.*[53,54] proposed also an interesting approach for estimating the vibrational effects on metal nuclear magnetic shieldings in metallic complexes. Their model is based on studying the influence of metal-ligand elongation on NMR shieldings using metal shielding/bond-length derivatives. By their coupling with DFT-based MD simulations it was possible to obtain a similar result of vibrational correction for Ti, V, Mn, and Co as using the standard perturbation theory approach (*e.g.* ref. 56).

7.2.7 Transferability

As mentioned already in Section 7.2.1.1 the zero-point vibrational corrections to hydrogen shieldings exhibit a unique feature – transferability. For example, the ZPVC for methyl hydrogens in methanol are almost the same as for methane, although they have quite a different electronic structure. This is exclusively a feature of the zero-point vibrational correction, since methyl hydrogens in methanol have equilibrium geometry values σ_H of about 2.5 ppm lower than in methane. Thus, it would be interesting to design a simple rule for estimates of ZPVCs to hydrogen shieldings in different functional groups.

Ruud *et al.*[37] analyzed the ZPVCs to σ_H for a test set of 38 molecules classified according to the nature of the functional group containing the hydrogen and provided a "rule-of-thumb" estimate of ZPVCs for hydrogens of almost any kind (see Table 7.7). Using their rule-of-thumb vibrational corrections they estimated the vibrationally averaged hydrogen shieldings in 15 non-polar molecules (mostly hydrocarbons) and compared them to shieldings vibrationally averaged by a standard quantum mechanical approach. The excellent correlation between these results (in terms of the chemical shifts relative to methane) is shown in Figure 7.1. Addition of both versions of vibrational corrections (QM and rule-of-thumb) reduces the RMS

Table 7.7 Transferable vibrational contributions (in ppm) to hydrogen shieldings of different hydrogen types.[37]

	ZPVC to σ_H
H_3C–H	−0.59
-CR_2H	−0.70 ± 0.11
=CRH	−0.46 ± 0.13
≡CH	−0.76 ± 0.01
H–CRO	−0.55 ± 0.06
RO–H	0.48 ± 0.13
RN–H_2	−0.18 ± 0.03
RCOO–H	−0.49 ± 0.06
Ar–H	−0.39 ± 0.06

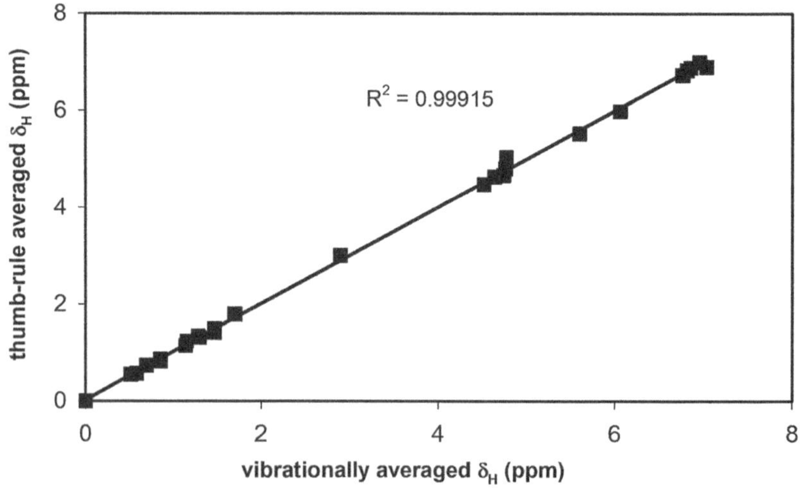

Figure 7.1 Correlation between zero-point vibrationally averaged σ_H calculated by the standard VPT2 procedure and estimates using the thumb-rule for the set of 15 non-polar molecules.[37]

deviation from experiment from 0.3 ppm for the equilibrium geometry results to 0.2 ppm for the vibrationally corrected values.

Contrary to the hydrogen shieldings, where the ZPVCs for a particular functional group are transferable between structurally very different systems, transferability of ZPVCs is observed for other atoms only in small groups of molecules or for structurally related functional groups. For example, in fluoromethanes (CH_3F, CH_2F_2, CHF_3, and CF_4) the transferable contribution for the ZPVC to fluorine shielding was estimated to -7.5 ± 1 ppm,[48] where the uncertainty of 1 ppm can be considered to be small compared to the several hundreds of ppm for the equilibrium geometry values of σ_F.

7.2.8 Methodological Aspects

7.2.8.1 Electron Correlation Effects

The majority of data and conclusions presented so far were based on uncorrelated calculations (*e.g.* Hartree–Fock) or did not treat electron correlation specifically. In this chapter a more detailed discussion of the effect of electron correlation on the ZPVC to NMR shieldings will be made. Since electron correlation effects are already negligible for the equilibrium geometry values of hydrogen shieldings, it is not surprising that the vibrational corrections to the hydrogen shieldings seem to be independent of electron correlation. Corrections obtained at the Hartree–Fock level compare, therefore, favorably with higher order methods. For example, the ZPV correction -0.59 ppm to σ_H in methane estimated at the Hartree–Fock level matches nicely the MCSCF result of -0.60 ppm.[8] The average deviation between MP2 and Hartree–Fock vibrational corrections to σ_H in CH_4, C_2H_2,

C_2H_4, C_2H_6, and C_6H_6 molecules was below 0.05 ppm.[37,40] Slightly exceptional to the rule, that correlation does not alter both hydrogen shielding as well as its vibrational correction,[21] seems to be ethyne. The correction to σ_H in ethyne calculated at the MCSCF level was −0.68 ppm while it was −0.76 ppm at the Hartree–Fock level. Thus, the chemical shift obtained at the correlated level was 1.81 ppm (related to CH_4), which is much closer to the experiment (1.58 ppm) than the Hartree–Fock shift of 1.15 ppm. A similar behavior was observed for another triple bond containing molecule, 1-butyne.[37]

Excellent agreement of the Hartree–Fock vibrational correction to ^{13}C shielding in CH_4 with the MCSCF results (−3.07 ppm *vs.* −3.20 ppm) implies that also carbon ZPVCs may be electron correlation-independent. However, selected carbon corrections calculated at several levels gathered in Table 7.3 (in Section 7.2.2) show the average RMS deviation of Hartree–Fock and MP2 results to be 0.73 ppm and the Hartree–Fock results to overestimate the MP2 results. Note that the MP2 results seem to be basis set dependent, since the data obtained with smaller pcS-2 basis set differ substantially from results obtained with the qz2p basis set. Selected coupled-cluster results indicate that the MP2 calculations underestimate the size of the vibrational corrections. Nevertheless MP2 ZPVCs are still considered to be a good compromise.

The analysis of 22 different vibrational corrections to ^{17}O nuclear shieldings (Table 7.5, in Section 7.2.4) calculated at the Hartree–Fock and MP2 levels revealed non-uniform changes in the ZPVCs if electron correlation is included.[45] Thus, namely the ZPVCs for alcoxy-oxygens in esters seem to be underestimated at the Hartree–Fock level, while most others are overestimated. The average RMS deviation of the Hartree–Fock ZPV corrections from MP2 values was 2.43 ppm for the whole test set. Pronounced exceptions are the N_2O and F_2O molecules, where electron correlation effects lead to a difference of approximately 7 ppm.

Prochnow *et al.* studied six different ^{15}N and five different ^{31}P zero-point vibrational corrections.[41] Comparison of the MP2 and CCSD(T) levels shows that MP2 underestimates the vibrationals corrections by RMSD of 2.66 ppm for ^{15}N and up to 1.5 ppm for ^{31}P shieldings (see Table 7.4 in Section 7.2.3). The exception is the PN molecule, for which strong electron correlation contributions to the ZPVC beyond the MP2 level were observed. The MP2 ZPVC even has the wrong sign. The maximum deviation of the results obtained at the Hartree–Fock level is in the order of 2 and 5 ppm for ^{15}N and ^{31}P shieldings, respectively, which means that the Hartree–Fock method provides ZPVC in closer agreement with CCSD(T) results for ^{15}N shieldings. Harding *et al.*[49] showed for the set of 30 vibrational corrections to ^{19}F shieldings that these corrections are underestimated at the Hartree–Fock level (average RMS 3.74 ppm) compared to the MP2 results, which is contrary to the ^{13}C or ^{17}O shieldings. Interestingly, the difference was much smaller for fluoro-hydrocarbons (about 1.16 ppm, Table 7.6), while for systems containing more heteroatoms it can increase more than five times.

High-level correlation methods are often impractical for larger systems due to the enormous computational time. Thus, density functional theory methods would seem to be a good compromise since they can partly account for electron correlation. However, precise predictions of nuclear magnetic shieldings at the DFT level are often hampered by the limited precision of current DFT functionals that are not well suited to describe magnetic phenomena including current density (see Tables 7.4–7.7). Higher-level methods enabling more precise treatment of electron correlation (*e.g.* coupled-cluster methods *etc.*) have then been used to estimate the DFT error in calculation of ZPVCs to shieldings.[26]

Summarizing, one can say that electron correlation can play an important role especially in systems containing multiple bonds or in aromatic molecules. However, corrections based on Hartree–Fock calculations give at least qualitative estimates. Electron correlation is negligible for vibrational corrections to hydrogen shieldings.

7.2.8.2 Basis Set Effects

It has been pointed out in many previous studies that the quality of NMR shieldings (in contrast to chemical shifts) depends strongly on the basis set used for their theoretical predictions, *e.g.* ref. 40, 43 and 57–59. Considering also the basis set dependence of anharmonic frequencies, for which often basis sets of the higher quality than double-ζ are needed for results close to the complete basis set limit (CBS),[60] one expects vibrational corrections to NMR shieldings to exhibit some dependence on the basis set. Dračínský *et al.*[26] studied the effect of the basis set size on the zero-point vibrational contributions to the hydrogen and carbon shieldings for CH_4, CH_3F, and CH_3Cl. Using several Pople-style basis sets as well as Dunning's aug-cc-pVTZ and aug-cc-pVQZ basis sets, they concluded that the dependence of the shielding derivatives on the basis set size is rather small. However, in some extreme cases as CH_4 the correction to σ_C can be reduced to $\pm 50\%$ when changing from the 6-31G to aug-cc-pVQZ, which was mostly due to the basis set dependence of the second-order derivative term as the first-order contributions were almost equal for all basis sets.

In the literature there are several suggestions for basis sets specifically tailored for the calculation of properties depending mainly on the near-nucleus region, *i.e.* nuclear magnetic shieldings. Among others, Jensen's polarization-consistent pcS-*n* basis set hierarchies[59] have been shown to provide accurate magnetic shieldings.[40,43] Based on calculations of vibrational corrections for a rather small test set of only nine small molecules (N_2, NH_3, CO, CO_2, CH_4, C_2H_2, C_2H_4, C_2H_6, and C_6H_6) at the DFT level with the BHandH functional with the pcS-2 and pcS-3 basis sets it was claimed that vibrational corrections for N_2, CO, CO_2, and NH_3 are well converged with these basis sets, while for hydrocarbons the difference between the pcS-2 and pcS-3 results indicated incomplete convergence. Thus, the basis set dependence of the vibrational corrections to N and O shieldings were considered to be negligible (*ca.* −0.1 ppm), while for C and H shieldings the changes reached *ca.* −4.1 ppm and −0.6 ppm, respectively.

Summarizing, one can conclude that, although the basis set convergence of vibrational corrections is usually fast, basis sets of at least a double-ζ quality are usually needed. Pople-style basis sets are less advisable for studying the convergence, as their hierarchy is not linear (although the 6-311++G** has been shown to provide good quality results of vibrational frequencies). Dunning's correlation consistent or Jensen's polarization-consistent pc-*n* basis sets are in this sense more recommended.

7.2.8.3 Relativistic Effects

The number of relativistic studies of heavy-element NMR shielding[61] is still significantly smaller than non-relativistic calculations and studies of relativistic effects in the calculation of vibrational corrections are even more rare.[55] Minaev *et al.*[62] investigated the sensitivity of the relativistic spin–orbit coupling contribution to light-atom NMR shieldings on vibrational motion. Lantto *et al.*[63] showed the importance of a full treatment of vibrational corrections to ^{13}C shieldings calculated with spin–orbit coupling included for explaining the isotope effects in CO_2, CS_2, CSe_2, and CTe_2. Furthermore, it was observed that ZPV corrections and a treatment of relativistic effects are necessary in order to reproduce qualitatively the halogen dependence of ^{13}C and ^{1}H shieldings in methyl halides.[64] However, there is to the best of our knowledge only one study treating in detail the relativistic character of the vibrational corrections.[65] Typically, NMR shieldings in systems containing heavy atoms are calculated with relativistic corrections to the equilibrium geometry value and the corrections originating from the anharmonicity are treated at the non-relativistic level. However, for example in HBr and HI the inclusion of the spin–orbit coupling reverses the sign of the vibrational correction compared to the non-relativistic value. This dramatic change is due to the coupling of the magnetic moment with the external magnetic field through the spin–orbit and Fermi-contact mechanisms. These contributions have oppositely directed geometry dependences compared to the non-relativistic contributions.[62]

Lantto *et al.*[65] employed the relativistic Breit–Pauli perturbational (BPPT) approach in order to estimate the finite-temperature vibrational effects on ^{129}Xe and ^{19}F shieldings in XeF_2. They observed that, unlike for non-relativistic shieldings, relativistic BPPT vibrational corrections estimated at the DFT level are rather independent of the choice of functional. Thus, the BLYP and B3LYP results for ^{129}Xe are almost identical. Moreover, the BHandHLYP functional, which was found to show poorer performance in non-relativistic vibrational corrections compared to CCSD(T) values, provides relativistic contributions to vibrational corrections to ^{129}Xe shieldings close to the results of functionals with lower admixtures of exact exchange. The relativistic contribution to the ZPVC is positive and about half the size of the non-relativistic ZPVC, which is negative. Thus, the total ZPVC is still negative. If the non-relativistic correction is calculated at the coupled-cluster level, the total vibrational correction decreases in absolute value. Note, that both Hartree–Fock and DFT data profit from an error cancellation and produce relativistic ZPVCs similar to the CCSD(T) level. As expected, vibrational effects are relatively more important for σ_F than for Xe. The ZPV correction represents *ca.* 1.5% of the ^{19}F equilibrium geometry value. It is noteworthy that unlike Xe there is a negative relativistic contribution to σ_F at the equilibrium geometry. Both non-relativistic and relativistic contributions to the vibrational correction are negative. The authors concluded finally that high-quality treatment of the nuclear motion could be reached combining *ab initio* shielding constant surfaces with DFT-estimated relativistic contributions. Using such an approach one can achieve almost quantitative agreement with the experimentally observed ^{19}F isotope shifts in XeF_2.

7.2.8.4 Expansion Terms Analysis

There is only a limited number of studies investigating systematically the relative importance of various derivative terms in the Taylor expansion of the vibrational Hamiltonian as well as the property surface.[20,21,26,66] Also only a few authors reflected on the effect of the full-quartic force field and higher order property derivatives than the standard second-order derivatives on ZPVCs of NMR shieldings.[24]

It has been shown that the effective geometry approach represents an effective and reliable tool for estimations of vibrational corrections to NMR shieldings.[8,24] Ruud *et al.*[37] observed using this approach that the shift in geometry accounts for almost 70% of the ZPV corrections to both for σ_C and σ_H in methane. For the hydrogens of the CH_3-group of methanol the situation changes (see Figure 7.2) as the contributions from the shift of geometry and from the harmonic property term are almost equal. Interestingly, in a closely related system, propane, the effect of the shift of the geometry is negligible for the hydrogens in the CH_3-group, while it amounts to ~30% of the total ZPVC in the CH_2-group. It is interesting that the transferability described in Section 7.2.7 holds (and accounts for *ca.* −0.7 ppm), even though the relative contributions to the ZPVC of the two terms vary considerably.

Figure 7.2 also shows an analysis for different types of aromatic hydrogens. The data are averaged over values obtained for toluene, phenol, aniline, and benzoic acid. Note, that all *ortho*-, *meta*-, and *para*-hydrogens exhibit a similar pattern – the contribution of the property derivatives is dominant and accounts for ~74% of the total ZPVC. Thus, it is completely opposite to the methane case. For the hydrogens in benzene the same contribution is only slightly smaller (68%) than in the substituted benzenes. Interestingly similar ratios as for the aromatic hydrogens are also obtained for the hydrogens in the methyl group in toluene. As one would expect, the situation changes dramatically for hydrogens attached to different heteroatoms. For the hydroxylic hydrogen of methanol, the correction due to the shift of the

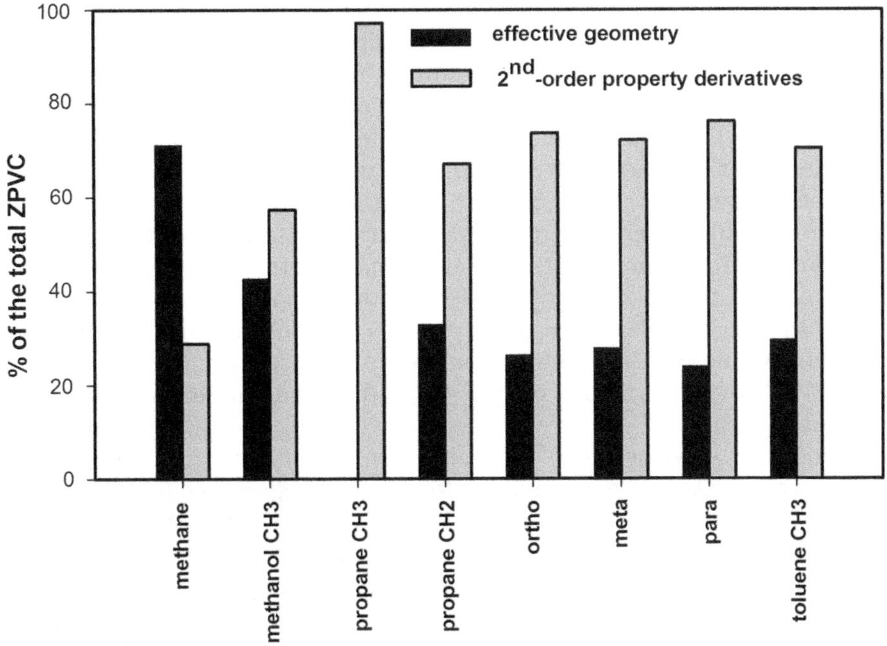

Figure 7.2 Contribution to the total σ_H ZPVC caused by the shift in geometry and by the second-order property derivatives.

geometry is positive (shielded by 0.49 ppm), while the second-order derivative of the property makes the hydrogen deshielded by about −0.24 ppm. The difference is even more significant for the phenolic hydrogen (0.80 ppm *vs.* −0.22 ppm). The effective geometry causes also positive correction to the shielding of the NH_2-group hydrogens in aniline (0.39 ppm), but the property second derivative contribution is −0.55 ppm and the total ZPVC is thus negative. Similarly for the carboxylic hydrogen in benzoic acid the effective geometry correction is positive (0.25 ppm) but smaller in absolute value than the property derivative correction (−0.81 ppm).

Dračínský *et al.*[26] investigated separately the effect of the terms in the expansion of the potential energy and of the shielding surface. For the shieldings in CH_4 and CH_3F they analyzed the importance of the expansion of the vibrational potential energy in the complete second-order treatment of the vibrational corrections. They observed substantial qualitative changes in the shielding, if the equilibrium geometry values are corrected by vibrational corrections obtained with only a harmonic vibrational wavefunction (−1.91 ppm and −3.95 ppm for σ_C and −0.24 ppm and −0.58 ppm for σ_H, respectively). This is in agreement with previous studies[37,45] and confirms the importance of the second derivatives of the shielding for the averaging, as the first derivative vanishes in the harmonic wavefunction limit. They concluded, therefore, that a useful estimate of the vibrational correction could often be obtained with the quadratic part of the molecular potential and second-order property derivatives only. However, it is noteworthy that the cubic terms in the expansion of the vibrational potential energy still change the ZPVC significantly. For example, for σ_C in CH_4 addition of cubic terms leads to changes as large as 80% of the harmonic contribution. Compared to the cubic correction, inclusion of the "semidiagonal" (force constants that have at least two identical indices) and fully quartic force fields led to negligible changes in ZPVC, although they could be important for more flexible systems or for excited vibrational states. Note that the total vibrational corrections for the same atom type are similar in CH_4 and CH_3F. However, the size of the individual contributions arising from particular terms in the potential energy expansion (harmonic, cubic, quartic) is different. While for CH_4 the harmonic term represents only $\pm$50% of the total ZPVC for both σ_C and σ_H, for CH_3F the harmonic term amounts to more than 95%.

Raynes and co-workers[20,21,66] went even beyond the aforementioned analyses and classified contributions to the shielding surface for ^{17}O and ^{1}H in H_2O, ^{13}C and ^{1}H in ethyne, and ^{13}C and ^{19}F in CH_3F not only as the first- and second-order but also as stretching, bending, or cross-terms according to the corresponding normal modes or curvilinear symmetry coordinates involved. For the ^{13}C shielding in CH_3F[66] the dominant contribution is the first-order stretching term providing more than 85% of the total -4.558 ppm vibrational correction. Two quadratic stretching terms are, however, larger than the first-order C–F stretching contribution. Note that while all stretching terms are negative, some bending contributions are also positive providing a total bending contribution of 0.705 ppm. Most of the bending contributions can be assigned to the H–C–H vibration. Cross-terms involving a combination of the stretching and bending vibrations were indicated to be minor with the total contribution of about 0.05 ppm. The total vibrational correction to the ^{19}F shielding is -9.425 ppm and is more than twice as large as the correction for ^{13}C. All stretching terms are again negative and first-order stretching dominates ($\sim$80%) again the total correction. Note, that all stretching terms (both first and second) involving C–H stretches are more than twice as large as the stretching terms involving C–F stretching. For the hydrogen shielding, the largest contribution comes from the vibration involving the bond containing the hydrogen of interest. Contrary to fully negligible cross-term effect (under 0.001 ppm), still significant contributions arise from the rest of the stretching (both first and second contributions) as well as bending terms but all fall below 0.1 ppm in the absolute value. Interestingly, the second-order stretching contributions for hydrogens are positive contrary to contributions for ^{13}C and ^{19}F. For the ^{17}O shielding in H_2O [20] again the negative first-order stretching contribution dominates with 75% of the -9.86 ppm total zero-point vibrational correction, followed by the also negative second-order stretching contribution ($\sim$37%), while the second-order bending contribution is positive and contributes only with 13%.[20] Ethyne,[21] finally, shows again a different pattern due to larger electron correlation effects (see Section 7.2.8.1).

Here the largest contribution to the vibrational corrections to the ^{13}C shieldings is the negative second-order bending contribution, which amounts to 82% of the total correction of −4.1 ppm, while stretching motions contribute with only 16%.

7.2.9 Temperature Effects and Isotopic Shifts

Temperature effects, which require Boltzmann averaging over excited rovibrational states, are most easily incorporated by the extension of the standard second-order perturbation theory expression by Toyama, Oka, and Morino.[11] Jameson and co-workers pioneered studies on temperature effects of shieldings in polyatomic systems in vacuum.[67,68] Considering the rather low levels of theory employed in the calculations (Hartree–Fock, small basis sets) the agreement with experiment is surprisingly good. They demonstrated for example that the large-amplitude inversion mode of NH_3 had to be modeled explicitly rather than by a perturbation expansion in order to reproduce experimental temperature dependence of the isotope shifts. However, for PH_3 the perturbational approach seemed to be sufficient due to the higher inversion barriers. The authors also found the shapes of the ^{31}P shielding surface in PH_3 and the ^{15}N shielding surface in NH_3 (calculated in terms of the symmetry coordinates) to be very similar. Nevertheless, according to several studies the thermal effect is usually an order of magnitude lower than the zero-point vibrational correction itself.[20,21,44,69,70] Thus, for example, if the CCSD(T) vibrational correction to ^{19}F in F_2 molecule is −30.87 ppm, then the temperature correction at room temperature is only −4.69 ppm. Adding both corrections to the equilibrium value of −197.53 ppm gives us −225.5 ppm, which is much closer to the experimental value of −232.8 ppm.[50] Moreover, estimated theoretical vibrational-temperature correction (−35.6 ppm) corresponds well with corrections reported by Jameson *et al.* (−40 ppm) obtained from gas-phase experiments.[50]

Vibrational averaging is also necessary in order to interpret changes in NMR chemical shifts caused by isotopic substitution.[20,21,26,66,70,71] For example, on the set of eight halogenmethane derivatives the secondary isotopic shift in the ^{13}C shieldings due to the hydrogen–deuterium exchange, as well as in the ^{1}H shieldings due to the ^{12}C–^{13}C exchange, was estimated.[26] The isotopic shift was very small (0.002 ppm) for the ^{1}H shieldings in all the halomethanes in good agreement with the earlier studies on H_2O,[20] OH^-,[71] and ethyne.[21] In the case of ^{13}C shieldings in the halomethanes the isotopic shift due to the hydrogen–deuterium exchange was observed to be −0.2 to −0.7 ppm with mean average deviation from experiment of 0.04 ppm again similar to the shifts in ethyne.[21] For the bromine derivatives, however, the calculated values were substantially overestimated probably due to a stronger relativistic effect. More importantly, the data clearly revealed the significance of the off-diagonal shielding derivatives for proper interpretation of (mainly non-hydrogen) isotopic shifts.[26]

7.2.10 Solvent Effects

Although it is out of the scope of this book, which concentrates on gas phase NMR only, we want to spend a few words on the effect of the solvent on the zero-point vibrational corrections to nuclear magnetic shieldings. As shown in previous sections, vibrational averaging can be a substantial part of the calculated NMR shielding. It is well known that solvent effects on equilibrium NMR shieldings are also often significant and thus one can assume that the effect of a solvent on the vibrational correction can be indispensable as well. There are only a few published studies dealing with the solvent effect on zero-point vibrational corrections. Kongsted *et al.*[72] concentrated on acetone (as an example of a molecule having a carbonyl group) and three diazene molecules in the gas phase and water. The authors observed using the continuum PCM solvent model and the KT3/6-311++G(2d,2p) level that the dielectric continuum increases slightly the magnitude of the ZPVC for carbon atoms of acetone. This follows the trend for equilibrium values of σ_C calculated in vacuum and in water. However, the increase of the ZPVC due to the solvent is minor (about 70% of the gas-phase ZPVC) compared to the increase of the equilibrium geometry value (~300% of the gas-phase equilibrium geometry value) and represents only 3% of the total solvent shift. Regarding the ZPVC on σ_O the authors found a sign change and decreased magnitude of the vibrational correction when going from vacuum to water. The ZPVC is approximately 12% of the total solvent shift, thus the effect on the equilibrium structure shielding dominates also for σ_O. Inclusion of the solvent in the calculation improved the agreement with experiment in both cases, even though the predicted change in σ due to the solvent it is still far from ideal ($\sigma_{exp} - \sigma_{calc.} = \sim 7$ ppm for ^{13}C and 31 ppm for ^{17}O of acetone). The main reason is said to be the inability of the PCM model to treat hydrogen bonding. In the case of pyrazine, pyrimidine, and pyridazine, the ZPVC to σ_N represents a substantial correction both in vacuum and in water (−5.15 to −7.47 ppm). On the other hand, when considering the total solvent shift, the ZPVC tends to cancel. As for acetone, including vibrational corrections improved the agreement with experiment even though the predicted solvent effect was underestimated.

Kaminský *et al.*[38] estimated the solvent effect on vibrational corrections to σ_C in C_{70} fullerene using first-principle molecular dynamic simulations performed on the BP86/def-SVP Born–Oppenheimer potential surface. They ran several microcanonical trajectories both in vacuum and in trichloroethane using the CPCM model. They obtained the vibrational corrections to σ_C as the difference between the average over 64 snapshots calculated at the BHandHLYP/IGLO-III level and the equilibrium geometry value. While for gas-phase simulations the corrections depend on carbon types (−1.4 ppm to −5.8 ppm), solvent effects made the corrections more uniform for all carbons (−1.2 ppm to −1.7 ppm). The solvent damps the molecular motion and hence makes the deshielding smaller and more uniform for all carbons. Note that using C_{60} as the secondary reference partly cancels the vibrational corrections, as the vibrational contributions for C_{60} are approximately 70% of the C_{70} value. The correction for chemical shifts is therefore rather minor: −0.6 and −0.2 for vacuum and trichloroethane, respectively.

7.2.11 Practical Aspects of ZPVC Calculations

In the last section we will discuss some practical aspects of the calculation of ZPVCs to nuclear magnetic shieldings.

Since the calculation of the shielding derivatives is still based on a numerical differentiation, the stability and accuracy is the key issue and the numerical differentiation step is a question. The nuclear shielding like spin–spin coupling constants are some of the properties most dependent on the geometry. According to several studies, where authors investigated the effect of the step length in the numerical differentiation on the total vibrational contribution, it appears that a reasonable numerical stability is achieved for steps longer than 0.05 bohr (0.026 Å)[8,73] and sometimes even larger.[74] This requirement is larger than the recommended step length of 0.0075 bohr (0.0040 Å) for the effective geometry. If the differentiation is performed in normal modes, it is often useful to choose a variable step length depending on $1/\sqrt{\omega}$ since for low-frequency modes a uniform step length results in too small changes in the Cartesian coordinates.

Imaginary frequencies can occur sometimes in the averaging approach using an expansion around an effective geometry. These were often associated with an internal rotation of some free-rotating group (*e.g.* $-CH_3$), which are often coupled to other vibrational modes. Therefore, one may argue that these intermolecular motions cannot be treated by a simple expansion around the equilibrium or effective geometry. Instead, a full description of the potential energy surface describing this internal motion should be considered. This would, however, increase the computational time tremendously. Baaden *et al.*[75] showed moreover that the rotation along the single bond causes a change of σ_H of about 1% of the electronic contribution and neglecting imaginary modes in ZPVC calculations thus represents only a negligible change in the total vibrational correction.

7.3 Examples of Vibrational Corrections to Coupling Constants

Calculations of rovibrational and temperature corrections to NMR spin–spin coupling constants (SSCCs) are still more the exception than the rule, although many authors acknowledge nowadays that with the accuracy of current high-level SSCC calculations, proper treatment of nuclear motion effects is necessary in order to further reduce the deviations from experiment. In the following we will present some illustrative examples for the calculation of rovibrational and temperature effects on SSCCs. We will hereby distinguish on one hand between calculations using vibrational wavefunctions obtained by the perturbation theory approach, as presented in Section 7.1, and by variational vibrational approaches and on the other hand between high-level wavefunction and density functional theory methods in the electronic structure calculations of the necessary coupling constant surfaces. Finally, we will comment on the recently emerging field of vibrational corrections to spin–spin coupling constants in molecules, where relativistic effects are important due to the presence of atoms with larger nuclear charges. Sample results for a few different types of SSCCs are given in Tables 7.8–7.10.

Table 7.8 Selected results for vibrational corrections to ^1J(XH) in various small molecules. Shown are the presumably most accurate results using correlated wavefunction methods and sample DFT results and for comparison also experimental values.

Coupling	Method	ZPVC/Hz	Total/Hz	% of Total
HD				
^1J(H–D)	FCI[76,a]	1.89	43.11	4.4
		(2.09 at 300 K)	(43.31 at 300 K)	(4.8)
	Exp.[76]		43.26 ± 6	
HF				
^1J(H–F)	RASSCF[88,b]	−26.90	499.5	−5.4
		(−29.70 at 300 K)	(496.7 at 300 K)	(−6.0)
	B3LYP[102,c]	−37.7	378.9	−10.0
	Exp.[92]		500 ± 20	
H_2O				
$^1J(^{17}O–H)$	SOPPA(CCSD)[19,d]	3.96	−77.59	− 5.1
		(4.34 at 300 K)	(−77.22 at 300 K)	(−5.6)
	CASSCF[89,e]	(4.58 at 300 K)	(−76.71 at 300 K)	− 6.0
	B3LYP[102,c]	5.2	−70.7	− 7.4
	Exp.[93]		−78.70 ± 0.02	
	Exp.[94]		−80.62 ± 0.06	
NH_3				
$^1J(^{15}N–^1H)$	SOPPA(CCSD)[5,f]	0.12	−61.968	−0.2
		(0.14 at 300 K)	(−61.947 at 300 K)	
$^1J(^{14}N–^1H)$[g]	BHandH[103,h]	−1.87	−71.01	
$^1J(^{15}N–^1H)$	Exp.[93]		−61.45 ± 0.03	
CH_4				
$^1J(^{13}C–^1H)$	SOPPA(CCSD)[17,i]	5.03	128.88	3.9
		(5.13 at 300 K)	(128.98 at 300 K)	(4.0)
	B3LYP[102,c]	5.3	137.9	3.8
	BHandH[103,h]	5.70	132.12	
	Exp.[95]		125.31 ± 0.01	
SiH_4				
^1J(Si–H)	SOPPA(CCSD)[23,j]	−7.59	−199.64	3.8
		(−7.83 at 298 K)	(−199.89 at 298 K)	(3.9)
	PBE0[104,k]	−3.9	−202.30	
	Exp.[96]		−202.5	
	Exp.[23]		−201.3 ± 0.4	

Table 7.8 (*Continued*)

Coupling	Method	ZPVC/Hz	Total/Hz	% of Total
Ethane				
$^1J(C–H)$	CCSD[77,l]	3.84	119.97	3.2
	BHandH[103,h]	5.26	130.59	4.0
	Exp.[97]		125.206	
Ethene				
$^1J(C–H)$	CCSD[77,l]	3.60	150.21	2.4
	CCSD[78,m]	3.96	156.29	2.5
	B3LYP[102,c]	5.1	170.4	3.0
	BHandH[103,h]	5.86	164.26	3.5
	Exp.[97]		156.302	
Ethyne				
$^1J(C–H)$	CCSD[77,l]	0.78	238.73	0.3
	CCSD[78,m]	2.56	248.89	1.0
	B3LYP[102,c]	4.7	276.4	1.7
	BHandH[103,h]	7.23	265.03	2.7
	Exp.[98]		247.56 ± 0.02 at 300 K	
Cyclopropene				
$^1J(C_1–H_1)$	CCSD[78,m]	3.49	227.07	1.5
	B3LYP[105,n]	4.5	244.2	1.8
	Exp.[99]		228.2	
$^1J(C_3–H_3)$	CCSD[78,m]	5.91	169.08	3.5
	B3LYP[105,n]	5.0	179.9	2.8
	Exp.[99]		167	

[a]SSCC surface: FCI/aug-pcJ-5 + 8s3p2d-1s1p1d (FC)/aug-pcJ-5 + 8s3p2d-1s (non-FC); FF: numerical FCI/aug-pV7Z potential energy curve.
[b]SSCC derivatives: RASSCF/ANO[6s5p4d3f/5s4p3d] + tight 3s/5s; FF: RASSCF/ANO[6s5p4d3f/5s4p3d].
[c]SSCC derivatives + FF: B3LYP/sHIII.
[d]SSCC derivatives: SOPPA(CCSD)/(17s7p5d2f/11s2p2d); FF: refined experimental.
[e]SSCC derivatives: CASSCF/[11s7p3d1f/9s3p1d]; FF: CCSD(T)/aug-cc-pVQZ.
[f]SSCC derivatives: SOPPA(CCSD)/(17s7p5d2f/11s2p2d); FF: CCSD(T) refined.
[g]Multiplied by γ(15N)/γ(14N).
[h]SSCC derivatives: BHandH/pcJ-3; FF: BHandH/pcJ-2.
[i]SSCC derivatives: SOPPA(CCSD)/[10s5p4d/6s2p]; FF: quadratic and cubic refined experimental.
[j]SSCC derivatives: SOPPA(CCSD)/aug-cc-pVTZ-Juc; FF: CCSD(T)/cc-pVQZ + one tight d function, refined.
[k]SSCC derivatives + FF: PBE0/6-311G**.
[l]SSCC derivatives: CCSD/dzp; FF: anharmonic CCSD/dzp.
[m]SSCC derivatives: CCSD/aug-cc-pVTZ-J; FF: CCSD(T)/cc-pVTZ.
[n]SSCC derivatives: B3LYP/HIIsu (FC) and B3LYP/HII (non-FC).

7.3.1 High-level Wavefunction Calculations on Small Molecules

High-level wavefunction calculations employing coupled cluster theory,[4,76–83] Møller–Plesset perturbation theory,[5,14–19,22,23,84–87] or multi-configurational self-consistent field approaches[88–91] in combination with one-electron basis sets specially optimized or at least tuned for the

Table 7.9 Selected results for vibrational corrections to $^1J(CC)$ in various small molecules. Shown are the presumably most accurate results using correlated wavefunction methods and sample DFT results.

Coupling	Method	ZPVC/Hz	Total/Hz	% of Total
Ethyne				
$^1J(C–C)$	CCSD[77,a]	−11.46	173.77	−6.6
	CCSD[78,b]	−10.62	176.57	−6.0
	B3LYP[102,c]	−9.3	195.8	4.7
	BHandH[103,d]	−2.83	201.95	1.4
	Exp.[98]		174.78 ± 0.02 at 300 K	
Ethene				
$^1J(C–C)$	CCSD[77,a]	0.55	67.53	0.8
	CCSD[78,b]	0.69	69.46	1.0
	B3LYP[102,c]	0.9	75.6	1.1
	BHandH[103,d]	2.06	77.91	2.6
	Exp.[97]		67.457	
Ethane				
$^1J(C–C)$	CCSD[77,a]	1.78	35.20	5.1
	BHandH[103,d]	1.78	36.53	4.9
	Exp.[97]		34.521	
Cyclopropene				
$^1J(C_1{=}C_2)$	CCSD[78,b]	−0.94	63.19	−1.5
	B3LYP[105,e]	0.2	66.7	0.3
$^1J(C_1–C_3)$	CCSD[78,b]	−0.01	9.77	−0.1
	B3LYP[105,e]	0.0	8.4	0.0
Cyclopropane				
$^1J(C–C)$	CCSD[77,a]	0.58	13.21	4.4
	B3LYP[105,e]	0.5	12.5	4.0
	Exp.[100]		12.4	

[a]SSCC derivatives: CCSD/dzp; FF: anharmonic CCSD/dzp.
[b]SSCC derivatives: CCSD/aug-cc-pVTZ-J; FF: CCSD(T)/cc-pVTZ.
[c]SSCC derivatives + FF: B3LYP/sHIII.
[d]SSCC derivatives: BHandH/pcJ-3; FF: BHandH/pcJ-2.
[e]SSCC derivatives: B3LYP/HIIsu (FC) and B3LYP/HII (non-FC).

calculation of spin–spin coupling constants are still very rare and generally restricted to small molecules. Mostly the vibrational second-order perturbation theory approach (VPT2) with expansion either around the equilibrium (see Section 7.1.2) or effective geometries (see Section 7.1.3) has been employed in these calculations. However, a few systems known to have large-amplitude vibrations have been investigated using variational treatments, and for diatomic molecules the vibrational Schrödinger equation is often solved numerically. In Tables 7.8–7.10, we have collected some representative examples listing both vibrationally and possibly temperature averaged coupling constants, vibrational corrections, and experimental values.[23,76,92–101] For the sake of later comparison also corresponding DFT results are included.[102–106]

Table 7.10 Selected results for vicinal ^{3}J(HH) in small hydrocarbons.

Coupling	Method	ZPVC/Hz	Total/Hz	% of Total
Ethyne				
^{3}J(H–H)	CCSD[77,b]	−0.37	9.50	−3.9
	CCSD[78,c]	−0.26	10.22	−2.5
	B3LYP[102,d]	−0.1	10.5	−1.0
	BHandH[103,e]	0.56	11.71	4.8
	Exp.[98]		9.62 ± 0.05 at 300 K	
Ethene				
cis-^{3}J(H–H)	CCSD[77,b]	0.90	11.79	7.6
	CCSD[78,c]	0.7	12.07	5.8
	B3LYP[102,d]	1.2	14.7	8.1
	BHandH[103,e]	1.40	15.80	8.9
	Exp.[97]		11.657	
trans-^{3}J(H–H)	CCSD[77,b]	1.38	18.02	7.7
	CCSD[78,c]	1.47	18.93	7.8
	B3LYP[102,d]	2.3	23.0	10.0
	BHandH[103,e]	2.38	22.85	10.4
	Exp.[97]		19.015	
Ethane				
^{3}J(H–H)[a]	CCSD[77,b]	0.39	7.65	5.1
	BHandH[103,e]	0.73		
	Exp.[97]		8.0	
Cyclopropene				
^{3}J(H_p–H_p)	CCSD[78,c]	0.12	1.86	6.6
	B3LYP[105,f]	0.3	2.1	14.3
^{3}J(H_o–H_p)	CCSD[78,c]	0.04	−1.96	−2.1
	B3LYP[105,f]	0.0	−2.3	0.0
Cyclopropane				
syn-^{3}J(H–H)	CCSD[77,b]	0.44	8.81	5.0
	B3LYP[105,f]	0.5	10.3	4.9
	Exp.[101]		9.0	
gauche-^{3}J(H–H)	CCSD[77,b]	0.39	5.11	7.6
	B3LYP[105,f]	0.6	6.4	9.4
	Exp.[101]		5.6	

[a]Average of *syn*- and *anti*-coupling.
[b]SSCC derivatives: CCSD/dzp; FF: anharmonic CCSD/dzp.
[c]SSCC derivatives: CCSD/aug-cc-pVTZ-J; FF: CCSD(T)/cc-pVTZ.
[d]SSCC derivatives + FF: B3LYP/sHIII.
[e]SSCC derivatives: BHandH/pcJ-3; FF: BHandH/pcJ-2.
[f]SSCC derivatives: B3LYP/HIIsu (FC) and B3LYP/HII (non-FC).

7.3.1.1 Using Vibrational Perturbation Theory

Raynes and co-workers[14–23] were among the first who studied systematically rovibrational and temperature effects on SSCCs. They employed the VPT2 approach with temperature corrections in combination with refined experimental force fields (FF) and electronic structure calculations of the SSCC derivatives using the second-order polarization propagator approximation with coupled cluster singles and doubles amplitudes, SOPPA(CCSD),[107] or predecessors and basis sets tuned for the calculation of SSCCs. The vibrational corrections were calculated in symmetry coordinates. In this way they studied CH_4,[15–18] H_2O,[19] HCCH,[22] and SiH_4.[18,23] To first order in the normal coordinates they studied also GeH_4 and SnH_4,[18] while for PbH_4 only the dependence on the Pb–H bond length has been investigated.[84] The results for vibrational and temperature corrections in water were afterwards confirmed by Casanueva *et al.*[89] employing sophisticated MCSCF wavefunctions and a comparable SSCC tuned basis set. The fluorine–hydrogen coupling in HF was studied by Åstrand *et al.*[88] at the RASSCF level with large basis sets and the effective geometry approach. The results for these simple XH_n hydrides are quite similar. The ZPVCs for $^1J(XH)$ couplings (X = C, Si, O, F) are in the order of 2 and 6% with temperature corrections less than 1 Hz. For methane, silane, and water this amounts to 4–8 Hz, while for HF it is almost 30 Hz due to the large value of the coupling. Interestingly vibrational averaging reduces the absolute value of the one-bond couplings in water and hydrogen–fluoride, while it increases it for the XH_4 molecules. For the $^2J(HH)$ couplings the ZPVC were calculated to be less than 1 Hz, which nevertheless can amount to as much as 9% as in the case of water, where averaging also reduces the absolute value of the geminal coupling constant contrary to methane. In silane the correction almost vanishes, but is negative. Inclusion of the vibrational corrections improves the agreement with experimental values with the exception of CH_4. The remaining differences from the experimental values are in the order of 1–3% or 1–4 Hz for the $^1J(XH)$ couplings and less than 1 Hz for the $^2J(HH)$ couplings again with the exception of methane, which was shown to be due to an insufficiently converged basis set in the calculation of the equilibrium geometry value.[107]

Apart from the simple hydrides only for very few other molecules have the vibrational corrections been calculated with correlated wavefunction methods. Sneskov and Stanton[77] as well as Faber and Sauer[78] studied some small hydrocarbons at the CCSD level with the VPT2 approach. In their study of ethyne, ethene, ethane, and cyclopropane, Sneskov and Stanton calculated both the SSCC derivatives and the anharmonic force field at the CCSD level with a rather small DZP basis set, while the equilibrium geometry values were obtained with a large basis set, but not one that was tuned for calculation of SSCCs. Faber and Sauer, on the other hand, employed the specialized SSCC basis set, aug-cc-pVTZ-J,[108] for the calculation of the coupling constant derivatives and calculated the cubic force field at the CCSD(T)/cc-pVTZ level. They studied also ethyne and ethene and in addition also cyclopropene and allene. For ethyne the final ZPV averaged results of Sneskov and Stanton reproduce the earlier results by Wigglesworth *et al.*[22] within 1 Hz for the ^2J and ^3J but differ significantly for the ^1J couplings due to a larger difference between the equilibrium geometry values. The ZPVCs for the $^1J(CC)$ and ^2J couplings are, however, very similar, while for the ^3J and $^1J(CH)$ couplings Sneskov and Stanton predict much smaller corrections.

For these couplings, the CCSD results of Faber and Sauer lie between those earlier results, suggesting that the discrepancy between the earlier results is due to both the basis set and the electronic structure methods. The ZPVCs for ethene are almost identical in the studies of Sneskov and Stanton and of Faber and Sauer, with the exception of the $^1J(CH)$ where the difference is 0.4 Hz. There is also a larger difference between the equilibrium geometry values for this and the $^1J(CC)$ coupling, which reflects the fact that Faber and Sauer employed a basis set tuned for the FC term of the coupling constants, whereas Sneskov and Stanton used a larger basis set, but one that was not optimized for SSCCs. Generalizing the results for these seven hydrocarbons, CH_4, C_2H_2, C_2H_4, C_2H_6, C_3H_4, and C_3H_6, one can state the ZPVCs are for the $^1J(CH)$ couplings $\sim$4 Hz or 2–4% of the total value, for the $^2J(HH)$ couplings less than 1.1 Hz and for the $^3J(HH)$ couplings less than 1.5 Hz or 5–8%. The ZPVCs for the $^1J(CC)$ and $^2J(CH)$ couplings are 1–2 Hz with the exception of ethyne, where these corrections are much larger in Hz but not in percentage of the total values and cyclopropene, where the ZPVCs to these couplings are smaller than 1 Hz. Adding the ZPVCs to the equilibrium geometry values improves the agreement with experiment for the ^1J and ^3J couplings, but not for the geminal couplings.

Finally, vibrational corrections have also been calculated with correlated wavefunction methods for two fluorine-containing molecules. Couplings to fluorine atoms are known to be difficult and require in general such high-level treatment. Jackowski *et al.*[79] measured and calculated $^1J(^{11}B^{19}F)$ in BF_3. The calculations were carried out at the CCSD level with basis set especially tuned for J coupling calculations. They did not perform a full ZPVC calculation but calculated the coupling constants at the temperature averaged geometry,[49] which corresponds to the main contribution in the effective geometry approach. They find a change of −7.7 Hz between the equilibrium and zero-point vibrationally averaged geometry value and another −0.74 Hz for the temperature dependence to 300 K, which amounts to $\sim$40% of the total value and strongly improves the agreement with experiment. The second molecule is difluoroethyne, whose $^3J(FF)$ coupling is a particular spectacular case due to its small size. Only calculations with appropriate basis sets, proper inclusion of vibrational corrections (at the CCSD/aug-ccJ-pVQZ level using a CCSD(T)/cc-pCVQZ cubic force field), and treatment of electron correlation at the highest possible level, here CCSD and CC3,[80] could give satisfactory agreement with experiment.[109] The vibrational corrections are 2–4 Hz, which is as large as the total coupling or the contribution from the triples excitation in the CC3 calculations.

7.3.1.2 *Using Variational Vibrational Methods*

The vibrational corrections to the SSCCs in diatomic molecules and in molecules with large amplitude motions like in NH_3 or in hydrogen-bonded systems are typically treated by solving the vibrational problem variationally. For the diatomic systems the rovibrational Schrödinger equation is solved numerically by the Numerov–Cooley[110,111] technique with numerical potential energy radial functions obtained either from the experimental spectroscopic constants or from high-level *ab initio* calculations. In this way perfect agreement with new experimental data has been obtained for the simplest molecule, HD, by Helgaker *et al.*[76] They carried out full configuration interaction (FCI) calculations both for the SSCCs radial function with a very large SSCC specialized basis set and for the potential energy radial function with an even larger basis set. Nevertheless the ZPVC differ by only ± 0.08 Hz from earlier predictions at lower levels of theory, CCSDPPA,[85] and CASSCF.[90] Using the same two methods also vibrational corrections to the couplings in CO and N_2 have been studied.[86,91] However, contrary to HD, for these two molecules the results are not yet in agreement either with experiment or with each other. The CCSDPPA calculations overestimate the experimental values, while the CASSCF calculations underestimate them and inclusion of vibrational corrections only worsens the agreement.

The SSCCs in NH_3, the prototypical molecule with a large-amplitude inversion mode, were studied by Yachmenev *et al.*[5] at the SOPPA(CCSD) level with a large, uncontracted basis set tuned for the calculation of SSCCs. For the rovibrational and temperature averaging they generated a SOPPA(CCSD) coupling constant surface of more than 2000 points, which was fitted to a fourth-order power series in internal coordinates, and employed fully coupled rovibrational wavefunctions, which were obtained with the TROVE Hamiltonian from a CCSD(T) potential energy surface refined by fitting to experimental vibrational energies.[112] At 300 K they found that the rovibrational correction to $^1J(^{15}N^1H)$ is with -0.2% rather small due to the near cancellation of a negative contribution from bending and a positive contribution from the own-bond stretching but in good agreement with experiment, while it amounts to 5% for $^1J(^1H^2H)$ in $^{15}NH_2D$. In order to assess the importance of the large-amplitude mode they also carried out the averaging with wavefunctions, which mimic the usually employed perturbation theory wavefunctions, and found that the perturbation theory approach significantly overestimates the correction to the geminal coupling constants and gives only half of the correction to the one-bond coupling. Furthermore, comparing the results from the fully coupled rovibrational wavefunctions with those from pure vibrational wavefunctions, the authors could determine the often ignored effect of rotations and their coupling with the vibrations for the temperature dependence of the coupling constants. It turned out that the rotational contributions have the opposite sign, but are almost as large as the vibrational ones, which implies that ignoring the rotational contributions overestimates the temperature dependence. Similarly, the dependence of the SSCCs in the hydronium ion H_3O^+ on the large-amplitude inversion and the O–H stretching mode have been studied[87] with the non-rigid bender formalism.[113] The reduced two-dimensional coupling constant surface was again obtained at the SOPPA(CCSD) level with the same basis set as in the previous calculations on H_2O.[19] The vibrational state specific values of the coupling constants and in particular of the geminal H–H coupling constant exhibit a large and non-monotonic dependence on the quantum number of the inversion mode leading also to a very non-linear temperature dependence at low temperatures. The zero-point vibrational corrections to the one- and two-bond couplings amount to 4% and 55% of the equilibrium geometry values – indication again of the importance of a proper treatment of the large amplitude inversion mode for such systems.

The SSCCs in the FHF^- was studied in detail by Hirata *et al.*[4] using the VSCF, VMP2, and VCI methods for vibrational averaging. They employed a CCSD potential energy surface. The coupling constant surfaces were obtained at the EOM-CCSD level with the (d-)aug-cc-pVXZ-su (X = D,T) basis sets as one-dimensional functions of the normal coordinates, *i.e.* no cross terms were included. Unlike the equilibrium value, the dependence of the vibrational corrections on the basis set employed in the calculation of the coupling constant surface was found to be almost negligible. The zero-point vibrational corrections to the one-bond and geminal couplings amount to 18 Hz (15%) and −43 Hz (25%), but they did not change on going from a VSCF to a VMP2 or VCI description. A harmonic treatment, however, completely failed, giving 34 Hz and −3 Hz, underestimating dramatically the correction to geminal coupling while predicting almost twice as large a correction to the one-bond coupling. In an earlier study, Del Bene *et al.*[81] obtained a similar vibrational correction for the F–F coupling (−41 Hz) using a two-dimensional vibrational model based on CCSD(T)/aug-cc-pVTZ and EOM-CCSD/qzp potential energy and coupling constant surfaces. This is despite the fact that the equilibrium value of Del Bene *et al.* differs by 25 Hz from that of Hirata *et al.*, which must be almost exclusively due to the different basis sets employed.

The effect of the dimer- and proton-stretching modes on the geminal $^{2h}J(^{15}N–^{15}N)$ across the hydrogen bond in the model hydrogen bonded complex CNH:NCH was studied by Jordan *et al.*[82] They reduced the vibrational problem to a two-dimensional problem including only the dimer- and proton-stretching modes for which they obtained anharmonic vibrational eigenfunctions. The two-dimensional potential energy surface was obtained at the MP2/aug-cc-pVTZ level while for the coupling constant surface the EOM-CCSD method was employed with the qzp,qz2p basis sets from Ahlrichs. They find that the zero-point vibrational correction is 0.68 Hz (~10%) and that several low-lying vibrationally excited states should be taken into account even at 298 K, lowering the vibrational correction to 0.38 Hz. In a similar study on $ClH:NH_3$ using the same methodology,[83] they reported a zero-point vibrational correction to $^{2h}J(^{35}Cl–^{15}N)$ of −1.6 Hz (21%). Again, low-lying vibrationally exited states played a role to the effect that the temperature averaged vibrational correction decreases with temperature to −1.3 Hz at 298 K.

7.3.2 DFT Calculations

Compared to correlated wavefunction methods, DFT based calculations have some advantages from a practical point of view, the most obvious being the lower formal scaling with system size. The lower computational cost makes it possible to carry out vibrational correction calculations on larger systems than what can be achieved using correlated methods. DFT also provides a suitable test bed for comparing different variational vibrational models, since performing the numerous calculations needed for higher order expansions of energy and property surfaces is much more feasible when using DFT. The lower accuracy of DFT compared to high level wavefunction calculations may be less of a concern, since the vibrational corrections are only concerned with differences, not absolute values of the SSCC. Still the fact that the number of calculations needed to obtain the parameters for calculating vibrational corrections also scales with system size means that even DFT calculations of vibrational corrections can become quite expensive.

7.3.2.1 Using Vibrational Perturbation Theory

Ruden *et al.*[102] studied the basis set dependence of vibrational corrections to the SSCCs of nine small molecules (H_2, HF, CO, N_2, H_2O, HCN, NH_3, CH_4, C_2H_2) using the B3LYP functional and the Huzinaga basis sets HX and HX–suX, X = II,III,IV. They found that the vibrational corrections depend somewhat on the choice of basis set, with for instance a change of up to 10% on going from HIII to HIV and a similar change on adding tight uncontracted s-functions. However, these effects largely cancel in most cases. It was mainly the Fermi-contact results that changed with basis sets, leading the authors to suggest that a smaller basis set could be used for the remaining terms. Since the vibrational corrections for the SSCCs in this study are an order of magnitude smaller that the equilibrium values, the basis set error on the vibrational correction should be at most a few percent of the total averaged SSCC and thus probably within the intrinsic error of common DFT functionals such as B3LYP. Indeed, comparing their results for the nine compounds as well for ethene and benzene to experimental values, it appears that adding vibrational corrections does not in general improve the agreement between B3LYP calculations and experiment. In a follow-up study allene and a number of small cyclic hydrocarbons[105] were treated using HIIsu2 for the Fermi-contact term and HII for the rest. The test set was further increased by Lutnæs *et al.*[106] to include also a few (hetero-)aromatic compounds, pyrrole, furan, thiophene, and benzene, though now only considering the Fermi-contact term. These papers have often been cited, *e.g.* in ref. 114 and 115, for the suggestion that vibrational corrections to ^{1}J(C–H) are usually 5 Hz. While this is true within 1 Hz for all but one SSCC in this study, is should be noted that they all involve very rigid systems and mostly sp^2 carbons. And even for these systems it appears to be difficult to generalize for the remaining types of SSCCs, for instance ^{3}J(H–H) appear to be between 0.1 and 0.6 Hz, but this does not match the results for ethene (1.2 and 2.3 Hz for the *cis* and *trans* couplings).

While the above results are on systems with only one conformer, it is important to keep in mind that many chemically relevant systems have several. Ideally all conformers should be treated in one vibrational model, but since the modes connecting conformers behave like multiwell potentials, this cannot be handled by the standard perturbation expansions, Section 7.1.1. Instead handling the conformers individually, *i.e.* ignoring tunneling between them, is an attractive option. This is exemplified in a study by Atieh *et al.*,[116] who calculated the geminal and vicinal H–H coupling constants in serine at the B3LYP/6-311++G** level of theory. Including only the ZPV corrections to the vibrational ground state of the lowest energy conformer did not improve the agreement with experiment, but when instead calculated as the Boltzmann average over the 22 conformers, reasonably good agreement with experiment was obtained with an RMSD of 0.7 Hz. Similar conformational considerations are used in the study of Esteban *et al.*,[117] who employed the VPT2 approach in the calculation of vibrational corrections to 70 vicinal proton–proton coupling constants in six mono-substituted and five 1,1-di-substituted ethanes, 3 mono-substituted cyclohexanes, 3 norbornane type molecules, and 11 three-membered ring compounds. They used B3LYP with double zeta type basis sets for the PES and with the locally dense BHH basis set[118] for the SSCCs, the latter consisting of an aug-cc-pVDZ basis set on non-coupled atoms, but a quite large basis on the coupled atoms. The vibrational corrections represent about 7% of the total couplings and are dominated by the harmonic contribution. A very good agreement with experiment was obtained by scaling the temperature-averaged vibrational correction by 0.8485. On the other hand, they found that comparing the scaled vibrationally averaged couplings with experimental values gave the same standard variation as comparing the equilibrium geometry values alone scaled by 0.9016. While some of the molecules in this study, in particular the ethanes, can be seen as having large amplitude modes, it appears that conformational averaging is sufficient in most cases when at least the populations of the different conformers are equal.

It has also been proposed to augment higher level single-point calculations with DFT vibrational corrections. This has for example been done by Helgaker *et al.*,[34] who employed VPT2 at the PBE/cc-pCVTZ level to correct their own CCSD/cc-pCVTZ(FC)/cc-pCVDZ calculations of the coupling constants of *o*-benzyne, which, due to its weak biradical character, has a low-lying excited triplet state. First of all they found that the PBE results at the PBE geometry were closest to the CCSD results at the experimental geometry. The ZPVC never become larger than 4 Hz and amounts in particular for the one-bond couplings to less than 3%, while for some of the geminal and vicinal C–C couplings and one vicinal C–H coupling ZPVC in the order of 3–4 Hz amount to much larger percentages due to the fact that the equilibrium values themselves are only of the order of 10 Hz.

The variations of the standard perturbation theory approaches, discussed in Section 7.1.7, have been used in combination with DFT. Woodford and Harbison[25] included contributions from the diagonal (in the normal mode coordinate system) cubic and quartic force constants, and they included thermal effects *via* a Boltzmann average of the vibrational states. They employed this approach to the NMR spectrum of 1′-imidazolyl-2′-desoxy-β-ribofuranose as a model system for purine nucleosides. From calculations at the B3LYP/6-311 + + G(2d,p) level they find ZPVCs of up to 5 Hz, where the largest values typically belong to the one-bond C–H couplings and therefore correspond to only a few percent. Percent-wise the largest ZPVCs (180% and 65%) were observed for the geminal C–H couplings to the two hydroxyl hydrogens of desoxyribofuranose. Finally, very large temperature corrections (~60% of the ZPVC) are predicted for the three one-bond C–H couplings of the imidazolyl ring. For this molecule a purely harmonic treatment fails not only quantitatively in most cases but also qualitatively. A more elaborate approach was presented by Dračínský *et al.*,[26] who included the full cubic and quartic force constants at the same level of perturbation theory in the expansion of the potential energy surface and employed a degeneracy-corrected perturbational formula. Using this approach they calculated vibrational corrections of coupling constants in methane- and halogen-substituted methanes $CH_{4-n}X_n$ as well as on three differently charged forms of alanine and for a model sugar. The calculations were carried out at the DFT/B3LYP level with the generic 6-311 + + G** basis set, *i.e.* without a basis set specialized for SSCC calculations. They find for CH_4 and CH_3F that the changes in the zero-point vibrational corrections due to the anharmonic part of the potential are not large, which is contrary to previous findings,[102] and that the effect of the quartic force field is negligible. In most cases the diagonal second derivatives lead to the largest contributions to the ZPVC by far, while for the vicinal hydrogen–hydrogen couplings in the model sugar the first and diagonal second derivatives contribute more evenly. The off-diagonal second derivatives, finally, only become important for isotope effects in the halogenated methanes. This approach was also applied by Kupka *et al.*[103] in a study on several small molecules previously studied by Ruden *et al.*[102] The force constants and the coupling constant derivatives were obtained at the BHandH/pcJ-2 and BHandH/pcJ-3 levels. They concluded, in agreement with Ruden *at al.*, that the vibrational corrections are not very sensitive to a change in the basis set giving changes of at most 6%, which will usually be quite insignificant compared to the total SSCC. That being said their results in some cases differ quite significantly from those which have been calculated previously using both wavefunction methods and B3LYP, so one may wonder whether BHandH is appropriate for this kind of calculation.

PBE0 together with the 6-311G** was *e.g.* employed by Rusakov and Krivdin[104] for the ZPVC to ^1J(Si–H) one-bond coupling constants in the series of halosilanes SiH_nX_{4-n} (X = F, Cl, Br, I). They combined them with higher-level equilibrium geometry values obtained, SOPPA, SOPPA(CC2), or SOPPA(CCSD) and a locally dense basis set (aug-cc-pVTZ-J for Si and H/6-311G** X) and even relativistic corrections. They find ZPVC in the order of −2.5 to −6.6 Hz (1 to 2% of the experimental values), while the relativistic corrections are between −1.7 and −8.4 Hz, *i.e.* of almost the same size. For SiH_4

they obtained −3.9 Hz for the ZPVC compared to the −7.59 Hz of the pure SOPPA(CCSD) calculation.[23]

Autschbach and co-workers[119] studied the vibrational corrections to the one-bond hydrogen-deuterium coupling constant in six transition metal (Ir, Os, Nb, Re, Ru) dihydrogen or dihydride complexes at the non-relativistic DFT level using non-specialized basis sets on the lighter atoms and effective core-potentials on the transition metal following earlier work on the $[Ir(dmpm)CpH_2]^{2+}$ complex by Gelabert *et al.*[120] The vibrational corrections were calculated at the VPT2 level augmented in the case of the $[Ir(dmpm)CpH_2]^{2+}$ complex with a new explicit approach for treating the hindered rotation of the cyclopentadienyl ring.[28] While the size of the coupling constant differs quite a lot between the different complexes, for five out of the six complexes the predicted ZPVC amount to more than 20% of the total couplings (or almost 5 Hz in the largest case) and the temperature dependence was also found to be significant.

7.3.2.2 Using Variational Vibrational Methods

Hansen *et al.*[3] compared the accuracy of the variational VSCF, VMP2, and VCI methods for calculations of zero-point and thermal averaging corrections to the SSCCs of a series of small molecules: N_2, CO, HF, H_2O, and C_2H_2. The calculations were carried out at the rather low level of DFT/B3LYP with the HIV-su4 basis set. For the type of molecules studied, *i.e.* without large amplitude modes, the standard perturbation theory approaches were found to perform well in comparison with the more accurate and expensive VCI approach. Indeed for none of the couplings in HCN, H_2O, CH_4, and C_2H_2 does the difference between the approaches exceed 10%, with a mean average deviation of only 0.15 Hz. Furthermore they found that in the VCI calculations the inclusion of more than two mode couplings in the SSCC surfaces, *i.e.* higher order (more than two) mixed derivatives of the SSCCs, is much less important than the mode-excitation level, *i.e.* carrying out VCI[2] or higher VCI calculations instead of VSCF. VCI was also compared to the perturbation theory approach in the previously mentioned study by Dračínský *et al.*[26] In most molecules in the study, differences between the VPT2 and VCI treatment are less than 0.1%, while large differences were observed for the cationic and anionic forms of alanine. Whether this difference is due to large amplitude motions (internal rotations) of alanine becoming more important in the charged forms or because DFT may have problems handling doublet states was not investigated further.

A study on the effect of the internal rotation of the trimethylsiloxy-group, on the SSCCs between the Si nucleus and the carbon nuclei in the benzene ring in three substituted silylated phenols was presented by Sychrovský *et al.*[121] They employed the rigid-bender formalism[122] with B3LYP/6-31G(d,p) and MP2/6-31G(d,p) potential surfaces as a function of the torsional angle while optimizing all other geometry parameters to obtain wavefunctions for the internal rotation states. While internal-rotation wavefunctions are neatly localized inside their potential well, at higher energies they will eventually exhibit free rotor behaviors. For MP2 this crossover is above 500 cm^{-1}, but using B3LYP it was already below 150 cm^{-1}, a difference which was diagnosed as being largely due to the failure of DFT at describing dispersion interactions. Using these energy surfaces they averaged the corresponding coupling constant torsional angle function obtained with B3LYP and the IGLOIII basis set. Carrying out averaging over the internal rotation reduced the percentage-wise deviation from experiment by almost a factor of 2, showing the importance of a proper treatment of such large-amplitude motions. However, the remaining deviations were on average still 37%.

7.3.3 Systems with Relativistic Effects

Quantum chemical calculations including relativistic effects are still a growing field and few studies include both these and vibrational corrections at the same time. One such example is the study of Autschbach *et al.*,[123] who included VPT2 vibrational and temperature corrections in their study of the one-bond ^{199}Hg–^{13}C coupling constant and anisotropy in methylmercury halides. Calculations were carried out with the zeroth-order regular approximation (ZORA) at the DFT/VWN level and using STO TZP basis sets, which for Hg was extended with tight *s*- and *p*-functions. With ~53 Hz for J and ~27 Hz for ΔJ or 6% and 3% of the predicted gas phase values, the vibrational corrections are not negligible but are relatively small. Bryce and Autschbach[124] included in their study of the one-bond coupling constants and coupling anisotropies of the 20 diatomic alkali metal halides also rovibrational corrections. The VPT2 corrections were calculated using ZORA-DFT with the PBE0 functional and the STO TZ2P basis set modified with four tight *s*-functions. Mostly the zero-point vibrational corrections were below 1 Hz with the exceptions of LiF (−2.5 Hz), LiI (1.5 Hz), CsF (−2.9 Hz), and CsBr (1.5 Hz). Percentage-wise in the majority the corrections were below 1% and the largest corrections were found for LiI and NaI with ~3.5%.

Using the VPT2 approach Cho *et al.*[125] calculated the temperature dependence of the ^{99}Tc–^{17}O coupling constant in $^{99}Tc(^{16}O_3)(^{17}O)^-$. They employed spin–orbit ZORA with the VWN functional with a TZ2P Slater type orbitals basis set augmented with extra four tight *s*-functions on Tc and the *s*-functions for O replaced by the corresponding from the QZ4P basis set. They observed changes of ~4 Hz (3.5%) in the coupling constant and could predict the correct, although too small, trend of the temperature dependence.

7.3.4 Isotope Effects

The vibrational corrections or in general the geometry dependence of the one-bond C–H couplings in methane and its isotopomers have also been the topic of several other studies[126–130] because it was experimentally observed that rather unusually the secondary isotope effect (on exchanging one of the other hydrogens by deuterium) is larger than the primary isotope effect, where the coupled hydrogen is replaced by deuterium.[131] Raynes *et al.* could show that it arises because the C–H coupling depends more on changes in the length of the bond to one of the other hydrogens than to the coupled hydrogen. They denoted this phenomena the *unexpected differential sensitivity* (UDS).[126] The same effect was also found for the one-bond carbon-hydrogen coupling in ethyne[22] and the Si–H coupling in silane.[23,127] Sauer and co-workers could finally show, by an analysis in terms of localized molecular orbital contributions, that a subtle balance in the sensitivity of the different localized molecular orbital contributions is responsible for the UDS in methane and silane.[128–130]

7.3.5 General Trends

Today most authors acknowledge the need for vibrational corrections for accurate calculations of SSCCs or when trying to reproduce experimental values. However, actual studies which include vibrational effects are still quite rare. A common compromise is to calculate properties at experimental geometries or at vibrationally averaged geometries; however, this will at most give the anharmonic contribution to the vibrational correction. In many cases the harmonic second derivative term is the larger contribution and predicting which of the two will be the dominating one for a particular SSCCs is non-trivial. Therefore, as a minimum, both of these contributions should be considered before considering a value to be vibrationally averaged and thus directly comparable to experimental values.

Nuclear spin–spin coupling constants are a tricky property to handle from a theoretical point of view and accurate calculation of single point values are demanding for both the basis set and the method employed. This means that cheaper methods, in particular DFT, often have intrinsic errors that are of similar size or larger than the vibrational corrections. As a consequence, a number of the mentioned studies did not show an improved agreement with experiment upon inclusion of vibrational effects. When calculating the vibrational corrections, however, this error largely cancels and therefore the vibrational corrections seem to depend much less on the method and basis set employed. Furthermore, for molecules without large amplitude motions the difference between vibrational corrections calculated *via* perturbation theory approach or *via* variational methods are often negligible. The remaining errors in the calculated values are therefore mostly due to errors in the method and basis set employed for the calculation at the equilibrium geometry. Nevertheless, there are cases, as can be seen in Tables 7.8–7.10, where the vibrational corrections depend significantly on the chosen calculation method. Some considerations of which electronic structure method and basis set to use are therefore still necessary.

In studies comparing the performance of various methods for the calculation of spin–spin coupling constants one should ideally include vibrational corrections for all methods. However, in order to avoid the cost of calculating vibrational corrections with all different methods, it has become popular to calculate only one correction at a lower level of theory and then apply it evenly to all calculated single point values or (essentially equivalent) to use this to "remove" vibrational effects from the experimental value leading to an *empirical* value equilibrium geometry value.[102] This raises a number of questions regarding consistency as effectively the vibrational average is taken with respect to the equilibrium geometry of the lower-level method used for the force field calculation whereas one would usually calculate the "equilibrium value" of the SSCC at a geometry obtained with a higher-level method. The vibrational corrections calculated on a different SSCC surface around a different geometry are then added to the high-level SSCC value. Due to the small size of the vibrational corrections to the total SSCC, however, the possible error induced by this inconsistency is probably acceptable and treating vibrational corrections as additive appears nevertheless to be a reasonable approach in practice.

References

1. T. Helgaker, S. Coriani, P. Jørgensen, K. Kristensen, J. Olsen and K. Ruud, *Chem. Rev.*, 2012, **112**, 543.
2. O. Christiansen, *Phys. Chem. Chem. Phys.*, 2007, **9**, 2942.
3. M. B. Hansen, J. Kongsted, D. Toffoli and O. Christiansen, *J. Phys. Chem. A*, 2008, **112**, 8463.
4. S. Hirata, K. Yagi, S. A. Perera, S. Yamazaki and K. Hirao, *J. Chem. Phys.*, 2008, **128**, 214305.
5. A. Yachmenev, S. N. Yurchenko, I. Paidarová, P. Jensen, W. Thiel and S. P. A. Sauer, *J. Chem. Phys.*, 2010, **132**, 114305.
6. C. W. Kern and R. L. Matcha, *J. Chem. Phys.*, 1968, **49**, 2081.
7. T. A. Ruden and K. Ruud, *Calculation of NMR and EPR Parameters: Theory and Applications*, ed. M. Kaupp, V. G. Malkin and M. Bühl, Wiley-VCH, Weinheim, 2004, ch. 10, p. 153.
8. K. Ruud, P.-O. Åstrand and P. R. Taylor, *J. Chem. Phys.*, 2000, **112**, 2668.
9. J. Watson, *Mol. Phys.*, 1968, **15**, 479.
10. H. H. Nielsen, *Rev. Mod. Phys.*, 1951, **23**, 90.
11. M. Toyama, T. Oka and Y. Morino, *J. Mol. Spectrosc.*, 1964, **13**, 193.
12. H. Batiz-Hernandez and R. A. Bernheim, *Prog. Nucl. Magn. Reson. Spectrosc.*, 1967, **3**, 63.
13. C. J. Jameson and H.-J. Osten, *J. Am. Chem. Soc.*, 1986, **108**, 2497.
14. J. Geertsen, J. Oddershede, W. T. Raynes and G. E. Scuseria, *J. Magn. Reson.*, 1991, **93**, 458.
15. W. T. Raynes, J. Geertsen and J. Oddershede, *Int. J. Quantum Chem.*, 1994, **52**, 153.
16. J. Geertsen, J. Oddershede, W. T. Raynes and T. L. Marvin, *Mol. Phys.*, 1994, **82**, 29.
17. R. D. Wigglesworth, W. T. Raynes, S. P. A. Sauer and J. Oddershede, *Mol. Phys.*, 1997, **92**, 77.

18. S. Kirpekar, T. Enevoldsen, J. Oddershede and W. T. Raynes, *Mol. Phys.*, 1997, **91**, 897.
19. R. D. Wigglesworth, W. T. Raynes, S. P. A. Sauer and J. Oddershede, *Mol. Phys.*, 1998, **94**, 851.
20. R. D. Wigglesworth, W. T. Raynes, S. P. A. Sauer and J. Oddershede, *Mol. Phys.*, 1999, **96**, 1595.
21. R. D. Wigglesworth, W. T. Raynes, S. Kirpekar, J. Oddershede and S. P. A. Sauer, *J. Chem. Phys.*, 2000, **112**, 736.
22. R. D. Wigglesworth, W. T. Raynes, S. Kirpekar, J. Oddershede and S. P. A. Sauer, *J. Chem. Phys.*, 2000, **112**, 3735, *J. Chem. Phys.*, 2001, **114**, 9192.
23. S. P. A. Sauer, W. T. Raynes and R. A. Nicholls, *J. Chem. Phys.*, 2001, **115**, 5994.
24. P. Åstrand, K. Ruud and D. Sundholm, *Theor. Chem. Acc.*, 2000, **103**, 365.
25. J. N. Woodford and G. S. Harbison, *J. Chem. Theory Comput.*, 2006, **2**, 1464.
26. M. Dračínský, J. Kaminský and P. Bouř, *J. Chem. Phys.*, 2009, **130**, 094106.
27. C. J. Jameson, A. C. de Dios and A. K. Jameson, *J. Chem. Phys.*, 1991, **95**, 1069.
28. B. C. Mort and J. Autschbach, *ChemPhysChem*, 2008, **9**, 159.
29. P. E. Maslen, D. Jayatilaka, S. M. Colwell, R. D. Amos and N. C. Handy, *J. Chem. Phys.*, 1991, **95**, 7409.
30. M. Ringholm, D. Jonsson, R. Bast, B. Gao, A. J. Thorvaldsen, U. Ekstrøm, T. Helgaker and K. Ruud, *J. Chem. Phys.*, 2014, **140**, 034103.
31. W. Schneider and W. Thiel, *Chem. Phys. Lett.*, 1989, **157**, 367.
32. P. Åstrand, K. Ruud and P. R. Taylor, *J. Chem. Phys.*, 2000, **112**, 2655.
33. T. A. Ruden, P. R. Taylor and T. Helgaker, *J. Chem. Phys.*, 2003, **119**, 1951.
34. T. Helgaker, O. B. Lutnæs and M. Jaszuński, *J. Chem. Theory Comput.*, 2007, **3**, 86.
35. O. B. Lutnæs, T. Helgaker and M. Jaszuński, *Mol. Phys.*, 2010, **108**, 2579.
36. A. A. Auer and J. Gauss, *Chem. Phys.*, 2009, **356**, 7.
37. K. Ruud, P. Åstrand and P. R. Taylor, *J. Am. Chem. Soc.*, 2001, **123**, 4826.
38. J. Kaminský, M. Buděšínský, S. Taubert, P. Bouř and M. Straka, *Phys. Chem. Chem. Phys.*, 2013, **15**, 9223.
39. A. A. Auer, J. Gauss and J. F. Stanton, *J. Chem. Phys.*, 2003, **118**, 10407.
40. T. Kupka, M. Stachów, M. Nieradka, J. Kaminský and T. Pluta, *J. Chem. Theory Comput.*, 2010, **6**, 1580.
41. E. Prochnow and A. A. Auer, *J. Chem. Phys.*, 2010, **132**, 064109.
42. H. Fukui, T. Baba, J. Narumi, H. Inomata, K. Miura and H. Matsuda, *J. Chem. Phys.*, 1996, **105**, 4692.
43. T. Kupka, M. Stachów, J. Kaminský and S. P. A. Sauer, *Magn. Reson. Chem.*, 2013, **51**, 482.
44. D. Sundholm, J. Gauss and A. Schäfer, *J. Chem. Phys.*, 1996, **105**, 11051.

45. A. A. Auer, *J. Chem. Phys.*, 2009, **131**, 024116.
46. J. Vaara, J. Lounila, K. Ruud and T. Helgaker, *J. Chem. Phys.*, 1998, **109**, 8388.
47. C. Benzi, O. Crescenzi, M. Pavone and V. Barone, *Magn. Reson. Chem.*, 2004, **42**, S57.
48. P. Åstrand and K. Ruud, *Phys. Chem. Chem. Phys.*, 2003, **5**, 5015.
49. M. E. Harding, M. Lenhart, A. A. Auer and J. Gauss, *J. Chem. Phys.*, 2008, **128**, 244111.
50. C. J. Jameson, A. K. Jameson and J. Honarbakhsh, *J. Chem. Phys.*, 1984, **81**, 5266.
51. A. Dransfeld, *Chem. Phys.*, 2004, **298**, 47.
52. K. A. Chernyshev and L. B. Krivdin, *Russ. J. Org. Chem.*, 2011, **47**, 355.
53. S. Grigoleit and M. Bühl, *J. Chem. Theory Comput.*, 2005, **1**, 181.
54. S. Grigoleit and M. Bühl, *Chem. - Eur. J.*, 2004, **10**, 5541.
55. M. Bühl, P. Imhof and M. Repisky, *Chem. Phys. Chem.*, 2004, **5**, 410.
56. M. Bühl, S. Grigoleit, H. Kabrede and F. T. Mauschick, *Chem. - Eur. J.*, 2006, **12**, 477.
57. M. Kaupp, M. Bühl and V. G. Malkin, *Calculation of NMR and EPR Parameters: Theory and Applications*, Wiley-VCH, Weinheim, 2004.
58. J. Kongsted, K. Aidas, K. V. Mikkelsen and S. P. A. Sauer, *J. Chem. Theory Comput.*, 2008, **4**, 267.
59. F. Jensen, *J. Chem. Theory Comput.*, 2008, **4**, 719.
60. M. Broda, A. Buczek, T. Kupka and J. Kaminský, *Vib. Spectrosc.*, 2012, **63**, 432.
61. See *e.g.* V. Arcisauskaite, J. I. Melo, L. Hemmingsen and S. P. A. Sauer, *J. Chem. Phys.*, 2011, **135**, 044306.
62. B. Minaev, J. Vaara, K. Ruud, O. Vahtras and H. Ågren, *Chem. Phys. Lett.*, 1998, **295**, 455.
63. P. Lantto, J. Vaara, A. M. Kantola, V. V. Telkki, B. Schimmelpfennig, K. Ruud and J. Jokisaari, *J. Am. Chem. Soc.*, 2002, **124**, 2762.
64. A. M. Kantola, P. Lantto, J. Vaara and J. Jokisaari, *Phys. Chem. Chem. Phys.*, 2010, **12**, 2679.
65. P. Lantto, S. Kangasvieri and J. Vaara, *J. Chem. Phys.*, 2012, **137**, 214309.
66. C. H. Gee and W. T. Raynes, *Chem. Phys. Lett.*, 2000, **330**, 595.
67. C. J. Jameson, A. K. Jameson, S. M. Cohen, H. Parker, D. Oppusunggu, P. M. Burrel and S. Wille, *J. Chem. Phys.*, 1981, **74**, 1608.
68. C. J. Jameson, A. C. de Dios and A. K. Jameson, *J. Chem. Phys.*, 1991, **95**, 9042.
69. D. Sundholm and J. Gauss, *Mol. Phys.*, 1997, **92**, 1007.
70. S. P. A. Sauer, V. Špirko, I. Paidarová and J. Oddershede, *Chem. Phys.*, 1994, **184**, 1.
71. S. P. A. Sauer, V. Špirko, I. Paidarová and W. P. Kraemer, *Chem. Phys.*, 1997, **214**, 91.
72. J. Kongsted and K. Ruud, *Chem. Phys. Lett.*, 2008, **451**, 226.
73. V. Barone, *J. Chem. Phys.*, 2005, **122**, 014108.
74. P. Daněček and P. Bouř, *J. Comput. Chem.*, 2007, **28**, 1617.
75. M. Baaden, P. Granger and A. Strich, *Mol. Phys.*, 2000, **98**, 329.

76. T. Helgaker, M. Jaszuński, P. Garbacz and K. Jackowski, *Mol. Phys.*, 2012, **110**, 2611.
77. K. Sneskov and J. F. Stanton, *Mol. Phys.*, 2012, **110**, 2321.
78. R. Faber and S. P. A. Sauer, *AIP Conf. Proc.*, 2016.
79. K. Jackowski, W. Makulski, A. Szyprowska, A. Antušek and M. Jaszuński, *Magn. Reson. Chem.*, 2009, **47**, 857.
80. R. Faber and S. P. A. Sauer, *Phys. Chem. Chem. Phys.*, 2012, **14**, 16440.
81. J. E. Del Bene, M. J. T. Jordan, S. A. Perera and R. J. Bartlett, *J. Phys. Chem. A*, 2001, **105**, 8399.
82. M. J. T. Jordan, J. S.-S. Toh and J. E. Del Bene, *Chem. Phys. Lett.*, 2001, **346**, 288.
83. J. E. Del Bene and M. J. T. Jordan, *J. Phys. Chem. A*, 2002, **106**, 5385.
84. S. Kirpekar and S. P. A. Sauer, *Theor. Chem. Acc.*, 1999, **103**, 146.
85. J. Oddershede, J. Geertsen and G. Scuseria, *J. Phys. Chem.*, 1988, **92**, 3056.
86. J. Geertsen, J. Oddershede and G. Scuseria, *J. Chem. Phys.*, 1987, **87**, 2138.
87. S. P. A. Sauer, C. K. Møller, H. Koch, I. Paidarová and V. Špirko, *Chem. Phys.*, 1998, **238**, 385.
88. P.-O. Åstrand, K. Ruud, K. V. Mikkelsen and T. Helgaker, *J. Chem. Phys.*, 1999, **110**, 9463.
89. J. Casanueva, J. San Fabián, E. Díez and A. L. Esteban, *J. Mol. Struct.*, 2001, **565**, 449.
90. O. Vahtras, H. Ågren, P. Jørgensen, H. J. Aa Jensen, S. B. Padkjær and T. Helgaker, *J. Chem. Phys.*, 1992, **96**, 6120.
91. O. Vahtras, H. Ågren, P. Jørgensen, T. Helgaker and H. J. Aa Jensen, *Chem. Phys. Lett.*, 1993, **209**, 201.
92. S. M. Bass, R. L. DeLeon and J. S. Muenter, *J. Chem. Phys.*, 1987, **86**, 4305.
93. R. E. Wasylishen and J. O. Friedrich, *Can. J. Chem.*, 1987, **65**, 2238.
94. N. M. Sergeyev, N. D. Sergeyeva, Y. A. Strelenko and W. T. Raynes, *Chem. Phys. Lett.*, 1997, **277**, 142.
95. A. Antušek, D. Kedziera, K. Jackowski, M. Jaszuński and W. Makulski, *Chem. Phys.*, 2008, **352**, 320.
96. E. A. V. Ebsworth and J. J. Turner, *J. Chem. Phys.*, 1962, **36**, 2628.
97. J. Kaski, P. Lantto, J. Vaara and J. Jokisaari, *J. Am. Chem. Soc.*, 1998, **120**, 3993.
98. K. Jackowski, M. Wilczek, M. Pecul and J. Sadlej, *J. Phys. Chem. A*, 2000, **104**, 5955.
99. H. Günther and H. Seel, *Org. Magn. Reson.*, 1976, **8**, 299.
100. H. Kalinowski, S. Berger and S. Braun, *Carbon-13 NMR Spectroscopy*, Wiley, Chichester, 1994.
101. H. Günther, *NMR Spectroscopy*, Wiley, Chichester, 1980.
102. T. A. Ruden, O. B. Lutnæs, T. Helgaker and K. Ruud, *J. Chem. Phys.*, 2003, **118**, 9572.
103. T. Kupka, M. Nieradka, M. Stachów, T. Pluta, P. Nowak, H. Kjær, J. Kongsted and J. Kaminský, *J. Phys. Chem. A*, 2012, **116**, 3728.

104. Yu. Yu. Rusakov and L. B. Krivdin, *Magn. Reson. Chem.*, 2013, **51**, 557.
105. T. A. Ruden, T. Helgaker and M. Jaszuński, *Chem. Phys.*, 2003, **296**, 53.
106. O. B. Lutnæs, T. A. Ruden and T. Helgaker, *Magn. Reson. Chem.*, 2004, **42**, S117.
107. T. Enevoldsen, J. Oddershede and S. P. A. Sauer, *Theor. Chem. Acc.*, 1998, **100**, 275.
108. P. F. Provasi, G. A. Aucar and S. P. A. Sauer, *J. Chem. Phys.*, 2001, **115**, 1324.
109. H. Bürger and S. Sommer, *J. Chem. Soc. Chem. Commun.*, 1991, 456.
110. B. Numerov, *Publ. Obs. Cent. Astrophys. Russ.*, 1933, **2**, 188.
111. J. Cooley, *Math. Comput.*, 1962, **15**, 363.
112. S. N. Yurchenko, J. Zheng, H. Lin, P. Jensen and W. Thiel, *J. Chem. Phys.*, 2005, **123**, 134308.
113. V. Špirko, *J. Mol. Spectrosc.*, 1983, **101**, 30.
114. S. N. Maximoff, J. E. Peralta, V. Barone and G. E. Scuseria, *J. Chem. Theory Comput.*, 2005, **1**, 541.
115. J. San Fabián, J. M. García de la Vega, R. Suardíaz, M. Fernández-Oliva, C. Pérez, R. Crespo-Oterod and R. H. Contreras, *Magn. Reson. Chem.*, 2013, **51**, 775.
116. Z. Atieh, A. R. Allouche, D. Graveron-Demilly and M. Aubert-Frécon, *Meas. Sci. Technol.*, 2011, **22**, 114015.
117. A. L. Esteban, E. Dıéz, M. P. Galache, J. San Fabián, J. Casanueva and R. H. Contreras, *Mol. Phys.*, 2010, **108**, 583.
118. E. Dıéz, J. Casanueva, J. San Fabián, A. L. Esteban, M. P. Galache, V. Barone, J. E. Peralta and R. H. Contreras, *Mol. Phys.*, 2005, **103**, 1307.
119. B. C. Mort and J. Autschbach, *J. Am. Chem. Soc.*, 2006, **128**, 10060.
120. R. Gelabert, M. Moreno, J. M. Lluch, A. Lledós and D. M. Heinekey, *J. Am. Chem. Soc.*, 2005, **127**, 5632.
121. V. Sychrovský, L. Benda, A. Prokop, V. Blechta, J. Schraml and V. Špirko, *J. Phys. Chem. A*, 2008, **112**, 5167.
122. V. Špirko, W. P. Kraemer and A. Cejchan, *J. Mol. Spectrosc.*, 1989, **136**, 340.
123. J. Autschbach, A. M. Kantola and J. Jokisaari, *J. Phys. Chem. A*, 2007, **111**, 5343.
124. D. L. Bryce and J. Autschbach, *Can. J. Chem.*, 2009, **87**, 927.
125. H. Cho, W. A. de Jong, B. K. McNamara, B. Rapko and J. E. Burgeson, *J. Am. Chem. Soc.*, 2004, **126**, 11583.
126. W. T. Raynes, J. Geertsen and J. Oddershede, *Chem. Phys. Lett.*, 1992, **197**, 516.
127. S. P. A. Sauer and W. T. Raynes, *J. Chem. Phys.*, 2000, **113**, 3121, *J. Chem. Phys.*, 2001, **114**, 9193.
128. S. P. A. Sauer and P. F. Provasi, *ChemPhysChem*, 2008, **9**, 1259.
129. P. F. Provasi and S. P. A. Sauer, *Phys. Chem. Chem. Phys.*, 2009, **11**, 3987, *Phys. Chem. Chem. Phys.*, 2010, **12**, 15132.
130. M. N. C. Zarycz, S. P. A. Sauer and P. F. Provasi, *J. Chem. Phys.*, 2014, **141**, 151101.
131. B. Bennett, W. T. Raynes and C. W. Anderson, *Spectrochim. Acta, Part A*, 1989, **45**, 821.

CHAPTER 8

Relativistic Calculations of Nuclear Magnetic Resonance Parameters

MICHAL REPISKY,* STANISLAV KOMOROVSKY, RADOVAN BAST AND KENNETH RUUD

Centre for Theoretical and Computational Chemistry (CTCC), Department of Chemistry, UiT The Arctic University of Norway, N-9037 Tromsø, Norway
*Email: michal.repisky@uit.no

8.1 Introduction

Relativistic effects are often considered to be important largely for heavy elements. However, this is not the case for nuclear magnetic resonance (NMR) parameters. This is in part due to the fact that the nuclear magnetic shielding tensors and the indirect nuclear spin–spin coupling tensors are governed to a large extent by the electron density near the nucleus (outer core-inner valence electrons), but also due to the fact that magnetic interactions can be considered relativistic in nature.[1] The coupling between the electron spin and the orbital motion of the electrons through the spin–orbit operator allows for new interaction mechanisms between the nuclear magnetic moments or between the nuclear magnetic moments and an externally applied magnetic field. Many of these new interaction mechanisms also grow rapidly with increasing nuclear charge, making these relativistic effects particularly important for molecules containing heavy elements. It has been

New Developments in NMR No. 6
Gas Phase NMR
Edited by Karol Jackowski and Michał Jaszuński

Published by the Royal Society of Chemistry, www.rsc.org

shown that the relativistic effects on the nuclear magnetic shielding tensors scale as $Z^{3.5}$ with respect to the nuclear charge Z.[2] However, also light elements in the vicinity of heavy elements can display large relativistic effects. Even the lightest element, hydrogen, can display dramatic effects arising from relativity, with hydrogen iodide[3,4] and mercury hydride complexes[5,6] being prime examples. This effect is often referred to as the heavy-atom effect on light atoms (HALA),[3,7] to be contrasted with the effect of a heavy element on the NMR properties of the heavy element itself, the HAHA effect.

Relativistic effects can be defined as the difference between the results of a relativistic calculation with a finite speed of light, $c \approx 137$ a.u., at a given level of electronic-structure theory, and the corresponding results obtained with $c \rightarrow \infty$. For heavy-element systems, there are no alternatives to a relativistic calculation, but also highly accurate calculations on light elements require a treatment of relativity.[8] Relativistic theory also offers a natural and consistent framework for describing magnetic properties and allows for a treatment of the whole periodic table on an equal footing. Relativistic effects can be studied perturbatively by *adding* relativistic corrections to a non-relativistic description, or by starting from a relativistic Hamiltonian with spin–orbit coupling included variationally from the start and selectively *removing* relativistic contributions using dedicated non-relativistic or scalar-relativistic Hamiltonians or by modifying the speed of light.

The last decade has seen a significant increase in new methodologies for calculating NMR properties at the relativistic two- and four-component levels of theory. These advances have been made possible in part by the development of computationally efficient two- and four-component electronic structure programs[9–12] as well as the development of techniques to calculate the NMR shielding and indirect spin–spin coupling tensors from these relativistic methodologies. These developments have been motivated by the importance of a full two- or four-component description in order to obtain accurate estimates for relativistic effects on molecular properties in general, and NMR parameters in particular.

Perturbation theory provides a convenient language for discussing relativistic effects on NMR parameters, but the formalism is complicated and in the case of indirect spin–spin coupling tensors, a complete computational protocol for calculating all the relativistic corrections to the indirect spin–spin coupling tensors still has not been realized. For this reason, we will not discuss results obtained using perturbation theory in this chapter.

In line with the topic of this book, we will put the main focus on calculations of relevance to gas phase NMR spectroscopy. In principle, most theoretical calculations can be considered to represent gas phase studies. However, we will restrict the discussion of calculations of relativistic effects either to focus on general trends in the effects of relativity, or limit ourselves to investigations of relevance to experimental gas phase NMR studies. We note that the use of relativistic methods for the study of NMR parameters is becoming routine even for fairly large molecules in complex environments, but we refer the reader to recent reviews covering these applications of relativistic effects on NMR parameters.[13–16]

In the next section, we present the basic models for including relativistic effects in quantum-chemistry calculations, together with the fundamentals of calculating NMR properties using relativistic methodology. We then briefly review calculations of relativistic effects on NMR properties, including both the NMR shielding tensor as well as the indirect spin–spin coupling tensor. Particular attention will be paid to the importance of relativistic effects for the determination of absolute shielding tensors. We end the chapter with some concluding remarks and an outlook.

8.2 Basic Theoretical Models of Relativistic Quantum Chemistry

In the following, we will provide a brief overview of the basic relativistic models for which we later will discuss the relativistic calculation of the NMR parameters. This will not only allow us to introduce the four-component Dirac–Coulomb–Breit Hamiltonian and families of different approximate two-component Hamiltonians, but it will also help us define the notation used throughout this chapter. Unless otherwise stated, we employ the Hartree system of atomic units.

8.2.1 Relativistic Four-component Hamiltonians

Relativistic calculations are in general computationally more costly than their non-relativistic (NR) counterparts, and for this reason most developments of relativistic methods to date have focused on independent-electron models, such as Hartree–Fock (HF) or density functional theory (DFT) methods. The most rigorous starting point for relativistic electronic-structure calculations is the four-component Hamiltonian. In contrast to the non-relativistic case, a closed-form expression for the relativistic many-electron Hamiltonian is not available. Instead, approximate Hamiltonians are constructed such that the one-electron Dirac operator $\hat{h}^{\mathrm{D}}$ is combined with an approximate expression for the two-electron interactions $\hat{g}$

$$\hat{H} = \sum_{i}^{N_e} \hat{h}^{\mathrm{D}}(i) + \frac{1}{2}\sum_{i \neq j}^{N_e} \hat{g}(i,j). \tag{8.1}$$

Here, within the Born–Oppenheimer approximation, $\hat{h}^{\mathrm{D}}$ describes a relativistic electron in a time-independent electrostatic field due to the fixed atomic nuclei[17,18]

$$\hat{h}^{\mathrm{D}} = (\beta - 1_{4\times4})c^2 + c(\vec{\alpha}\cdot\vec{p}) + \hat{V}1_{4\times4} = \begin{bmatrix} \hat{V}1_{2\times2} & c(\vec{\sigma}\cdot\vec{p}) \\ c(\vec{\sigma}\cdot\vec{p}) & (\hat{V} - 2c^2)1_{2\times2} \end{bmatrix}, \tag{8.2}$$

where c is the speed of light, $\vec{p} = -i\vec{\nabla}$ is the quantum-mechanical momentum operator, and $\hat{V}$ is the potential energy of interaction of the electron with an external electrostatic field. The two-by-two or four-by-four subscript in eqn (8.2) indicates the multicomponent character of the scalar operators. In 1928, Dirac introduced four new dynamical variables,[17,18] the so-called Dirac matrices

$$\vec{\alpha} = \begin{bmatrix} 0_{2\times 2} & \vec{\sigma} \\ \vec{\sigma} & 0_{2\times 2} \end{bmatrix}, \quad \beta = \begin{bmatrix} 1_{2\times 2} & 0_{2\times 2} \\ 0_{2\times 2} & -1_{2\times 2} \end{bmatrix}, \tag{8.3}$$

in order to formulate the quantum-mechanical expression for the relativistic energy of an electron, where the differentiations with respect to space and time variables appear linearly. The Dirac matrices account for spin in the relativistic theory and are in the standard representation composed of the unit matrix and three Pauli spin matrices

$$\sigma_x = \begin{bmatrix} 0 & 1 \\ 1 & 0 \end{bmatrix}, \quad \sigma_y = \begin{bmatrix} 0 & -i \\ i & 0 \end{bmatrix}, \quad \sigma_z = \begin{bmatrix} 1 & 0 \\ 0 & -1 \end{bmatrix}. \tag{8.4}$$

The one-electron Dirac operator has various interesting properties, of which the most relevant for us is its energy spectrum. In the absence of a static potential $\hat{V}$, the solution of the time-independent Dirac wave equation (in this paragraph, we have chosen to make the m-dependence explicit)

$$\left[\beta mc^2 + mc(\vec{\alpha}\cdot\vec{p})\right]\psi = E\psi, \tag{8.5}$$

gives a continuous spectrum of scattering states lying in two disjoint energy intervals $(-\infty, -mc^2\rangle$ and $\langle mc^2, \infty)$. Note that the presence of an attractive interaction potential $\hat{V}$ in eqn (8.5), such that its expectation values $\langle\hat{V}\rangle$ are within $0 > \langle\hat{V}\rangle > -2mc^2$, gives rise to a countable set of *discrete electronic bound states* in the region $mc^2 > E > -mc^2$.[19] These bound states are of relevance for relativistic quantum chemistry. In general, the eigenstates ψ in eqn (8.5) are four-component complex vector functions, so-called Dirac spinors

$$\psi = \begin{bmatrix} \psi^{L} \\ \psi^{S} \end{bmatrix} = \begin{bmatrix} \psi^{L\alpha} \\ \psi^{L\beta} \\ \psi^{S\alpha} \\ \psi^{S\beta} \end{bmatrix}, \tag{8.6}$$

which are often expressed *via* the large (ψ^{L}) and small components (ψ^{S}). The names reflect their relative size – the large-component spinor is of the order $O(c^0)$ and gives a dominant contribution to the positive-energy solutions, whereas the small component spinor is a factor c smaller (and *vice versa* for the negative-energy solutions).

Practical calculations in the framework of Dirac–Hartree–Fock or Dirac–Kohn–Sham theory are, however, performed with a different Hamiltonian from the one provided in eqn (8.2). For reasons discussed below, it is convenient to transform the small component ψ^{S} of the Dirac spinor into the pseudo-large component $\tilde{\psi}^{L}$ according to

$$\begin{bmatrix} \psi^{L} \\ \psi^{S} \end{bmatrix} = \begin{bmatrix} 1_{2\times2} & 0_{2\times2} \\ 0_{2\times2} & \frac{1}{2c}(\vec{\sigma}\cdot\vec{p}) \end{bmatrix} \begin{bmatrix} \psi^{L} \\ \tilde{\psi}^{L} \end{bmatrix}. \tag{8.7}$$

This non-unitary transformation was first proposed by Kutzelnigg,[20] and leads to a one-electron Dirac equation with a modified metric[21]

$$\begin{bmatrix} \hat{V} & \hat{T} \\ \hat{T} & \frac{1}{4c^2}(\vec{\sigma}\cdot\vec{p})\hat{V}(\vec{\sigma}\cdot\vec{p}) - \hat{T} \end{bmatrix} \begin{bmatrix} \psi^{L} \\ \tilde{\psi}^{L} \end{bmatrix} = E \begin{bmatrix} 1_{2\times2} & 0_{2\times2} \\ 0_{2\times2} & \frac{1}{2c^2}\hat{T} \end{bmatrix} \begin{bmatrix} \psi^{L} \\ \tilde{\psi}^{L} \end{bmatrix}, \tag{8.8}$$

where the kinetic energy operator appears in its familiar NR form

$$\frac{1}{2}(\vec{\sigma}\cdot\vec{p})(\vec{\sigma}\cdot\vec{p}) = \frac{p^2}{2} = \hat{T}. \tag{8.9}$$

The use of the modified Dirac equation eqn (8.8) offers several advantages compared to the parent Dirac equation:

(i) In the modified Dirac equation, the large and the pseudo-large components have the same symmetry, and thus the same primitive basis $\{X\}$ can be used for the representation of the modified Dirac spinors eqn (8.7)

$$\begin{bmatrix} \psi^{L} \\ \tilde{\psi}^{L} \end{bmatrix} = \sum_{\mu} \begin{bmatrix} X_{\mu}C^{L}_{\mu} \\ X_{\mu}\tilde{C}^{L}_{\mu} \end{bmatrix}, \tag{8.10}$$

where $\mathbf{C}^{L}$ and $\tilde{\mathbf{C}}^{L}$ are the two-component complex expansion coefficients. This also leads to a one-to-one ratio between the matrix dimension of the different components, thereby reducing the memory requirements in computer implementations. However, care must be taken with the use of *contracted* basis sets, since the large and the pseudo-large component possess a different radial distribution at any finite speed of light, in particular in the vicinity of heavy nuclei. This implies that the contraction coefficients for the components will differ.

(ii) The modified Dirac equation guarantees, in *any* finite basis representation, the correct non-relativistic limit ($c\rightarrow\infty$) for the kinetic energy, provided that eqn (8.10) is used for the spinor expansion. In contrast, the analogous representation of the parent Dirac spinors produces shortcomings in the computed kinetic energy, which persists even in the non-relativistic limit.[20] Schwarz and Wallmeier related this problem to an inadequate representation of the small components and the primary mechanism for variational failure observed in the early days of relativistic molecular calculations.[22] Moreover, eqn (8.8) only contains inverse powers of the speed of light, and the non-relativistic limit for the positive-energy solutions can therefore be obtained numerically, provided that a finite-sized model is used for the nuclei. In practice, the non-relativistic limit can be achieved by performing calculations with an increased speed of light; however, care must be taken for very high values of c, as numerical instabilities in this case can emerge in the computer implementation due to the symmetric orthonormalization step, which contains the factor $\sqrt{c}$ in the numerator. However, the increase of c by a factor 10–15 leads in our experience to stable and "converged" non-relativistic results.

(iii) The modified Dirac equation is the point of departure for the derivation of various quasi-relativistic (two-component) Hamiltonians.[23] Moreover, it is also a natural starting point for separating the scalar and spin-dependent contributions. For this, we follow the work of Dyall[21] and invoke the Dirac identity

$$(\vec{\sigma}\cdot\vec{p})\hat{V}(\vec{\sigma}\cdot\vec{p}) = \vec{p}\hat{V}\cdot\vec{p} + i\vec{\sigma}\cdot(\vec{p}\hat{V}\times\vec{p}), \tag{8.11}$$

in order to obtain the spin-free and spin-dependent terms of the modified Dirac Hamiltonian in eqn (8.8)

$$\tilde{\hat{h}}^{D} = \begin{bmatrix} \hat{V} & \hat{T} \\ \hat{T} & \frac{1}{4c^2}\vec{p}\hat{V}\cdot\vec{p} - \hat{T} \end{bmatrix} + \begin{bmatrix} 0_{2\times2} & 0_{2\times2} \\ 0_{2\times2} & \frac{1}{4c^2}i\vec{\sigma}\cdot\vec{p}\hat{V}\times\vec{p} \end{bmatrix}. \tag{8.12}$$

Note that the entire spin dependence in the modified Dirac Hamiltonian arises from the second term involving the potential energy of the small component. By omitting this term, one can perform the four-component calculation in a spin–orbit-free mode making use of real algebra only. The leading-order relativistic contributions $O(c^{-2})$ to the spin-dependent term are associated with the so-called spin–orbit (SO) interaction that can be further divided into one-electron and two-electron parts based on the interaction mechanisms involved. As a demonstration, we can consider the one-electron spin–orbit interaction arising from the nuclear Coulomb attraction $\hat{V} = -\sum_K \frac{Z_K}{r_K}$, assuming for simplicity a point-charge nuclear model. This is typically the dominant SO contribution proportional to the charge of nucleus Z

$$\frac{1}{4c^2}i\vec{\sigma}\cdot\vec{p}\hat{V}\times\vec{p} = \frac{1}{4c^2}\sum_K Z_K\frac{\vec{\sigma}\cdot(\vec{r}_K\times\vec{p})}{r_K^3} = \frac{1}{4c^2}\sum_K Z_K\frac{\vec{\sigma}\cdot\vec{l}_K}{r_K^3}. \tag{8.13}$$

Here, $\vec{r}_K = \vec{r} - \vec{R}_K$, and $\vec{l}_K$ is the orbital angular momentum of an electron with respect to the nuclear position $\vec{R}_K$. In addition, the two-electron parts of the Hamiltonian give rise to various two-electron contributions to the spin–orbit interaction – the Coulomb interaction gives the spin–same–orbit interaction, whereas the Breit interaction between the electrons is responsible for the spin–other–orbit and electron spin–spin interactions. The explicit form of the two-electron interactions shall be discussed later in this chapter.

Finally, the algebraic form of the modified one-electron Dirac equation eqn (8.8) suitable for practical implementations reads

$$\begin{bmatrix} \mathbf{V} & \mathbf{T} \\ \mathbf{T} & \frac{1}{4c^2}\mathbf{W} - \mathbf{T} \end{bmatrix} \begin{bmatrix} \mathbf{C}^{\mathrm{L}} \\ \tilde{\mathbf{C}}^{\mathrm{L}} \end{bmatrix} = E \begin{bmatrix} \mathbf{S} & \mathbf{0} \\ \mathbf{0} & \frac{1}{2c^2}\mathbf{T} \end{bmatrix} \begin{bmatrix} \mathbf{C}^{\mathrm{L}} \\ \tilde{\mathbf{C}}^{\mathrm{L}} \end{bmatrix}. \tag{8.14}$$

Here, we facilitate the finite basis expansion in terms of scalar functions X [eqn (8.10)], where **S, T, V, W** are matrices defined as

$$\begin{aligned} S_{\mu\nu} &= \langle X_\mu \mid X_\nu \rangle, \quad T_{\mu\nu} = \frac{1}{2}\langle X_\mu | p^2 | X_\nu \rangle, \\ V_{\mu\nu} &= \langle X_\mu | \hat{V} | X_\nu \rangle, \quad W_{\mu\nu} = \langle (\vec{\sigma}\cdot\vec{p})X_\mu | \hat{V} | (\vec{\sigma}\cdot\vec{p})X_\nu \rangle. \end{aligned} \tag{8.15}$$

The error in the total energies introduced by the ansatz eqn (8.7) was shown by Stanton and Havriliak[24] to be of order $O(c^{-4})$. Note that eqn (8.14) can be obtained without imposing the spinor transformation when the Dirac eigenstates are expanded directly in the so-called restricted kinetically balanced (RKB) basis[24]

$$\begin{bmatrix} \psi^{\mathrm{L}} \\ \psi^{\mathrm{S}} \end{bmatrix} = \sum_\mu \begin{bmatrix} X_\mu & 0_{2\times 2} \\ 0_{2\times 2} & \frac{1}{2c}X_\mu^{\mathrm{RKB}} \end{bmatrix} \begin{bmatrix} C_\mu^{\mathrm{L}} \\ \tilde{C}_\mu^{\mathrm{L}} \end{bmatrix}, \tag{8.16}$$

where

$$X_\mu^{\mathrm{RKB}} = \vec{\sigma}\cdot\vec{p}X_\mu. \tag{8.17}$$

When solving eqn (8.14), conventional non-relativistic basis sets of Slater or Gaussian type are commonly used. However, the use of such basis sets in relativistic calculations is justified only in combination with a finite-sized nuclear model because the use of the more common point-charge model induces a weak singularity in the electronic wavefunction at the nucleus, which cannot be described by conventional basis sets. A more thorough analysis of different nuclear models is given in ref. 25.

We have so far not considered the two-electron term $\hat{g}(i, j)$ in eqn (8.1). A closed-form expression for the Lorentz-invariant two-electron interaction is not known and the instantaneous (non-relativistic) Coulomb interaction is considered as the first approximation to the electron–electron repulsion

$$\hat{g}^{\mathrm{C}}(i,j) = r_{ij}^{-1}1_{4\times 4}. \tag{8.18}$$

The frequency-independent Breit interaction[26] $\hat{g}^{\mathrm{B}}(i, j)$ can be added in order to account for leading-order $O(c^{-2})$ relativistic corrections to the two-electron interaction[26,27]

$$\hat{g}^{\mathrm{B}}(i,j) = \hat{g}^{\mathrm{Gaunt}}(i,j) + \hat{g}^{\mathrm{gauge}}(i,j) = -\frac{1}{2r_{ij}}\left\{\vec{\alpha}_i\cdot\vec{\alpha}_j + \frac{(\vec{\alpha}_i\cdot\vec{r}_{ij})(\vec{\alpha}_j\cdot\vec{r}_{ij})}{r_{ij}^2}\right\}. \tag{8.19}$$

The Dirac–Coulomb–Breit (DCB) Hamiltonian is nowadays considered as a rigorous representation of many-electron systems in the framework of quantum chemistry. However, because of the complicated form of the two-electron Breit operator, in particular its second gauge term, most relativistic four-component molecular calculations are based on more approximate Hamiltonians, such as the Dirac–Coulomb (DC) and Dirac–Coulomb–Gaunt (DCG) Hamiltonians, respectively.

Because of its computational effectiveness, density functional theory (DFT) is one of the most popular ways of including electron correlation effects in calculations. The starting point for defining the exchange-correlation energy of a generalized-gradient-approximation (GGA) functional is

$$E^{xc} = \int \varepsilon^{xc}\left(n, s, \vec{\nabla} n, \vec{\nabla} s\right) dV, \tag{8.20}$$

where n is the electron density and s the spin density. In non-relativistic theory, the spin and spatial degrees of freedom are completely decoupled and a quantization axis for the spin angular momentum can be chosen independently of the molecular orientation. Conventionally, the quantization axis is chosen along the z-axis, then the spin density in eqn (8.20) has only one non-zero component $s = s_z$. This approach is termed as *collinear* DFT formulation. However, when spin–orbit coupling is included, the collinear approach breaks the rotational invariance of the energy, and this is a non-physical feature when performing a molecular DFT calculation. In addition to this problem, van Wüllen[28] demonstrated that the collinear approach in general is not able to recover the full spin polarization. To circumvent these problems, the so-called *non-collinear* approach can be invoked.[28,29] In this approach, a more general definition of the spin density and the corresponding spin polarization is considered by using the norm of the spin vector $s = |\vec{s}|$. The same distinction between collinear and non-collinear approach applies to the exchange–correlation kernel and its derivatives, needed for property calculations.[2,30–32] In two-component theories, the spin density is defined as an inner product of the wavefunction with the Pauli matrices $\vec{s} = \langle\vec{\sigma}\rangle$; however, the definition of the spin density in the four-component relativistic domain remains unresolved. This is related to the fact that the energy functional (8.20) should also depend on the current density.[33,34]

8.2.2 Relativistic Two-component Hamiltonians

The relatively high computational cost of four-component relativistic calculations, typically associated with the use of complex algebra and two distinct basis sets, has motivated the development of less expensive two-component (quasi-relativistic) methods. Since the Dirac equation represents a system of two coupled equations, one can in principle obtain the same positive-energy spectrum from a two-component equation, provided that we know a unitary transformation $\hat{U}$ that block-diagonalizes (decouples) the parent four-component Hamiltonian

$$\hat{U}^{\dagger}\begin{bmatrix}\hat{h}^{\mathrm{LL}} & \hat{h}^{\mathrm{LS}}\\ \hat{h}^{\mathrm{SL}} & \hat{h}^{\mathrm{SS}}\end{bmatrix}\hat{U}=\begin{bmatrix}\hat{h}^{(+)} & 0\\ 0 & \hat{h}^{(-)}\end{bmatrix},\quad \hat{U}^{\dagger}\hat{U}=1. \tag{8.21}$$

Similarly, the unitary transformation brings the four-component spinors in a form where the upper two components are non-zero for the positive-energy solutions (+), and the lower two for the negative-energy solutions (−)

$$\begin{bmatrix}\psi^{(+)}\\ 0\end{bmatrix}=\hat{U}^{\dagger}\begin{bmatrix}\psi_{+}^{\mathrm{L}}\\ \psi_{+}^{\mathrm{S}}\end{bmatrix}, \tag{8.22a}$$

$$\begin{bmatrix}0\\ \psi^{(-)}\end{bmatrix}=\hat{U}^{\dagger}\begin{bmatrix}\psi_{-}^{\mathrm{L}}\\ \psi_{-}^{\mathrm{S}}\end{bmatrix}. \tag{8.22b}$$

Here, $\psi^{\mathrm{L}}_{+/-}$ and $\psi^{\mathrm{S}}_{+/-}$ refer to eigenfunctions of the original Dirac Hamiltonian.

This approach goes back to the work of Foldy and Wouthuysen[35] (FW), where the authors proposed an exponential ansatz for the unitary operator. For a free-particle (FP) Dirac Hamiltonian (8.5), this transformation makes the system of equations fully decoupled, where

$$\hat{U}=\hat{U}^{\mathrm{FPFW}}=\begin{bmatrix}\hat{A} & \frac{1}{c}\hat{A}\hat{B}\\ \frac{1}{c}\hat{A}\hat{B} & -\hat{A}\end{bmatrix}, \tag{8.23}$$

with

$$\hat{A}\equiv\sqrt{\frac{\hat{e}_p+1}{2\hat{e}_p}},\quad \hat{B}=\frac{1}{\hat{e}_p+1}\vec{\sigma}\cdot\vec{p},\quad \hat{e}_p=\sqrt{1+\frac{p^2}{c^2}}. \tag{8.24}$$

Unfortunately, in the presence of a Coulomb potential, the FW transformation based on the expansion parameter c^{-1} produces operators that are singular and cannot be used in variational calculations.

The Douglas–Kroll (DK) method[36] instead builds up a sequence of unitary transformations, $\hat{U}=(\hat{U}_0\hat{U}_1\hat{U}_2\ldots)$ based on an expansion in $\hat{V}/c^2$, where each $\hat{U}_i$ successively eliminates the leading-order off-diagonal term from the transformed Hamiltonian. The very first step in the DK procedure is associated with the free-particle Foldy–Wouthuysen transformation, $\hat{U}_0=\hat{U}^{\mathrm{FPFW}}$, while subsequent operators follow the ansatz, $\hat{U}_n=\hat{W}_n+\sqrt{1+\hat{W}_n^2}$ with an anti-Hermitian $\hat{W}$ operator. Alternative parameterizations to the

Douglas–Kroll approach have been proposed by Nakajima and Hirao[37] and Wolf *et al.*[38] Finally after reaching the desired order, the remaining off-diagonal terms of the transformed Hamiltonian are neglected. In practice, one has to deal with matrix elements depending non-linearly on the momentum operator. Hess[39] proposed to evaluate these terms in the momentum space, making the Douglas–Kroll (DK) approach rather straightforward to implement and commonly referred to as the Douglas–Kroll–Hess (DKH) method. In most implementations, the transformation is performed up to second order, although a few groups have developed techniques for higher orders.[37,38] Using modern computer algorithms, van Wüllen reported a generalized DKH transformation scheme up to sixth order,[40] Reiher and Wolf presented an infinite-order approach with computational results reported for DKH14.[41,42] Recently, Peng and Hirao introduced a DKH-based algorithm with polynomial cost to order 100.[43]

An alternative method to obtain high-order decoupling transformations was suggested by Barysz *et al.*[44–46] The authors analyzed the higher-order coupling terms in the free-particle transformed Hamiltonian in powers of c^{-1}, rather than the external potential, as in the DKH method. To eliminate the off-diagonal terms in the power series expansion, however, they proposed an iterative procedure. As a result, a *single* step was needed to perform the decoupling transformation to the desired order.

A modified procedure for finding the unitary transformation $\hat{U}$ was suggested by Heully *et al.*[47] The authors proposed a general expression for $\hat{U}$ in terms of a coupling operator $\hat{R}$, which in the notation of Kutzelnigg[48] reads

$$\hat{U} = \hat{U}^{\text{Heully}} = \hat{W}_1 \hat{W}_2,$$
$$\hat{W}_1 = \begin{bmatrix} 1 & -\hat{R}^\dagger \\ \hat{R} & 1 \end{bmatrix}, \quad \hat{W}_2 = \frac{1}{\sqrt{1+\hat{R}^\dagger\hat{R}}} \begin{bmatrix} 1 & 0 \\ 0 & 1 \end{bmatrix}. \tag{8.25}$$

Here, $\hat{W}_1$ ensures decoupling of the parent four-component Hamiltonian, whereas $\hat{W}_2$ provides a re-normalization of the wavefunctions. Once an *exact* expression for $\hat{R}$ is known, the decoupling transformation can be done in a single step and the resulting eigenvalue equation of the form

$$\hat{h}^{(+)}\psi^{(+)} = E\psi^{(+)},$$
$$\hat{h}^{(+)} = \frac{1}{\sqrt{1+\hat{R}^\dagger\hat{R}}}\left[\hat{V} + c(\vec{\sigma}\cdot\vec{p})\hat{R} + c\hat{R}^\dagger(\vec{\sigma}\cdot\vec{p}) + \hat{R}^\dagger\left(\hat{V} - 2c^2\right)\hat{R}\right]\frac{1}{\sqrt{1+\hat{R}^\dagger\hat{R}}}, \tag{8.26}$$

is fully characterized by two-component quantities with the electronic spectra equivalent to the original Dirac Hamiltonian. As pointed out by Iliaš and Saue,[49] a change of metric in eqn (8.26) gives an eigenvalue equation which is equivalent to Dyall's normalized elimination of the small component (NESC) method[50,51] (in the x-representation)

$$\psi^{(+)} = \sqrt{1+\hat{R}^{\dagger}\hat{R}}\psi^{\text{NESC}},$$

$$\left[\hat{V} + c(\vec{\sigma}\cdot\vec{p})\hat{R} + c\hat{R}^{\dagger}(\vec{\sigma}\cdot\vec{p}) + \hat{R}^{\dagger}\left(\hat{V} - 2c^2\right)\hat{R}\right]\psi^{\text{NESC}} = E\left(1+\hat{R}^{\dagger}\hat{R}\right)\psi^{\text{NESC}}. \tag{8.27}$$

The importance of the approach proposed by Heully and co-workers lies, however, in the fact that the general $\hat{R}$-parameterization of $\hat{U}$ (8.25) determines four key relations:

(i) between the wavefunction components, as derived from eqn (8.22),

$$\psi^{\text{S}}_{+} - \hat{R}\psi^{\text{L}}_{+} = 0, \tag{8.28a}$$

$$\psi^{\text{L}}_{-} + \hat{R}^{\dagger}\,\psi^{\text{S}}_{-} = 0. \tag{8.28b}$$

(ii) between the Hamiltonian elements, as derived from eqn (8.21),

$$\hat{h}^{\text{SL}} + \hat{h}^{\text{SS}}\,\hat{R} - \hat{R}\,\hat{h}^{\text{LL}} - \hat{R}\,\hat{h}^{\text{LS}}\,\hat{R} = 0, \tag{8.29a}$$

$$\hat{h}^{\text{LS}} - \hat{h}^{\text{LL}}\,\hat{R}^{\dagger} + \hat{R}^{\dagger}\,\hat{h}^{\text{SS}} - \hat{R}^{\dagger}\,\hat{h}^{\text{SL}}\,\hat{R}^{\dagger} = 0. \tag{8.29b}$$

Note that the matrix formulation and algebraic solution of these equations is the central idea of modern quasi-relativistic methods, named generically as the exact two-component (X2C) method. Before we proceed, however, let us first discuss the analytic form of $\hat{R}$.

A closed analytic expression for the *exact* coupling $\hat{R}$ between the large and small components is known and can be derived from the Dirac equation by employing an elimination-of-the-small-component (ESC) technique

$$\hat{R} = \frac{1}{2c}\left[1 + \frac{E-\hat{V}}{2c^2}\right]^{-1}(\vec{\sigma}\cdot\vec{p}). \tag{8.30}$$

Due to the complicated nature of the exact $\hat{R}$ operator which involves an explicit energy (*state-specific*) dependence, a number of approximate two-component approaches have been devised in the last decades. We will briefly discuss these methods in terms of the two-component eqn (8.26) and (8.30) with an *approximate* $\hat{R}$. For a more comprehensive survey, the reader is referred to standard textbooks[52,53] or recent review articles.[23,54,55]

The (Breit–)Pauli (BP) Hamiltonian represents a common starting point for discussing the leading-order relativistic corrections $O(c^{-2})$ to the energy. In this approach, the coupling operator $\hat{R}$ in eqn (8.30) is approximated by

$$\hat{R}^{\text{Pauli}} = \frac{1}{2c}(\vec{\sigma}\cdot\vec{p}), \tag{8.31}$$

and inserted into the two-component expression [eqn (8.26)], retaining terms to order $O(c^{-2})$. In the most complete picture, the transformed Hamiltonian $h^{(+)}$ in eqn (8.26) is derived from the many-electron Dirac–Coulomb–Breit Hamiltonian and is referred to as the full Breit–Pauli Hamiltonian. A comprehensive compilation of contributions to the BP Hamiltonian is, however, beyond the scope of the present chapter, and we refer instead the interested reader to the book by Moss.[56] Here, we only provide the Pauli eigenvalue equation

$$\left[\hat{T}+\hat{V}-\frac{1}{8c^2}p^4+\frac{1}{8c^2}\left(\Delta\hat{V}\right)+\frac{1}{4c^2}\vec{\sigma}\cdot\left\{\left(\vec{\nabla}\hat{V}\right)\times\vec{p}\right\}\right]\psi^{\mathrm{P}}=E\psi^{\mathrm{P}}, \tag{8.32}$$

where the third, fourth, and fifth terms of the Pauli Hamiltonian arise from the leading-order one-electron relativistic corrections, namely mass–velocity (MV), Darwin, and the Pauli spin–orbit operator. The MV term is associated with the relativistic mass increase of the electron, whereas the Darwin operator is a correction term associated with the creation (interference) of electron-positron pairs in the close vicinity of electrons, the so-called "Zitterbewegung". Irrespective of the nature of the wavefunction, the Pauli Hamiltonian is not suitable for variational molecular electronic structure calculations because: (i) the mass–velocity and Darwin terms are not bounded from below[57] (ii) the Darwin term exhibits strong singularities due to a higher-order derivative of the nuclear potential where the Taylor expansion of $(1-x)^{-1}$ with $x=\dfrac{\hat{V}-E}{2c^2}$ is not mathematically justified, as x does not necessarily lie within the radius of convergence, $\left|\dfrac{\hat{V}-E}{2c^2}\right|<1$.

To avoid numerical problems with the Pauli Hamiltonian, the so-called regular expansion has been proposed.[58] Now, $\hat{R}$ in eqn (8.30) is rearranged first

$$\hat{R}=\frac{c}{2c^2-\hat{V}}\left[1+\frac{E}{2c^2-\hat{V}}\right]^{-1}(\vec{\sigma}\cdot\vec{p}), \tag{8.33}$$

followed by a power series expansion in $E(2c^2-\hat{V})^{-1}$. The radius of convergence, $\left|\dfrac{E}{2c^2-\hat{V}}\right|<1$ becomes more desirable as the potential appears in the denominator. This justifies the regular expansion not only to valence regions where $\hat{V}$ is small compared to c^2, but also to regions close to the nucleus, where $\hat{V}$ typically acquires large negative values, suggesting that the expansion remains valid for almost all quantum-chemical applications. The zeroth-order regular approximation[59] (ZORA) method is obtained by retaining in eqn (8.33) the zeroth-order expansion term only

$$\hat{R}^{\mathrm{ZORA}}=\frac{1}{2c}K(\vec{r})(\vec{\sigma}\cdot\vec{p}),\quad K(\vec{r})=\left(1-\frac{\hat{V}}{2c^2}\right)^{-1}, \tag{8.34}$$

and omitting the renormalization term in eqn (8.26)

$$\left[\frac{1}{2}(\vec{\sigma}\cdot\vec{p})K(\vec{r})(\vec{\sigma}\cdot\vec{p})+\hat{V}\right]\psi^{\mathrm{ZORA}}=E\psi^{\mathrm{ZORA}}. \tag{8.35}$$

All leading-order relativistic terms were recovered in the infinite-order regular approximation[60] (IORA) method by including a non-unit metric missing in the previous ZORA equation, thus providing an energy that is correct to $O(c^{-2})$. Both the ZORA and the IORA methods can be used in variational calculations and, in particular, the ZORA approach has found many practical applications. For $\hat{V}\ll 2c^2$, the ZORA kinematic factor $K(\vec{r})$ can be reduced to $K(\vec{r})\approx 1+\hat{V}/(2c^2)$, which identifies the leading-order ZORA relativistic corrections as the Darwin term and the spin–orbit term, but not the mass–velocity term that is present in the Pauli Hamiltonian. Moreover, the SO term can also include the two-electron spin-same-orbit contributions in a mean-field fashion when the potential $\hat{V}$ is replaced by the mean-field Kohn–Sham potential. On the other hand, the ZORA Hamiltonian does not fulfill the electric gauge transformation, which means that a change in the potential by a constant Δ, *i.e.* $\hat{V}\rightarrow\hat{V}+\Delta$, does not lead to the equivalent change in energy, $E\rightarrow E+\Delta$. To resolve the problem, $\hat{V}$ is typically replaced by a frozen core or model potentials, the latter designed from superposed atomic potentials. Moreover, the ZORA core orbitals in heavy atoms give substantial errors compared to the DKH2 approach due to the fact that a power series expansion of $[1+E/(2c^2-\hat{V})]^{-1}$ in $E/(2c^2-\hat{V})$ is not justified for very large E. To partially resolve the problem with the energy of the core orbitals as well as the missing renormalization term in eqn (8.35), a scaled ZORA approach has been introduced.[61]

In contrast to previous approaches that rely on finding a closed analytic expression for the decoupling operator $\hat{R}$, a very appealing alternative method, the so-called *exact two-component* (X2C) approach, has been developed in recent years. The method takes advantage of the formalism proposed by Heully,[47] but instead of solving eqn (8.28) and (8.29) in operator form, X2C solves the equations directly in the matrix representation. This allows us to obtain an exact decoupling for the parent four-component one-electron equation in a single step. A predecessor to the modern X2C implementation was a two-step X2C approach, developed by Jensen and Iliaš[62] and inspired by the work of Barysz *et al.*[45,46] The formalism was later improved towards a one-step approach independently by Kutzelnigg and Liu[63,64] and Iliaš and Saue.[49] The exact two-component Hamiltonian at the matrix level has also been reported by Liu and Peng[65] within the density functional theory framework.

8.3 Relativistic Quantum Chemical Models for NMR Parameters

Within the Born–Oppenheimer approximation and closed-shell molecular systems, the NMR shielding tensor can be calculated as the leading-order response of the total energy to an external magnetic field $\vec{B}$ and the nuclear magnetic moment of the Kth nucleus $\vec{\mu}^K$

$$\sigma_{uv}^{K} = \left.\frac{\mathrm{d}^2 E(\vec{B}, \vec{\mu}^K)}{\mathrm{d}B_u \mathrm{d}\mu_v^K}\right|_{\vec{B}=\vec{\mu}^K=0}, \tag{8.36}$$

whereas the reduced indirect nuclear spin–spin coupling tensor involves the energy derivatives with respect to the magnetic moments of two nuclei K and L

$$K_{uv}^{KL} = \left.\frac{\mathrm{d}^2 E(\vec{\mu}^K, \vec{\mu}^L)}{\mathrm{d}\mu_u^K \mathrm{d}\mu_v^L}\right|_{\vec{\mu}^K=\vec{\mu}^L=0}. \tag{8.37}$$

The total energy depends on $\vec{B}$, $\vec{\mu}^K$, and $\vec{\mu}^L$ *via* the vector potential, $\vec{A}=\vec{A}_0+\vec{A}_K+\vec{A}_L$, satisfying the Coulomb gauge, $\vec{\nabla}\cdot\vec{A}=0$, where the vector potentials are defined as

$$\vec{A}_0 = \frac{1}{2}\vec{B}\times\vec{r}_0, \quad \vec{r}_0 = \vec{r}-\vec{R}_0, \tag{8.38}$$

$$\vec{A}_K = \frac{\vec{\mu}^K \times \vec{r}_K}{r_K^3}, \quad \vec{r}_K = \vec{r}-\vec{R}_K. \tag{8.39}$$

Here, $\vec{r}$, $\vec{R}_0$ and $\vec{R}_K$ refer to coordinates of the electron, the gauge origin, and the nucleus K, respectively. For heavier elements, the point model for the magnetic moment (8.39) should be replaced by a more physical finite-sized model

$$\vec{A}_K = -\vec{\mu}^K \times \vec{\nabla} \iiint \frac{G(|\vec{R}-\vec{R}_K|)}{|\vec{r}-\vec{R}|}\,\mathrm{d}^3\vec{R}, \tag{8.40}$$

where G is a function representing the finite magnetic moment distribution. It is convenient to use a Gaussian distribution, although other models can also be used. Relativistic calculations using a finite-sized distribution for the magnetic moment have been presented for hyperfine couplings,[66] NMR shielding[67] and indirect spin–spin coupling tensors.[68] In general, finite magnetic moment effects have no significant impact on the calculated NMR shielding tensors, but the situation is very different for indirect spin–spin coupling constants. In the latter case, the effects are in general of the order of about 5%, but larger differences (20%) have been reported for one-bond Hg–Hg couplings.[68] For indirect spin–spin coupling tensors, finite-sized nuclear magnetic moment models are strongly recommended, both for improving basis set convergence and for increasing numerical stability of the implementation.

Different methods for relativistic calculations of NMR parameters have been developed in recent years, building on different levels of approximation for the energy expression. The conceptually most straightforward approach for calculating NMR parameters is to treat both the magnetic and relativistic energy corrections perturbatively, including the leading-order relativistic operators arising from the two-component Breit–Pauli (BP) molecular Hamiltonian. An important shortcoming of these methods is the fact that the BP Hamiltonian is not variationally stable.[69,70] Furthermore, a perturbational treatment of its singular operators is limited to the leading orders, and this may not always be sufficient for obtaining accurate NMR parameters. Although the former shortcoming can be avoided using special techniques such as a frozen core approximation[71] or using the direct perturbation theory approach of Kutzelnigg,[72] the latter disadvantage is inherent to perturbation techniques and cannot be avoided. Furthermore, perturbation calculations beyond lowest-order contributions will require a large number of perturbation equations to be solved, often involving complicated integrals.[73–75] However, at lower order, these perturbational methods provide an excellent interpretation tool of the origins of the relativistic corrections to the NMR properties formulated in a non-relativistic context.

As in the previous section, our focus will be on methods that are capable of treating magnetic resonance parameters of both light and heavy nuclei from first principles – that is, methods that describe the relativistic effects variationally. These methods can be divided into two main categories: Quasi-relativistic (two-component) methods and fully relativistic (four-component) methods. The most popular two-component methods for prediction of NMR parameters are currently the zeroth-order regular approximation (ZORA),[76] the Douglas–Kroll–Hess method (DKH),[67] and recently also the exact two-component (X2C) approach.[77] At the four-component level, there exist implementations of NMR parameters at the Hartree–Fock level of theory involving the unrestricted kinetic balance formalism.[78,79] Most developments, however, have been done within the DFT formalism utilizing various magnetic balance (MB) concepts such as simple magnetic balance,[80] restricted magnetic balance,[81–84] or a transformation proposed by Kutzelnigg,[85,86] details of which will be discussed later.

The expression for the bilinear energy derivative eqn (8.36) and (8.37) can be written generally for a single determinant method (HF or DFT) as

$$E^{11} = \langle \varphi_i^{00} | h^{11} | \varphi_i^{00} \rangle + \langle \varphi_i^{00} | h^{01} | \varphi_i^{10} \rangle + \langle \varphi_i^{10} | h^{01} | \varphi_i^{00} \rangle, \tag{8.41}$$

where superscripts 00, 10, 01 indicate the order of the Taylor expansion with respect to two different perturbations, φ is a MO and h is a one-electron operator. Here and in the following, we assume implicit summation over repeated indices where i, j will be used for occupied orbitals, a, b for virtual, and p, q for general molecular orbitals (MOs). Note that eqn (8.41) is independent of the level of approximation used for treating relativistic effects as long as these are included variationally. The first term in eqn (8.41) (called the diamagnetic contribution) is easily evaluated in terms of the unperturbed (field-free) MOs (φ_i^{00}) and operators that are bilinear in the vector potentials (h^{11}). The last two terms (called the paramagnetic contributions) require the evaluation of the linear response function (φ_i^{10}), which is usually the most computationally demanding step. It is convenient to expand φ_i^{10} in the solutions of the unperturbed system

$$\varphi_i^{10} = \beta_{pi}\,\varphi_p^{00}. \tag{8.42}$$

In the case of a perturbation-independent basis set expansion, eqn (8.42) can be reduced to a summation over virtual MOs

$$\varphi_i^{10} = \beta_{ai}\,\varphi_a^{00}, \tag{8.43}$$

with the expansion coefficients obtained from perturbation theory as

$$\beta_{ai} = \frac{\langle \varphi_a^{00} | F^{10} | \varphi_i^{00} \rangle}{\varepsilon_i^{00} - \varepsilon_a^{00}}, \tag{8.44}$$

where ε^{00} are orbital energies. The response of the Fock operator F^{10} consists of both one-electron terms and contributions from the exchange-correlation and HF kernels. Due to the kernels, eqn (8.44) must be solved iteratively in quantum-chemical calculations. It is worth noting that in the case of non-relativistic NMR shielding calculations involving pure DFT functionals and closed-shell systems, the response of the Fock operator will only contain the one-electron terms, and in this case eqn (8.44) can be solved directly without imposing an iterative algorithm. However, this is no longer the case if the spin–orbit interaction is included in the Hamiltonian.[87] For instance in hydrogen iodide, the contribution of the exchange-correlation kernel to the Fock operator is responsible for more than 3 ppm of the total ^{1}H NMR shielding.[87]

The magnetic perturbation operators in eqn (8.41) will strictly speaking vanish in the non-relativistic limit. This is a manifestation of the fact that in the non-relativistic limit there are no magnetic fields, as for instance in the electric limit, $|E| \gg c\,|B|$, only the expression corresponding to Gauss' law in Maxwell's equations remains non-zero.[88] Nevertheless, it is customary to refer to the first non-vanishing terms (in a $1/c$ expansion) as non-relativistic operators. These operators can be written as

$$h^{11} = \frac{1}{c^2}\vec{A}_1 \cdot \vec{A}_2, \tag{8.45}$$

$$h^{01} = \frac{1}{c}\vec{A}_2 \cdot \vec{p} + \frac{i}{2c}\vec{\sigma} \cdot (\vec{p} \times \vec{A}_2), \tag{8.46}$$

where $\vec{A}_1$ and $\vec{A}_2$ correspond to either $\vec{A}_0$ and $\vec{A}_K$ for the NMR shielding tensor or $\vec{A}_K$ and $\vec{A}_L$ for the indirect spin–spin coupling tensor. Inserting eqn (8.45) and (8.46) into eqn (8.41), the non-relativistic expressions for the NMR parameters are recovered, involving both diamagnetic and paramagnetic contributions.

A similar decomposition into dia- and paramagnetic terms is not obvious in the relativistic framework due to the absence of operators that are quadratic in the vector potentials, since the magnetic-field-dependent operator in four-component theory is only linear in the vector potential

$$h^{01} = \vec{\alpha} \cdot \vec{A}_2. \tag{8.47}$$

As such, the second-order energy correction (8.41) consists only of a paramagnetic term and thus only the last two terms in eqn (8.41) survive. In the non-relativistic (NR) limit, both the diamagnetic and paramagnetic terms must nevertheless be recovered. This apparent conundrum can be traced to the fact that there is a conceptual difference between the relativistic and non-relativistic paramagnetic term. In relativistic theory, the summation over virtual MOs (8.43) contains both positive- and negative-energy states. Virtual positive energy states correspond in the NR limit to the standard virtual MOs of NR theory. It is therefore no surprise that the diamagnetic term is recovered *via* the negative-energy summation in the paramagnetic term. Using perturbation theory, it can be shown that

$$\frac{\left\langle \varphi_i^{00,\mathrm{rel}} \middle| h^{01} \middle| \varphi_{\bar{a}}^{00,\mathrm{rel}} \right\rangle \left\langle \varphi_{\bar{a}}^{00,\mathrm{rel}} \middle| F^{10} \middle| \varphi_i^{00,\mathrm{rel}} \right\rangle}{\varepsilon_i^{00} - \varepsilon_{\bar{a}}^{00}} + c.c. \tag{8.48}$$

$$= \frac{1}{c^2} \langle \varphi_i^{00,\mathrm{NR}} | \vec{A}_1 \cdot \vec{A}_2 | \varphi_i^{00,\mathrm{NR}} \rangle + O(c^{-4}), \tag{8.49}$$

where $\bar{a}$ indicates summation over negative-energy states. This relation was initially analyzed for one-electron systems by Sternheim[89] and later extended to many-electron systems by Pyykkö.[90] Aucar *et al.*[91] showed on a series of small XH_2 molecules using relativistic polarization propagators at the RPA level that the diamagnetic contribution arises exclusively from the pp-block (positronic-positronic) of the principal propagator, *i.e.* positive to negative energy virtual excitations.

Without taking any special measures, the negative-energy summation in the relativistic paramagnetic term (8.48) will display a very poor basis-set convergence, even for atoms.[92] Both the problem of a "missing" diamagnetic term and the poor basis set convergence are manifestations of the insufficient basis set representation for the coupling between the large and the small components of the Dirac four-vectors (φ^{L} and φ^{S}). Even in the NR limit, the coupling between the components requires a magnetic field-dependent operator

$$\varphi_p^{\mathrm{S}} \approx \frac{1}{2c} \left(\vec{\sigma} \cdot \vec{p} + \frac{1}{c} \vec{\sigma} \cdot \vec{A} \right) \varphi_p^{\mathrm{L}}. \tag{8.50}$$

This implies that even if the state-of-the-art kinetic balance condition (8.17) is invoked, the basis set is not sufficient to represent the magnetic balance coupling (8.50), and in practice NMR calculations require extensive basis sets even for molecules containing only light elements.

There exist several approaches which take the magnetic balance coupling (8.50) between the large and small components of the Dirac four-vectors into consideration. We will briefly comment on three of the most computationally and theoretically reliable techniques.

8.3.1 External Field-dependent Unitary Transformation (EFUT)

Kutzelnigg[85] proposed a unitary transformation that decouples the large and small components of the wavefunction such that only the RKB basis set is required to properly describe the wavefunction. This transformation is designed in the spirit of the Foldy–Wouthuysen transformation (8.23) in the presence of a magnetic field. Using this transformation, Dirac spinors can be parameterized up to linear terms in the vector potential in the following manner

$$\varphi_p^{\mathrm{EFUT}} = \begin{bmatrix} 1 & \frac{1}{2c^2}\vec{\sigma}\cdot\vec{A} \\ -\frac{1}{2c^2}\vec{\sigma}\cdot\vec{A} & 1 \end{bmatrix} \begin{bmatrix} 1 & 0 \\ 0 & \frac{1}{2c}\vec{\sigma}\cdot\vec{p} \end{bmatrix} \begin{bmatrix} X_\mu C^L_{\mu p} \\ X_\mu C^S_{\mu p} \end{bmatrix}. \tag{8.51}$$

The first successful implementation and numerical assessment of the different approaches based on Kutzelnigg's transformation was presented by Xiao *et al.*[86] on the NMR shielding tensor calculations. The authors concluded that only an external magnetic-field-dependent operator should be transformed, leading to the so-called EFUT approach. Transforming also the operator $\vec{\alpha}\cdot\vec{A}_K$, where $\vec{A}_K$ is the vector potential generated by the Kth nucleus, will result in singular operators (first noted as numerical instabilities by Visscher)[93] that makes methods based on Kutzelnigg's transformation unsuitable for calculations of indirect spin–spin coupling tensors, even in conjunction with a finite-sized model for the magnetic moment distribution (8.40).

8.3.2 Restricted Magnetic Balance (RMB)

In order to introduce magnetic fields in the Dirac Hamiltonian eqn (8.2), it is common to use the principle of minimal coupling

$$\vec{p} \rightarrow \vec{p} + \frac{1}{c}\vec{A}. \tag{8.52}$$

Keeping in mind that the RKB condition is crucial for describing the proper balance between the large and the small components of the wavefunction in the absence of magnetic fields (8.16), it is very intuitive to use the minimal coupling substitution (8.52) when balancing basis set for the wavefunction components in the presence of magnetic fields (8.50). This is a central idea of the so-called restricted magnetically balanced (RMB) approach, developed and implemented by Komorovský and co-workers.[87] The Dirac spinor in the RMB basis can be expressed as

$$\varphi_p^{\mathrm{RMB}} = \begin{bmatrix} 1 & 0 \\ 0 & \dfrac{1}{2c}(\vec{\sigma}\cdot\vec{p} + \dfrac{1}{c}\vec{\sigma}\cdot\vec{A}) \end{bmatrix} \begin{bmatrix} X_\mu C_{\mu p}^L \\ X_\mu C_{\mu p}^S \end{bmatrix}. \tag{8.53}$$

A clear advantage of the RMB-based method compared to EFUT or sMB (*vide infra*) is its applicability to the calculation of indirect nuclear spin–spin coupling tensors, as first demonstrated by Repiský *et al.*[94]

8.3.3 Simple Magnetic Balance (sMB)

Olejniczak *et al.*[80] proposed an elegant but conceptually different method from EFUT and RMB for dealing with the magnetic balance problem in NMR shielding tensor calculations. The simple magnetic balance method is in many aspects similar to RMB but without the need to implement new integrals. The main underlying idea of the sMB method is connected to the properties of Gaussian basis functions. When employing London atomic orbitals (see the discussion below), acting with the RMB condition (8.53) on a Gaussian basis function centered on the Kth nucleus G^K will result in a linear combination of other Gaussian basis functions. These functions can be generated by gradient operators and are therefore referred to as an unrestricted kinetically balanced basis set (UKB)

$$\left(\vec{p} + \frac{1}{c}\vec{A}_{0K}\right)G^K = \sum_\gamma c_\gamma G_\gamma^K, \tag{8.54}$$

$$G_\gamma^K \in \{\nabla_x\, G^K\} \cup \{\nabla_y\, G^K\} \cup \{\nabla_z\, G^K\}, \tag{8.55}$$

where $\vec{A}_{0K}$ is the vector potential generated by an external magnetic field (8.38) with the gauge origin centered on nucleus K. The sMB approach can be divided into two steps. In the first step, perturbation-free MOs are obtained using the RKB basis. Then, when solving the response equations (8.44), the UKB basis set (8.55) is used instead. However, for the UKB complement of the perturbation-free MO coefficients, one-electron energies are not available and are therefore approximated by $-2c^2$. This can be expected to be a good approximation, although it is difficult to estimate its error. The main disadvantage of the sMB method is its formulation for calculating indirect spin–spin coupling tensors, since a straightforward extension of the procedure in a finite basis would not lead to correct NR expressions, which is a necessary condition for any reliable relativistic technique.

8.3.4 Other Methods for Solving the Magnetic Balance Problem

- Prior to the sMB method, the UKB basis set was often used for both perturbation-free and response calculations.[78,91] Although the use of UKB will provide better basis set convergence compared to the RKB basis alone, it may suffer from linear dependencies in the basis set. More importantly, the UKB basis will not span the same space, as dictated by the magnetic balance condition (8.50) for methods with gauge-origin dependence problem. Moreover methods utilizing UKB basis are more computationally involved compared to methods which included magnetic balance explicitly.
- Two more methods with explicit magnetic balance have been proposed, the orbital decomposition approach (ODA)[95] and the Sternheim decomposition (SD).[96] Although these methods will recover the non-relativistic limit in finite basis sets correctly, they exhibit poorer basis set convergence compared to EFUT, RMB, and sMB methods. Extending the arguments of Stanton and Havriliak[97] to the case of different magnetic balance conditions, it can be shown that EFUT, RMB, and sMB exhibit in the worst case variational instability of order $O(c^{-4})$, whereas ODA and SD will exhibit instability of the order $O(Z^4c^{-4})$ (where Z is the atomic number). Therefore, the heavier the system studied, the poorer the basis set convergence.

Utilizing proper magnetic balance between the upper and lower components of the Dirac four-component wavefunction will result in basis set requirements comparable to those of non-relativistic methods. However, note that the matrices in four-component relativistic approaches are 4^2 larger than in the corresponding one-component case. A further improvement of the basis set convergence as well as solving the gauge-origin problem of the NMR shielding results can be obtained by means of London orbitals. The London atomic orbitals (LAOs) are defined as[98]

$$X_\mu^{\mathrm{LAO}} = \omega_\mu^B X_\mu, \quad \omega_\mu^B = \exp\left\{-\frac{i}{2c}\left[\vec{B} \times \left(\vec{R}_\mu - \vec{R}_0\right)\right] \cdot \vec{r}\right\}. \tag{8.56}$$

Here, $\vec{R}_\mu$ is the position of the nucleus where basis function X_μ is centered and $\vec{R}_0$ is the global gauge origin. London orbitals are often referred to as gauge-including atomic orbitals (GIAOs), and as a result of the ansatz (8.56), every basis function is moved from the global gauge origin to the position of nucleus where the basis function is centered. The rationalization of the ansatz (8.56) follows from the fact that in the NR theory of an atomic, one-electron system, the London phase factor ω^B describes the first-order magnetic field dependence of the wavefunction. This follows from the fact that for atomic systems the angular momentum operator commutes with the perturbation-free Hamiltonian, and at the same time is responsible for the interaction with the external magnetic field. In contrast, these effects are described in relativistic theory with different operators. The leading-order term of the first-order magnetic field dependence of the relativistic wavefunction is again the London orbital, with additional terms of order $O(c^{-2})$, which justifies the use of LAOs in relativistic theory. However, slower basis set convergence should be expected for heavy-atom NMR shieldings. In practice, basis sets of triple-zeta quality are in most cases sufficient to obtain DFT results converged with respect to basis set completeness.[99]

In relativistic theories, the London phase factor eqn (8.56) is applied on top of the corresponding magnetic balance Γ defined either by eqn (8.51) or eqn (8.53)

$$\varphi_p^{\mathrm{LAO,rel}} = \begin{bmatrix} \omega_\mu^B & 0 \\ 0 & \omega_\mu^B \end{bmatrix} \begin{bmatrix} \Gamma^{LL} & \Gamma^{LS} \\ \Gamma^{SL} & \Gamma^{SS} \end{bmatrix} \begin{bmatrix} X_\mu C_{\mu p}^L \\ X_\mu C_{\mu p}^S \end{bmatrix}. \tag{8.57}$$

Although the four-component MOs are still gauge-origin dependent, the final NMR shielding expressions will be gauge-origin independent. Including the GIAO phase factor requires the evaluation of new integrals, which will on one hand slow down the calculations, but on the other hand significantly improve the convergence of NMR results towards the basis set limit. As a consequence, any reliable modern method for calculating NMR shielding constants must utilize GIAOs. Except for the sMB method, where London orbitals are implicitly included, the combination of magnetic balance with the London phase factor can be developed and implemented separately. For Kutzelnigg's transformation, the work was done by Cheng *et al.*[83] and in the case of the RMB method, independently by Cheng *et al.*[83] and Komorovský *et al.*[84]

The four-component MOs in eqn (8.57) can be decomposed into three parts

$$\varphi_p^{10} = \varphi_p^{10,\omega} + \varphi_p^{10,m} + \varphi_p^{10,r}. \tag{8.58}$$

Substituting this formal decomposition into eqn (8.41) and noting that in the relativistic theory only a linear operator in the vector potential is present (8.47), the following three contributions will arise

$$E^{11} = \left\langle \varphi_i^{10,m} \middle| \vec{\alpha} \cdot \vec{A}_2 \middle| \varphi_i^{00} \right\rangle \tag{8.59}$$

$$+ \left\langle \varphi_i^{10,r} \middle| \vec{\alpha} \cdot \vec{A}_2 \middle| \varphi_i^{00} \right\rangle \tag{8.60}$$

$$+ \left\langle \varphi_i^{10,\omega} \middle| \vec{\alpha} \cdot \vec{A}_2 \middle| \varphi_i^{00} \right\rangle + c.c. \tag{8.61}$$

The third contribution in eqn (8.61) originates from the London phase factor and will only appear in NMR shielding calculations. The first contribution in eqn (8.59) arises from the magnetic balance of the four-component MOs [eqn (8.51) or (8.53)] and will in the NR limit recover the diamagnetic contribution for any finite basis set. This is a consequence of the folding of the magnetic balance operators in the magnetic field-dependent part of the four-component Hamiltonian eqn (8.47). As a result, bilinear operators in the vector potentials are recovered also in relativistic theory. Finally, the second contribution in eqn (8.60) is the relativistic counterpart of the paramagnetic contribution, since the regular part of the response MOs $\varphi_i^{10,r}$ is expanded in the perturbation-free MOs (8.42). The summation involved in eqn (8.42) includes all MOs (occupied, negative, and positive energy virtuals), but this time the summation over negative-energy states will only give a minor contribution because the diamagnetic term has already been projected out and now is part of the first contribution in eqn (8.59). Note that the separation into dia- and paramagnetic terms in the relativistic theories is unique only in the spirit of the NR limit. Since the EFUT and RMB approaches handle the magnetic balance differently, the distribution of relativistic effects between dia- and paramagnetic terms [eqn (8.59)–(8.61)] is different, and only the final NMR shielding or spin–spin coupling constants should be compared.

8.4 Examples of Relativistic Effects on NMR Parameters

We will close the chapter by giving some examples of calculations of relativistic effects on the NMR shielding and spin–spin coupling constants. Our goal is not to give a comprehensive review of all relativistic calculations of NMR parameters, but rather to give examples of studies that will help illustrate the importance of relativistic corrections for the shielding and spin–spin coupling constants. We will in particular try to illustrate the factors that govern relativistic effects so that a feeling for when relativistic effects can be expected to be important can be gained.

Whereas a perturbational analysis of the relativistic corrections to the NMR parameters often can provide a very nice qualitative insight into the factors that determine the importance of the relativistic corrections,[7,73–75] we will restrict our examples to calculations using two- or four-component relativistic methods. This partly illustrates that such methods have now reached a level of maturity and computational efficiency such that there is no need to use non-relativistic calculations if there is reason to believe relativistic effects will be important for the calculated NMR parameters. As such, four-component methods are today a convenient and reliable tool for modeling and understanding NMR shielding and indirect spin–spin coupling tensors of systems containing heavy elements. Despite the complexity of the formalism, modern implementations allow calculations on systems containing up to 100 atoms.[100–103]

We will also limit ourselves to examples relevant in the context of gas phase NMR spectroscopy, and will largely ignore examples for which solvent effects are important or have been included. However, we emphasize that the computational methodology to a very large extent is similar for gas phase and solvent calculations. For a more complete review of the literature on studies that include relativistic effects in the calculation of NMR parameters, we refer to recent reviews of the field.[13–16]

In our discussion of nuclear magnetic shielding constants, we will focus on a very important consequence of relativistic effects, namely the determination of absolute shielding tensors. Whereas in the non-relativistic domain there is a close relation between spin-rotation tensors and the paramagnetic contribution to the shielding tensor,[104,105] this relation breaks down when relativistic effects are taken into account.[106,107] This has significant consequences for the experimental determination of absolute shielding tensors, even for fairly light elements, and we will give some examples of studies of absolute shielding tensors when considering this breakdown.

8.4.1 Nuclear Magnetic Shielding Constants and Chemical Shifts

When discussing relativistic effects in general, an important distinction is made between scalar relativistic effects and spin–orbit effects. One reason for this, as discussed in the previous section, are the properties of the operators involved, where the scalar relativistic effects in general do not involve the electronic spin coordinates and account for the changes in the electron density in the vicinity of the nucleus, whereas the spin–orbit effects include coupling to the spin coordinates of the electron and in general require two- or four-component methods. The spin–orbit corrections will split the degeneracy of atomic and molecular orbitals through the coupling of the electron spin with the orbital motion of the electrons. The spin–orbit interactions will thus also directly affect the electron density in the valence regions.

In the case of NMR, we measure the interaction of the nuclear magnetic moments with local magnetic fields arising from different sources, either from an external apparatus (NMR shielding) or from magnetic moments of other nuclei (indirect nuclear spin–spin coupling). In all these cases, the spin of the electron can couple to these sources and thus be a mediator for the interaction mechanisms. This is obvious in the case of the indirect spin–spin coupling constants, where the Fermi-contact and spin–dipolar contributions involve the electronic spin degrees of freedom for coupling the nuclear magnetic moments. In the case of the nuclear shielding tensor, the spin–orbit operator creates new coupling mechanisms, in which the external magnetic field induces interactions with the orbital magnetic moments of the electrons. Through the spin–orbit operator, this induced magnetic moment can couple to the electronic spin degrees of freedom, which in turn can couple to the nuclear magnetic moments through the Fermi-contact and spin–dipolar interaction mechanisms. The close relation between this relativistic contribution to the shielding constant and the indirect spin–spin coupling constants have been discussed by Pyykkö *et al.*[7]

The spin–orbit effects are the most striking effects of relativity on nuclear magnetic shielding constants in the sense that they significantly influence

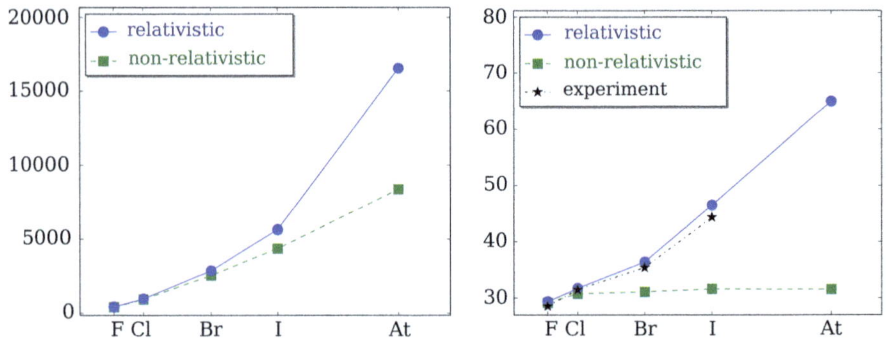

Figure 8.1 Isotropic NMR shieldings constants (in ppm) of the heavy atom (left) and the hydrogen (right) in the HX (X = F, Cl, Br, I, At) series. Calculated relativistic (four-component using mDKS-RMB-GIAO method) and non-relativistic DFT results were obtained with the BP86 functional and Dyall's cvqz basis set (for computational details, see ref. 2). Experimental absolute shielding constants were obtained from chemical shifts published in ref. 109 and four-component relativistic calculations for the CH_4 as reference compound (31.0 ppm).

the shielding also of light elements in the vicinity of heavy elements. A common set of molecules that demonstrate these relativistic effects very convincingly are the hydrogen halides, for which the relativistic corrections to the hydrogen chemical shieldings are larger than the normal span of the chemical shift of the proton. We show in Figure 8.1 the relativistic and non-relativistic shielding constants in the HX (X = F, Cl, Br, I, At) series. Focusing on the hydrogen shieldings, we first of all note that relativistic effects are mandatory in order to reproduce even qualitatively the experimental trends in the shielding constants. Even for fairly light elements such as chlorine and bromine, there are non-negligible effects from the presence of the heavy element due to spin–orbit interactions. This effect is often referred to as the heavy-atom effect on the light atoms.[3]

An interesting example of the use of relativistically calculated NMR chemical shifts to help characterize molecular structure, is the case of Breitfussin A and B, a secondary metabolite of an Arctic hydrozoan with a novel molecular structure.[108] Despite the use of a multitude of experimental methods, the structure of the oxazole part of the molecule could only be resolved through a combined experimental and theoretical NMR approach, for which the presence of bromine and iodine mandated the inclusion of relativistic effects in order to elucidate the ^{13}C chemical shifts. Full four-component Dirac–Kohn–Sham calculations were performed on this molecule with 34 atoms, illustrating the powers of currently available four-component relativistic calculations.

Turning now our attention to the heavy-element shielding in the hydrogen halides, these have been extensively studied and are known to display strong relativistic effects. Indeed, it has been shown that the relativistic effects on the halogen shielding increase approximately as $Z^{3.5}$, where Z is the nuclear charge of the halogen.[2] This scaling is visible in the plot of the heavy-atom shielding with and without relativistic effects in Figure 8.1. In contrast to the hydrogen shieldings, the relativistic corrections to the heavy-atom shielding have a dominating contribution from the electron density close to the nucleus, although the spin–orbit corrections are not negligible either. As such, these effects are much more atomic in nature and, as a consequence, some partial cancellation will occur in the chemical shifts.

When performing highly accurate calculations of shielding tensors for comparison to high-quality NMR data, high-level non-relativistic calculations performed at the CCSD(T) level of theory are often combined with corrections from vibrational effects and relativistic corrections. In most cases, the vibrational corrections are calculated non-relativistically. However, it is important to realize that because the spin–orbit operator couples to the Fermi-contact interaction, the geometry dependence of the relativistic spin–orbit correction can be substantial and, in the case of hydrogen iodide, the vibrational correction to the shielding tensor changes sign compared to the non-relativistic level when relativistic effects are included also in the vibrational averaging procedure.[110]

Spin–orbit effects are included both in two- and four-component calculations. The majority of such calculations have been performed with the two-component spin–orbit ZORA Hamiltonian. Whereas such calculations in general provide very good results for chemical shifts, the situation is different in the case of the absolute shielding tensor itself. One illustration of this is the gas phase NMR study of the absolute shielding tensor of ^{119}Sn in tetramethyltin by Makulski.[111] An experimental absolute shielding of 2172 ± 200 ppm was determined by the author through the use of the Flygare relation.[112] The result was in only fair agreement with non-relativistic density-functional theory calculations that gave a shielding of 2523 ppm.[113] Scalar relativistic effects would seem to improve this result, being 2283 ppm using the BP86 functional and the ZORA Hamiltonian, but this result deteriorated when including spin–orbit effects in the ZORA approach, giving a result of 2749 ppm. Subsequent four-component calculations differed significantly from the SO-ZORA results, but going in the opposite direction of experiment, being 3199 ppm.[107] This clearly shows that relativistic effects are significant for heavy-element shieldings (about a 50% increase of the non-relativistic value) and that there is a need to go beyond the SO-ZORA Hamiltonian when considering absolute shielding constants. The origin of the large discrepancy between the four-component shielding tensor and experiment can be traced to the breakdown of the relation connecting shielding tensors to spin-rotation tensors,[107] and we will return to this in the next section.

Recently, Sun *et al.* presented an implementation of nuclear magnetic shielding tensors at the X2C level of theory, and presented the first pilot calculations.[77] The preliminary results look very promising, and may suggest that the calculation of shielding tensors in the X2C framework can be a very reliable and fast route to nuclear magnetic shielding tensors at the two-component level. However, the implementation appears not to have been extensively used.

Simultaneously, great strides have been made in the efficient computation of NMR shielding tensors at the full relativistic four-component level of theory, for both Hartree–Fock[78] and density functional theory.[80,81,84] We have in particular demonstrated the efficiency of four-component relativistic Dirac–Kohn–Sham calculations for the evaluation of NMR shielding constants for organometallic complexes[103] and organic molecules with heavy halogens.[101] These studies illustrate that four-component relativistic theory at the DFT level today can rival the size of molecular systems that can be handled by non-relativistic methods, but with an accuracy that will be consistently better than those of the corresponding non-relativistic methods because of the exact handling of relativistic effects. What is currently missing from the perspective of accurate gas phase NMR studies of shielding constants is electron correlation methods such as the very successful coupled-cluster singles and doubles with perturbative triples [CCSD(T)] method for calculating shielding constants at the two- or four-component levels of theory. This is, from the point of view of high-accuracy calculations of NMR shielding constants of molecules in the gas phase, an important area of developments. Work in this direction is being pursued by Gauss and co-workers.[114,115]

8.4.1.1 *Absolute NMR Shielding and Spin-rotation Tensors*

By considering the non-relativistic expressions, Flygare[104,105] suggested that the nuclear spin-rotation (NSR) tensors could be used to determine, in combination with theoretical calculations, the absolute nuclear magnetic shielding tensors using the relation between the electronic part of the spin-rotation tensor $\mathbf{M}^{K,\mathrm{e1}}$ and the paramagnetic contributions to the shielding tensor

$$\boldsymbol{\sigma}_K^{\mathrm{para}} = \frac{2\pi}{\hbar}\frac{m_\mathrm{p}}{m_\mathrm{e}}\frac{10^9}{g_K}\mathbf{M}^{K,\mathrm{el}}\,\mathbf{I} = \boldsymbol{\sigma}_K^{\mathrm{SR}} - \frac{2\pi}{\hbar}\frac{m_\mathrm{p}}{m_\mathrm{e}}\frac{10^9}{g_K}\mathbf{M}^{K,\mathrm{nuc}}\mathbf{I}, \tag{8.62}$$

where m_p and m_e are the proton and electron mass, respectively, g_K the nuclear g-value of nucleus K, $\hbar$ the reduced Planck constant, $\mathbf{I}$ the moment of inertia tensor, and $\boldsymbol{\sigma}_K^{\mathrm{SR}}$ refers to the NSR tensor in ppm, whereas $\mathbf{M}^{K,\mathrm{nuc}}$ is its nuclear contribution in kHz.

The nuclear spin-rotation tensor measures the coupling between the nuclear magnetic moment and the small magnetic moment induced by the rotation of the molecular framework. Indeed, this induced magnetic moment is in the non-relativistic limit proportional to the magnetic moment induced by an external magnetic field. The nuclear spin-rotation tensors are an important source of experimental data, as they can be measured with high accuracy in rotational microwave spectra. Considering the non-relativistic relation between the spin-rotation and the nuclear shielding tensors and the fact that these measurements are conducted in the gas phase, the spin-rotation tensors measured in rotational spectra have in the past been one of the most important sources of experimental information on absolute shielding tensors of molecules in the gas phase.

The relation in eqn (8.62), which we will refer to as the *first Flygare relation*, provides a theoretical connection between absolute shielding tensors and nuclear spin-rotation tensors. Flygare and Goodisman[112] suggested that a more useful way to determine absolute shielding constants experimentally would be to use the relation

$$\sigma_K \approx \sigma_K^{\mathrm{SR}} + \sigma_K^{\mathrm{FA}}, \tag{8.63}$$

where only experimentally available data are used to derive an absolute shielding tensor, σ_K^{FA} being the free-atom NMR shielding tensor. Equation (8.63) stems from the fact that the diamagnetic contribution to the isotropic NMR shielding can be approximated by the sum of free-atom isotropic NMR shielding and the nuclear contribution to the spin-rotation tensor [second term on the right-hand side of eqn (8.62)]. In combination with accurately measured experimental nuclear spin-rotation tensors, the *second Flygare relation* (8.63) has been an important route for determining experimental absolute nuclear magnetic shielding tensors,[104,112] as these are difficult to determine directly in an NMR experiment. Flygare's relation eqn (8.62) has been extensively used to determine semi-experimental absolute shielding tensors, see *e.g.* ref. 116–118 for some recent examples.

In the relativistic domain, the magnetic moment operator arising from the external magnetic field has the form

$$\vec{\mu} = -\frac{1}{2}\vec{r}_0 \times c\vec{\alpha}. \tag{8.64}$$

It was questioned in the literature[1,79] that the coupling of the electronic state and the rotational state should be described by different operators in relativistic domain. Thus, a straightforward relation between the interaction with an external magnetic field and the rotation of a molecule is not possible. However, it was shown only recently by Aucar *et al.*[106] that the operator of this coupling is actually the total angular momentum

$$\vec{J}_e = \vec{L} + \frac{1}{2}\vec{\Sigma}. \tag{8.65}$$

The operators in eqn (8.64) and (8.65) are fundamentally different, since whereas the magnetic moment eqn (8.64) couples large and small components of the wavefunction, the total angular momentum eqn (8.65) couples large–large and small–small components of the four-component wavefunction

$$\vec{\alpha} = \begin{bmatrix} 0 & \vec{\sigma} \\ \vec{\sigma} & 0 \end{bmatrix}, \quad \vec{\Sigma} = \begin{bmatrix} \vec{\sigma} & 0 \\ 0 & \vec{\sigma} \end{bmatrix}. \tag{8.66}$$

This fundamental difference will lead to a dramatic breakdown of Flygare's relation eqn (8.62) for heavy elements.[2]

The first calculations of NSR tensors following the theory of Aucar *et al.*[106] appeared in early 2013, when Malkin *et al.* presented a first study of the absolute shielding of ^{119}Sn in SnH_4, $Sn(CH_3)_4$, and $SnCl_4$.[107] The work was motivated by a combined experimental and theoretical study of the gas phase NMR parameters of tetramethyltin by Makulski,[111] in which it was observed that the absolute shielding calculated using the SO-ZORA Hamiltonian was in poorer agreement with experiment than with the spin–orbit-free ZORA results. This effect was further accentuated at the four-component level, with differences as large as 1000 ppm being observed.[107]

An interesting observation from the study of Malkin *et al.* is that for all three systems, SnH_4, $Sn(CH_3)_4$, and $SnCl_4$, the relativistic correction to the Flygare relation eqn (8.62) was almost constant at 1000 ppm. A similar correction of 1000 ppm can also be found in the SnX (X = O, S, Se, and Te) series.[119] The reason for this surprising additivity can be understood from a perturbational analysis of the relativistic corrections to the nuclear magnetic shielding and nuclear spin-rotation constants.[106,119] There are five contributions that only appear for the nuclear shielding constant, and four of these contributions sample regions of the electron density close to the nucleus by sampling the electron density at the nucleus $[\delta(\vec{r}_K)]$ or close to the nucleus (r_K^{-n}, $n \geq 2$). The last term involves the orbital angular momentum around the nucleus of interest and will vanish for molecules with spherical symmetry or have only one non-vanishing component for linear molecules. Thus, for highly symmetric molecules, the relativistic corrections arising from the breakdown of Flygare's relation have a strongly atomic character.

Jaszuński and co-workers have revised a number of absolute shieldings based on the breakdown of eqn (8.62), either based on experimental nuclear spin-rotation constants or the use of highly accurate coupled-cluster calculations. These studies include a series of monofluorides XF (X = B, Al, Ga, In, and Tl)[120] and hexafluorides XF_6 (X = S, Se, Te, Mo, W)[121] as well as three transition metal monocarbonyls XCO (X = Ni, Pd, Pt).[99] A combined theoretical and experimental study of the H^{35}Cl and H^{37}Cl isotopes, in which the combined use of gas phase NMR data and high-level coupled-cluster calculations and relativistic corrections following eqn (8.63) enabled the accuracy of the magnetic dipole moments of ^{35}Cl and ^{37}Cl to be improved, the revised magnetic dipole moments being $0.821721(5)\mu_N$ and $0.683997(4)\mu_N$, respectively (μ_N being the nuclear magneton).

Despite the breakdown of Flygare's relation eqn (8.62), it can nevertheless be a useful tool for providing benchmark results for absolute shielding constants for lighter elements against which highly accurate *ab initio* calculations can be benchmarked, as done for instance in a recent study by Helgaker *et al.* of the absolute shielding tensor of ^{33}S.[116] However, relativistic effects are still present in the experimental spin-rotation data and, in order to create a truly non-relativistic absolute shielding tensor, the relativistic corrections must be removed from the experimental spin-rotation constant before the Flygare relation is applied. Komorovsky *et al.*[117] showed that in the case of ^{33}S absolute shielding tensors, subtracting the relativistic corrections to the nuclear spin-rotation tensors increased the isotropic shielding by 1.9 ppm, giving an experimental absolute shielding of 718.3(4.6) ppm, largely resolving a previously noted discrepancy with highly accurate CCSD(T) results calculated for $T = 300$ K of 719.0 ppm. Nevertheless, the estimated relativistic absolute shielding is 742.9(4.6) ppm,[117] highlighting that the breakdown of eqn (8.62) has significant effects on the absolute shielding tensor even for such a light nucleus as ^{33}S.

8.4.2 Indirect Nuclear Spin–Spin Coupling Constants

As experimental studies of indirect spin–spin coupling constants in the gas phase are much more limited than for the shielding constants, also the literature on two- and four-component calculations of indirect spin–spin coupling constants is much more limited than for the shielding constants. There are multiple reasons for this. One is of course the more limited experimental interest, in particular in the gas phase, but also the much higher computational costs, since in general three response equations [see eqn (8.44)] along each Cartesian direction have to be determined for each NMR-active nucleus. An additional factor is the fact that the indirect spin–spin coupling constants are often dominated by operators that involve the electron spin (as the indirect spin–spin coupling constants couple the nuclear magnetic moments through the electron spin density), which often cannot be reliably calculated at the Hartree–Fock level of theory due to triplet instabilities, and thus DFT or electron-correlated approaches are needed, also for relativistic calculations.

For almost all one-bond indirect spin–spin coupling constants, their isotropic values, which are the only ones that can be detected for a freely rotating gaseous sample, are dominated by the Fermi-contact interaction. This interaction couples the two nuclear magnetic moments through the spin density at the two nuclei. Other mechanisms, and in particular the coupling of the nuclear magnetic moments through the orbital motion of the electrons, the paramagnetic spin–orbit operator, become more important for spin–spin couplings across multiple bonds. In contrast, the anisotropic component is instead dominated by cross terms of the Fermi-contact operator and the spin-dipolar operator. This will lead to potentially very different relativistic effects on the isotropic and anisotropic spin–spin coupling constants. These differences between the isotropic and anisotropic components of the spin–spin coupling constants are to a large extent dictated by the symmetry of these operators, the Fermi-contact operator being fully isotropic and the spin-dipolar operator fully anisotropic.[76] Introducing the relativistic spin–orbit operator will have the consequence that the electron spin will no longer be a good quantum number, mixing the spin and orbital components. We can thus expect that the inclusion of the relativistic spin–orbit operator, as done in two- and four-component theories, can potentially change the behavior of the isotropic and anisotropic parts of the indirect spin–spin coupling tensors.

The first calculations of indirect spin–spin coupling constants at the four-component Dirac–Hartree–Fock level of theory were presented by Visscher *et al.*[79,122] who studied the one- and two-bond couplings in the hydrogen halides and a set of carbon-group hydrides. The first four-component DFT implementation of indirect spin–spin coupling constants also involving RMB condition was reported by Repiský *et al.*[94] Autschbach and Ziegler implemented the spin–orbit ZORA approach.[76] They found that in some systems, such as plumbane, the spin–orbit effects partially cancel the scalar relativistic, demonstrating that the inclusion of only scalar relativistic effects may actually give an incorrect account of the importance of the relativistic effects. Spin–orbit effects were also found to be essential for the TlX series (X = F, Cl, Br, I), as scalar relativistic effects can give wrong answers for couplings involving heavy *p*-block elements for which the Fermi-contact and paramagnetic spin–orbit contributions are large already in the non-relativistic limit.

In addition to being observable in NMR spectra, indirect spin–spin coupling constants can also be observed in molecular beam and stimulated Raman experiments.[124,125] Bryce *et al.* performed a comparison of scalar and two-component spin–orbit ZORA for the calculation of indirect spin–spin coupling constants in a series of interhalogen dimers and compared these to experimental observations.[123] We have collected some of their results in Table 8.1. Agreement with experiment is in general found to be satisfactory, although rather significant differences remain. Some of these deviations no doubt arise from the approximations in the ZORA Hamiltonian, the choice of basis and limitations in the functional, as well as the lack of vibrational corrections, which can be quite significant for indirect spin–spin coupling constants.[126] An interesting observation from the results of Bryce *et al.* is that the mixed Fermi-contact–spin dipolar contribution accounts for about 20–25% of the isotropic indirect spin–spin coupling constants. As this contribution would be zero in the absence of spin–orbit effects, it clearly demonstrates the importance of including spin–orbit effects in the calculation of indirect spin–spin coupling constants.

Moncho and Autschbach presented a very comprehensive benchmark of one-, two-, three-, and four-bond indirect spin–spin coupling constants in 47 different molecules with a variety of heavy elements (W, Pt, Hg, Tl, and

Table 8.1 Experimental and spin–orbit ZORA indirect spin–spin coupling constants calculated using the ZORA V basis and the BP86 functional. Results taken from ref. 123.

Molecule	J_{iso} Exp.	J_{iso} SO-ZORA	J_{aniso} Exp.	J_{aniso} SO-ZORA
$^{35}Cl^{19}F$	840	969	−907	−1143
$^{35}Cl^{81}Br$		711		−788
$^{35}Cl^{127}I$		678		−801
$^{81}Br^{19}F$	5240	5648	−6306	−6420
$^{81}Br^{127}I$		3993		−4538
$^{127}I^{19}F$	5730	4908	−5856	−6223

Pb).[127] In addition to testing the effects of a continuum solvation model on the spin–spin coupling constants, they also benchmarked the performance of purely scalar as well as spin–orbit relativistic effects as described by the (SO)-ZORA Hamiltonians. The performance of pure and hybrid exchange–correlation functionals were also investigated. Somewhat surprisingly, SO-ZORA did not lead to any overall improvement in the agreement with available experimental data. For Pb and W, including spin–orbit corrections actually increased the standard deviation. In contrast, hybrid functional (PBE0) was found in general to perform better than the pure DFT functional (PBE), and this effect was particularly noteworthy for the SO-ZORA calculations. Overall, the best performance was obtained using the SO-ZORA Hamiltonian with the PBE0 functional. Clearly, it would be interesting to perform a similar benchmark study at the four-component level of theory in order to assess the quality of the SO-ZORA Hamiltonian for calculating NMR indirect spin–spin coupling constants. A first step in this direction was presented by Demissie *et al.*,[103] who performed a combined theoretical and experimental study of the chemical shifts and indirect spin–spin coupling constants for a series of tungsten complexes using four-component relativistic DFT with RMB, using both pure (BP86) and hybrid (B3LYP) functionals. They obtained notably better agreement with the experimental data when using the B3LYP functional, corroborating the findings of Moncho and Autschbach.

8.5 Concluding Remarks

In this chapter, we have discussed different two- and four-component methods for calculating the NMR shielding and indirect spin–spin coupling tensors. By including spin–orbit effects variationally, we ensure an accurate and reliable account of the most significant relativistic effects on the NMR parameters. We have discussed various approximations for the two- and four-component Hamiltonians as well as their implementation for the calculation of the NMR shielding and indirect spin–spin coupling tensors. Although two- and four-component relativistic calculation of NMR parameters is still rather young compared to the calculation of NMR parameters at the non-relativistic level of theory, the field is in rapid development, both in terms of the computational efficiency of the two- and four-component relativistic methods as well as in terms of an understanding of what are the appropriate Hamiltonians and computational approaches needed for the calculation of NMR parameters at the relativistic level of theory. By a few illustrative examples, we have shown that relativistic effects cannot be ignored if there are heavy elements present in a molecule, even for as light elements as those from the third row of the periodic table. If such heavier elements are present, we have also demonstrated that the effects of relativity in these cases are as significant for the near-lying light elements as for the heavy elements themselves. This is particularly the case for the shielding tensor.

Of particular interest for accurate measurements of gas phase NMR spectra is the determination of absolute shielding tensors, and through this the determination of nuclear magnetic moments. We have shown that the recent development of a relativistic theory for the nuclear spin-rotation tensor has undermined the commonly used approach for determining absolute shielding tensors by relating the spin-rotation constants to the nuclear shielding tensor, the effects being noticeable even in the water molecule. Indeed, the interaction of a nuclear magnetic moment with the magnetic moment induced by the molecular rotation and the magnetic moment induced by an external magnetic field has been shown to have different physical origins in relativistic theory. Although clearly making the determination of absolute shielding tensors even more challenging, it remains possible to derive semi-experimental absolute shielding scales by using relativistic four-component calculations, though the input from the theoretical calculations increases significantly compared to the non-relativistic case, and thus it can be questioned to what extent such absolute shielding tensors can still be considered "experimental".

The advances made in two- and four-component relativistic theory and the importance of relativistic effects on NMR shielding and indirect spin–spin coupling tensors will in few years lead to a situation that these calculations will be the *de facto* standard for NMR calculations on molecules containing heavy elements. We believe these developments will further boost the importance of NMR as a tool for understanding molecular structure and intermolecular interactions, both in gas phase and in solution.

Acknowledgements

This work has been supported by the Research Council of Norway through a Centre of Excellence Grant and project grants (Grant No. 179568, 214095, 177558, and 191251).

References

1. T. Saue, *Adv. Quantum Chem.*, 2005, **48**, 383–405.
2. S. Komorovsky, M. Repisky, E. Malkin, T. B. Demissie and K. Ruud, *J. Chem. Theory Comput.*, 2015, **11**, 3729–3739.
3. P. Pyykkö, A. Görling and N. Rösch, *Mol. Phys.*, 1987, **61**, 195–205.
4. I. Morishima, K. Endo and T. Yonezawa, *J. Chem. Phys.*, 1973, **59**, 3356–3364.
5. M. Kaupp and O. L. Malkina, *J. Chem. Phys.*, 1998, **108**, 3648–3659.
6. J. Vaara, O. L. Malkina, H. Stoll, V. G. Malkin and M. Kaupp, *J. Chem. Phys.*, 2001, **114**, 61–71.
7. M. Kaupp, O. L. Malkina, V. G. Malkin and P. Pyykkö, *Chem. – Eur. J.*, 1998, **4**, 118–126.
8. T. Helgaker, W. Klopper and D. P. Tew, *Mol. Phys.*, 2008, **106**, 2107–2143.

9. DIRAC, a relativistic ab initio electronic structure program, Release DIRAC14 (2014), written by T. Saue, L. Visscher, H. J. Aa. Jensen and R. Bast. with contributions from V. Bakken, K. G. Dyall, S. Dubillard, U. Ekström, E. Eliav, T. Enevoldsen, E. Faßhauer, T. Fleig, O. Fossgaard, A. S. P. Gomes, T. Helgaker, J. K. Lærdahl, Y. S. Lee, J. Henriksson, M. Iliaš, Ch. R. Jacob, S. Knecht, S. Komorovský, O. Kullie, C. V. Larsen, H. S. Nataraj, P. Norman, G. Olejniczak, J. Olsen, Y. C. Park, J. K. Pedersen, M. Pernpointner, R. di Remigio, K. Ruud, P. Sałek, B. Schimmelpfennig, J. Sikkema, A. J. Thorvaldsen, J. Thyssen, J. van Stralen, S. Villaume, O. Visser, T. Winther and S. Yamamoto (see http://www.diracprogram.org).
10. ReSpect, version 3.4.2 (2015), Relativistic Spectroscopy DFT program of authors M. Repisky, S. Komorovsky, V. G. Malkin, O. L. Malkina, M. Kaupp, K. Ruud, with contributions from R. Bast, U. Ekström, M. Kadek, S. Knecht, L. Konecny, E. Malkin, I. Malkin-Ondik (see www.respectprogram.org).
11. E. J. Baerends, T. Ziegler, J. Autschbach, D. Bashford, A. Bérces, F. M. Bickelhaupt, C. Bo, P. M. Boerrigter, L. Cavallo, D. P. Chong, L. Deng, R. M. Dickson, D. E. Ellis, M. van Faassen, L. Fan, T. H. Fischer, C. Fonseca Guerra, M. Franchini, A. Ghysels, A. Giammona, S. J. A. van Gisbergen, A. W. Götz, J. A. Groeneveld, O. V. Gritsenko, M. Grüning, S. Gusarov, F. E. Harris, P. van den Hoek, C. R. Jacob, H. Jacobsen, L. Jensen, J. W. Kaminski, G. van Kessel, F. Kootstra, A. Kovalenko, M. V. Krykunov, E. van Lenthe, D. A. McCormack, A. Michalak, M. Mitoraj, S. M. Morton, J. Neugebauer, V. P. Nicu, L. Noodleman, V. P. Osinga, S. Patchkovskii, M. Pavanello, P. H. T. Philipsen, D. Post, C. C. Pye, W. Ravenek, J. I. Rodríguez, P. Ros, P. R. T. Schipper, H. van Schoot, G. Schreckenbach, J. S. Seldenthuis, M. Seth, J. G. Snijders, M. Solà, M. Swart, D. Swerhone, G. te Velde, P. Vernooijs, L. Versluis, L. Visscher, O. Visser, F. Wang, T. A. Wesolowski, E. M. van Wezenbeek, G. Wiesenekker, S. K. Wolff, T. K. Woo and A. L. Yakovlev, ADF2014, SCM, Theoretical Chemistry, Vrije Universiteit, Amsterdam, The Netherlands, http://www.scm.com.
12. W. Liu, F. Wang and L. Li, *J. Theor. Comput. Chem.*, 2003, **2**, 257–272.
13. J. Vaara, *Phys. Chem. Chem. Phys.*, 2007, **9**, 5399–5418.
14. J. Autschbach, in *Relativistic Methods for Chemists*, ed. M. Barysz and Y. Ishikawa, Springer, 2010, p. 521.
15. J. Autschbach, *J. Chem. Phys.*, 2012, **136**, 150902.
16. W. Liu, *Phys. Rep.*, 2014, **537**, 59.
17. P. Dirac, *Proc. R. Soc. London, Ser. A*, 1928, **117**, 610–624.
18. P. Dirac, *Proc. R. Soc. London, Ser. A*, 1928, **118**, 351–361.
19. I. P. Grant and H. M. Quiney, *Phys. Rev. A: At., Mol., Opt. Phys.*, 2000, **62**, 022508.
20. W. Kutzelnigg, *Int. J. Quantum Chem.*, 1984, **25**, 107–129.
21. K. G. Dyall, *J. Chem. Phys.*, 1994, **100**, 2118–2127.
22. W. Schwarz and H. Wallmeier, *Mol. Phys.*, 1982, **46**, 1045–1061.

23. T. Saue, *ChemPhysChem*, 2011, **12**, 3077–3094.
24. R. E. Stanton and S. Havriliak, *J. Chem. Phys.*, 1984, **81**, 1910–1918.
25. L. Visscher and K. G. Dyall, *At. Data Nucl. Data Tables*, 1997, **67**, 207–224.
26. G. Breit, *Phys. Rev.*, 1929, **34**, 553–573.
27. J. Gaunt, *Proc. R. Soc. London, Ser. A*, 1929, **122**, 513–532.
28. C. van Wüllen, *J. Comput. Chem.*, 2002, **23**, 779–785.
29. L. M. Sandratskii, *Adv. Phys.*, 1998, **47**, 91–160.
30. F. Wang and T. Ziegler, *J. Chem. Phys.*, 2004, **121**, 12191–12196.
31. J. Gao, W. Zou, W. Liu, Y. Xiao, D. Peng, B. Song and C. Liu, *J. Chem. Phys.*, 2005, **123**, 054102.
32. R. Bast, H. J. Aa. Jensen and T. Saue, *Int. J. Quantum Chem.*, 2009, **109**, 2091–2112.
33. H. Eschrig and V. D. P. Servedio, *J. Comput. Chem.*, 1999, **20**, 23–30.
34. T. Saue and T. Helgaker, *J. Comput. Chem.*, 2002, **23**, 814–823.
35. L. L. Foldy and S. A. Wouthuysen, *Phys. Rev.*, 1950, **78**, 29–36.
36. M. Douglas and N. M. Kroll, *Ann. Phys.*, 1974, **82**, 89–155.
37. T. Nakajima and K. Hirao, *J. Chem. Phys.*, 2000, **113**, 7786–7789.
38. A. Wolf, M. Reiher and B. A. Hess, *J. Chem. Phys.*, 2002, **117**, 9215–9226.
39. B. A. Hess, *Phys. Rev. A: At., Mol., Opt. Phys.*, 1985, **32**, 756–763.
40. C. van Wüllen, *J. Chem. Phys.*, 2004, **120**, 7307–7313.
41. M. Reiher and A. Wolf, *J. Chem. Phys.*, 2004, **121**, 2037–2047.
42. M. Reiher and A. Wolf, *J. Chem. Phys.*, 2004, **121**, 10945–10956.
43. D. Peng and K. Hirao, *J. Chem. Phys.*, 2009, **130**, 044102.
44. M. Barysz, A. J. Sadlej and J. G. Snijders, *Int. J. Quantum Chem.*, 1997, **65**, 225–239.
45. M. Barysz and A. J. Sadlej, *J. Chem. Phys.*, 2002, **116**, 2696–2704.
46. D. Kedziera and M. Barysz, *J. Chem. Phys.*, 2004, **121**, 6719–6727.
47. J.-L. Heully, I. Lindgren, E. Lindroth, S. Lundqvist and A.-M. Martensson-Pendrill, *J. Phys. B: At. Mol. Phys.*, 1986, **19**, 2799–2815.
48. W. Kutzelnigg, *J. Chem. Phys.*, 1999, **110**, 8283–8294.
49. M. Iliaš and T. Saue, *J. Chem. Phys.*, 2007, **126**, 064102.
50. K. G. Dyall, *J. Chem. Phys.*, 1997, **106**, 9618–9628.
51. D. Cremer, W. Zou and M. Filatov, *Wiley Interdiscip. Rev.: Comput. Mol. Sci.*, 2014, **4**, 436–467.
52. K. G. Dyall and K. Faegri Jr., *Introduction to Relativistic Quantum Chemistry*, Oxford University Press, USA, New York, 2007.
53. M. Reiher and A. Wolf, *Relativistic Quantum Chemistry: The Fundamental Theory of Molecular Science*, John Wiley & Sons, 2014.
54. M. Iliaš, V. Kellö and M. Urban, *Acta Phys. Slovaca*, 2010, **60**, 259–391.
55. W. Liu, *Mol. Phys.*, 2010, **108**, 1679–1706.
56. R. E. Moss, *Advanced Molecular Quantum Mechanics*, Chapman and Hall, London, 1973.
57. J. Almlöf and O. Gropen, *Relativistic Effects in Chemistry*, John Wiley & Sons, New York, 1996, vol. 8, pp. 203–244.
58. C. Chang, M. Pelissier and P. Durand, *Phys. Scr.*, 1986, **34**, 394–404.

59. E. V. Lenthe, E. J. Baerends and J. G. Snijders, *J. Chem. Phys.*, 1993, **99**, 4597–4610.
60. K. G. Dyall and E. van Lenthe, *J. Chem. Phys.*, 1999, **111**, 1366–1372.
61. E. van Lenthe, E. J. Baerends and J. G. Snijders, *J. Chem. Phys.*, 1994, **101**, 9783–9792.
62. H. J. Aa. Jensen and M. Iliaš, unpublished.
63. W. Kutzelnigg and W. Liu, *J. Chem. Phys.*, 2005, **123**, 241102.
64. W. Liu and W. Kutzelnigg, *J. Chem. Phys.*, 2007, **126**, 114107.
65. W. Liu and D. Peng, *J. Chem. Phys.*, 2006, **125**, 044102.
66. E. Malkin, M. Repiský, S. Komorovský, P. Mach, O. L. Malkina and V. G. Malkin, *J. Chem. Phys.*, 2011, **134**, 044111.
67. R. Fukuda, M. Hada and H. Nakatsuji, *J. Chem. Phys.*, 2003, **118**, 1015–1026.
68. J. Autschbach, *ChemPhysChem*, 2009, **10**, 2274–2283.
69. R. E. Moss and H. P. Trivedi, *Mol. Phys.*, 1979, **38**, 1611–1620.
70. J. D. Morrison and R. E. Moss, *Mol. Phys.*, 1980, **41**, 491–507.
71. G. Schreckenbach and T. Ziegler, *Int. J. Quantum Chem.*, 1997, **61**, 899–918.
72. W. Kutzelnigg, *Z. Phys.*, 1989, **11**, 15–28.
73. P. Manninen, P. Lantto, J. Vaara and K. Ruud, *J. Chem. Phys.*, 2003, **119**, 2623–2637.
74. P. Manninen, K. Ruud, P. Lantto and J. Vaara, *J. Chem. Phys.*, 2005, **122**, 114107.
75. P. Manninen, K. Ruud, P. Lantto and J. Vaara, *J. Chem. Phys.*, 2006, **124**, 149901.
76. J. Autschbach and T. Ziegler, *J. Chem. Phys.*, 2000, **113**, 9410–9418.
77. Q. Sun, W. Liu, Y. Xiao and L. Cheng, *J. Chem. Phys.*, 2009, **131**, 081101.
78. M. Ilias, T. Saue, T. Enevoldsen and H. J. Aa. Jensen, *J. Chem. Phys.*, 2009, **131**, 124119.
79. L. Visscher, T. Enevoldsen, T. Saue, H. J. Aa. Jensen and J. Oddershede, *J. Comput. Chem.*, 1999, **20**, 1262–1273.
80. M. Olejniczak, R. Bast, T. Saue and M. Pecul, *J. Chem. Phys.*, 2012, **136**, 014108.
81. S. Komorovský, M. Repiský, O. L. Malkina, V. G. Malkin, I. M. Ondik and M. Kaupp, *J. Chem. Phys.*, 2008, **128**, 104101.
82. M. Repiský, S. Komorovský, O. L. Malkina and V. G. Malkin, *Chem. Phys.*, 2009, **356**, 236–242.
83. L. Cheng, Y. Xiao and W. Liu, *J. Chem. Phys.*, 2009, **131**, 244113.
84. S. Komorovský, M. Repiský, O. L. Malkina and V. G. Malkin, *J. Chem. Phys.*, 2010, **132**, 154101.
85. W. Kutzelnigg, *Phys. Rev. A: At., Mol., Opt. Phys.*, 2003, **67**, 032109.
86. Y. Xiao, W. Liu, L. Cheng and D. Peng, *J. Chem. Phys.*, 2007, **126**, 214101.
87. S. Komorovský, M. Repiský, O. L. Malkina, V. G. Malkin, I. Malkin Ondík and M. Kaupp, *J. Chem. Phys.*, 2008, **128**, 104101.
88. G. Manfredi, *Eur. J. Phys.*, 2013, **34**, 859–871.

89. M. M. Sternheim, *Phys. Rev.*, 1962, **128**, 676–677.
90. P. Pyykkö, *Chem. Phys.*, 1983, **74**, 1–7.
91. G. A. Aucar, T. Saue, L. Visscher and H. J. Aa. Jensen, *J. Chem. Phys.*, 1999, **110**, 6208–6218.
92. J. Vaara and P. Pyykkö, *J. Chem. Phys.*, 2003, **118**, 2973–2976.
93. L. Visscher, *Adv. Quantum Chem.*, 2005, **48**, 369–381.
94. M. Repiský, S. Komorovský, O. L. Malkina and V. G. Malkin, *Chem. Phys.*, 2009, **356**, 236–242.
95. Y. Xiao, D. Peng and W. Liu, *J. Chem. Phys.*, 2007, **126**, 081101.
96. L. Cheng, Y. Xiao and W. Liu, *J. Chem. Phys.*, 2009, **130**, 144102.
97. R. E. Stanton and S. Havriliak, *J. Chem. Phys.*, 1984, **81**, 1910–1918.
98. F. London, *J. Phys. Radium.*, 1937, **8**, 397–409.
99. T. B. Demissie, M. Jaszuński, E. Malkin, S. Komorovský and K. Ruud, *Mol. Phys.*, 2015, **113**, 1576–1584.
100. S. Komorovsky, M. Repisky, K. Ruud, O. L. Malkina and V. G. Malkin, *J. Phys. Chem. A*, 2013, **117**, 14209.
101. T. B. Demissie, M. Repisky, S. Komorovsky, J. Isaksson, J. S. Svendsen, H. Dodziuk and K. Ruud, *J. Phys. Org. Chem.*, 2013, **26**, 679–687.
102. T. B. Demissie, M. Repisky, H. Liu, K. Ruud and P. M. Kozlowski, *J. Chem. Theory Comput.*, 2014, **10**, 2125–2136.
103. T. B. Demissie, N. Kostenko, S. Komorovsky, M. Repisky, J. Isaksson, A. Bayer and K. Ruud, *J. Phys. Org. Chem.*, 2015, **28**, 723–731.
104. W. H. Flygare, *Chem. Rev.*, 1974, **74**, 653–687.
105. W. H. Flygare, *J. Chem. Phys.*, 1964, **41**, 793–800.
106. I. A. Aucar, S. S. Gómez, M. C. Ruiz de Azúa and C. G. Giribet, *J. Chem. Phys.*, 2012, **136**, 204119.
107. E. Malkin, S. Komorovsky, M. Repisky, T. B. Demissie and K. Ruud, *J. Phys. Chem. Lett.*, 2013, **4**, 459–463.
108. K. Ø. Hanssen, B. Schuler, A. J. Williams, T. B. Demissie, E. Hansen, J. H. Andersen, J. Svenson, K. Blinov, M. Repisky, F. Mohn, G. Meyer, J.-S. Svendsen, K. Ruud, M. Elyashberg, L. Groos, M. Jaspars and J. Isaksson, *Angew. Chem., Int. Ed.*, 2012, **51**, 12238.
109. W. G. Schneider, H. J. Bernstein and J. A. Pople, *J. Chem. Phys.*, 1958, **28**, 601–607.
110. B. F. Minaev, J. Vaara, K. Ruud, O. Vahtras and H. Ågren, *Chem. Phys. Lett.*, 1998, **295**, 455–461.
111. W. Makulski, *J. Mol. Struct.*, 2012, **1017**, 45–50.
112. W. H. Flygare and J. Goodisman, *J. Chem. Phys.*, 1968, **49**, 3122–3125.
113. A. C. de Dios, *Prog. Nucl. Magn. Reson. Spectrosc.*, 1998, **29**, 229.
114. L. Cheng and J. Gauss, *J. Chem. Phys.*, 2011, **135**, 244104.
115. L. Cheng, J. Gauss and J. F. Stanton, *J. Chem. Phys.*, 2013, **139**, 054105.
116. T. Helgaker, J. Gauss, G. Cazzoli and C. Puzzarini, *J. Chem. Phys.*, 2013, **139**, 244308.
117. S. Komorovsky, M. Repisky, E. Malkin, K. Ruud and J. Gauss, *J. Chem. Phys.*, 2015, **142**, 091102.

118. C. Puzzarini, G. Cazzoli, M. E. Harding, J. Vazquez and J. Gauss, *J. Chem. Phys.*, 2015, **142**, 124308.
119. S. Komorovsky, M. Repisky, T. B. Demissie, E. Malkin and K. Ruud, *Nuclear Spin-rotation and Shielding Constants in the SnX (X = O, S, Se, Te) Series*, unpublished.
120. M. Jaszuński, T. B. Demissie and K. Ruud, *J. Phys. Chem. A*, 2014, **118**, 9588–9595.
121. K. Ruud, T. B. Demissie and M. Jaszuński, *J. Chem. Phys.*, 2014, **140**, 194308.
122. T. Enevoldsen, L. Visscher, T. Saue, H. J. Aa. Jensen and J. Oddershede, *J. Chem. Phys.*, 2000, **112**, 3493–3498.
123. D. L. Bryce, R. E. Wasylishen, J. Autschbach and T. Ziegler, *J. Am. Chem. Soc.*, 2002, **124**, 4894–4900.
124. A. Yokozeki and J. S. Muenter, *J. Chem. Phys.*, 1980, **72**, 3796–3804.
125. J. P. Wallerand, F. du Burck, B. Mercier, A. N. Goncharov, M. Himbert and C. Bordé, *Eur. Phys. J. D*, 1999, **6**, 63–76.
126. T. A. Ruden, O. B. Lutnæs, T. Helgaker and K. Ruud, *J. Chem. Phys.*, 2003, **118**, 9572–9581.
127. S. Moncho and J. Autschbach, *J. Chem. Theory Comput.*, 2010, **6**, 223–234.

CHAPTER 9

High-resolution Spectra in PHIP

RODOLFO H. ACOSTA,* IGNACIO PRINA AND LISANDRO BULJUBASICH

FaMAF-Universidad Nacional de Córdoba, IFEG-CONICET, 5016 Córdoba, Argentina
*Email: racosta@famaf.unc.edu.ar

9.1 Introduction

Nuclear Magnetic Resonance (NMR) is one of the most powerful analytical techniques used for materials characterization at a microscopic level. The application of NMR in science and technology includes chemistry, biology, food research and quality control, environmental studies of plants and soils. Determination of pore structures has a great impact in the oil industry and medicine. Additionally Magnetic Resonance Imaging (MRI) is perhaps the most powerful diagnosis technique used in medicine in modern days. Despite all the power of NMR, there is a major drawback in its application that is the poor inherent sensitivity of the signals that can be detected. This fundamental insensitivity originates from the minuscule size of nuclear magnetic moments, which results in an exceedingly small equilibrium nuclear spin polarization even in high magnetic fields. Traditionally, NMR has dealt with excitation and detection of nuclear spin angular momentum in systems in thermal equilibrium with an external static magnetic field. The intensity of the NMR signal is proportional to the population difference of quantum states, which is driven by the difference in energy levels and is given by $\gamma B_0/k_B T$, where γ is the nuclear gyromagnetic ratio, B_0 is the external

New Developments in NMR No. 6
Gas Phase NMR
Edited by Karol Jackowski and Michał Jaszuński

Published by the Royal Society of Chemistry, www.rsc.org

magnetic field intensity, T the absolute temperature and k_B Boltzmann's constant. The excess population can be described by the polarization P. For instance, for spin-1/2 nuclei, P is the population difference between the two energy states over the whole spin population. A sample of water at room temperature, placed in a magnetic field of 4.7 T has a polarization for ^1H that amounts to $P = 1.6 \times 10^{-5}$. This amounts to a population difference of only 1 in 62 500 for protons. The most common strategies to overcome this small polarization are the use of higher magnetic fields or low temperature probes. However, in many applications temperature is not a variable, as for instance in physiological studies.

An alternative approach is to drive the system into a metastable state where the population difference is externally increased, namely hyperpolarization. Different methods have been developed in the last years such as laser polarization (LP) of noble gases *via* optical pumping,[1–5] dynamic nuclear polarization (DNP),[6–8] Chemically Induced DNP (CIDNP),[9,10] or Parahydrogen Induced Polarization (PHIP),[11–14] among others. Hyperpolarization has a particular impact on gas phase NMR, where the low density of gases renders even lower signals as compared to the liquid state. LP noble gases (^3He and ^{129}Xe) are particularly suitable to be used in medical applications, mainly in human lung imaging. Helium is a perfectly inert gas that can be inhaled in large quantities without adverse effects. The solubility in blood is negligible and polarizations up to 70% have been achieved.[15,16] Xenon can be dissolved in liquids, and perfuse to enter the blood mainstream, and can be used to obtain detailed information of the tissue of the lung.[17,18] Many research and clinical applications involve the measurement of changes in relaxation times to probe local amounts of oxygen,[19,20] restricted diffusion for assessment of emphysema[21–23] or asthma,[24] among many others.[25–28] Another outstanding feature of gases is the high diffusion coefficient as compared to liquids, which is particularly useful to probe long distances in porous media.[29,30] By determination of the diffusion coefficient at short and long measurement times the surface-to-volume ratio of the system and the tortuosity, a quantity directly related to the system's transport properties, can be respectively determined,[29,30] and information on nanotubes can be obtained by single file diffusion.[31]

Hyperpolarization with PHIP involves a chemical reaction, where protons originally forming part of the parahydrogen (p-H_2) molecules are deposited into an unsaturated precursor before the NMR signal acquisition, resulting in a product molecule with a specific hyperpolarized site. The hydrogenation reaction can be carried out either at the same high magnetic field where the NMR experiment is performed, widely known as PASADENA (parahydrogen and synthesis allow dramatically enhanced nuclear alignment) protocol,[11] or at low magnetic fields as in the ALTADENA (Adiabatic Longitudinal Transport After Dissociation Engenders Net Alignment) protocol.[12] In this chapter we will restrict the discussion to PASADENA. The main feature of a spectrum acquired under this condition is the antiphase character of the signal, associated with the presence of longitudinal two-spin order terms which are initially present in the density operator once the chemical reactions have ceased.[32,33] As the coupling constants are in the order of a few Hz, even a slight linebroadening will result in partial peak cancellation. The effect of partial cancellation not only diminishes the signal intensity, but also introduces a deformation of a spectrum, where the separation of the antiphase peaks becomes much larger than the actual *J*-coupling values.

Spin echoes are usually used for the refocusing of magnetic field inhomogeneities; however, evolution due to *J*-couplings is not affected by the refocusing 180° pulse. Spin echoes in combination with *J*-coupling delays have been successfully applied for PHIP-MRI[34,35] and chemical reaction monitoring with low-field time domain NMR.[36]

In liquids, the application of a Carr–Purcell–Meiboom–Gill (CPMG) sequence will render a decay that is modulated by the evolution under *J*-couplings, namely, acquisition of a *J*-spectrum.[37] In this chapter we describe the performance of *J*-spectroscopy in PHIP. Two main aspects are considered: on one hand, partial peak cancelling is removed due to the enhanced resolution of a *J*-spectrum, usually in the order of 0.1 Hz; on the other hand, the evolution of the density operators steaming from PHIP under this multipulse sequence differs substantially from operators that arise from thermal polarization. This results in a frequency separation of both types of signals, even in situations where a resonance from a thermally polarized species overlaps with a hyperpolarized one. We refer to this method as parahydrogen discriminated-PHIP or PhD-PHIP.[38,39] The chapter is organized as follows: first the basic aspects of PHIP are reviewed. Then, the relevant aspects of *J*-spectroscopy are summarized and the particular aspects of PhD-PHIP are described. Validation of the method under different experimental situations is presented for liquids. While all the results are limited to the weak coupling regime, usually found at high magnetic fields, it must be noted that frequency separation and partial peak cancelling has also been shown at low and inhomogeneous magnetic fields (0.5 Tesla) in the strong coupling regime,[40] and could readily be applied for the study of gases. Finally, we present simulations for the performance of the method on gas phase NMR. We restrict our discussion to the most commonly used reaction of propylene into propane, which occurs upon hydrogenation with p-H_2. There are two main results that could render this experiment very valuable. One is the possibility to achieve highly resolved spectra for hyperpolarized gases. However, the principal result is the possibility to detect hydrogenation even in much diluted or slowly reacting systems. This can be very valuable in the field of catalyst research, where PHIP has proven to be an ideal probe.[41]

9.2 Parahydrogen Induced Polarization (PHIP)

There are in the literature several articles, reviews and books dealing with the basics of PHIP with more or less detail (see for instance ref. 42–46).

Nevertheless, on behalf of self-consistency, we include here a short summary of PHIP and its more remarkable features.

9.2.1 Brief Description of p-H_2

The hydrogen molecule can be simply described from the quantum mechanical point of view by considering the orbital motion of the electrons, the electron spin, the oscillation of the nuclei, the nuclear rotational state, and the orientation of the nuclear spins. The assumption that the overall wavefunction can be factorized in five independent wavefunctions is made. It can be demonstrated that the first three functions are symmetric with respect to the nuclei.[47] Thus, the symmetry of the hydrogen wavefunctions is given by the symmetry of the rotational and the nuclear spin functions. Only the combination of a symmetric and an anti-symmetric function is consistent with Pauli's principle. In this way, a division between the isomers of hydrogen is formed: parahydrogen, p-H_2, is composed by a combination of the symmetric rotational states and the anti-symmetric nuclear spin states (a singlet, with total spin $S=0$); orthohydrogen, o-H_2, has symmetric nuclear spin functions (a triplet with $S=1$) and anti-symmetric rotational states.[42,47] As transitions between rotational states with different symmetry involve transitions between singlet and triplet nuclear spin states, which are symmetry forbidden, the proportion of ortho- and parahydrogen is quasi-stable independently of the temperature.[42,46,47] However, in 1929 Bonhoffer and Harteck discovered that in the presence of a catalyst, thermodynamic equilibrium is rapidly achieved, providing a simple method for the manipulation of the parahydrogen fraction.[48] At room temperature 25% of the molecules are in the para-state, whereas in the presence of a catalyst (charcoal, for instance) the amount of p-H_2 increases with decreasing temperature. For instance, at 77 K the proportion is 50%. If the gas is removed from the presence of the catalyst and subsequently warmed up, gas enriched in the para-state is obtained. In the absence of paramagnetic impurities, the excess of molecules in parahydrogen will be converted to orthohydrogen very slowly. The lapse of time depends on the properties of the gas reservoir walls and can vary from hours to days. In any case the enriched state is stable enough to perform several NMR experiments.

9.2.2 ALTADENA and PASADENA

In 1985 Weitekamp and co-workers realized that the p-H_2 singlet state could be transformed into observable NMR signal. This inspired the acronym PASADENA. The idea was published in 1986[11] and the experimental demonstration was reported a few months later.[49] Since then, hydrogenations with many catalysts and complexes were reported, helping the rapid growth of the technique (see ref. 44 for a thorough historical view). In 1988, Pravica and Weitekamp introduced a new approach which produces enhanced NMR signal qualitatively different from those obtained in PASADENA experiments. The technique was denominated ALTADENA.[12]

Although the acronyms are helpful in rapidly differentiating between the two types of experiments, the more general acronym PHIP (Parahydrogen Induced Polarization) is customarily used to refer to both experiments.

The fundamental differences between PASADENA and ALTADENA are schematically summarized in Figure 9.1. For the sake of simplicity, let us first limit the analysis to the cases where the former p-H_2 atoms form a weakly coupled two-spin system after the hydrogenation reaction. In PASADENA experiments the reaction and the NMR experiment are both performed in the presence of a high magnetic field. This produces a spectrum consisting of two antiphase doublets originated from the longitudinal two-spin order terms ($\propto I_1^z I_2^z$) present in the density operator right after the hydrogenation. In the case of ALTADENA the chemical reaction is carried out at low magnetic fields (typically earth field or similar) and a physical transport of the sample to the high magnetic field for the NMR experiment is performed. This results in a spectrum displaying two out-of-phase peaks according to the term $\propto I_1^z I_2^z + 1/2(I_1^z - I_2^z)$, the latter arising from the adiabatic transport through different magnetic field strengths.[42,50] In fact, the factor 1/2 is exclusively related to the adiabatic character of the transport, and will change if the sample is moved to the magnet in a non-adiabatic way.

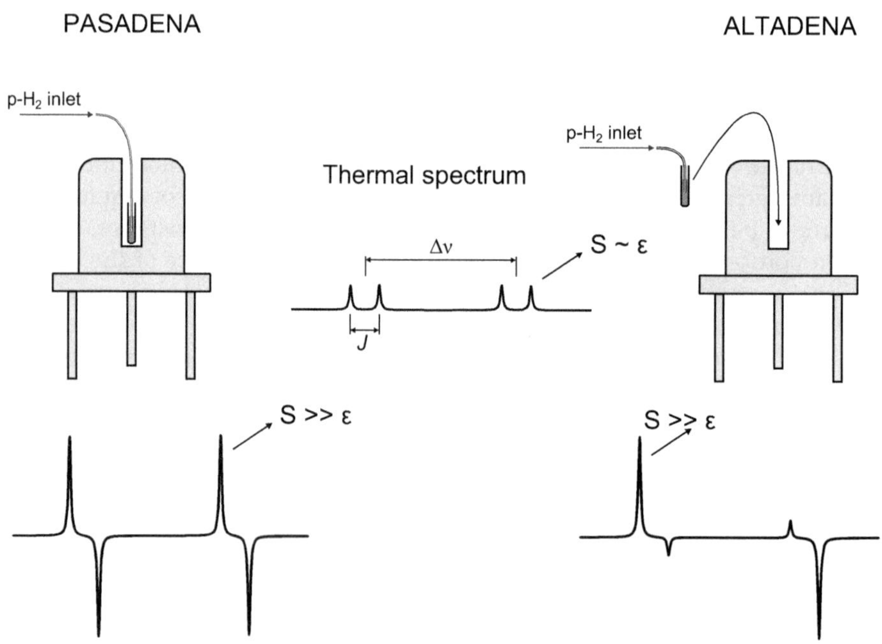

Figure 9.1 Scheme of PHIP where the chemical reaction is performed at different magnetic fields, showing the main characteristics of the antiphase spectra for two weakly interacting spins. PASADENA is carried out at high magnetic fields and the spectrum consists of two antiphase doublets. ALTADENA consists of hydrogenation at low magnetic fields and the spectrum at high fields consists of two out-of-phase peaks.

When the p-H_2 atoms are coupled to another spin within the product molecule, the situation changes. In PASADENA, the hyperpolarization will be circumscribed at the two-spin system, unless one of the former p-H_2 is strongly coupled to a third spin. In the latter case, the hyperpolarization will be partially transferred to the third spin, and two spin-orders between all pairs of spins will contribute to the NMR spectrum. In contrast, as the chemical reaction in ALTADENA is conducted at low magnetic field where all spins are strongly coupled (including heteronuclei as ^{13}C, ^{15}N, *etc.*), after the adiabatic transport the spectrum will display hyperpolarization in all the sites. Moreover, the spectra obtained in that case will be markedly different if the sample's travel to the magnet is non-adiabatic.[46,51]

9.2.3 Hydrogenation

In order to perform a PHIP experiment an exclusive requirement must be fulfilled: both atoms comprising a single p-H_2 molecule must be transferred pairwise to the target molecule. In this way the spin correlation of the singlet is conserved in the product, leading to hyperpolarization. It is mandatory to mention that there are examples when only one hydrogen atom is transferred to a product molecule, showing hyperpolarization,[52,53] but these are considered as exceptions. In general the pairwise addition can be considered as a restrictive condition.[41]

Most of the PHIP experiments concerning spectroscopy and imaging found in the literature are performed using homogeneous catalysts, usually transition metal complexes, and the target molecules are dissolved in a liquid phase.[35,42,43,45,46,54] This is not a minor point, as PHIP applications to clinical MRI are hindered by the difficulty of catalyst separation. One approach to circumvent this obstacle is the use of heterogeneous catalytic reactions,[41] where the possibility of implementing a continuous hyperpolarization scheme by a flow of substrates through a bed reactor is present. However, the most used supported metal catalysts are not compatible with the pairwise addition. Instead, such a reaction is most likely conducted adding somehow randomly hydrogen atoms to the target molecules. Despite this weakness, the feasibility of PHIP experiments in heterogeneous reactions (HET-PHIP) has been largely demonstrated by Koptyug and coworkers.[55–57]

To the best of our knowledge, there are no reports of PHIP experiments to hyperpolarize gases with inhomogeneous catalysis. In contrast, the use of heterogeneous reactions is ideal for gases, thus extending the list of samples that can be hyperpolarized. The first demonstration of HET-PHIP in gas-phase was reported by Koptyug *et al.* back in 2007,[55] those results being successfully extended to the MRI field.[34,58,59] A full description of the advances in HET-PHIP is out of the scope of this chapter; for a detailed chronology see ref. 41 and references therein.

For the particular model sample presented in this chapter, hydrogenation of hexyne into hexene will be considered (Figure 9.2 (A)), resulting in a

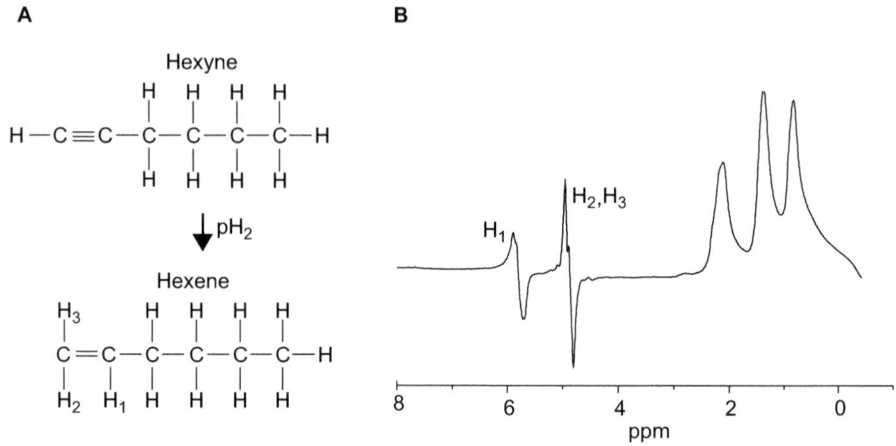

Figure 9.2 (A) Scheme of the hydrogenation of hexyne into hexene where the relevant protons are labelled. (B) Typical NMR spectrum of Hexene obtained with the PASADENA protocol at 7 T.

weakly coupled three-spin system labelled as H_1, H_2, and H_3, where the former are the original p-H_2 protons. The spectrum (Figure 9.2 (B)) shows a typical anti-phase signal of PHIP and no hyperpolarized signal is transferred to the other protons as the experiment is in the PASADENA condition. A 45° rf pulse was used to acquire the FID in order to maximize the PHIP signals. In this case, the reaction was performed inside the magnet and enriched hydrogen was bubbled directly into the sample using a Festo® hose and a needle valve to control the gas flow. The sample was prepared under controlled nitrogen atmosphere in 10 mm NMR tubes and consists of a solution of 0.62 g of hexyne, 5.2 g of acetone-d6, and 0.016 g of the catalyst (rhodium complex: CAS 79255-71-3).

9.3 *J*-Spectroscopy

In the protocols described above it is customary to acquire the signal after a single 45° radiofrequency pulse. As an alternative one could implement a stroboscopic acquisition of the signal during a train of refocusing pulses resembling the CPMG pulse sequence. In favorable circumstances, this technique produces a remarkable enhancement in resolution through the elimination of the effects of magnetic field inhomogeneities at the middle of the pulse separation. In this section we present the theoretical basis behind the acquisition and data processing to render a so-called *J*-spectrum or, more generally speaking, a spin-echo-spectrum.

9.3.1 Theoretical Background

The present discussion is restricted to an isotropic liquid, where the direct dipole–dipole interactions average out and only *J*-couplings are considered.

Additionally, the sample is supposed free of chemical exchange. Under these conditions, the system can be fully described by the isotropic part of the chemical shift and the isotropic *J*-coupling Hamiltonians, written in the rotating frame as:[33,60]

$$\mathscr{H} = 2\pi \sum_{j} \nu_j I_j^z + 2\pi \sum_{j<k} J_{jk} \mathbf{I}_j \cdot \mathbf{I}_k, \tag{9.1}$$

where ν_j represents the chemical shift of the *j*-th spin within a molecule and J_{jk} the coupling constant between the *j*-th and the *k*-th spins, both in units of Hz.

Let us start the analysis with a thorough description of the evolution of an initial density operator $\rho(0)$ during the formation of a single echo. In the following we denote by t_E the *echo time* and by P_X the refocusing pulse (180°) applied in the *x*-direction of the rotating frame, as indicated in Figure 9.3 (this choice of direction is arbitrary and the rest of the analysis is unaffected by this particular selection).

This single echo sequence is composed, therefore, of an initial free evolution period under the Hamiltonian $\mathscr{H}$ of duration $t_E/2$, followed by a 180° pulse and a second free evolution of duration $t_E/2$. With this time-independent Hamiltonian, the density operator at any time *t* is related to the initial density operator by:

$$\rho(t) = \mathscr{U}(t)\rho(0)\mathscr{U}^{\dagger}(t)\,; \quad \mathscr{U}(t) = \exp(-i\mathscr{H}t) \tag{9.2}$$

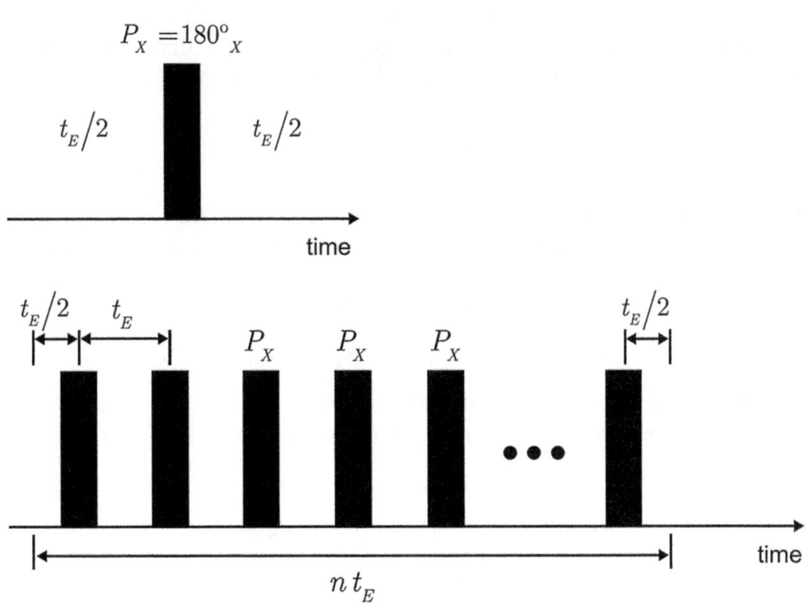

Figure 9.3 Scheme of the pulse sequence with the relevant notation.

where $\mathcal{U}$ is the usual time-evolution operator, or propagator.[33,60] For a single echo we can write:

$$\mathcal{U}(t_E) = \mathcal{U}(t_E/2)P_X\mathcal{U}(t_E/2) \tag{9.3}$$

By inserting an identity at the right-hand side of eqn (9.3) and rearranging terms we get:

$$\mathcal{U}(t_E) = \mathcal{U}(t_E/2)P_X\mathcal{U}(t_E/2)P_{-X}P_X = \mathcal{U}(t_E/2)\tilde{\mathcal{U}}(t_E/2)P_X \tag{9.4}$$

The new operator $\tilde{\mathcal{U}}$ is nothing but the rotated version of the propagator, which results in:

$$\tilde{\mathcal{U}}(t_E/2) = P_X\mathcal{U}(t_E/2)P_{-X} = P_X\exp(-i\mathcal{H}t_E/2)P_{-X} = \exp(-iP_X\mathcal{H}P_{-X}t_E/2) \tag{9.5}$$

Equivalently, we define $\tilde{\mathcal{H}}$ as the rotated version of $\mathcal{H}$,

$$\tilde{\mathcal{H}} = P_X\mathcal{H}P_{-X} = -2\pi\sum_j \nu_j I_j^z + 2\pi\sum_{j<k} J_{jk}\mathbf{I}_j \cdot \mathbf{I}_k \tag{9.6}$$

The effect of the refocusing pulses is the sign inversion of the Hamiltonian terms linear on the spin operators, whereas the terms bilinear on spin operators remain unchanged (the term is scalar and therefore invariant under rotations). Combining eqn. (9.4)–(9.6) yields:

$$\mathcal{U}(t_E) = \exp(-i\mathcal{H}t_E/2)\exp(-i\tilde{\mathcal{H}}t_E/2)P_X \tag{9.7}$$

It is worthy to remark that eqn (9.7) is valid for a general system of N-1/2 spins fulfilling the restrictions enumerated above.

Additionally, if the system is weakly coupled, meaning that any pair of spins satisfies the condition $|\nu_j - \nu_k| \gg |J_{jk}|$, we can go further with the calculations. In such a situation, the Hamiltonian takes the form

$$\mathcal{H}_\mathrm{w} = 2\pi\sum_j \nu_j I_j^z + 2\pi\sum_{j<k} J_{jk} I_j^z I_k^z \tag{9.8}$$

where only the z-part of the J-coupling Hamiltonian (denoted by $\mathcal{H}_\mathrm{w}^J = 2\pi\sum_{j<k} J_{jk} I_j^z I_k^z$), which commutes with the chemical shift Hamiltonian, is retained. The propagator in the case of weak coupling is

$$\mathcal{U}_\mathrm{w}(t_E) = \exp(-i\mathcal{H}_\mathrm{w}t_E/2)\exp(-i\tilde{\mathcal{H}}_\mathrm{w}t_E/2)P_X \tag{9.9}$$

The weak Hamiltonians before and after the rotation commute, $[\mathcal{H}_\mathrm{w}, \tilde{\mathcal{H}}_\mathrm{w}] = 0$, therefore the exponential functions in eqn (9.9) can be unified to yield:

$$\mathcal{U}_\mathrm{w}(t_E) = \exp\{-i(\mathcal{H}_\mathrm{w} + \tilde{\mathcal{H}}_\mathrm{w})t_E/2\}P_X \tag{9.10}$$

Given that $\mathcal{H}_\mathrm{w} + \tilde{\mathcal{H}}_\mathrm{w} = 2\mathcal{H}_\mathrm{w}^J$, we obtain

$$\mathcal{U}_\mathrm{w}(t_E) = \exp(-i\mathcal{H}_\mathrm{w}^J t_E)P_X \equiv \mathcal{U}_\mathrm{w}^J(t_E)P_X \tag{9.11}$$

If a second echo is added to the sequence right after the first one, the corresponding propagator is

$$\mathcal{U}_{\mathrm{w}}(t_E)\mathcal{U}_{\mathrm{w}}(t_E) = \mathcal{U}^J_{\mathrm{w}}(t_E)P_X\mathcal{U}^J_{\mathrm{w}}(t_E)P_X = \mathcal{U}^J_{\mathrm{w}}(2t_E)P_XP_X \tag{9.12}$$

This calculation can be straightforwardly generalized to n-echoes (see Figure 9.2), obtaining

$$\underbrace{\mathcal{U}_{\mathrm{w}}(t_E)\ \ldots\ \mathcal{U}_{\mathrm{w}}(t_E)}_{n\ \text{times}} = \mathcal{U}^J_{w}(nt_E)\underbrace{P_X\ \ldots\ P_X}_{n\ \text{times}} \tag{9.13}$$

In practice, during a CPMG-like sequence,[61] where the exciting pulse is applied 90° shifted from the refocusing pulses, the term corresponding to the pulses can be excluded from the propagator of eqn (9.13). In this example in particular, the excitation pulse will be aligned with the y-direction in the rotating frame, placing the magnetization aligned to the x-direction. Thus, the evolution with $P_X \ldots P_X$ will not affect the state. Therefore, we can consider the propagator to be

$$\underbrace{\mathcal{U}_{\mathrm{w}}(t_E)\ \ldots\ \mathcal{U}_{\mathrm{w}}(t_E)}_{n\ \text{times}} = \mathcal{U}^J_{\mathrm{w}}(nt_E) \tag{9.14}$$

This means that the intensities of the echoes are modulated only by the evolution of a weak J-coupling Hamiltonian. Figure 9.4 shows numerical simulations on a two-spin system in order to exemplify the latter result. When both spins are magnetically equivalent all the echoes have the same shape, their maximum intensities being modulated by an exponential function, as usually observed for instance in water. Even if both spins have different chemical shifts but no J-coupling is present, the intensities collected at the middle of the successive echoes still exhibit a mono-exponential decay, although the echo shapes appear modulated since they are formed by two FIDs face-to-face. In the case of a two-spin system weakly coupled, the train of pulses will form echoes with their maximum intensities modulated by the evolution described in eqn (9.14).

At this point it is worthy to shortly review the historical perspective. The idea of performing a Fourier transform of the data collected at the top of the echoes in a pulse train was first introduced in the 1960s. In 1961, Powles and Hartland[62] reported theoretically and experimentally a method for determining the homonuclear indirect coupling (J-coupling) in liquid samples by a multipulse sequence. They provided closed formulas for the first three echoes and showed that in the case of simple molecules under the condition of weak coupling the result for the n-th echo can be inferred to be a simple evolution with a J-coupling Hamiltonian (as expressed in eqn (9.14)). The authors used the term "collapsed spectrum" for this kind of result, given that in this weak coupling limit, the evolution with chemical shift vanishes and the multiplets associated with the J-coupling Hamiltonians literally collapse at the center of the spectral window.[62]

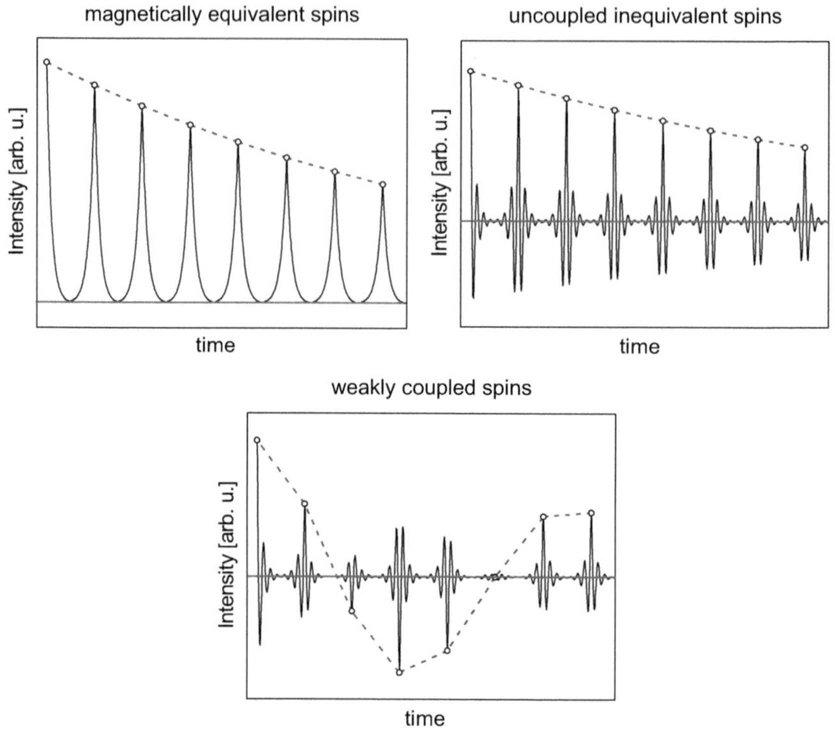

Figure 9.4 Simulated time evolution of the tops of the echoes in a CPMG sequence for magnetically equivalent spins, uncoupled inequivalent spins and weakly coupled spins.

A few years later the results were extended to a variety of larger spin systems with the help of the density operator formalisms.[63,64] In 1971, Ray Freeman and D. Hill[37] made the distinction between "*J*-spectroscopy", limited to the collection of spectra obtained with a CPMG-pulse sequence under certain restrictions, and "spin-echo spectroscopy" which embraces all the other cases, where one or more restrictions are violated. In order to render a *J*-spectrum the system and the spectrometer must fulfil the following conditions: (i) The magnetic field strength (*i.e.* B_0) must be intense enough to ensure the weak coupling condition for every pair of coupled spins, *i.e.* $|\nu_j - \nu_k| \gg |J_{jk}|, \forall_{j,k}$. Otherwise, the top of the echoes will be affected by evolutions with the chemical shift Hamiltonian; (ii) the echo time must be sufficiently long compared to the inverse of the smallest chemical shift difference within the sample, *i.e.* $t_E \gg (\nu_j - \nu_k)^{-1}, \forall_{j,k}$; deviation from this condition shifts the frequencies in the spectrum; (iii) the 180° pulses must be carefully adjusted; pulses deviated from this condition will produce fake resonance lines in the spectrum; (iv) no extra physical mechanism able to produce the echo modulations must be present; this includes direct dipole–dipole interactions (usually averaged out in isotropic liquids), chemical exchange, *etc.*

When the four conditions are fulfilled simultaneously, the absence of chemical shift modulation at the top of the echoes produces a spectrum centered at zero frequency with only *J*-coupling information. Figure 9.5 (A)–(B) shows numerical simulations included to highlight the similarities and differences between the well-known NMR-spectroscopy and *J*-spectroscopy. The calculations were performed in a two-spin system with the following

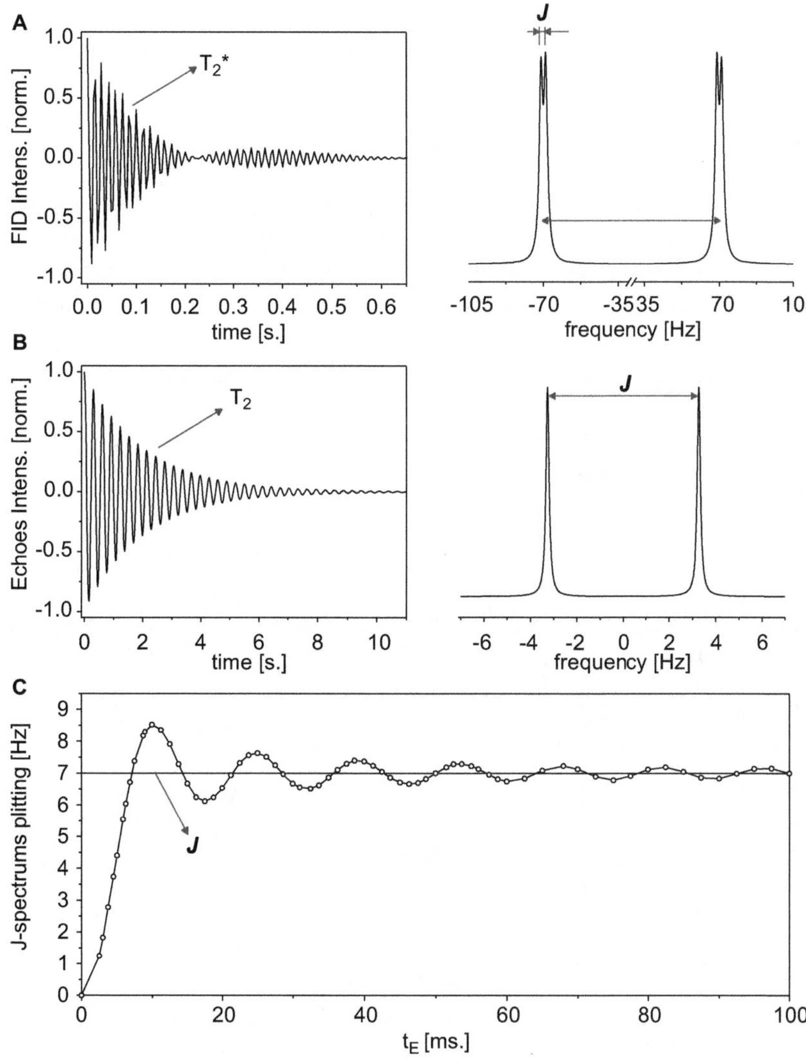

Figure 9.5 (A) Simulated time evolution for a weakly coupled spin pair after a single pulse with the resulting spectrum. Chemical shift and *J*-coupling splittings are observed. (B) Evolution of the signal acquired on the echo centres of a CPMG sequence. The spectrum shows the *J*-coupling information. (C) Frequency splitting in the *J*-spectra simulated for different echo times. Only for long echo times can the true *J*-coupling constant be obtained.

parameters: $B_0 = 7$ T, $\Delta\nu = 140$ Hz, $J = 7$ Hz, $t_E = 20$ ms, a magnetic field inhomogeneity corresponding to a 2 Hz line width (*i.e.* $T_2^* \sim 160$ ms) and $T_2 = 2$ s. The difference between T_2 and T_2^*, usually of one order of magnitude or more in isotropic liquids, is responsible for the resolution enhancement, as observed in Figure 9.4 (the resolution in the *J*-spectrum is $(\pi T_2)^{-1} = 0.16$ Hz). In Figure 9.5 (C), the peak difference in the *J*-spectrum is plotted against the echo time to illustrate the consequences of failing to fulfil condition (ii). An oscillation around $J = 7$ Hz is present, and the separation between peaks converges to the real value for long echo times.

As stated above, the more general case is the spin-echo spectroscopy. Perhaps the most likely situation consists in the violation of condition (i), *e.g.* when at least one pair of nuclei in the molecule are not in the weak coupled regime. The frequencies observed in the spin-echo spectrum will no longer be associated only with linear combinations of the *J*-coupling constants. These kinds of spectra also present a strong dependence on the echo time. To illustrate the case, we present numerical simulations performed on a three-spin system in the weak and strong coupling regime (see Figure 9.6).

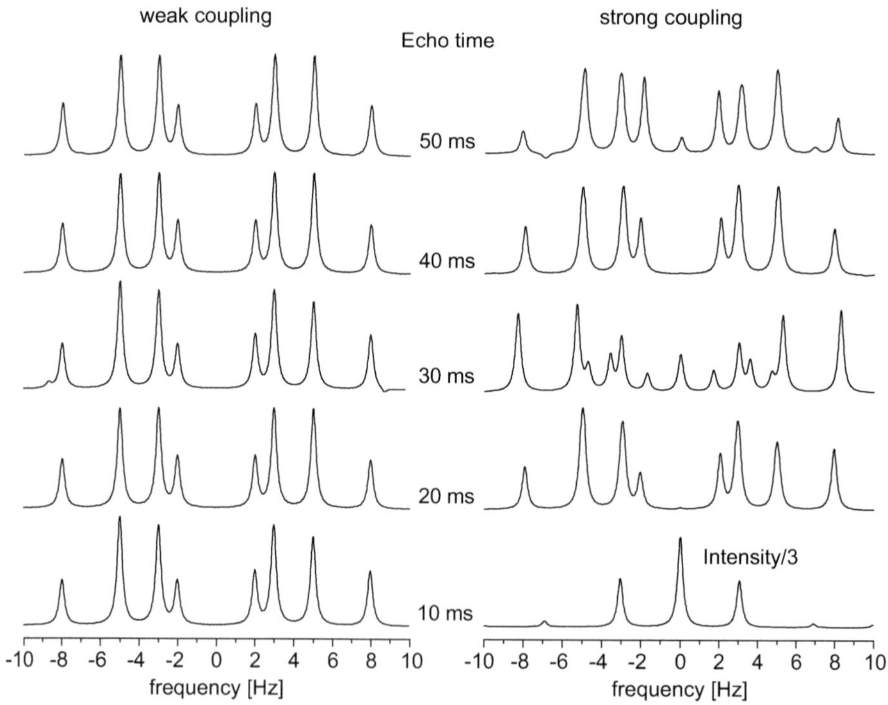

Figure 9.6 Numerical simulations for a three-spin system in the weak (left) and strong (right) coupling regime with the following parameters: $J_{12} = 10$ Hz, $J_{13} = 6$ Hz, $J_{23} = 0$ Hz, and magnetic fields that satisfy $\Delta\nu_{12}/J_{12} = 100$ (weak) and $\Delta\nu_{12}/J_{12} = 5$ (strong). Under strong coupling the spectra strongly depend on the echo time. All spectra are normalized to their individual intensity to enable comparison.

Labelling the spins from 1–3, the parameters used in the simulations were: $J_{12} = 10$ Hz, $J_{13} = 6$ Hz, $J_{23} = 0$ Hz, and two magnetic fields were assumed to induce chemical shifts that satisfy $\Delta\nu_{12}/J_{12} = 100$ and $\Delta\nu_{12}/J_{12} = 5$, for the weak and strong coupling, respectively. From the figure it can be clearly observed that while the *J*-spectra are echo time independent in the range covered in the calculations, the spin-echo spectra shapes substantially change and different frequency lines appear associated with the modulations produced by the chemical shifts.

9.3.2 Partial *J*-Spectra

Either in weak or strong coupling cases, when dealing with a multi-spin system, a *J*-spectrum can be difficult to interpret, mainly due to the interference of several multiplets within a small frequency range. In the simulated three-spin system with only two non-zero couplings already eight resonant lines appear in a range of 20 Hz in the *J*-spectrum whereas the spin-echo spectra consist of a number of lines ranging from 3 to 13 depending on the echo time. A method to simplify the analysis was proposed by Ray Freeman and D. Hill in the same article,[37] consisting of the acquisition of the echoes maxima with a digital filter, centered at the chemical shift of the desired multiplet. Given an *N*-spin system it is possible to perform *N* independent experiments with different digital filters and render *N* simplified subspectra. The authors proposed the names partial *J*-spectrum or partial spin-echo spectrum for every one of those subspectra.

The method can be easily understood with the help of Figure 9.7. In the left panel the ethanol molecule is displayed along with a simulated NMR spectrum with enough resolution to clearly discern the multiplets. The rest of the figure shows simulated and experimental *J*-spectra where the triplet and quartet collapsed at zero frequency (central panel) and partial *J*-spectra with the digital filter centered at both multiplets individually. The experimental *J*-spectrum presents a discrepancy in the quartet intensities coming from off-resonance effects because the rf pulses were applied on-resonance on the triplet. As two individual experiments are carried out to acquire the partial *J*-spectra, these can be performed on-resonance, thus the intensities of the experimental data match the simulated ones.

9.3.3 Technical Considerations

Liquid state NMR is usually carried out in 5 mm OD NMR tubes, in which the sample length exceeds the dimension of the sensitive coil volume. In this way, magnetic field distortions due to changes in the magnetic susceptibility that arise from water/air interfaces are removed. As an alternative, Shigemi plugs to match susceptibilities may be used. In the case of *J*-spectroscopy, phase accumulation due to magnetic field inhomogeneities is removed at the echo centers. Here the key role is the homogeneity of the rf field, which is necessary in order to achieve 180° rotations in the magnetization on each

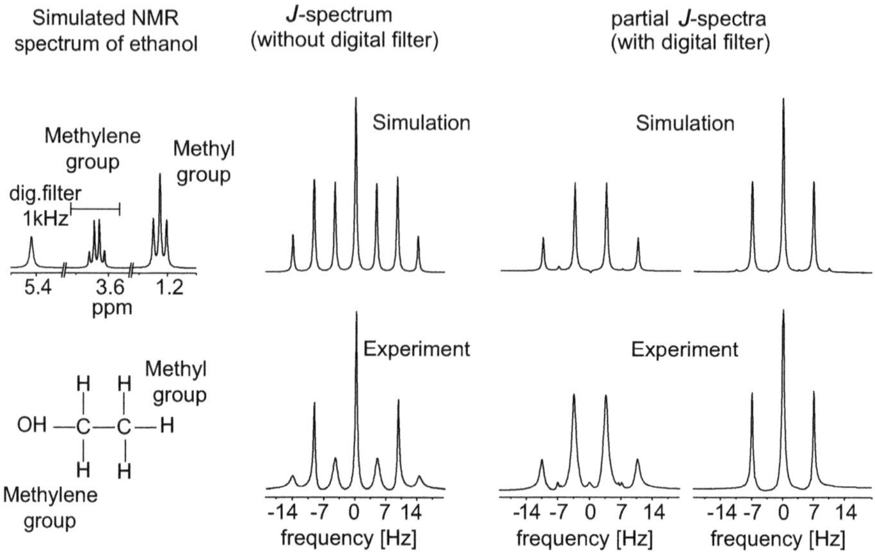

Figure 9.7 (left) Simulated high-resolution spectrum of ethanol and sketch of the molecule. (centre) Simulated and experimental *J*-spectra, where all resonances collapse to zero frequency. The discrepancy of the intensities in the experimental data is due to off-resonance effects in the 180° pulses in the methyl group. (right) Partial *J*-spectra, two on-resonance individual *J*-spectra are acquired for each group.
Adapted with permission from ref. 38. Copyright 2013 Elsevier.

echo. The results shown in all the experiments presented in this chapter were obtained by using 10 mm OD NMR sample tubes with a sample length of half of the birdcage radiofrequency coil. A standard 10 mm liquid probe from Bruker was used in all the experiments. Additionally, the phase cycling $yy\bar{y}\bar{y}$ was used for the 180° pulses in order to reduce stimulated echo artifacts on the echo train decay.[38,65,66]

The acquisition of partial *J*-spectra was carried out by using the digital filters built into a Bruker Avance II 300 console. Most modern NMR equipment is designed to perform a digital data acquisition with an oversampling rate. Depending on the manufacturer and spectrometer version, acquisition is continuous or stroboscopic. In any case, between the command to start acquisition and the real data sampling, most equipment acquires a set of data points which are used to calculate the digital filter. In general narrow filters are needed, for instance we used 1 kHz filter widths, corresponding to an acquisition time, or dwell time, of dw = 250 μs. Particular care must be taken in order to ensure that the top of each echo is correctly sampled in this condition. Here we chose to start acquisition after the 90° pulse in order to compensate for the filter calculation time only once during the pulse sequence. The receiver was then blanked during the 180° pulses and unblanked to record the signal. In this way the only condition that must be fulfilled is that the quantity $t_E/4dw$ should be an integer.

9.4 *J*-Spectroscopy in PHIP (PhD-PHIP)

A method aimed to remove magnetic field inhomogeneities and thus enhancing the spectral resolution is always welcome. This gains more relevance in PHIP, as the antiphase signals are particularly vulnerable through the possible partial peak cancellation (see Figure 9.1). The decrease of signal is particularly pronounced in PASADENA experiments, as the splitting of peaks with opposite intensities is proportional to the homonuclear *J*-coupling constants, typically of few Hz.[38,67] In ALTADENA, on the other hand, the separation is rather proportional to the chemical shift differences. Given this distinction, in the rest of the chapter we will be dealing with PASADENA.

Inhomogeneities might result from an inhomogeneous B_0 itself or be induced by the presence of elements within the probe that perturb the polarizing field. This is a common scenario in PASADENA experiments, where a gas delivery setup (either bubbling or through permeable membranes) is present. The use of a CPMG to acquire highly resolved multiplets opens the possibility for the detection of PHIP signals in a broader variety of situations, including special setups or complex samples.

Signals that give rise to antiphase or inphase spectra evolve markedly differently during a train of refocusing pulses. As explained in this section, the difference between both spectra is a shift of half of the spectral width. This provides a mechanism to discriminate between thermal and hyperpolarized contributions in the resulting *J*-spectrum, which is denoted by the acronym PhD-PHIP (Parahydrogen Discriminated-PHIP).[39] In what follows we give an insight into the theoretical basis of the method, including experimental demonstrations.

9.4.1 Theoretical Basis

In order to present a complete description of the mechanism behind the signal separation in PhD-PHIP, some assumptions are made. For simplicity, and with no loss of generality, we base the analysis on an isolated two-spin system in the weak coupling regime. We consider the evolutions of density operators, under the product operator formalism, corresponding to a thermally polarized sample as well as of an operator right after a hydrogenation with parahydrogen in a high magnetic field. The pulse sequence consists of an exciting 45°_{y} pulse followed by a train of 180°_{x} pulses separated by t_E.

For thermally polarized spins, the initial density operator in the high field approximation is given by:[33]

$$\rho_{Th}(0) \propto (I_1^z + I_2^z), \tag{9.15}$$

which can be expressed as:

$$\rho_{Th}(0^+) \propto (I_1^x + I_2^x) \tag{9.16}$$

after the application of the excitation pulse. At the top of the echoes only evolutions with the truncated *J*-coupling Hamiltonian, $\mathscr{H}^J_w = 2\pi J_{12} I_1^z I_2^z$, are present. The state of the system immediately before the application of the first 180° pulse is:

$$\rho_{Th}(t_E/2) \propto (I_1^x + I_2^x)\cos(\pi J_{12} t_E/2) + (2I_1^y I_2^z + 2I_1^z I_2^y)\sin(\pi J_{12} t_E/2) \quad (9.17)$$

The first term commutes with a 180°_x pulse whereas the term $(2I_1^y I_2^z + 2I_1^z I_2^y)$ changes its sign twice (*i.e.* I_k^z and I_j^y invert their signs simultaneously leaving an unchanged product). Therefore, $\rho_{Th}(t_E/2)$ is unaltered by the refocusing pulse. The subsequent evolution yields a state at the echo center equal to:

$$\rho_{Th}(t_E) \propto (I_1^x + I_2^x)\cos(\pi J_{12} t_E) + (2I_1^y I_2^z + 2I_1^z I_2^y)\sin(\pi J_{12} t_E), \quad (9.18)$$

which produces the following detectable magnetization:

$$M_{Th}(t_E) \propto \cos(\pi J_{12} t_E). \quad (9.19)$$

Following the same reasoning, the density operator at the top of the second echo results in:

$$\rho_{Th}(2t_E) \propto (I_1^x + I_2^x)\cos(\pi J_{12} 2t_E) + (2I_1^y I_2^z + 2I_1^z I_2^y)\sin(\pi J_{12} 2t_E), \quad (9.20)$$

with its corresponding signal

$$M_{Th}(2t_E) \propto \cos(\pi J_{12} 2t_E). \quad (9.21)$$

Extending the calculations to the *n*-th echo, we obtain

$$M_{Th}(nt_E) \propto \cos(\pi J_{12} n t_E), \quad (9.22)$$

i.e. the real part of the top of the echoes is modulated by a cosine function. This is depicted in Figure 9.8 (gray vectors), where the evolutions are schematized based on the rules of the product operators formalisms (for further details see ref. 33, 60, 68).

In the case of PHIP under PASADENA conditions, where the former parahydrogen atoms are transferred to a molecule forming an AX system, the density operator results in:

$$\rho_{Ph}(0) \propto 2I_1^z I_2^z. \quad (9.23)$$

After the 45°_y rf pulse, we have

$$\rho_{Ph}(0^+) \propto (2I_1^x I_2^z + 2I_1^z I_2^x). \quad (9.24)$$

The first evolution period under the weak coupling Hamiltonian produces

$$\rho_{Ph}(t_E/2) \propto (I_1^y + I_2^y)\sin(\pi J_{12} t_E/2) + (2I_1^x I_2^z + 2I_1^z I_2^x)\cos(\pi J_{12} t_E/2). \quad (9.25)$$

At this point the difference between thermally polarized spins and PHIP becomes clear: the sign of both terms in eqn (9.25) change under the action of a 180°_x pulse. The first term, linear on spin operators, changes its sign as usual. The second term, bilinear, also changes because I_k^z changes and I_j^x commutes with the pulse, yielding the product changed (see black vectors in Figure 9.8). Thus, right after the pulse we have,

$$\rho_{Ph}(t_E/2^+) \propto -(I_1^y + I_2^y)\sin(\pi J_{12} t_E/2) - (2I_1^x I_2^z + 2I_1^z I_2^x)\cos(\pi J_{12} t_E/2) \quad (9.26)$$

which further evolves into

$$\rho_{Ph}(t_E) \propto -(I_1^y + I_2^y)\sin(\pi J_{12}t_E) - (2I_1^x I_2^z + 2I_1^z I_2^x)\cos(\pi J_{12}t_E), \quad (9.27)$$

at the position of the first echo. The echo amplitude is,

$$M_{Ph}(t_E) \propto -i\sin(\pi J_{12}t_E). \quad (9.28)$$

Notice that the echo will be formed 90° shifted compared to the thermal case (*i.e.* in the imaginary channel), as usually encountered in single pulse PHIP experiments.[42,46] After the following evolution with $\mathscr{H}_{\mathrm{w}}^{J}$, the density operator prior to the application of the second 180° pulse is:

$$\rho_{Ph}(3t_E/2) \propto -(I_1^y + I_2^y)\sin(\pi J_{12}3t_E/2) - (2I_1^x I_2^z + 2I_1^z I_2^x)\cos(\pi J_{12}3t_E/2) \quad (9.29)$$

Once again, the refocusing pulse changes the global sign of the density operator (see point 5 in Figure 9.8),

$$\rho_{Ph}(3t_E/2^+) \propto (I_1^y + I_2^y)\sin(\pi J_{12}3t_E/2) + (2I_1^x I_2^z + 2I_1^z I_2^x)\cos(\pi J_{12}3t_E/2) \quad (9.30)$$

At the position of the second echo, we then have

$$\rho_{Ph}(2t_E) \propto (I_1^y + I_2^y)\sin(\pi J_{12}2t_E) + (2I_1^x I_2^z + 2I_1^z I_2^x)\cos(\pi J_{12}2t_E), \quad (9.31)$$

corresponding to a signal intensity

$$M_{Ph}(2t_E) \propto +i\sin(\pi J_{12}2t_E). \quad (9.32)$$

Based on these calculations, the intensity of the *n*-th echo will be

$$M_{Ph}(nt_E) \propto i(-1)^n \sin(\pi J_{12}2t_E). \quad (9.33)$$

This imposes a sign alternation of the successive echoes, in addition to the modulation due to the *J*-coupling. The differences between thermal and PHIP become evident in the numerical simulations of Figure 9.9 (A), where the echo intensities for a CPMG train applied to an AX two-spin system are plotted against time. The odd-even effect on alternating echoes is modulated by the *J*-coupling evolution.

Digitalization of time domain signals renders a string of n numbers, representing a total acquisition time of $T = ndw$. In *J*-spectroscopy the dwell time is given by the echo time t_E. When the Fast Fourier Algorithm is applied to such a string, another string of length n is obtained (if zero filling is not considered), representing frequencies, with the maximum value being $\nu_M = 1/t_E$ and spectral resolution $1/T$.[69] In the following lines, the detailed calculations concerning the FFT of discrete signals originated from a train of echoes are outlined, in order to show the differences in the frequency domain.

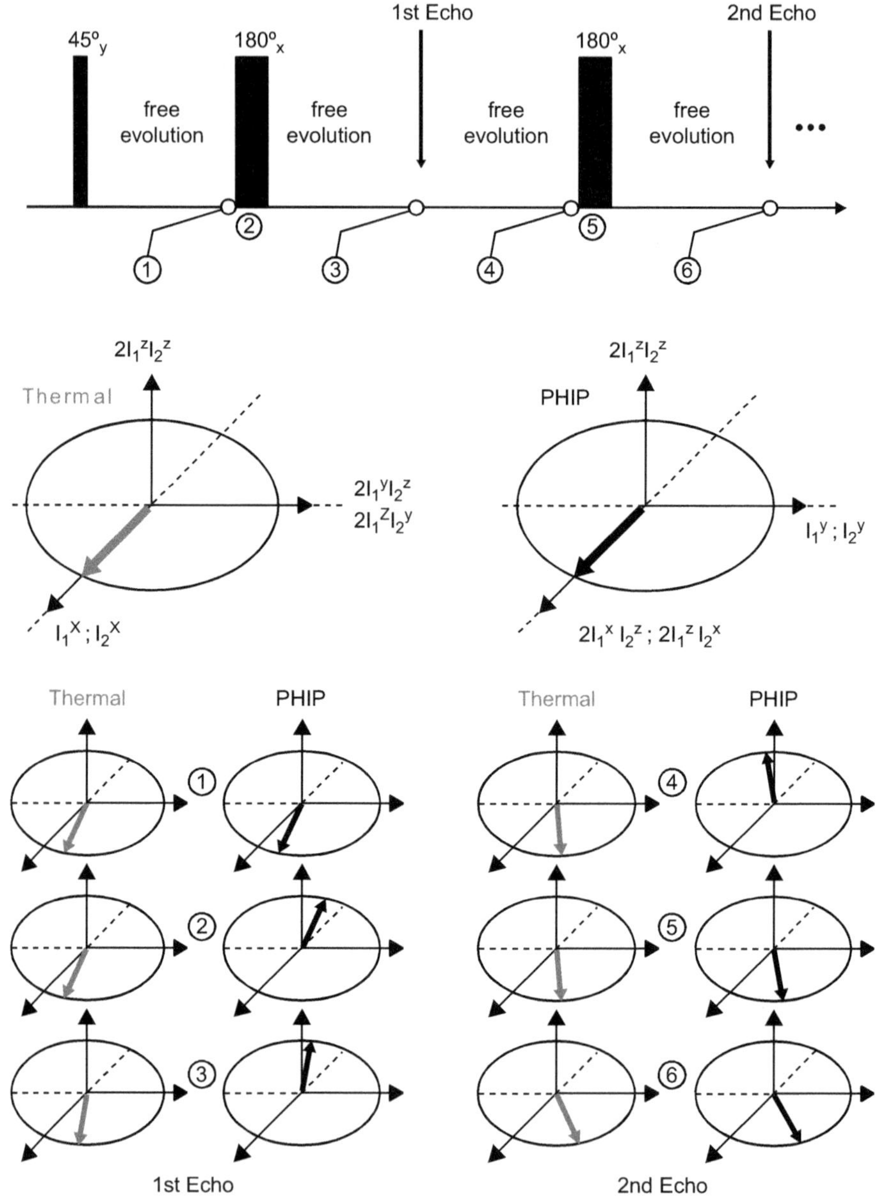

Figure 9.8 (Top) Pulse sequence and considered time events. (Center) Subspaces spanned by the product of the spin operators corresponding to scalar couplings; (Bottom) Evolution of a single spin operator during the formation of two consecutive echoes, where the phase alternation on PHIP signals is depicted.
Adapted with permission from ref. 39. Copyright 2013 American Chemical Society.

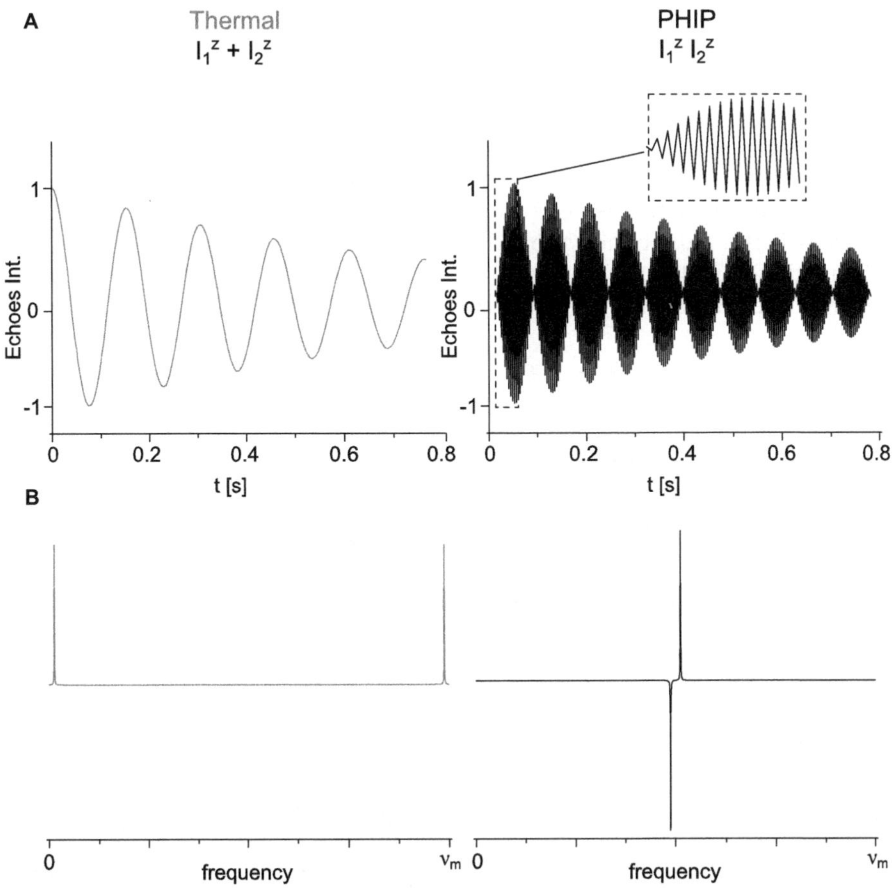

Figure 9.9 (A) Simulated evolution of the real part of the thermal and PHIP signals at the top of the echoes during a CPMG sequence. (B) Spectra obtained without a frequency shift on the FFT.
Reprinted with permission from ref. 39. Copyright 2013 American Chemical Society.

Let us assume a continuous signal $f(t)$ to be discretized in equally spaced points, with separation t_E. With the help of a "Dirac comb" function, the discrete version of $f(t)$ is given by

$$f_D(t) = f(t) \sum_{j=0}^{\infty} \delta(t - jt_E), \tag{9.34}$$

where δ denotes the Dirac delta function. The Fourier transform of this discrete function is

$$f_D(\nu) = F(\nu) \odot \sum_{j=0}^{\infty} \delta\left(\nu - \frac{j}{t_E}\right) = \sum_{j=0}^{\infty} F\left(\nu - \frac{j}{t_E}\right), \tag{9.35}$$

where $\odot$ denotes a convolution. The frequency response consists of a series of replicates separated by $1/t_E$ (*i.e.* the spectral width). Therefore the echo time must be carefully set in order to prevent aliasing. A common mathematical operation included in the software of most NMR spectrometers is a shift of the spectrum by $n/2$ points and reassignment of the frequencies to range from $-\nu_M/2$ to $+\nu_M/2-1$. Once such a function is applied to signal from thermal spins, a *J*-spectrum as shown in Figure 9.5 (B) is obtained, whereas if the shift is not applied the *J*-spectrum appears as shown in the left-hand side of Figure 9.9 (B).

On the other hand, when dealing with PHIP signals we can write

$$f_D(t) = f(t)\sum_{j=0}^{\infty}(-1)^j\delta(t - jt_E), \tag{9.36}$$

which can be split in two odd and even terms as:

$$f_D(t) = f(t)\sum_{k=0}^{\infty}\delta(t - 2kt_E) - f(t)\sum_{k=0}^{\infty}\delta(t - (2k+1)t_E). \tag{9.37}$$

Defining $\tau = t - t_E$ the above expression can be rewritten as

$$f_D(t) = f(t)\sum_{k=0}^{\infty}\delta(t - 2kt_E) - f(\tau + t_E)\sum_{k=0}^{\infty}\delta(\tau - 2kt_E). \tag{9.38}$$

Invoking the Shift Theorem (*i.e.* $F[f(t+t_0)] = F(\nu)e^{-i\nu t_0}$), eqn (9.38) becomes

$$F_D(\nu) = \sum_{k=0}^{\infty}F\left(\nu - \frac{k}{2t_E}\right) - e^{-i\nu t_E}\sum_{k=0}^{\infty}F\left(\nu - \frac{k}{2t_E}\right) = (1 - e^{-i\nu t_E})\sum_{k=0}^{\infty}F\left(\nu - \frac{k}{2t_E}\right). \tag{9.39}$$

Given that $(k/2t_E) = (k/t_E) - (k/2t_E)$ we arrive at the final form

$$F_D(\nu) = (1 - e^{-i\nu t_E})\sum_{k=0}^{\infty}F\left(\left(\nu + \frac{k}{2t_E}\right) - \frac{k}{t_E}\right). \tag{9.40}$$

Comparing eqn (9.35) and (9.40) it becomes clear that, besides a phase factor, both expressions differ in a frequency shift by an amount $1/2t_E$, *i.e.* half of the spectral width. This influences the result of the final spectrum as reflected in the simulations of Figure 9.9 (B), where the spectral lines appear at the center of the *J*-spectrum in PHIP, whereas thermal contributions are placed in the borders. This natural separation provides a mechanism to distinguish between signals originated from PHIP or thermally polarized spins and is the basis of PhD-PHIP.

9.4.2 Experimental Results

The performance of PhD-PHIP is not at all restricted to AX spin systems, and the idea can be extended to larger spin systems so far as the

J-spectroscopy conditions are considered, although a theoretical description is somehow cumbersome. To experimentally validate the method, we have used the hydrogenation of 1-hexyne into 1-hexene, shown in Figure 9.10 (A), where only the more relevant hydrogen atoms are labelled. In order to experimentally simulate a condition where a thermal signal cancels an antiphase peak, a few drops of CH_2Cl_2 were incorporated into the sample. In this way a large thermally polarized peak appears near the chemical shift of the site occupied for one of the former p-H_2 protons, H_a, in the product molecule as shown in Figure 9.10 (B). The horizontal line shows the width and central position of the digital filter to be used in the PhD-PHIP. A partial-*J*-spectrum of H_a acquired with $t_E = 8$ ms is shown in Figure 9.10 (C), where the frequency separation is clearly observed.

By comparing this *J*-spectrum with one acquired in a pure sample it can be observed that, regardless of an artifact in the center of the spectral windows, both results are in good agreement (see Figure 9.11 (A) and (B)).[39] At this point it must be clarified that these spectra correspond to single shot acquisitions, which is the most common experiment carried out when dealing with the hyperpolarized samples.

One of the potentials of the method is in the study of on-going chemical reactions. In this case a competition between hyperpolarized signals with thermal ones originated from molecules that have already reacted will unavoidably be present. This situation is shown in Figure 9.12 for a composed solution of 0.05 g of 1-hexyne (reactant), 0.1 g of hexane (product), 1.3 g of acetone-d_6, and 0.01 g of rhodium catalyst (CAS 79255-71-3). Prior to the incorporation of p-H_2 to the system, the spectrum shown in Figure 9.12 (A) is obtained, where the digital filter window used for the acquisition of the *J*-spectrum is sketched. On the right column, the thermal signals for H_a appear on the borders of the spectral window. After the fifth bubbling-detection cycle, the antiphase signal of Figure 9.12 (B) is obtained, where the negative antiphase peak is clearly distorted by the presence of thermal signals corresponding to unreacted educt molecules and product molecules from former hydrogenations that have already relaxed to a thermal state. The *J*-spectrum from the right panel shows that the antiphase signals are symmetric, as all thermal contributions move to the borders of the frequency window. In Figure 9.12 (C) the spectrum obtained after 20 bubbling-detection cycles is shown. In the spectrum (left panel), the antiphase character of the PHIP signals is barely observed and the signal has noticeably decreased due to sample evaporation. As the length of the sample has changed, the homogeneity of the magnetic field varied from the initial condition due to magnetic susceptibility changes resulting in a broader spectrum. However, the *J*-spectrum, even if less intense, is clearly undistorted with respect to the initial experiment. This simple experiment shows that PhD-PHIP is able to detect hydrogenations even in very diluted systems or in situations where very low polarization is achieved.

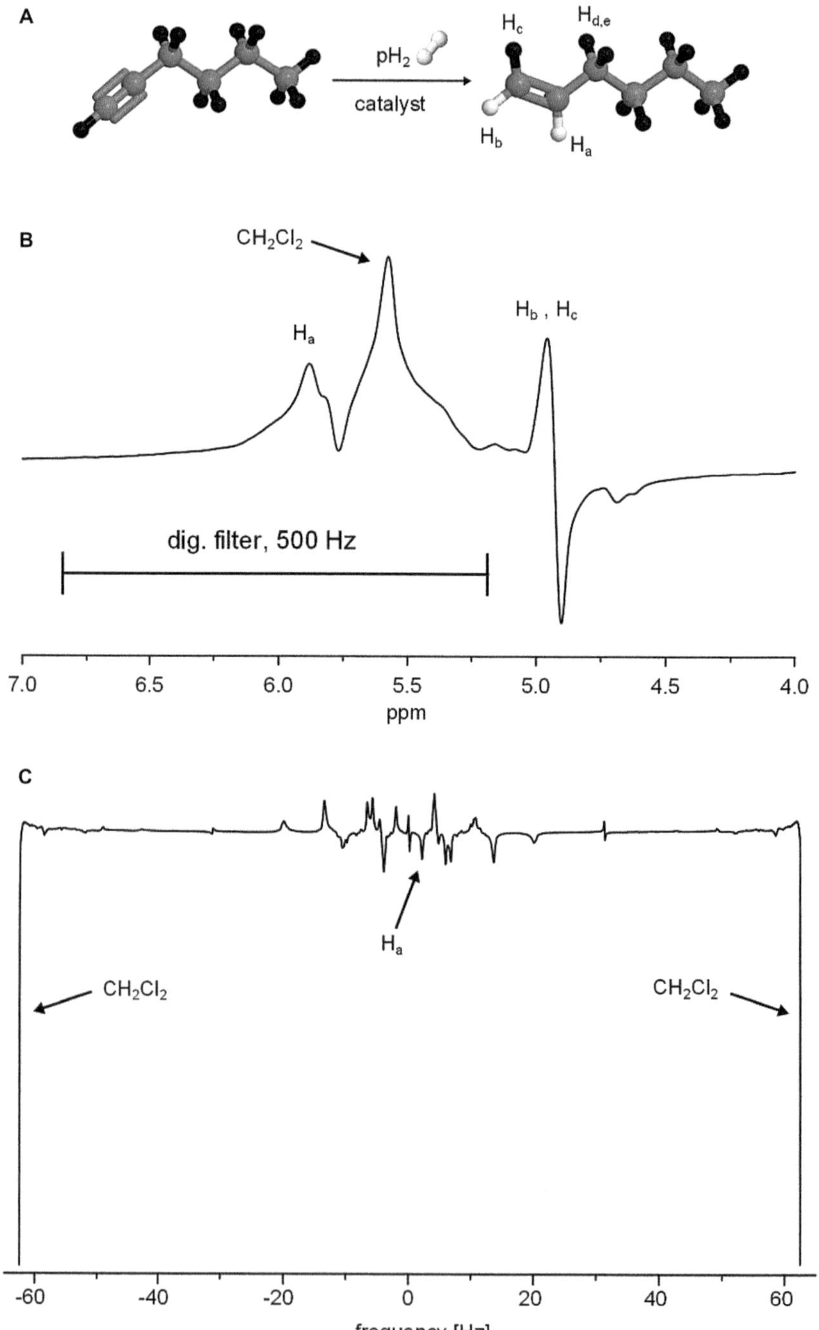

Figure 9.10 (A) Scheme of pairwise hydrogenation of hexene into hexyne. (B) NMR spectrum for a 45° excitation pulse with the presence of CH_2Cl_2. (C) *J*-spectrum acquired with a 500 Hz digital filter.
Reprinted with permission from ref. 39. Copyright 2013 American Chemical Society.

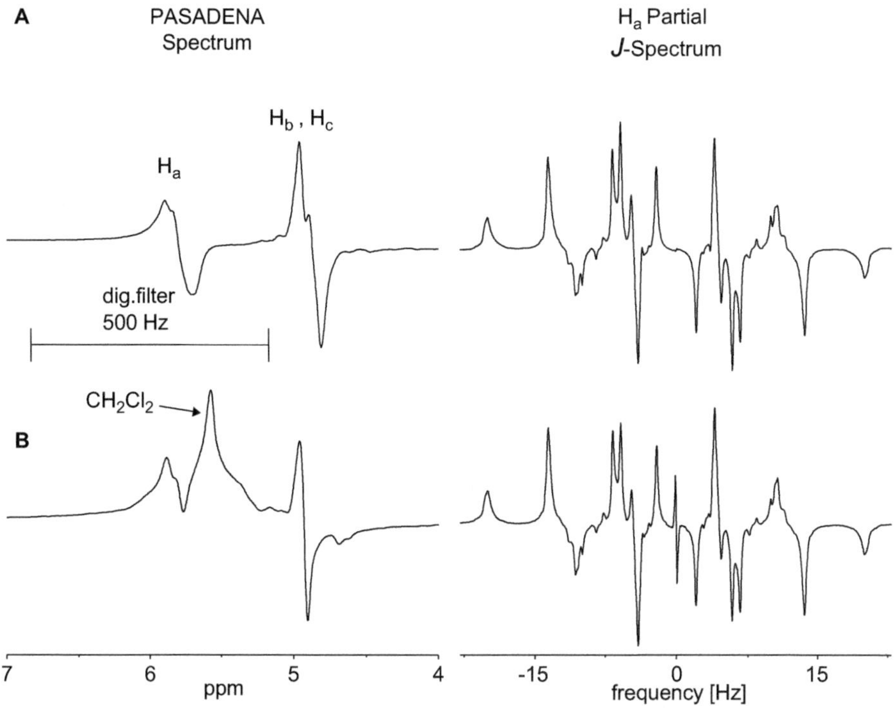

Figure 9.11 Comparison of the relevant spectral region of hyperpolarized (A) hexene and (B) hexene/CH_2Cl_2 mixture. (Left) ^1H spectra after a 45° excitation pulse, (right) PhD-PHIP spectra.
Reprinted with permission from ref. 39. Copyright 2013 American Chemical Society.

9.5 PhD-PHIP in Gases

We now turn our attention to analyzing the possibility of applying PhD-PHIP for the study of hyperpolarized gases. Simulations are shown for the most popular case of hyperpolarized propane gas or HP propane.[55] Simulations of propane spectra with thermal and PHIP initial density operators using all the spins in the product molecule at 7 Tesla are considered. In this magnetic field, the difference in chemical shift is much higher than the *J*-coupling between them, that is, the weak coupling limit can be considered.

The typical chemical reaction with a H_2 molecule is shown in Figure 9.13 (A); the double bond is reduced and the former H_2 atoms become part of a methylene group (H_{2a}) and of a methyl group (H_{1X}) in the propane molecule. Simulations of a propane spectrum with a line width of Lw = 1 Hz are shown in Figure 9.13 (B); here all the couplings in the molecule can be observed: seven peaks for H_{2A} and three for H_{1X}. Reactions with normal hydrogen and parahydrogen are presented where each spectrum is normalized to its own FID. In the case of PHIP in the PASADENA protocol, the characteristic

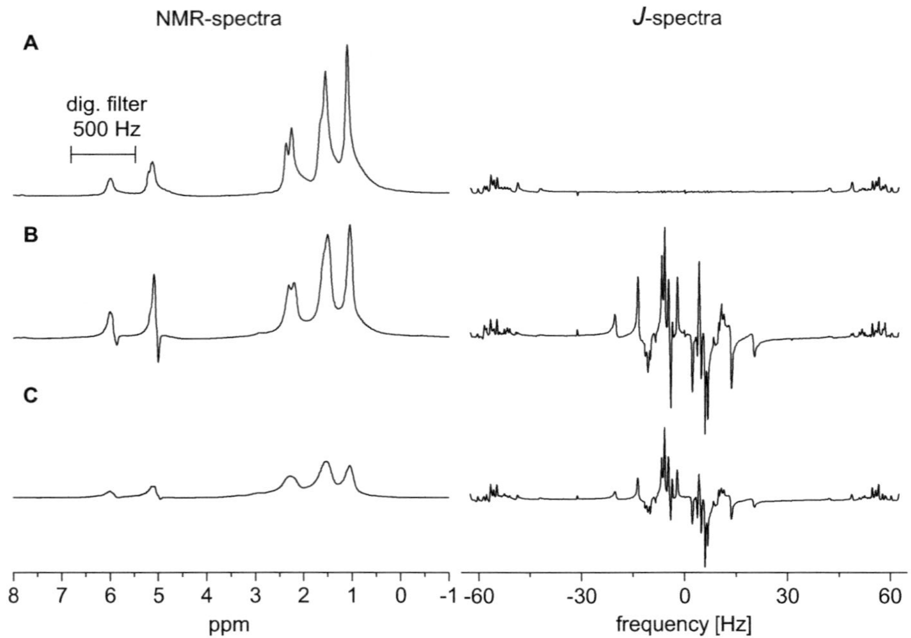

Figure 9.12 Left column: ^{1}H spectra after a 45° excitation pulse. Right column: Partial *J*-spectra. (A) Spectra obtained for 66% hexene and 33% hexyne. (B) Spectra after hydrogenation with *p*-H_2. (C) Spectra obtained after 20 bubbling/measurement cycles. A considerable amount of sample has evaporated, changing the shimming of the sample. Even though most of hexyne has reacted increasing the proportion of thermal hexene, the low amount of hyperpolarized samples is clearly observed with PhD-PHIP.
Reprinted with permission from ref. 76. Copyright 2013 Elsevier.

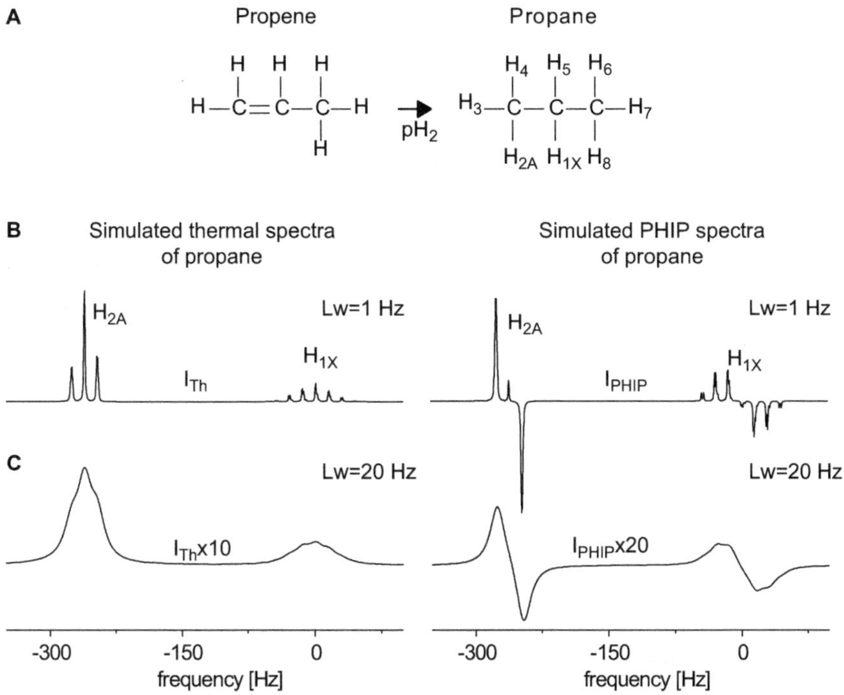

Figure 9.13 (A) Scheme of the pairwise hydrogenation of propene into propane. (B) Simulated spectra for a thermally polarized and PHIP sample of propane with high resolution. (C) Influence of line broadening in the spectra; in the case of PHIP partial peak cancelling is observed.

antiphase signal is obtained. In gas phase NMR, diffusion plays an important role in the line width of the spectrum as even for high magnetic field homogeneity, the long distances that a molecule can diffuse during the signal acquisition will render a phase accumulation over a significant volume. In the experimental results in the literature, line widths of *ca.* 20 Hz are observed.[34,41,55,59,70] In order to simulate this experimental condition, a line broadening was introduced in the simulated data; Figure 9.13 (C) shows the effect of the line broadening: the loss of the *J*-coupling information is accompanied by a reduction of the spectrum intensity, which is of course more pronounced for PHIP.[38]

In the previous section we showed that the use of PhD-PHIP renders highly resolved multiplets even with poor magnetic field homogeneity or very low reaction rates. An extreme case was presented for ethanol, where the shimming of the system was systematically changed and the partial *J*-spectra resulted unmodified.[38] This is mainly due to the refocusing of phase accumulation due to local gradients by the 180° pulses. In the case of gases particular care must be taken due to the high diffusion coefficients, where the echo can be completely suppressed as in the case of ^{3}He.[71,72] In the case of propane, in a very simplified approach, we can consider a local linear gradient to be responsible for the observed line widths such that, for a line width of 20 Hz in a 10 mm NMR tube, the following relation is satisfied: $\frac{\gamma}{2\pi}g=\frac{20\ \text{Hz}}{0.01\ \text{m}}$. The envelope of the time domain evolution in a CPMG sequence follows the well-known equation:[73]

$$I(g,nt_E)=I_0\,\mathrm{e}^{-\frac{nt_E}{T_2}}\,\mathrm{e}^{-\frac{1}{3}\left(\frac{\gamma}{2\pi}gnt_E\right)^2 D\,nt_E}. \tag{9.41}$$

During the pulse sequence, the effective relaxation time is then:

$$\frac{1}{T_{2,\text{eff}}}=\frac{1}{T_2}+\frac{1}{3}D\left(\frac{\gamma}{2\pi}gt_E\right)^2, \tag{9.42}$$

where for gases in the molar range of propane typically, $D\sim10^{-6}\ \text{m}^2\,\text{s}^{-1}$. Under these conditions, loss of resolution is negligible. For instance, for an echo time of 50 ms and considering $T_2=2$ s, an effective $T_{2,\text{eff}}$ value of 1.97 s is obtained, which is less than 1% lower. Changing the echo time between 1 ms and 100 ms will slightly change $T_{2,\text{eff}}$ in 0.0004% to 4%, respectively. In this way, line widths of less than 0.2 Hz could be obtained in free hyperpolarized gases.

A second aspect to be considered is for gas embedded in porous media. In this case, relaxation times will be considerably small compared with the free gas due to surface interactions.[73] Line widths of 15 Hz have been reported for gas confined in a catalyst bed,[34] thus, taking an estimate value of $T_2=200$ ms, a *J*-spectrum with a line width of 1 Hz can be obtained. Naturally, the effects of diffusion are greatly reduced in a confined gas and the apparent diffusion coefficient will depend on the cavity size, pressure, temperature, and dilution with other species.[74,75] Even in these very complex scenarios, and with effective relaxation times as short as a few hundred milliseconds, PhD-PHIP promises to be a very important tool in the study of hyperpolarized gases.

Simulations of *J*-spectra for propane with thermal and PHIP operators were carried out considering the eight spins in the molecule for an arbitrary echo time of $t_E=8$ ms and an effective transverse decay of 200 ms. Even though the resolution enables the discrimination of different resonances, and the antiphase characteristic of PASADENA is clearly observed, the resulting spectra are rather featureless due to the collapse of the chemical shift (see Figure 9.14 (A)). This is easily solved by using partial *J*-spectroscopy. In effect, the simulations of partial *J*-spectra shown in Figure 9.14 (B)–(C) show the expected resonances. Note that all the plots are normalized to the thermal and PHIP *J*-spectrum respectively. Here, two independent simulations must be carried out in order to detect each group independently. Additionally, single shot experiments were considered, thus the small artifacts in PhD-PHIP arise from off-resonant effects on the 180° pulses.

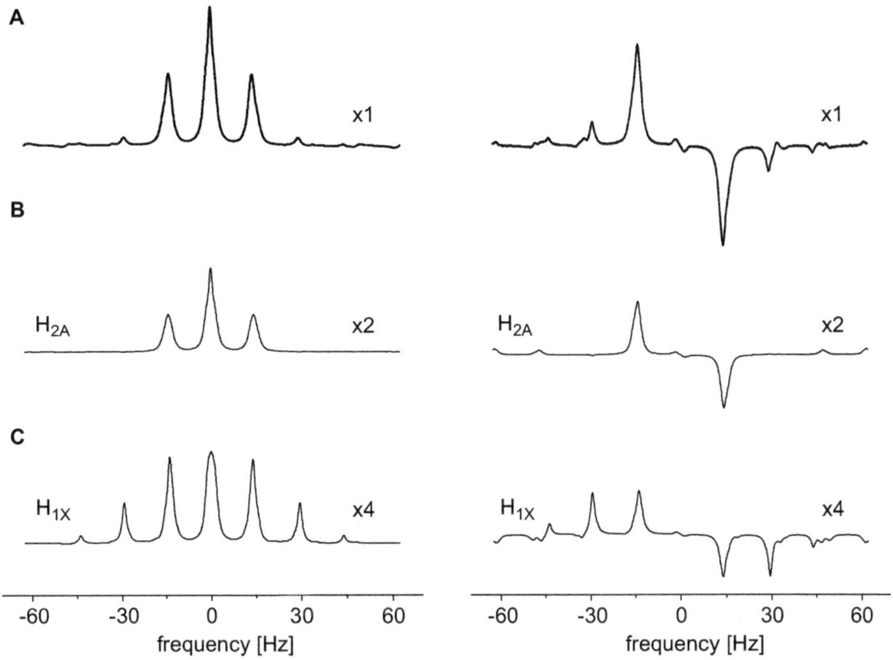

Figure 9.14 Simulated spectra of propane acquired with a CPMG sequence. Left column: thermally polarized. Right column: PHIP. (A) *J*-spectra where many resonances are not clearly observed as all the information collapses to zero frequency. (B) Partial *J*-spectra corresponding to acquisition of the methyl group of propane. (C) Partial *J*-spectra for the hyperpolarized methylene group.

9.6 Summary

The use of PhD-PHIP to enhance *J*-coupling determination and partial peak cancelling suppression appears to be a useful experimental approach to the study of catalyst reactions in liquids and gases. Application of a CPMG pulse sequence to obtain a *J*-spectrum with a sharp line width that is determined by T_2 relaxation rather than T_2^*, is applied in PHIP experiments. The particular evolution of the density matrix corresponding to an initial state derived from PHIP renders an odd-even effect on the successive echoes that differs from thermal density operators. This causes a natural spectral separation of both types of signals that can be exploited to suppress antiphase signal cancellation due to the overlap with thermal signals.

Several demonstrations of the technique have been presented in liquids. These include high-resolution spectra acquisition, suppression of signal cancellation into composite systems with frequency overlap of thermal and PHIP signals, and removal of thermal signals arising from on-going chemical reactions. Simulations of the performance of the method in the gas phase are presented in this book for the first time. As PhD-PHIP is able to detect very low reaction rates, it can readily be applied in the very extensive field of catalysts research, where gas-phase PHIP has shown to be an excellent work test bench. In summary, PhD-PHIP has potential for the study of chemical reactions in gas phase NMR and seems to be an excellent complement to the well-established techniques currently in use.

Acknowledgements

We would like to acknowledge the financial support received from CONICET, FoNCyT, SeCyT-UNC, and the Partner Group for NMR Spectroscopy with High Spin Polarization with the Max Planck Institute for Polymer Research, Mainz, Germany.

References

1. M. A. Bouchiat, T. R. Carver and C. M. Varnum, *Phys. Rev. Lett.*, 1960, **5**, 373.
2. S. Appelt, A. B.-A. Baranga, C. J. Erickson, M. V. Romalis, A. R. Young and W. Happer, *Phys. Rev. A: At., Mol., Opt. Phys.*, 1998, **58**, 1412.
3. F. D. Colegrove, L. D. Schearer and G. K. Walters, *Phys. Rev.*, 1963, **132**, 2561.
4. G. Eckert, W. Heil, M. Meyerhoff, E. W. Otten, R. Surkau, M. Werner, M. Leduc, P. J. Nacher and L. D. Schearer, *Nuclear Instruments & Methods in Physics Research Section A – Accelerators Spectrometers Detectors and Associated Equipment*, 1992, **320**, 53.
5. E. Stolz, M. Meyerhoff, N. Bigelow, M. Leduc, P. J. Nacher and G. Tastevin, *Appl. Phys. B: Lasers Opt.*, 1996, **63**, 629.
6. A. Abragam and M. Goldman, *Rep. Prog. Phys.*, 1978, **41**, 395.
7. K. H. Hausser and D. Stehlik, *Adv. Magn. Reson.*, 1968, **3**, 79.
8. J. H. Ardenkjær-Larsen, B. Fridlund, A. Gram, G. Hansson, L. Hansson, M. H. Lerche, R. Servin, M. Thaning and K. Golman, *Proc. Natl. Acad. Sci.*, 2003, **100**, 10158.
9. R. G. Lawler, *J. Am. Chem. Soc.*, 1967, **89**, 5519.
10. J. Bargon, *Helv. Chim. Acta*, 2006, **89**, 2082.
11. C. R. Bowers and D. P. Weitekamp, *Phys. Rev. Lett.*, 1986, **57**, 2645.
12. M. G. Pravica and D. P. Weitekamp, *Chem. Phys. Lett.*, 1988, **145**, 255.
13. T. C. Eisenschmid, R. U. Kirss, P. P. Deutsch, S. I. Hommeltoft, R. Eisenberg, J. Bargon, R. G. Lawler and A. L. Balch, *J. Am. Chem. Soc.*, 1987, **109**, 8089.
14. C. R. Bowers and D. P. Weitekamp, *J. Am. Chem. Soc.*, 1987, **109**, 5541.
15. E. Babcock, B. Chann, T. G. Walker, W. C. Chen and T. R. Gentile, *Phys. Rev. Lett.*, 2006, **96**, 083003.
16. E. J. R. van Beek, J. Schmiedeskamp, J. M. Wild, M. N. J. Paley, F. Filbir, S. Fichele, F. Knitz, G. H. Mills, N. Woodhouse, A. Swift, W. Heil, M. Wolf and E. W. Otten, *Eur. Radiol*, 2003, **13**, 2583.
17. M. S. Albert, G. D. Cates, B. Driehuys, W. Happer, B. Saam, C. S. Springer Jr and A. Wishnia, *Nature*, 1994, **370**, 199.

18. Y. V. Chang, J. D. Quirk, I. C. Ruset, J. J. Atkinson, F. W. Hersman and J. C. Woods, *Magn. Reson. Med.*, 2014, **71**, 339.
19. A. J. Deninger, B. Eberle, M. Ebert, T. Grossmann, W. Heil, H. U. Kauczor, L. Lauer, K. Markstaller, E. Otten, J. Schmiedeskamp, W. Schreiber, R. Surkau, M. Thelen and N. Weiler, *J. Magn. Reson.*, 1999, **141**, 207.
20. K. K. Gast, A. Biedermann, A. Herweling, W. G. Schreiber, J. Schmiedeskamp, E. Mayer, C. P. Heussel, K. Markstaller, H. U. Kauczor and B. Eberle, *Eur. J. Radiol.*, 2008, **18**, 530.
21. X. J. Chen, L. W. Hedlund, H. E. Moller, M. S. Chawla, R. R. Maronpot and G. A. Johnson, *Proc. Natl. Acad. Sci.*, 2000, **97**, 11478.
22. M. Salerno, E. E. de Lange, T. A. Altes, J. D. Truwit, J. R. Brookeman and J. P. Mugler, *Radiology*, 2002, **222**, 252.
23. D. A. Yablonskiy, A. L. Sukstanskii, J. C. Leawoods, D. S. Gierada, G. L. Bretthorst, S. S. Lefrak, J. D. Cooper and M. S. Conradi, *Proc. Natl. Acad. Sci.*, 2002, **99**, 3111.
24. L. L. Tsai, R. W. Mair, M. S. Rosen, S. Patz and R. L. Walsworth, *J. Magn. Reson.*, 2008, **193**, 274.
25. Z. Liu, T. Araki, Y. Okajima, M. Albert and H. Hatabu, *Eur. J. Radiol.*, 2014, **83**, 1282.
26. R. H. Acosta, P. Blümler, K. Münnemann and H. W. Spiess, *Prog. Nucl. Magn. Reson. Spectrosc.*, 2012, **66**, 40.
27. M. S. Conradi, B. T. Saam, D. A. Yablonskiy and J. C. Woods, *Prog. Nucl. Magn. Reson. Spectrosc.*, 2006, **48**, 63.
28. S. J. Kadlecek, K. Emami, M. C. Fischer, M. Ishii, J. Yu, J. M. Woodburn, M. NikKhah, V. Vahdat, D. A. Lipson, J. E. Baumgardner and R. R. Rizi, *Prog. Nucl. Magn. Reson. Spectrosc.*, 2005, **47**, 187.
29. R. W. Mair, D. G. Cory, S. Peled, C. H. Tseng, S. Patz and R. L. Walsworth, *J. Magn. Reson.*, 1998, **135**, 478.
30. R. W. Mair, M. S. Rosen, R. Wang, D. G. Cory and R. L. Walsworth, *Magn. Reson. Chem.*, 2002, **40**, S29.
31. C. Y. Cheng and C. R. Bowers, *J. Am. Chem. Soc.*, 2007, **129**, 13997.
32. S. Aime, R. Gobetto, F. Reineri and D. Canet, *J. Magn. Reson.*, 2006, **178**, 184.
33. M. H. Levitt, *Spin Dynamics: Basics of Nuclear Magnetic Resonance*, John Wiley & Sons, Chichester, England, Hoboken, NJ, 2008.
34. L. S. Bouchard, S. R. Burt, M. S. Anwar, K. V. Kovtunov, I. V. Koptyug and A. Pines, *Science*, 2008, **319**, 442.
35. J. F. Dechent, L. Buljubasich, L. M. Schreiber, H. W. Spiess and K. Munnemann, *Phys. Chem. Chem. Phys.*, 2012, **14**, 2346.
36. T. Ratajczyk, T. Gutmann, S. Dillenberger, S. Abdulhussaein, J. Frydel, H. Breitzke, U. Bommerich, T. Trantzschel, J. Bernarding, P. C. M. M. Magusin and G. Buntkowsky, *Solid State Nucl. Magn. Reson.*, 2012, **43–44**, 14.
37. R. Freeman and H. D. W. Hill, *J. Chem. Phys.*, 1971, **54**, 301.
38. L. Buljubasich, I. Prina, M. B. Franzoni, K. Munnemann, H. W. Spiess and R. H. Acosta, *J. Magn. Reson.*, 2013, **230**, 155.

39. I. Prina, L. Buljubasich and R. H. Acosta, *J. Phys. Chem. Lett.*, 2013, **4**, 3924.
40. I. Prina, L. Buljubasich and R. H. Acosta, *J. Magn. Reson.*, 2015, **251**, 1.
41. K. Kovtunov, V. Zhivonitko, I. Skovpin, D. Barskiy and I. Koptyug, in *Hyperpolarization Methods in NMR Spectroscopy*, ed. L. T. Kuhn, Springer, Berlin Heidelberg, 2013, vol. 338, pp. 123–180.
42. J. Natterer and J. Bargon, *Prog. Nucl. Magn. Reson. Spectrosc.*, 1997, **31**, 293.
43. D. Canet, C. Aroulanda, P. Mutzenhardt, S. Aime, R. Gobetto and F. Reineri, *Concepts Magn. Reson., Part A*, 2006, **28A**, 321.
44. C. R. Bowers, in *Encyclopedia of Magnetic Resonance*, ed. R. K. Harris, John Wiley & Sons, Ltd, Chichester, UK, 2007.
45. R. A. Green, R. W. Adams, S. B. Duckett, R. E. Mewis, D. C. Williamson and G. G. R. Green, *Prog. Nucl. Magn. Reson. Spectrosc.*, 2012, **67**, 1.
46. L. Buljubasich, M. B. Franzoni and K. Münnemann, *Top. Curr. Chem.*, 2013, **338**, 33.
47. L. Pauling and E. B. Wilson Jr., *Introduction to Quantum Mechanics, with Applications to Chemistry*, McGraw-Hill, 1935.
48. K. F. Bonhoeffer and P. Harteck, *Die Naturwissenschaften*, 1929, **17**, 182.
49. C. R. Bowers and D. P. Weitekamp, *J. Am. Chem. Soc.*, 1987, **109**, 5541.
50. L. Buljubasich, M. B. Franzoni, H. W. Spiess and K. Münnemann, *J. Magn. Reson.*, 2012, **219**, 33.
51. S. E. Korchak, K. L. Ivanov, A. V. Yurkovskaya and H. M. Vieth, *Phys. Chem. Chem.*, 2009, **11**, 11146.
52. A. B. Permin and R. Eisenberg, *J. Am. Chem. Soc.*, 2002, **124**, 12406.
53. D. J. Fox, S. B. Duckett, C. Flaschenriem, W. W. Brennessel, J. Schneider, A. Gunay and R. Eisenberg, *Inorg. Chem.*, 2006, **45**, 7197.
54. F. Reineri, A. Viale, G. Giovenzana, D. Santelia, W. Dastru, R. Gobetto and S. Aime, *J. Am. Chem. Soc.*, 2008, **130**, 15047.
55. I. V. Koptyug, K. V. Kovtunov, S. R. Burt, M. S. Anwar, C. Hilty, S. I. Han, A. Pines and R. Z. Sagdeev, *J. Am. Chem. Soc.*, 2007, **129**, 5580.
56. K. V. Kovtunov, V. V. Zhivonitko, A. Corma and I. V. Koptyug, *J. Phys. Chem. Lett.*, 2010, **1**, 1705.
57. K. V. Kovtunov, I. E. Beck, V. I. Bukhtiyarov and I. V. Koptyug, *Angew. Chem. Int. Ed.*, 2008, **47**, 1492.
58. L.-S. Bouchard, K. V. Kovtunov, S. R. Burt, M. S. Anwar, I. V. Koptyug, R. Z. Sagdeev and A. Pines, *Angew. Chem.*, 2007, **119**, 4142.
59. V.-V. Telkki, V. V. Zhivonitko, S. Ahola, K. V. Kovtunov, J. Jokisaari and I. V. Koptyug, *Angew. Chem. Int. Ed.*, 2010, **49**, 8363.
60. R. R. Ernst, G. Bodenhausen and A. Wokaun, *Principles of Nuclear Magnetic Resonance in One and Two Dimensions*, Oxford, 1994.
61. S. Meiboom and D. Gill, *Rev. Sci. Instrum.*, 1958, **29**, 688.
62. J. G. Powles and A. Hartland, *Proc. Phys. Soc.*, 1961, **77**, 273.
63. A. Allerhand, *J. Chem. Phys.*, 1966, **44**, 1.
64. T. Tokuhiro and G. Fraenkel, *J. Chem. Phys.*, 1968, **49**, 3998.

65. T. Gullion, D. B. Baker and M. S. Conradi, *J. Magn. Reson.*, 1969, **1990**(89), 479.
66. M. B. Franzoni, R. H. Acosta, H. M. Pastawski and P. R. Levstein, *Philos. Trans. R. Soc. A*, 2012, **370**, 4713.
67. D. Graafen, S. Ebert, O. Neudert, L. Buljubasich, M. B. Franzoni, J. F. Dechent and K. Münnemann, in *Annual Reports on NMR Spectroscopy*, ed. A. W. Graham, Academic Press, 2014, vol. 82, pp. 167–215.
68. O. W. Sørensen, G. W. Eich, M. H. Levitt, G. Bodenhausen and R. R. Ernst, *Prog. Nucl. Magn. Reson. Spectrosc.*, 1984, **16**, 163.
69. P. T. Callaghan, *Principles of Nuclear Magnetic Resonance Microscopy*, Clarendon Press, 1993.
70. K. V. Kovtunov, D. A. Barskiy, A. M. Coffey, M. L. Truong, O. G. Salnikov, A. K. Khudorozhkov, E. A. Inozemtseva, I. P. Prosvirin, V. I. Bukhtiyarov, K. W. Waddell, E. Y. Chekmenev and I. V. Koptyug, *Chemistry*, 2014, **20**, 11636.
71. P. P. Zanker, J. Schmidt, J. Schmiedeskamp, R. H. Acosta and H. W. Spiess, *Phys. Rev. Lett.*, 2007, **99**, 263001.
72. P. P. Zänker, J. Schmidt, J. Schmiedeskamp, R. H. Acosta and H. W. Spiess, *Chem. Phys. Lett.*, 2009, **481**, 137.
73. P. T. Callaghan, *Translational Dynamics and Magnetic Resonance: Principles of Pulsed Gradient Spin Echo NMR*, Oxford University Press, Oxford, New York, 2011.
74. T. M. de Swiet, *J. Magn. Reson. B*, 1995, **109**, 12.
75. L. Agulles-Pedrós, R. H. Acosta, P. Blümler and H. W. Spiess, *J. Magn. Reson.*, 2009, **197**, 56.
76. L. Buljubasich, I. Prina, M. B. Franzoni and R. H. Acosta, *Microporous Mesoporous Mater.*, 2016, **205**, 21.

CHAPTER 10

Optical Hyperpolarization of Noble Gases for Medical Imaging

TADEUSZ PAŁASZ[a] AND BOGUSŁAW TOMANEK*[b,c]

[a] Marian Smoluchowski Institute of Physics, Faculty of Physics, Astronomy and Applied Computer Science of the Jagiellonian University, Stanisława Łojasiewicza 11, 30-348 Kraków, Poland; [b] University of Alberta, Department of Oncology, Division of Medical Physics, 3-12 University Terrace, Edmonton, Alberta T6G 2T4, Canada; [c] Henryk Niewodniczański Institute of Nuclear Physics, Polish Academy of Sciences, Radzikowskiego 152, 30-342 Kraków, Poland

*Email: tomanek@ualberta.ca

10.1 Introduction

The Magnetic Resonance Imaging (MRI) technique was proposed in the early 1970s and soon after found clinical applications to observe *in vivo* soft tissues containing water molecules. Introduction of MRI was possible due to much earlier work on nuclear magnetic resonance phenomena. Although contrasting in standard – proton-based MRI is different from hyperpolarized noble gas MRI – the same physical principles apply to both techniques. Currently the number of clinical applications of hyperpolarized noble gases MRI is steadily increasing. Image quality, unlike in standard MRI, depends, however, on efficient gas polarization produced by dedicated gas polarizers. The first MR images of mouse lungs using hyperpolarized ^{129}Xe gas were

New Developments in NMR No. 6
Gas Phase NMR
Edited by Karol Jackowski and Michał Jaszuński

Published by the Royal Society of Chemistry, www.rsc.org

demonstrated by Albert *et al.* in 1994.[1] The first MR images of the human lungs filled with polarized ^{3}He were reported in 1996.[2–4]

Lungs must be filled with an MR "sensitive" gas such as ^{3}He or ^{129}Xe noble gas to obtain an MR image. These gases have one-half nuclear spin, similarly to protons (nuclei of hydrogen atoms) commonly used in standard MR imaging due to the high content of water in the human body, thus providing a strong MR signal. The density of gases in the lungs, at body temperature and at atmospheric pressure or close to it, is three orders of magnitude lower than the density of hydrogen in soft tissues containing mostly water molecules. This low density, thus low MR signal, must be compensated with nuclear polarization, providing an excess of excited spins much larger than at the thermal equilibrium. Techniques such as optical pumping of atoms and polarization transfer from electrons to nuclei (with spin or metastability exchange) are used for that purpose. These methods allow to increase nuclear polarization by four to five orders of magnitude in comparison to the thermal polarization in the magnetic field (1.5–3 T) produced by a standard medical MRI scanner. While there are other techniques of polarization than spin exchange optical pumping (SEOP) and metastability exchange optical pumping (MEOP), such as dynamic nuclear polarization (DNP), para-hydrogen induced polarization (PHIP), or a "brute force" technique, only SEOP and MEOP methods will be discussed in this chapter as they are the most common. The reader interested in MEOP without using elevated ^{3}He pressure and high magnetic fields is referred to recent reviews.[5–13]

10.2 Boltzmann Equilibrium Polarization and Hyperpolarization

According to quantum mechanics, atomic nuclei comprising an odd number of protons or an odd number of neutrons have non-zero nuclear spin. Magnetic moments μ of these nuclei associated with the spin of the protons and neutrons are randomly oriented in space but in the presence of an external (static) magnetic field B_0 these moments align along the direction of this field.

From a quantum mechanics perspective, the ^{1}H, ^{3}He, and ^{129}Xe nuclei with nuclear spin of one-half (quantum number $I = 1/2$) can occupy only two quantum spin states in the external magnetic field B_0. These quantum states can be represented by spins precession (a classical interpretation of quantum states) around the magnetic field direction (low energy state $E_\uparrow$) and opposite to the field (high energy state $E_\downarrow$). These two energy states are described by the magnetic quantum numbers $m_\mathrm{I} = \pm 1/2$. Their energy difference is equal to

$$\Delta E = E_\downarrow - E_\uparrow = |\gamma|\hbar B_0, \tag{10.1}$$

where the magnetogyric ratio γ depends on the nucleus. For protons (nuclei of ^{1}H atoms) the magnetogyric ratio $\gamma_\mathrm{H} \cong 2.67 \cdot 10^{8}$ rad s^{-1} T^{-1} and its value is equal to the ratio of the intrinsic magnetic dipole moment to the spin angular momentum. The magnetogyric ratios of 1/2 spin noble gases nuclei are: $-2.04 \cdot 10^8$ rad s^{-1} T^{-1} and $-7.4 \cdot 10^7$ rad s^{-1} T^{-1} for ^{3}He and ^{129}Xe respectively. Note their negative values, unlike the positive value of the hydrogen magnetogyric ratio. The spins rotate around the magnetization B_0 with the so-called Larmor frequency ω_0.

$$\omega_0 = \gamma B_0 \tag{10.2}$$

The sign of the magnetogyric ratio determines the direction of the precession. The spin and the magnetic moment are in opposite directions to each other when γ is negative. Because the magnetogyric ratio expresses the ratio between the observed angular frequency and the strength of the magnetic field its value is often provided in MHz T^{-1}. For example: $\frac{\gamma_H}{2\pi} \cong 42.58$ MHz T^{-1}, $\frac{\gamma_{He}}{2\pi} \cong 32.43$ MHz T^{-1}, $\frac{\gamma_{Xe}}{2\pi} \cong 11.78$ MHz T^{-1} for protons ^{3}He and ^{129}Xe, respectively. Most clinical MRI systems image protons, and thus operate at the frequency corresponding to the proton Larmor frequency (*e.g.* 63.87 MHz at 1.5 T). For noble gases MRI, a medical scanner has to be equipped with multi-frequency modules (*e.g.* broadband radio-frequency (RF) amplifiers, receiving channels, *etc.*) and radio-frequency coils tuned to the Larmor frequency of the imaged gas. For instance, an MRI scanner operating at 1.5 T with a standard RF proton coil tuned to 63.87 MHz should have additional RF coils resonating at 48.65 MHz for ^{3}He imaging and at 17.67 MHz for ^{129}Xe.

The net magnetization produced by a sample is the vector sum of all individual magnetic moments in the external magnetic field. The phases of the individual magnetic moments are random in both energy states. At room temperature and with no external magnetic field, both energy states have approximately the same number of spins. The Zeeman energy splitting between two energy levels of the nuclear spin one-half is equal to $|\gamma|\hbar B_0$ (see eqn (10.1)) and is a small fraction of the thermal energy $k_B T$ of the spin assembly.

Let $N_\uparrow$ denote the number of spins in the $m_I = 1/2$ state and $N_\downarrow$ denote the number of spins in the $m_I = -1/2$ state, then the ratio

$$\frac{N_\uparrow}{N_\downarrow} = e^{-\frac{|\gamma|\hbar B_0}{2k_B T}} \approx 1 - \frac{|\gamma|\hbar B_0}{2k_B T} \tag{10.3}$$

exists, where h is Planck's constant ($h \cong 6.63 \cdot 10^{-34}$ Js), $\hbar = \frac{h}{2\pi}$, and k_B is Boltzmann's constant ($k_B \cong 1.38 \cdot 10^{-23}$ J K^{-1}). In thermal equilibrium the absolute nuclear spin polarization can be defined as

$$P = \frac{N_\uparrow - N_\downarrow}{N_\uparrow + N_\downarrow} \tag{10.4}$$

and

$$P = \tanh \frac{|\gamma|\hbar B_0}{2k_B T} \approx \frac{|\gamma|\hbar B_0}{2k_B T}. \tag{10.5}$$

Eqn (10.5) gives $P \approx 5 \cdot 10^{-6}$ for an assembly of protons in a water sample at the human body temperature ($\sim 37\ ^\circ C$) and in the magnetic field of 1.5 T. An increase of the magnetic field B_0 increases the net magnetization of a given sample and thus the MR signal. For this reason modern clinical MR scanners are equipped with magnets producing magnetic fields up to 3 T, while human research systems reach 7 T. The magnitude of the magnetization M can be calculated as a product of the nuclear magnetic moment, nuclei density, and nuclear spin polarization. The magnetization of a sample with N nuclear spins ($N = N_\uparrow + N_\downarrow$) is then given by

$$M = \frac{1}{2} N \gamma \hbar P. \tag{10.6}$$

The ratio between the number of hydrogen nuclei in the water sample and the number of noble gas nuclei in the same volume is $\frac{N_H}{N_{He}} \approx 10^3$. In addition, lower than for protons, the magnetogyric ratio γ of helium and xenon further decreases the MR signal. The magnetization is proportional to γ^2 (eqn (10.5) and (10.6)) and in normal conditions and at 1.5 T its value is a few orders of magnitude smaller for noble gas than for water protons:

$$\frac{M_{He}}{M_H} \approx 2 \cdot 10^{-4} \text{ and } \frac{M_{Xe}}{M_H} \approx 2.7 \cdot 10^{-5}.$$

In other words, the MR signal from thermally polarized noble gas is about five orders of magnitude lower than that of water. This huge difference can, however, be compensated by hyperpolarization of gases using SEOP or MEOP processes. It should be noted that there are very few water molecules in lungs, preventing proton lung MR imaging, but allowing HP gas imaging.

10.3 Spin Exchange Optical Pumping of ^{3}He and ^{129}Xe

The SEOP process is challenging due to an exceptionally large energy gap between the optical and nuclear spin energy levels. Let us consider a typical SEOP polarizer utilizing rubidium vapor for the optical pumping. In this case the spin orientation of rubidium electrons is transferred to the noble gas nuclei (see Figure 10.1).

For SEOP, the frequency of light emitted by a high-power laser is tuned to the specific transition in alkali metal atoms. A laser diode (or a laser diode bar) is used as a typical light source for the SEOP process. This kind of laser provides linear light polarization. A quarter-wave plate ($\lambda/4$ retarder) converts linearly to circularly polarized light. An optical system (lenses and mirrors or beam splitters) shapes the light beam passing through the glass cell, where optical pumping and spin exchange processes take place. The cell and the output of the gaseous system carrying polarized gas must be placed in a homogeneous magnetic field, parallel to the direction of the light beam. Both the gas inlet and outlet are mounted on the opposite sides of the cell.

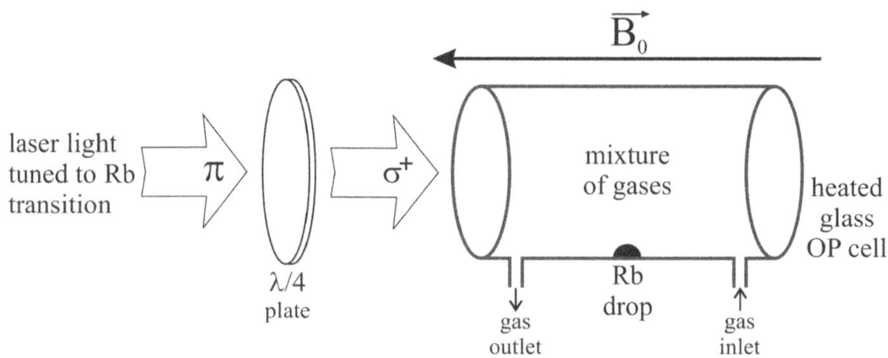

Figure 10.1 Diagram of the SEOP system. Linearly polarized (π) laser light passes through the quarter-wave plate ($\lambda/4$) and then, as circularly clockwise (σ^+) polarized light, illuminates rubidium atoms vaporized from the alkali metal droplet inside the heated glass cell. Optically pumped rubidium atoms are mixed with noble gas (^{3}He or ^{129}Xe with buffer gases). The optical pumping (OP) process and spin orientation transfer take place in the homogeneous magnetic field B_0 produced by a set of coils surrounding the OP cell, tubes, and polarized gas storage containers in the cell output (not shown).

The valves (not shown in Figure 10.1) allow both the continuous and the batch gas flow mode. The cell is filled with pure noble gas for ^{3}He SEOP polarization and with a mixture of noble and buffer gases (*e.g.* ^{4}He and nitrogen) for ^{129}Xe polarization. A small amount of alkali metal (usually rubidium) is placed inside the optical pumping cell and heated to vaporize.

To enable the optical pumping of rubidium atoms laser light of wavelength $\lambda = 794.7$ nm is used (photons of energy $E = \frac{hc}{\lambda} \cong 1.56$ eV conformed to rubidium D_1 transition). According to eqn (10.1), the nuclear spin energy transition is equal to $\Delta E \cong 10^{-10}$ eV for ^{129}Xe gas placed in the magnetic field of $B_0 = 2$ mT. Therefore nuclear polarization with the use of light is a two-step process. Firstly, laser light is used to polarize the valence electronic spins of rubidium atoms (or other alkali atoms such as potassium or cesium). Absorbed photons are reemitted by the alkali atoms and, as a result of the optical pumping, populations of energy states are changed.[14–16] The energy gap between the optically pumped states is in range of tens of nano-electrono-volts (depending on the Zeeman splitting, *i.e.* energy gap is equal to $\Delta E \cong 2 \cdot 10^{-8}$ eV in case of ^{85}Rb $F = 2$ of $5^2S_{1/2}$ state, see Figure 10.2). Then, in the next step, the electronic spin polarization is transferred to the noble gas nuclei. It can be realized by the spin exchange mediated by the Fermi contact in van der Waals molecules – stable clusters consisting of the alkali metal atom and the noble gas atom coupled together by the van der Waals forces. The alkali atom electronic spin polarization can also be transferred to the noble gas nuclei in a different way, by the binary collisions between these atoms.[5,6]

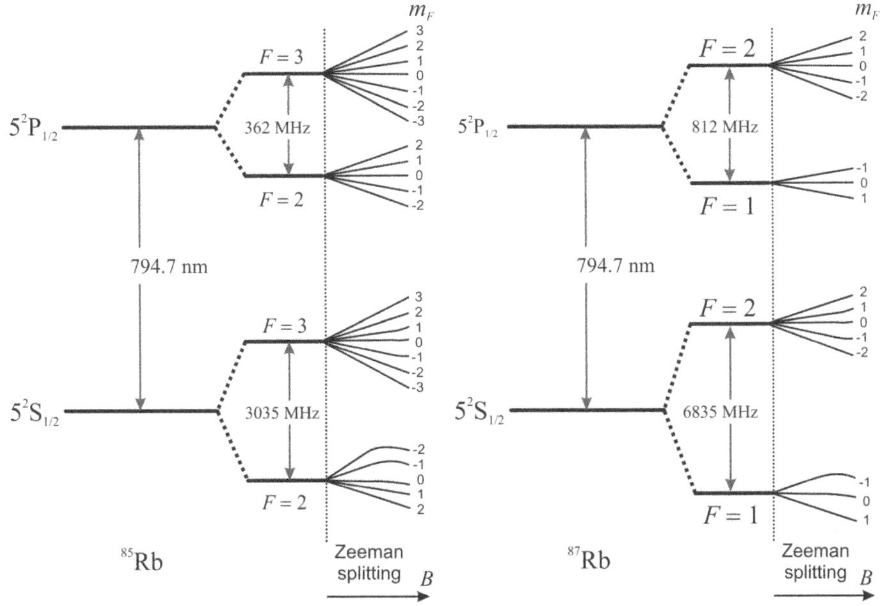

Figure 10.2 A rubidium atomic structure of two stable isotopes with Zeeman splitting in the external magnetic field. D_1 transition between the ground level $5^2S_{1/2}$ and the first excited level $5^2P_{1/2}$ has wavelength of 794.7 nm.

10.3.1 Optical Pumping of Alkali Metal Atoms

Electron spins are manipulated with light in the optical pumping process. The angular momentum of photons can be transferred with high efficiency to the atoms. The optical pumping is possible in the atomic system with at least two energy levels; however, the lower atomic level must have at least two sublevels of the fine or hyperfine structure. The selected lower and upper atomic energy levels are coupled by the optical transition. Circularly polarized and resonant to this transition light selectively depopulates one of the lower state sublevels. The most suitable alkali metal for SEOP is rubidium (only one valence electron). The rubidium atomic structure is presented in Figure 10.2. Only the ground level $5^2S_{1/2}$ and the first excited level $5^2P_{1/2}$ with the hyperfine structure are shown. Rubidium has two stable isotopes: ^{85}Rb (natural abundance 72.17%) and ^{87}Rb (27.83%), has low-temperature melting point (39.3 °C), and is easy to vaporize (at 160 °C rubidium vapor pressure is close to 1 Pa).[17,18] Relatively inexpensive, high-power laser diode bars and spectrally narrowed high-power lasers corresponding to rubidium transitions from the ground level are commercially available. Such lasers operate at the rubidium D_1 transition line (emitting light with the wavelength of 794.7 nm, or the corresponding frequency of 377 THz).

A two-level system composed of rubidium $5^2S_{1/2}$ and $5^2P_{1/2}$ energy levels is presented in Figure 10.3. The alkali atom has one valence electron, with the spin angular momentum $s = 1/2$. The rubidium ground state has the orbital

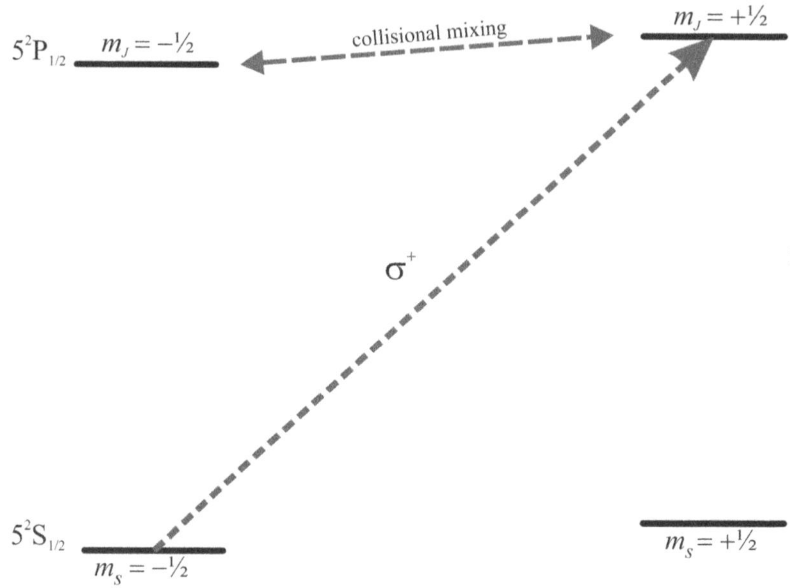

Figure 10.3 A scheme of the optical pumping process. The $m_J = +1/2$ level is excited by resonant and circularly (σ^+) polarized laser light. Collisional mixing between the upper states is caused by the helium atoms. Collisions with nitrogen gas activate quenching to the lower state to prevent radiation trapping.

angular momentum $L = 0$ and thus the total angular momentum of this state, which combines both the spin and orbital angular momentum, is equal to 1/2. The spin multiplicity is equal to $2s + 1 = 2$ and corresponds to its two quantum states. The first excited energy level $5^2P_{1/2}$ has angular momentum equal to 1/2. The ground energy level $5^2S_{1/2}$ consists of the two sublevels $m_S = \pm 1/2$. Likewise, the upper state has also two sublevels $m_J = \pm 1/2$. In a weak magnetic field (a few mT within an optical cell of a typical SEOP polarizer), these sublevels have slightly different energies.

In the optical pumping the ensemble of rubidium atoms is illuminated by laser light tuned to the energy transition between the $5^2S_{1/2}$ and the $5^2P_{1/2}$ rubidium electron states. The linear polarization of light emitted by the laser diode becomes circular after passing through a quarter-wave plate ($\lambda/4$). The $\lambda/4$ plate shifts the phase between two perpendicular polarization components of the light wave. The circular polarization may be referred to as a clockwise (σ^+) or anti-clockwise (σ^-) polarization, depending on the rotation direction of the electric field. This direction can be controlled by rotating the quarter-wave plate. The oscillating electric field of light induces transitions between the ground and the excited atomic states. A σ^+ polarized photon carries spin $s = 1$ and has the spin projection $m_S = 1$ in the magnetic field B_0. The angular momentum conservation law imposes the selection rule: the total change of $\Delta m = +1$ with the transition from $m_S = -1/2$ to $m_J = +1/2$ is compensated by the absorption of the σ^+circularly polarized photon. By absorption of such photons, the valence electrons in $m_s = -1/2$ state are promoted to $m_J = +1/2$ state of the excited $5^2P_{1/2}$ energy level. Photons with the spin projection $m_s = 1$ cannot be absorbed by the electrons in $m_s = +1/2$ state of the system. But, of course, a similar process can be induced by circularly σ^- polarized laser light (photons spin projection $m_s = -1$) promoting the $m_s = +1/2 \rightarrow m_J = -1/2$ transition with the total change of $\Delta m = -1$.

Rubidium atoms cannot remain in the excited state for a long time. The natural line width (FWHM – half width at half maximum of absorption line profile) of the described transition in ^{87}Rb is equal to $\Gamma \cong 2\pi \cdot 5.75$ MHz corresponding to 27.7 ns of the $5^2P_{1/2}$ state lifetime. The excited state decays to the ground state by the spontaneous emission of light (see Figure 10.4).

For an isolated rubidium atom, probability of the decay from $m_J = +1/2$ to $m_s = -1/2$ (transition σ^- in Figure 10.4) is twice as big as that of decay to $m_s = +1/2$ (transition π with emission of linearly polarized photon). For optical pumping of the $m_s = +1/2$ state, this more likely return to the state $m_s = -1/2$ is not desired because it depletes population of the $m_s = +1/2$ state. This undesired process would depopulate the $m_s = +1/2$ state. Therefore it is necessary to introduce a noble gas. The collisions of excited rubidium atoms with the noble gas atoms redistribute the populations of excited state sublevels $m_J = \pm 1/2$.

For example, in the SEOP method used for ^{3}He production, excited rubidium atoms collide with the noble gas atoms many times during their lifetime due to high ^{3}He density. In the ^{129}Xe SEOP process, ^{4}He atoms

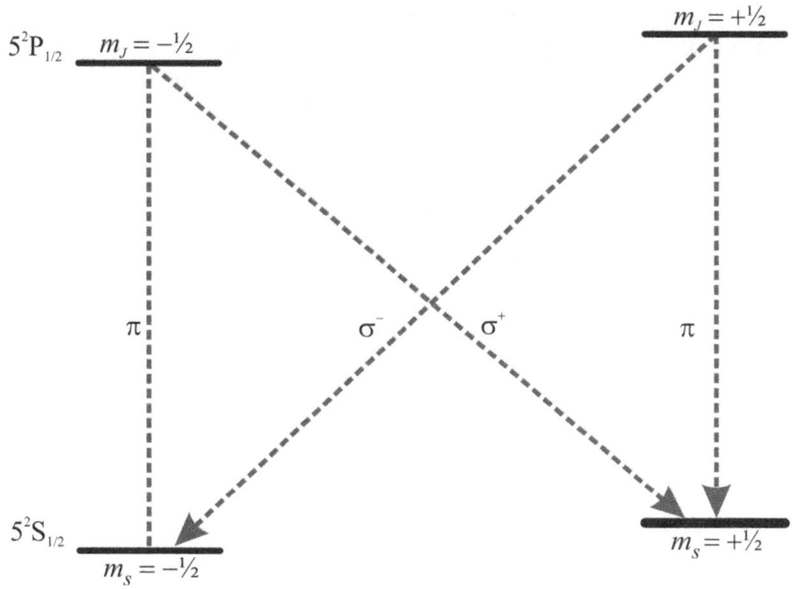

Figure 10.4 Rubidium optical decay *via* spontaneous emission of light.

added to the gas mixture at high enough pressure behave as a buffer gas colliding with excited rubidium atoms. With collisional mixing of the excited state sublevels probabilities of returning to both of the ground state sublevels become identical. With σ^+ polarized laser light continuously pumping the $m_J = +1/2$ excited sublevel, the $m_s = -1/2$ ground sublevel depopulates and the rubidium valence electrons are polarized to the spin-up state.

Light emitted in the spontaneous decay process propagates in any direction. Its wavelength is resonant to $5^2S_{1/2}$ – $5^2P_{1/2}$ transition and emitted unpolarized photons can be absorbed by other rubidium atoms. If the average free path of these photons is much shorter than the dimension of the optical pumping cell, the photons are trapped inside the cell. This process is called *radiation trapping* and reduces the final polarization of valence electrons. The decay from the excited state without emission of photons can restrict this process. The non-radiative decay occurs during collisions of the excited rubidium atoms with other molecules *e.g.* nitrogen (N_2). Therefore N_2 quenching becomes the main source of de-excitation and eliminates radiation trapping. For this process to occur nitrogen density must be higher than 0.1 amagat.[19,20] (The amagat is a unit of ideal gas density and is defined as the number of atoms per unit volume at pressure of 101 325 Pa and temperature of 0 °C.)

Both the laser light intensity and light spectral profile (line width) have direct impact on the rubidium ground state polarization.

Let us define the rubidium vapor ground state polarization as

$$P_{\mathrm{Rb}} = \frac{Pm_{s\uparrow} - Pm_{s\downarrow}}{2}, \tag{10.7}$$

where $Pm_{s\uparrow}$ and $Pm_{s\downarrow}$ are the occupation probabilities of $m_s = +1/2$ and $m_s = -1/2$ sublevels, respectively. P_{Rb} denotes longitudinal polarization, along the magnetic field B_0. In alkali vapor in thermal equilibrium, no transverse magnetization is produced by the optical pumping process. The interactions between light and valence electron spins can be expressed by the optical Bloch equation. The dynamics of polarization is then described as:

$$\frac{\mathrm{d}P_{\mathrm{Rb}}}{\mathrm{d}t} = -\frac{P_{\mathrm{Rb}} - P_{\mathrm{Rb}}^{\mathrm{eq}}}{T_r}, \tag{10.8}$$

where equilibrium polarization (the value of which depends on the contradictory processes of creation and destruction of polarization, characterized by the optical pumping rate P_{OP} and the spin destruction rate Γ_{SD}) is given by

$$P_{\mathrm{Rb}}^{\mathrm{eq}} = \frac{1}{2}\frac{P_{\mathrm{OP}}}{P_{\mathrm{OP}} + \Gamma_{\mathrm{SD}}} \tag{10.9}$$

and T_r is the effective relaxation time.[16,21]

There are a number of rubidium relaxation processes, described by the spin destruction rate Γ_{SD} causing the relaxation: collisions between alkali atoms (redistributing the angular momentum between the ground-state sublevels), atoms collisions with buffer gases, and collisions with a cell's wall (mostly due to paramagnetic impurities of glass, but generally depending on the composition of the cell wall). Obviously Γ_{SD} is proportional to alkali vapor density. The optical pumping rate

$$P_{OP} = \frac{\Omega_R^2}{4\Gamma_{OD}} \frac{\Gamma_{OD}^2}{\Gamma_{OD}^2 + \Delta^2} \tag{10.10}$$

is a function of the optical Rabi frequency Ω_R, the optical coherence decay Γ_{OD}, and the offset Δ of the laser frequency ω_L from the rubidium resonance frequency ω_{Rb} expressed as $\Delta = \omega_L - \omega_{Rb}$. The Rabi frequency characterizes the strength of the coupling between the atoms and resonant light. The squared Rabi frequency Ω_R^2 and thus the optical pumping rate is proportional to the incident light intensity. The optical coherence decay is equal to the half of the alkali absorption line width. This line is usually broadened by the collisions and pressure. Polarization of the alkali atoms increases exponentially with time according to the equation

$$P_{Rb}(t) = P_{Rb}^{eq}\left(1 - e^{-\frac{t}{T_r}}\right) \tag{10.11}$$

with the effective relaxation rate $1/T_r$.

To obtain the high polarization, the requirement

$$\Gamma_{SD} \ll P_{OP} \tag{10.12}$$

should be fulfilled (see eqn (10.9)). Therefore high power and narrow band lasers are preferable for the spin exchange optical pumping.

10.3.2 Spin Exchange between Optically Pumped Alkali Metal Atoms and Noble Gas Nuclei

The alkali metal electronic polarization, produced in the optical pumping process, can be transferred to nuclear spins of noble gases by atomic collisions. Electronic spins of polarized rubidium atoms interact with nuclear spins of helium or xenon atoms through hyperfine interactions. The coupling strength depends on the distance between atoms. Rubidium polarization decreases during the spin exchange process but it is constantly replenished by the optical pumping. Dynamics of noble gas nuclear spin polarization P can be expressed as:[6,21]

$$\frac{dP}{dt} = \gamma_{SE}(P_{Rb} - P) - \Gamma_{SR}P, \tag{10.13}$$

where γ_{SE} is the spin-exchange rate including processes contributing to the nuclear polarization growth. Spin-exchange or spin destruction are the interactions occurring in the gas mixture. The nuclear spin relaxation rate $\Gamma_{SR} = \frac{1}{T_1}$ depends on the collisions of atoms with the cell wall, dipole–dipole coupling, and collisions with gas impurities. The increase of the noble gas polarization P during the pumping process, assuming zero polarization at the beginning ($t = 0$), can be calculated from the equation:

$$P = P_{Rb} \frac{\gamma_{SE}}{\gamma_{SE} + \Gamma_{SR}} \left(1 - e^{-(\gamma_{SE} + \Gamma_{SR})t}\right). \qquad (10.14)$$

The increase of the nuclear polarization within several gas relaxation time constants $T_1 = \frac{1}{\Gamma_{SR}}$ (arbitrarily chosen and adequate for ^3He),[8] calculated from eqn (10.14) are presented in Figure 10.5.

In the steady state, the alkali atoms electron polarization transferred to noble gas nuclear spins is determined by:

$$P = P_{Rb} \frac{\gamma_{SE}}{\gamma_{SE} + \Gamma_{SR}}. \qquad (10.15)$$

To achieve high noble gas polarization P the spin-exchange rate should hugely exceed the nuclear spin relaxation rate, $\gamma_{SE} \gg \Gamma_{SR}$. Because γ_{SE} is proportional to the rubidium gas density, it can easily be increased by thermal heating of the cell. Therefore SEOP process is usually performed at temperatures between 100 and 160 °C (see the diagram of rubidium vapor

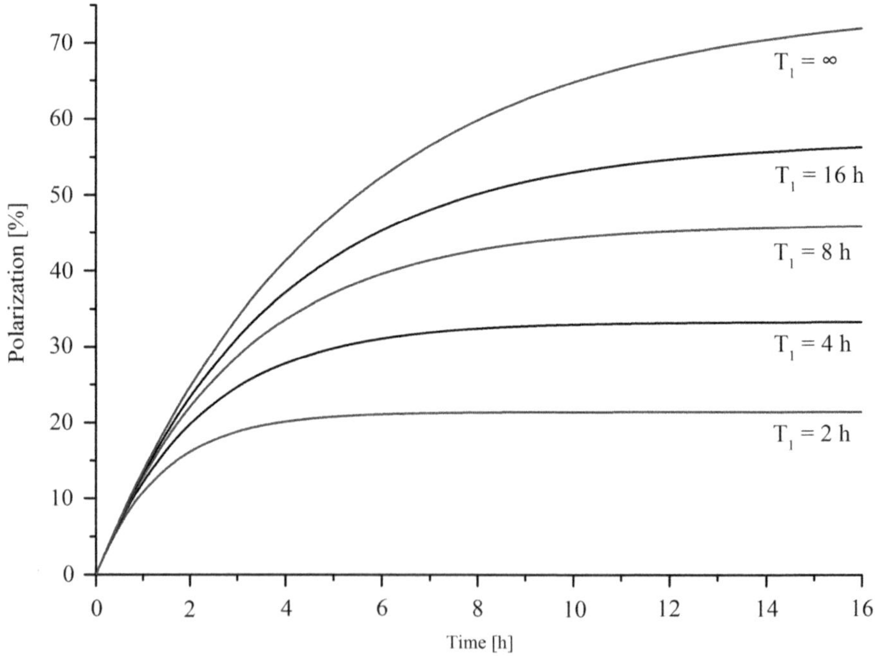

Figure 10.5 The theoretical increase of noble gas polarization P within time for several values of noble gas (^3He) relaxation time constant T_1. Calculations are based on eqn (10.14) with arbitrary and fixed rubidium polarization $P_{Rb} = 0.75$ and spin exchange rate $\gamma_{SE} = 0.2\ h^{-1}$.

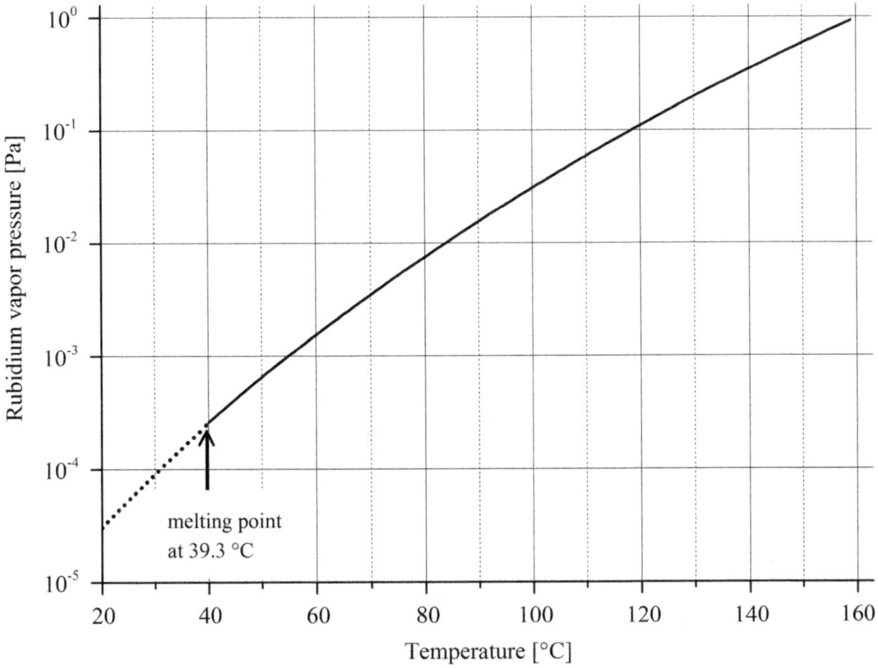

Figure 10.6 Vapor pressure of ^{87}Rb (logarithmic scale) *vs.* temperature calculated from the model equation.[17,18]

pressure *vs.* temperature in Figure 10.6). At rubidium vapor partial pressure above 0.1 Pa spin-exchange rate is sufficiently high.

In a well-illuminated part of the cell most of the alkali atoms are optically pumped into the non-absorbing level of the ground state and the electron spin polarization reaches a very high value. In contrast, spin polarization is nearly zero in the dark sections of the cell. A laser beam with the Gaussian intensity profile illuminates more intensively the central part of the cell. The Beer–Lambert law describes attenuation of the resonant laser light traveling through dense alkali metal vapor.[22] The light intensity decreases along the direction of propagation in the cell (z-axis) as well as transversely (along radius r) and therefore the alkali metal polarization should be considered in terms of a cylindrical coordinate system $P_{Rb}(r, \phi, z)$. In addition, the density of alkali vapor is usually inhomogeneous within the volume of the externally heated optical pumping cell. The temperature gradient along glass walls or optical windows can also be produced by hot air or oil bath. In addition, high-power laser light heats atoms inside the cell. Finally, the spin exchange rate γ_{SE} is also a function of (r, ϕ, z) and we can rewrite eqn (10.15) as:

$$P = P_{Rb}(r, \varphi, z)\frac{\gamma_{SE}(r, \varphi, z)}{\gamma_{SE}(r, \varphi, z) + \Gamma_{SR}}. \quad (10.16)$$

Due to the long relaxation time of noble gases and their diffusion within the cell, the spin polarization of ^{3}H or ^{129}Xe is nearly the same throughout the cell.

The polarization transfer from the polarized alkali metal atomic electrons to the noble gas nuclei can occur during two types of atomic interactions: two-body binary collisions in the very short time scale[23,24] and a van der Waals molecules formation.[25] In the former the angular momentum can be transferred from the electron spins of the optically pumped atoms to the nuclear spins of ^{3}He. Bouchiat *et al.* observed this effect[26] in ^{3}He used as buffer gas for the optical pumping of rubidium vapor. The Overhauser nuclear polarization effect causes dipolar interactions between polarized alkali atoms and nuclei of noble gas at considerably higher pressure than that of the alkali vapor. Because of low rubidium concentration the effect resembles relaxation of optically polarized impurity. The process of two-body binary collisions has low efficiency and the exchange times are very long (order of tens of hours). In the experiment performed by Bouchiat *et al.* optically pumped rubidium vapor under partial pressure of 0.13 Pa generated polarization of the ^{3}He sample at $2.84 \cdot 10^{5}$ Pa within $\sim 10^{5}$ seconds. The rubidium atoms were optically polarized by the light resonant with D_1 transition. As a light source the authors used a lamp with rubidium atoms excited by a radiofrequency discharge. Light was passing through an interference filter (to reject the other spectral line – D_2, emitted from excited rubidium atoms along with the desired D_1 line) and circularly polarized (by a linear polarizer and a quarter-wave plate). Since this first experiment, efficiency of ^{3}He SEOP has improved with the use of laser light for optical pumping and it is widely exploited nowadays.[27]

The binary collisions used for spin exchange typically take place between partners characterized by a big disparity in mass, like tiny ^{3}He and heavy Rb, and provide significant contribution to the ^{129}Xe SEOP at high gas pressure. Because helium atoms have weak van der Waals attraction to rubidium atoms and do not create a molecule, this spin exchange mechanism is not utilized in ^{3}He SEOP. The idea of nuclear polarization of ^{3}He induced by the binary collision with the dipolar exchange is presented in Figure 10.7. The up and down arrows indicate orientations (related to the external magnetic field) of rubidium valence electron and ^{3}He nuclear spin before and after a collision.

In 1978, Grover[28] reported that the spin exchange method can also be used for isotopes of xenon and krypton. In the case of nuclei of heavy noble gases the spin exchange rates are dominated by the interactions in van der Waals molecules.[29] A molecule consisting of an alkali metal atom and a noble gas atom is created in a three-body collision. This process is schematically illustrated in Figure 10.8. A rubidium atom and a xenon atom collide in the presence of a nitrogen molecule and form a weakly bound Rb–Xe van der Waals molecule. The third body – a nitrogen molecule – carries away the binding energy of the van der Waals molecule. The Rb–Xe molecule evolves freely until it is broken up by a collision with another nitrogen molecule.

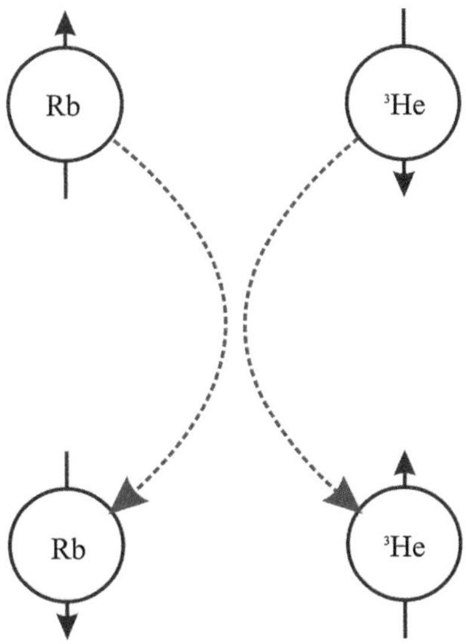

Figure 10.7 Schematic concept of the spin exchange during the two-body binary collision.

The Rb–Xe molecule half-lifetime is relatively long ($\sim 10^{-9}$ s at gas pressure of a few kPa), much longer than the time needed for the binary collision ($\sim 10^{-12}$ s).[6,30] It is also long enough for the nuclear spin of the xenon atom to flip-flop due to the weak coupling between the electron spin of the rubidium atom and the rotational angular momentum of the Rb–Xe van der Waals molecule.[31,32] In this spin exchange process only a small fraction of the spin orientation is transferred from the alkali atom to the noble atom nucleus.[25] The remaining (more than 90%) is lost to the rotational motion of the Rb–Xe van der Waals molecule.

10.3.3 Relaxation Processes

The crucial issue in gas polarization is relatively short longitudinal relaxation of the noble gas spins in both the optical pumping and the storage cells. This short relaxation time limits the ultimate achievable polarization and restricts storage time of polarized gas. The longitudinal relaxation processes can be classified into two categories: intrinsic (generated by collisions between noble gas atoms with impurities in gas and in the case of ^{129}Xe SEOP, caused by xenon dimers formation) and extrinsic (caused by magnetic field gradients and collisions with paramagnetics in glass walls). The formation of stable xenon van der Waals dimers occurs at low gas pressure.[33] Xenon dimers are created during the three-body collisions and their molecular lifetime lasts till the next collision with another atom. For a gas mixture, the relaxation rate of the Rb–Xe van der Waals molecules is independent of total gas density.[34] The SEOP in xenon is most efficient at

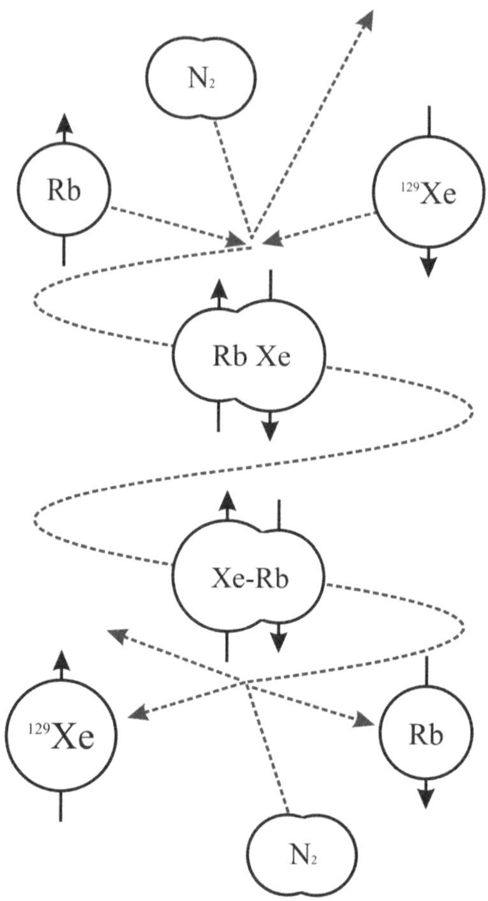

Figure 10.8 Schematic concept of the spin exchange during the three-body van der Waals molecule formation.

low concentrations of gas (usually 1–2%) because of the predominant contribution of the collisions with other xenon atoms to the relaxation.[24,30] Thus, it is said "Xenon is its own worst enemy".[13] The amount of transient xenon dimers created in the process of the binary collisions is negligible at low gas pressure. The extrinsic relaxation processes are mostly caused by the magnetic field gradients, while collisions with paramagnetic impurities in glass walls add to shortening the relaxation time. The wall relaxation is pressure independent but strongly depends on the surface-to-volume ratio (cell geometry). This type of relaxation can be reduced in the high magnetic field, using a glass with lower content of iron ions or reducing coating of the cell walls. The surfaces coating (silane or siloxane based) can extend available time for the storage of hyperpolarized gases by preventing gas diffusion to glass and the interactions with paramagnetic impurities.[35]

For optical pumping of the rubidium D_1 transition high-power laser diode arrays (LDA) are used as a light source.[36,37] These lasers (GaAlAs) have relatively high power output (up to several tens of watts) but emit a broad spectrum (up to a few nanometers) bandwidth. Many SEOP polarizers run at increased gas pressure to broaden the D_1 transition to improve photon absorption.[38] Therefore, a gas mixture used in xenon polarizers contains a large fraction of helium. However, this high percentage of noble gas would depolarize rubidium electron spins too much, therefore the modern SEOP polarizers are equipped with the narrow-line lasers.[39,40]

There are a few types of external laser resonators attached to the diode bars. As the dispersive element a diffraction grating is used most frequently. A typical laser diode array consists of a dozen or so emitters arranged in a straight line. The light emitted from the laser diode elements has a typical diffraction-limited divergence angle (40–50°) in the direction perpendicular to the array. There is also a multimode divergence angle ($\sim$10°) parallel to the array. Light from each diode has to be collimated, reflected off a diffraction grating at a specific angle, and directed back onto the diode with sufficient efficiency. The external cavity is usually arranged in so-called Littrow configuration in which the blaze angle is chosen such that diffraction angle and incidence angle are the same. The diffracted beam is back-reflected from the grating into the direction of the incident beam. As a result, the spectral power of light generated by the external-cavity setup is approximately a factor of 10 greater than that of the free-running array. For example, the spectral power from a 20 W laser diode array can reach 250 W THz^{-1} and exceeds that of a typical 100 W array by a factor of 2. The initial line width of the laser diode bar can be narrowed more than 20 times and attain FWHM below 0.1 nm. The latest generation of the SEOP systems incorporate volume Bragg gratings (VBG or volume holographic gratings – VHG).[41–43] Such volume gratings reflect only a spectrally narrow part of the initial spectrum back onto the laser diodes and force them to emit at this particular wavelength.

Delivery of polarized xenon from the optical pumping cell can be accomplished with a cold trap, consisting of a glass spiral tube. The tube allows separation of noble gas (Xe) from gas mixture (*e.g.* helium, nitrogen). It is immersed in the liquid nitrogen bath to freeze xenon while the gas mixture flows through the polarizer outlet. This accumulation process takes about 30 min.[44] A strong permanent magnet around the cold trap is used to conserve frozen xenon polarization by keeping its spin states separated. When the polarization process is finished polarized gas is sublimated from frozen xenon with hot water flowing around the glassware and then xenon is expanded directly to a plastic bag (made from polyvinyl fluoride – PVF, commercially named Tedlar®) and transported directly to an MR scanner.

Spin-exchange optical pumping of ^{3}He can be also performed with a rubidium–potassium mixture of alkali metal vapors or with pure potassium alone.[45,46] In this so-called Hybrid-SEOP rubidium atoms absorb light from a standard 795 nm laser and then collisionally polarize potassium (K) atoms. Spin-exchange collisions between K and ^{3}He atoms occur with much greater efficiency than in Rb–^{3}He collisions. As a result, transfer of the angular momentum to the ^{3}He increases by a factor of 10 or more. The highest helium polarization rate was obtained with K–Rb vapor density ratio between 2 and 6.[47] SEOP with the pure potassium (without rubidium vapor) is also possible with the use of high power, spectrally narrowed lasers operating at 770 nm (potassium D_1 transition).[48]

10.4 Metastability Exchange Optical Pumping of ^{3}He

The metastability exchange optical pumping method was proposed by Colegrove *et al.* in 1963.[49] The theoretical aspects of this method had been intensively investigated ever since,[50] but, at that pioneering time, optical pumping in helium was performed using a weak light source from a helium lamp, thus low polarization was obtained (order of a few percent). Because laser sources operating at a wavelength suitable for MEOP were developed in the 1980s, much higher nuclear polarizations of ^{3}He have been obtained since then. In addition, a theoretical model of MEOP with the restriction of the low pumping light intensity was proposed in 1985.[51] High nuclear polarization of ^{3}He can be currently obtained by optical pumping of ^{3}He metastable state (*i.e.* a particular excited state of ^{3}He atom that has a longer lifetime than the ordinary excited states) and the transfer of angular momentum to the ground state by collisions with metastability exchange. The MEOP technique provides very high nuclear polarization (up to 70%) with very good photon efficiency. Usually, for each photon absorbed in the optical pumping process, one ^{3}He nucleus is polarized. High-power fiber lasers suitable for this method are now commercially available. The only drawback of the standard MEOP is the limited range of operating pressures (hundreds of Pa). However, this range can be extended to several tens of kPa by performing MEOP in a high magnetic field,[52] such as used in 1.5 T human MR scanners,[53–55] and up to 4.7 T and possibly even higher in experimental MRI systems.[56]

10.4.1 Optical Pumping of ^{3}He and Metastability Exchange

The key process in the MEOP method is the net transfer of angular momentum from absorbed light to electron spins of ^{3}He atoms in excited states. The structure of low-lying energy states of ^{3}He is presented in Figure 10.9.

The 1^1S_0 ground level is a singlet spin state with no orbital angular momentum. The total electronic angular momentum of ^{3}He in the ground state 1^1S_0 is zero ($J = 0$). The nuclear spin one-half results in two magnetic sublevels ($m_F = \pm 1/2$) of the ground state. The level structure of the excited states is determined from the total Hamiltonian including hyperfine interactions. The optical transition between 1^1S_0 and 2^3S_1 energy states is strictly forbidden. In order to populate the 2^3S_1 metastable state and perform the

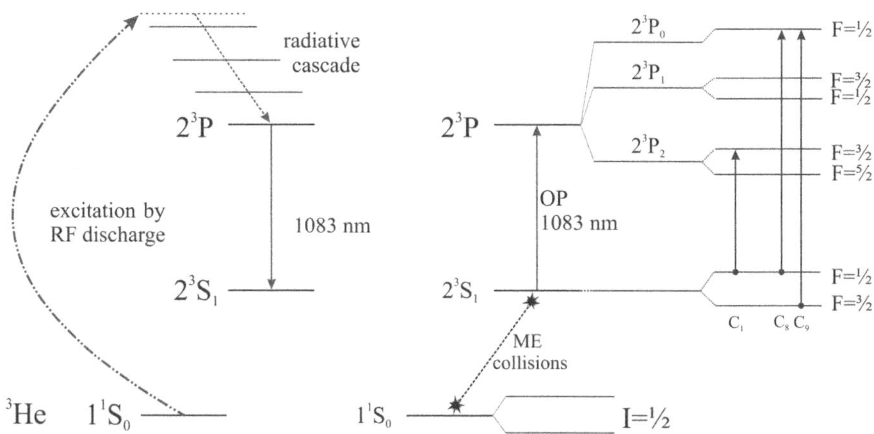

Figure 10.9 Populating the 2^3S_1 energy level of ^3He by the RF discharge and radiative cascade from the higher energy levels. The direct transition from the 1^1S_0 to the 2^3S_1 state is forbidden. Optical pumping (OP) of 2^3S_1 states (F = 3/2 or F = 1/2) by using the circularly polarized light with wavelength 1083 nm tuned to one of the 2^3P states is shown. In the low magnetic field the transitions C_8 (2^3S_1, F = 1/2 – 2^3P_0, F = 1/2) and C_9 (2^3S_1, F = 3/2 – 2^3P_0, F = 1/2) are most effective. The nuclear spin polarization of ^3He increases due to the metastability exchange (ME) collisions between atoms in 2^3S_1 and 1^1S_0 states.

optical pumping process, a weak radio-frequency discharge is used to sustain plasma in helium gas. This continuous RF discharge produces ions and, by electron collisions, causes population of the highly excited states of helium atoms. Then, in the radiative cascade, the 2^3S_1 metastable state is populated. Atoms in this metastable state are minority species in plasma with a typical density of 10^{11} cm^{-3} compared to $2.6 \cdot 10^{16}$ cm^{-3} of density of helium atoms at 100 Pa in typical optical pumping conditions. The structure of the energy states can be explained by hyperfine interactions. The first excited state 2^3S_1 has two hyperfine sublevels $F = 1/2$ and $= 3/2$. The next excited state 2^3P has the fine and hyperfine structure with five sublevels: 2^3P_0 ($F = 1/2$), 2^3P_1 ($F = 1/2$, $F = 3/2$), and 2^3P_2 ($F = 3/2$, $F = 5/2$). The optical transitions between 2^3S_1 and 2^3P states correspond to the resonant line of wavelength $\lambda = 1083$ nm. C_i (where $i = 1\ldots9$) refer to the transitions by increasing energy: from the 2^3S_1, F = 1/2 → 2^3P_2, $F = 3/2$ transition, indicated as C_1, to the 2^3S_1, F = 3/2 → 2^3P_0, $F = 1/2$ transition (C_9). The transition between 2^3S_1, $F = 1/2$ and 2^3P_2, $F = 5/2$ states is forbidden in the zero magnetic field.[57]

Let us consider ^3He energy level structure in the presence of a weak magnetic field. In the case of electronic triplet state of ^3He, the nuclear spin $I = 1/2$ splits two hyperfine sublevels of the 2^3S_1 state into six magnetic sublevels. The same way, the five hyperfine sublevels of the 2^3P state are split into 18 magnetic sublevels. The C_8 and C_9 transitions are spectrally well resolved at a low magnetic field and are frequently used for optical pumping.

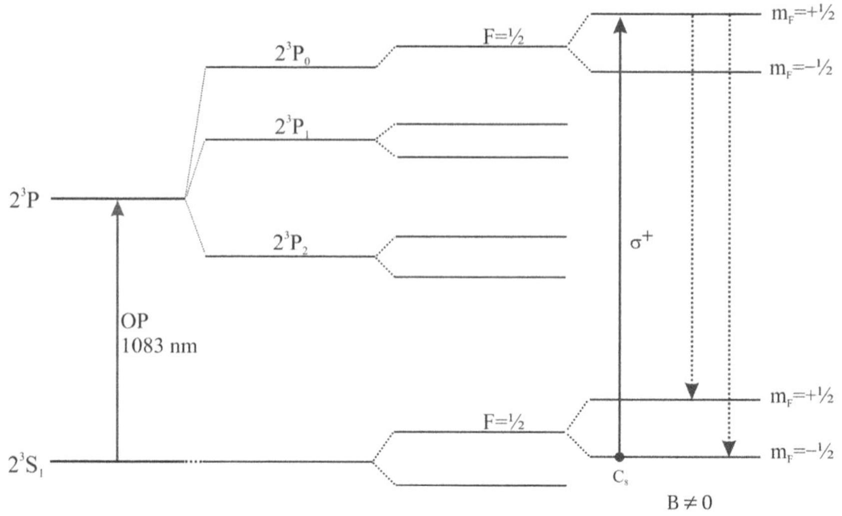

Figure 10.10 Schematic diagram of the optical pumping using C_8 component of the 2^3S_1, $F=1/2-2^3P_0$, $F=1/2$ transition in ^{3}He. In the presence of a weak magnetic field, clockwise circularly polarized light (σ^+) causes transition with angular momentum selection rule $\Delta m_F=+1$. As a result of optical pumping the $m_F=-1/2$ state is depopulated and most population is transferred to the $m_F=+1/2$ state.

The other transitions (C_1...C_7) are not well resolved because of the broadening due to the Doppler effect. A schematic diagram of the optical pumping using the C_8 component is presented in Figure 10.10.

In the presence of a weak magnetic field, the absorption of clockwise circularly polarized light (σ^+) causes transition from 2^3S_1, $m_F=-1/2$ state to 2^3P_0, $m_F=+1/2$ state, because of the angular momentum selection rule $\Delta m_F=+1$. A spontaneous reemission from 2^3P_0, $m_F=+1/2$ state to both sublevels of 2^3S_1 state occurs. The continuous depletion of the $m_F=-1/2$ sublevel results in higher population of the $m_F=+1/2$ sublevel. This optical pumping process increases the net magnetization of the total angular momentum of the ^{3}He atoms in the 2^3S_1 metastable state.

During so-called metastability exchange collisions a polarized metastable atom (labeled as $^3He_\uparrow(2^3S_1)$, see expression (10.17)) and a non-polarized ^{3}He atom in the ground state $^3He(1^1S_0)$ exchange their electronic states. In this collision the nuclear polarization remains unaffected. The initially polarized metastable atom is now in the ground state with the polarized nucleus $^3He_\uparrow(1^1S_0)$, and the second atom is excited from the ground to the metastable state $^3He(2^3S_1)$. The amount of the metastable atoms in a sample is not reduced by the metastability exchange collisions.

$$^3He_\uparrow(2^3S_1)+{}^3He(1^1S_0)\rightarrow{}^3He_\uparrow(1^1S_0)+{}^3He(2^3S_1) \tag{10.17}$$

A schematic of the MEOP system is presented in Figure 10.11. Linearly polarized (π) laser light of wavelength 1083 nm passes through the

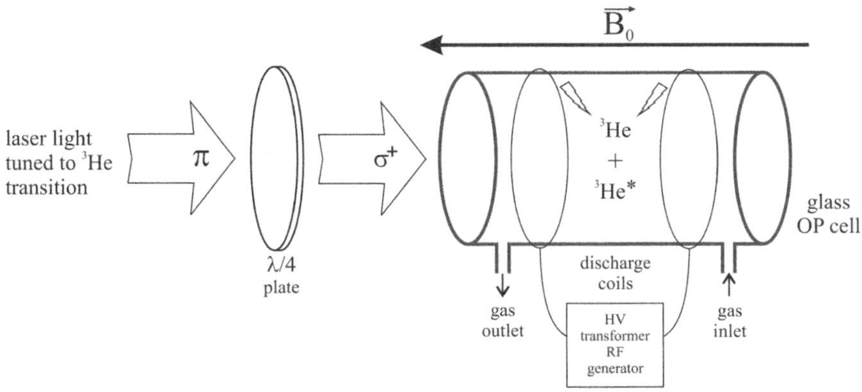

Figure 10.11 Diagram of the MEOP system. Linearly polarized (π) laser light passes through the quarter-wave plate ($\lambda/4$) and then as a clockwise circularly (σ^+) polarized light illuminates the optical pumping (OP) cell with ^{3}He* atoms in the metastable state.

quarter-wave plate ($\lambda/4$) and then, as a clockwise circularly (σ^+) polarized light, illuminates the optical pumping (OP) cell with a mixture of ^{3}He atoms in the ground state and a small amount of ^{3}He* atoms in the metastable state. A weak RF discharge in helium gas is produced by a set of external antennas connected to the matching transformer powered by an RF generator with an amplifier. The helium metastable state density depends on discharge intensity and can be controlled by changing RF power transferred to the antennas. The ^{3}He spectra of the clockwise (σ^+) and anti-clockwise (σ^-) circularly polarized light are indistinguishable, thus purely circular polarization of the pumping laser light must be applied. The MEOP method provides the highest steady state polarization for ^{3}He pressure lower than 100 Pa. Polarization production rates and general efficiency of MEOP depend mainly on the RF discharge intensities and on the used laser. Nuclear polarizations of the order of 80% were obtained thanks to the use of dedicated lasers (such as ytterbium fiber laser, diode laser oscillator coupled to an ytterbium-doped fiber amplifier, ytterbium-doped fiber oscillator coupled to ytterbium-doped fiber amplifier).[58,59]

10.4.2 Compression of Polarized ^{3}He

The main disadvantage of the standard MEOP technique is low ^{3}He gas pressure ($\sim$100 Pa) required for the most efficient conditions, *i.e.* low magnetic field of a few mT. When polarizing ^{3}He at higher pressure (exceeding 500 Pa), the obtained steady state nuclear polarization is much lower. Therefore, for any medical application, an additional, polarization-preserving compression is necessary. This higher gas density can be achieved by a mechanical compression which poses a number of technical challenges such as avoiding any objects causing relaxation and maintaining high purity of gas.[60] In early experiments the Toepler compressor was used.[61] In this system, inside a glass tube a mercury piston was activated periodically to evacuate ^3He from the pumping cell and to compress it in a storage cell. A piston compressor made of titanium was used in the consequent experiments. Such a solution has been used until now in a large-scale helium polarizer (Institute of Physics, Johannes Gutenberg University, Mainz) and in construction of a compact helium polarizer facility.[62] Such a polarization-conserving piston compressor is driven by hydraulics and can achieve final gas pressure up to 600 kPa. In the first step polarized ^3He is compressed into a buffer cell and then, after achieving the desired amount, is pumped to a detachable transport cell. Instead of a very efficient but rather complicated piston compressor a peristaltic compressor suitable for relaxation-free compression of polarized gas can be used. In this solution a pumping rotor equipped with a few pressing rollers and a tube with polarized helium are placed in a depressurized chamber with a lower than atmospheric pressure. This vacuum prevents the tube from squeezing under atmospheric pressure while compressing ^3He in the optical pumping cell at low pressure.[63,64] The pumping rotor is driven by an engine placed far away from the region of homogeneous magnetic field created around OP and polarized gas storage cells.

10.4.3 MEOP at High Magnetic Field and Elevated Pressures

At pressures above 500 Pa the steady state polarization of ^3He produced in MEOP decreases rapidly and its value reaches only a few percent at 4 kPa or higher pressure. There are two main relaxation channels in this case:

(1) The ionizing Penning collisions (see expression (10.18)) shorten the lifetime of ^3He atoms in the metastable 2^3S_1 state. This process induces an unfavorable ratio between the metastable state and the ground state populations of helium atoms.

$$He(2^3S_1) + He(2^3S_1) \rightarrow He(1^1S_0) + He^+(1^1S_0) + e^- \qquad (10.18)$$

The incidence of Penning collisions increases non-linearly with gas pressure and effects the efficiency of MEOP.[65]

(2) At higher pressure incidence of three-body collisions (10.19) increases with creation of metastable helium molecules $He_2(2^3S_1)$.

$$He(2^3S_1) + 2He(1^1S_0) \rightarrow He_2(2^3S_1) + He(1^1S_0) \qquad (10.19)$$

The rate of creation of such molecules increases with the square of pressure and their diffusion increases linearly with gas pressure.[52] Collisions between nuclear polarized helium atoms (in the ground state) and metastable helium molecules are similar to the metastable exchange collisions. As a result nuclear angular momentum can be dissipated in the rotational states of the helium molecule by the spin–orbit coupling.[66,67]

In the standard MEOP conditions low magnetic field (a few mT) has negligible influence on the structure of helium atomic states. Fine and hyperfine interactions play a more significant role than interactions between electronic and nuclear spins with the magnetic field. In practice, the optical pumping can be performed with both clockwise and anti-clockwise circularly polarized light with similar results. At a higher magnetic field the structure of 2^3S and 2^3P levels and therefore transitions between them are significantly modified. At the exact value of $B=0.1602$ T, the first energy crossing of 2^3S eigenstates appears between $F=3/2$, $m_F=3/2$ (referred to as A_4, where six sublevels $A_1 \ldots A_6$ of 2^3S are labeled by their increasing order of energy)[57] and $F=1/2$, $m_F=-1/2$ (A_5). At even higher magnetic field the energy levels are mostly determined by m_S and m_L, otherwise F is not an adequate quantum number. However, the relation $m_L+m_I+m_S=m_F$ is still fulfilled. Eighteen eigenstates of 2^3P level ($B_1 \ldots B_{18}$) are also shifted, effecting very complicated structure of the transitions. Furthermore, the spectra depend on the light polarization (circular σ^+, circular σ^-, or linear). This is a significant benefit of hyperfine decoupling, because it is no longer essential to use extremely pure circular polarization of light for the optical pumping.

The first MEOP experiments at high magnetic field were performed in 2001.[66] At $B=0.1$ T and gas pressure of 4 kPa polarization of ^{3}He was two times larger in comparison to the previous low magnetic field experiments. This promising result was followed by subsequent experiments. In the magnetic field of 1.5 T the helium spectra are stretched out to over 170 GHz with some overlapping components at room temperature (due to the Doppler broadening and collisions in a medium such as helium plasma inside an OP cell). Therefore, for a given circular polarization of pumping light, the absorption spectrum consists of a few main components appearing as unresolved pairs or quartets. One set of two unresolved transitions was found as the most efficient component for MEOP in the high magnetic field.[68] Then, an accurate optical method to measure the nuclear polarization of ^{3}He atoms in the ground state was implemented.[69–71] In this method, collective spin temperature between the metastable and ground states is established by metastable exchange collisions. Absorption of the weak probe laser beam is used to measure the relative populations of the two hyperfine sublevels of the metastable state that are not utilized in optical pumping process. Because the metastable sublevels have to be continuously populated by plasma discharge in the OP cell, this method can be used to monitor the dynamics of MEOP and relaxation processes. Thereafter, the non-standard MEOP method was extended to 6.7 kPa at 2 T with 51% polarization.[54] Finally, systematic research was performed in magnetic fields up to 4.7 T (see Figure 10.12) and up to 26 kPa of gas pressure.[56]

At low ^{3}He pressure and in low magnetic field (standard MEOP conditions), plasma distribution is uniform in the entire OP cell volume. Under elevated gas pressure and in high magnetic field plasma is non-uniformly distributed in the OP cell. Generated in helium discharge plasma appear close to the glass surface, in the proximity of the antenna mounted outside

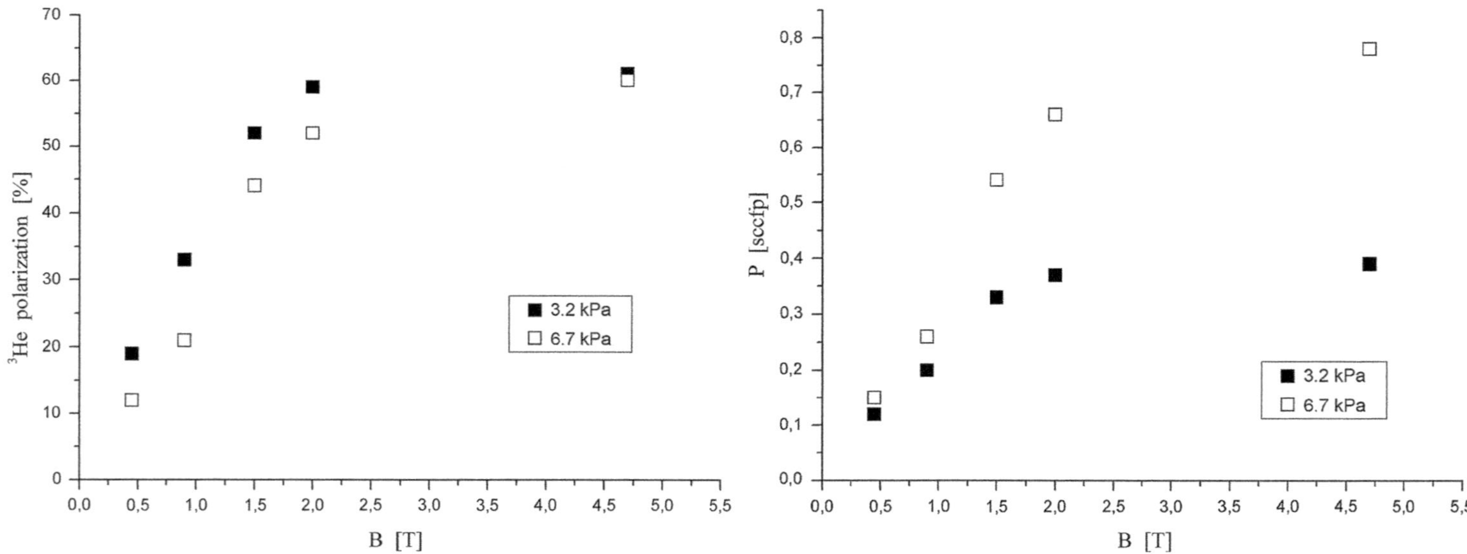

Figure 10.12 (Left) Steady-state ^{3}He polarization as a function of magnetic field B in the 3.2 kPa (black squares) and 6.7 kPa (open squares) sealed cells at fixed pump laser power (0.5 W) and weak RF discharge. (Right) The same ^{3}He polarization was expressed in "sccfp" (*standard cubic centimeter of fully polarized gas*) to show the great improvement of MEOP efficiency in non-standard conditions. The amount of polarized helium equivalent to the standard cubic centimeters of fully polarized gas was calculated as: $P[\text{sccfp}] = \dfrac{\text{gas pressure in the cell}}{\text{atmospheric prassure}} \cdot \text{cell volume}[\text{cm}^3] \cdot P[\%]$.

of the OP cell. At higher gas pressure, the two- and three-body collisions producing metastable helium molecules significantly decrease the population of metastable helium atoms. As a result, efficiency of the MEOP process is dramatically reduced in regions with no plasma. In addition, there is only partial overlapping of a standard Gaussian laser beam profile and population of atoms in the metastable state. Thus a large fraction of the pumping laser power is wasted. In order to optimize efficiency of the MEOP process in non-standard conditions, the spatial shape of the pumping laser beam profile should match the annular plasma distribution. The annular profile of the laser beam can be created by a pair of axicons. The axicon is a refractive, cylindrically symmetric, optical element also called a conical prism. A pair of axicons can transform the Gaussian laser beam into an annular beam without loss of light power. The intensity distribution of the generated beam can be continuously modified by changing the incident Gaussian beam diameter and the distance between the two conical prisms. It was shown that, with the optical pumping light intensity distribution matching the metastable atoms distribution, it is possible to polarize ^{3}He atoms with high field MEOP at pressures exceeding 25 kPa (a quarter of atmospheric pressure!) while maintaining nuclear polarization of 20% or more.[72] Driven by these promising results the high field ^{3}He polarizer was designed to fit inside the MRI radio-frequency receiving coil that is used in human chest MR imaging. The MRI coil with the polarizer (consisting of gas handling system, table with mounted optical elements, glass cells, gas compression, and storage units) can be mounted on the patient bed inside a standard 1.5 T medical scanner.[55,73]

10.5 Summary

In this chapter the two most common polarization techniques of noble gases were presented: spin exchange optical pumping and metastability exchange optical pumping. Both methods use circularly polarized light to change distribution of the atomic magnetic moments in the optical pumping process. In the SEOP technique, alkali metal atoms are first optically pumped, and then, due to collisions, exchange spins polarization with ^{3}He or ^{129}Xe atoms. In the MEOP technique, metastable ^{3}He atoms are optically pumped, and then, due to collisions, exchange metastability providing polarization of the ground ^{3}He state.

Modifications of these methods delivering higher noble gases polarization, including hybrid SEOP or MEOP in standard conditions and in elevated pressure and high magnetic fields, were presented as well.

There are various noble gases polarizers. Their efficacy depends on power and spectral width of the lasers needed for optical pumping. Recent progress in laser technology allows construction of more and more powerful lasers (up to hundreds of watts) and thus more efficient optical pumping. The laser diode bars are currently the most frequently used for SEOP. For the construction of the external resonators used in such narrowband lasers, volume holographic grating is used. Fiber optics lasers are applied to produce light of 1083 nm wavelength used in the MEOP technique.

A very promising, although not yet commonly accepted, is application of a MEOP polarizer, working in a high magnetic field and at elevated pressure. This system can produce hyperpolarized ^{3}He inside a magnet used for MRI. This configuration rectifies issues associated with gas compression and gas transportation. Furthermore, the speed of the polarized gas production is a huge advantage of MEOP over SEOP polarizers. However, the current high costs of ^{3}He prevent its development and widespread implementation, although recycling systems can reduce its effective costs.

Polarizing systems allowing production of large quantities of hyperpolarized ^{3}He or ^{129}Xe for scientific and medical applications have also been constructed[74,75] followed by commercial production of very efficient SEOPs. Finally, it should be noted that a description of a home-made, automated ^{129}Xe hyperpolarizer is currently available as open source.[76,77]

References

1. M. S. Albert, G. D. Cates, B. Driehuys, W. Happer, B. Saam, C. S. Springer Jr and A. Wishnia, *Nature*, 1994, **370**, 199.
2. H. U. Kauczor, D. Hofmann, K. F. Kreitner, H. Nilgens, R. Surkau, W. Heil, A. Potthast, M. V. Knopp, E. W. Otten and M. Thelen, *Radiology*, 1996, **201**(2), 564.
3. M. Ebert, T. Grossmann, W. Heil, W. E. Otten, R. Surkau, M. Leduc, P. Bachert, M. V. Knopp, L. R. Schad and M. Thelen, *Lancet*, 1996, **347**, 1297.
4. J. R. MacFall, H. C. Charles, R. D. Black, H. Middleton, J. C. Swartz, B. Saam, B. Driehuys, C. Erickson, W. Happer, G. D. Cates, G. A. Johnson and C. E. Ravin, *Radiology*, 1996, **200**(2), 553.
5. W. Happer, *Lect. Notes Phys.*, 1997, **499**, 121.
6. T. G. Walker and W. Happer, *Rev. Mod. Phys.*, 1997, **69**(2), 629.
7. T. Chupp and S. Swanson, *Adv. At., Mol., Opt. Phys.*, 2001, **45**, 41.
8. J. C. Leawoods, D. A. Yablonskiy, B. Saam, D. S. Gierada and M. S. Conradi, *Concepts Magn. Reson.*, 2001, **13**(5), 277.
9. B. M. Goodson, *J. Magn. Reson.*, 2002, **155**, 157.
10. X. Wu, T. Nishino and H. Liu, in *Biomedical Photonics*, ed. Tuan Vo-Dinh, 2003, CRC Press, Boca Raton, London, New York, Washington, D.C., ch. 27.
11. A. M. Oros and N. J. Shah, *Phys. Med. Biol.*, 2004, **49**, R105.
12. J. Becker, J. Bermuth, M. Ebert, T. Grossmann, W. Heil, D. Hofmann, H. Humblot, M. Leduc, E. W. Otten, D. Rohe and R. Surkau, *Nucl. Instrum. Methods Phys. Res., Sect. A*, 1998, **402**, 327.
13. B. M. Goodson, Hyperpolarizing nuclear spins of noble gases via spin-exchange optical pumping: Principles and Practice for Enhancing NMR and MR, *COST Annual Meeting: Spin Hyperpolarisation in NMR and MRI*, Zürich, Switzerland June 27, 2014.

14. A. Kastler, *Nobel Lecture*, December 12, 1966; *Nobel Lectures, Physics 1963–1970*, Elsevier Publishing Company, Amsterdam, 1972, p. 186.
15. W. Happer and B. S. Mathur, *Phys. Rev.*, 1967, **163**(1), 12.
16. W. Happer, *Rev. Mod. Phys.*, 1972, **44**(2), 169.
17. D. A. Steck, Rubidium 85 D Line Data, available online at http://steck.us/alkalidata (revision 2.1.6, 20 September 2013).
18. D. A. Steck, Rubidium 87 D Line Data, available online at http://steck.us/alkalidata (revision 2.1.4, 23 December 2010).
19. A. Fink, D. Baumer and E. Brunner, *Phys. Rev. A: At., Mol., Opt. Phys.*, 2005, **72**, 053411.
20. A. Fink and E. Brunner, *Appl. Phys. B: Lasers Opt.*, 2007, **89**, 65.
21. T. G. Walker, *J. Phys.: Conf. Ser.*, 2011, **294**, 012001.
22. M. S. Freeman, K. Emami and B. Driehuys, *Phys. Rev. A: At., Mol., Opt. Phys.*, 2014, **90**, 023406.
23. C. C. Bouchiat, M. A. Bouchiat and L. C. L. Pottier, *Phys. Rev.*, 1969, **181**(1), 144.
24. M. A. Bouchiat, J. Brossel and L. C. Pottier, *J. Chem. Phys.*, 1972, **56**(7), 3703.
25. W. Happer, E. Miron, S. Schaefer, D. Schreiber, W. A. van Wijngaarden and X. Zeng, *Phys. Rev. A: At., Mol., Opt. Phys.*, 1984, **29**(6), 3092.
26. M. A. Bouchiat, T. R. Carver and C. M. Varnum, *Phys. Rev. Lett.*, 1960, **5**(8), 373.
27. E. Babcock, B. Chann, T. G. Walker, W. C. Chen and T. R. Gentile, *Phys. Rev. Lett.*, 2006, **96**, 083003.
28. B. C. Grover, *Phys. Rev. Lett.*, 1978, **40**(6), 391.
29. N. D. Bhaskar, W. Happer and T. McClelland, *Phys. Rev. Lett.*, 1982, **49**(1), 25.
30. M. E. Wagshul and T. E. Chupp, *Phys. Rev. A: At., Mol., Opt. Phys.*, 1994, **49**(5), 3854.
31. S. Appelt, A. Ben-Amar Baranga, C. J. Erickson, M. V. Romalis, A. R. Young and W. Happer, *Phys. Rev. A: At., Mol., Opt. Phys.*, 1998, **58**(2), 1412.
32. S. Appelt, T. Ünlü, K. Zilles, N. J. Shah, S. Baer-Lang and H. Halling, *Appl. Phys. Lett.*, 1999, **75**(3), 427.
33. B. C. Anger, G. Schrank, A. Schoeck, K. A. Butler, M. S. Solum, R. J. Pugmire and B. Saam, *Phys. Rev. A: At., Mol., Opt. Phys.*, 2008, **78**, 043406.
34. B. Chann, E. Babcock, L. W. Anderson and T. G. Walker, *Phys. Rev. A: At., Mol., Opt. Phys.*, 2002, **66**, 032703.
35. S. R. Breeze, S. Lang, I. Moudrakovski, C. I. Ratcliffe, J. A. Ripmeester, B. Simard and G. Santyr, *J. Appl. Phys.*, 1999, **86**, 4040.
36. M. E. Wagshul and T. E. Chupp, *Phys. Rev. A: At., Mol., Opt. Phys.*, 1989, **40**(8), 4447.
37. W. J. Cummings, O. Häusser, W. Lorenzon, D. R. Swenson and B. Larson, *Phys. Rev. A: At., Mol., Opt. Phys.*, 1995, **51**(6), 4842.
38. M. V. Romalis, E. Miron and G. D. Cates, *Phys. Rev. A: At., Mol., Opt. Phys.*, 1997, **56**(6), 4569.

39. B. Chann, I. Nelson and T. G. Walker, *Opt. Lett.*, 2000, **25**(18), 1352.
40. B. Chann, E. Babcock, L. W. Anderson, T. G. Walker, W. C. Chen, T. B. Smith, A. K. Thompson and T. R. Gentile, *J. Appl. Phys.*, 2003, **94**(10), 6908.
41. P. Nikolau, N. Whiting, N. A. Eschmann, K. E. Chaffee, B. M. Goodson and M. J. Barlow, *J. Magn. Reson.*, 2009, **197**(2), 249.
42. N. Whiting, P. Nikolaou, N. A. Eschmann, B. M. Goodson and M. J. Barlow, *J. Magn. Reson.*, 2011, **208**, 298.
43. N. Whiting, P. Nikolaou, N. A. Eschmann, M. J. Barlow, R. Lammert, J. Ungar, W. Hu, L. Vaissie and B. M. Goodson, *Appl. Phys. B: Lasers Opt.*, 2012, **106**, 775.
44. M. S. Rosen, T. E. Chupp, K. P. Coulter, R. C. Welsh and S. D. Swanson, *Rev. Sci. Instrum.*, 1999, **70**(2), 1546.
45. A. Ben-Amar Baranga, S. Appelt, M. V. Romalis, C. J. Erickson, A. R. Young, G. D. Cates and W. Happer, *Phys. Rev. Lett.*, 1998, **80**(13), 2801.
46. E. Babcock, I. Nelson, S. Kadlecek, B. Driehuys, L. W. Anderson, F. W. Hersman and T. G. Walker, *Phys. Rev. Lett.*, 2003, **91**(12), 123003.
47. W. C. Chen, T. R. Gentile, T. G. Walker and E. Babcock, *Phys. Rev. A: At., Mol., Opt. Phys.*, 2007, **75**, 013416.
48. G. Wang, W. Shao and E. W. Hughes, *Phys. Rev. A: At., Mol., Opt. Phys.*, 2003, **68**, 065402.
49. F. D. Colegrove, L. D. Schearer and G. K. Walters, *Phys. Rev.*, 1963, **132**(6), 2561.
50. R. L. Gamblin and T. R. Carver, *Phys. Rev.*, 1965, **138**(4A), A946.
51. P. J. Nacher and M. Leduc, *J. Phys.*, 1985, **46**, 2057.
52. P.-J. Nacher, E. Courtade, M. Abboud, A. Sinatra, G. Tastevin and T. Dohnalik, *Acta Phys. Pol., B*, 2002, **33**(8), 2225.
53. M. Abboud, A. Sinatra, X. Maître, G. Tastevin and P.-J. Nacher, *Europhys. Lett.*, 2004, **68**(4), 480.
54. A. Nikiel, T. Pałasz, M. Suchanek, M. Abboud, A. Sinatra, Z. Olejniczak, T. Dohnalik, G. Tastevin and P.-J. Nacher, *Eur. Phys. J.: Spec. Top.*, 2007, **144**, 255.
55. G. Collier, T. Pałasz, A. Wojna, B. Głowacz, M. Suchanek, Z. Olejniczak and T. Dohnalik, *J. Appl. Phys.*, 2013, **113**, 204905.
56. A. Nikiel-Osuchowska, G. Collier, B. Głowacz, T. Pałasz, Z. Olejniczak, W. P. Węglarz, G. Tastevin, P.-J. Nacher and T. Dohnalik, *Eur. Phys. J. D*, 2013, **67**, 200.
57. E. Courtade, F. Marion, P.-J. Nacher, G. Tastevin, K. Kiersnowski and T. Dohnalik, *Eur. Phys. J. D*, 2002, **21**, 25.
58. T. R. Gentile, M. E. Hayden and M. J. Barlow, *J. Opt. Soc. Am. B*, 2003, **20**(10), 2068.
59. G. Tastevin, S. Grot, E. Courtade, S. Bordais and P.-J. Nacher, *Appl. Phys. B: Lasers Opt.*, 2004, **78**, 145.
60. T. R. Gentile, G. L. Jones, A. K. Thompson, R. R. Rizi, D. A. Roberts, I. E. Dimitrov, R. Reddy, D. A. Lipson, W. Gefter, M. D. Schnall and J. S. Leigh, *Magn. Reson. Med.*, 2000, **43**, 290.

61. W. Heil, G. Eckert, M. Leduc, M. Meyerhoff, P. J. Nacher, E. W. Otten, Th. Proschka, L. D. Schearer and R. Surkau, in *High Energy Spin Physics*, ed. W. Meyer *et al.*, Springer-Verlag, Berlin Heidelberg, 1991, vol. 2, p. 178.
62. C. Mrozik, O. Endner, C. Hauke, W. Heil, S. Karpuk, J. Klemmer and E. W. Otten, *J. Phys.: Conf. Ser.*, 2011, **294**, 012007.
63. P.-J. Nacher, Peristaltic compressors suitable for relaxation-free compression of polarized gas, *U.S. Pat. No.* US6655931B2, 2003.
64. G. Collier, M. Suchanek, A. Wojna, K. Cieślar, T. Pałasz, B. Głowacz, Z. Olejniczak and T. Dohnalik, *Opt. Appl.*, 2012, **XLII**(1), 223.
65. L. D. Schearer, *Phys. Rev. Lett.*, 1969, **22**, 629.
66. E. Courtade, PhD thesis, Université Pierre et Marie Curie – Paris VI, 2001, http://tel.archives-ouvertes.fr/tel-00001447.
67. G. Tastevin, B. Głowacz and P.-J. Nacher, *J. Low Temp. Phys.*, 2010, **158**, 339.
68. M. Abboud, A. Sinatra, G. Tastevin, P.-J. Nacher and X. Maître, *Laser Phys.*, 2005, **15**(4), 475.
69. E. Stoltz, M. Meyerhoff, N. Bigelow, M. Leduc, P.-J. Nacher and G. Tastevin, *Appl. Phys. B: Lasers Opt.*, 1996, **63**, 629.
70. E. Stoltz, B. Villard, M. Meyerhoff and P.-J. Nacher, *Appl. Phys. B: Lasers Opt.*, 1996, **63**, 635.
71. K. Suchanek, M. Suchanek, A. Nikiel, T. Pałasz, M. Abboud, A. Sinatra, P.-J. Nacher, G. Tastevin, Z. Olejniczak and T. Dohnalik, *Eur. Phys. J.: Spec. Top.*, 2007, **144**, 67.
72. T. Dohnalik, A. Nikiel, T. Pałasz, M. Suchanek, G. Collier, M. Greńczuk, B. Głowacz and Z. Olejniczak, *Eur. Phys. J.: Appl. Phys.*, 2011, **54**, 20802.
73. T. Dohnalik, B. Głowacz, Z. Olejniczak, T. Pałasz, M. Suchanek and A. Wojna, *Eur. Phys. J.: Spec. Top.*, 2013, **222**, 2103.
74. F. W. Hersman, I. C. Ruset, S. Ketel, I. Muradian, S. D. Covrig, J. Distelbrink, W. Porter, D. Watt, J. Ketel, J. Brackett, A. Hope and S. Patz, *Acad. Radiol.*, 2008, **15**(6), 683.
75. F. W. Hersman, D. W. Watt, I. C. Ruset, J. H. Distelbrink and J. Ketel, *Phys. Procedia*, 2013, **42**, 171.
76. P. Nikolaou, A. M. Coffey, L. L. Walkup, B. M. Gust, N. Whiting, H. Newton, S. Barcus, I. Muradyan, M. Dabaghyan, G. D. Moroz, M. S. Rosen, S. Patz, M. J. Barlow, E. Y. Chekmenev and B. M. Goodson, *Proc. Natl. Acad. Sci.*, 2013, **110**(35), 14150.
77. P. Nikolaou, A. M. Coffey, L. L. Walkup, B. M. Gust, N. Whiting, H. Newton, I. Muradyan, M. Dabaghyan, K. Ranta, G. D. Moroz, M. S. Rosen, S. Patz, M. J. Barlow, E. Y. Chekmenev and B. M. Goodson, *Magn. Reson. Imaging*, 2014, **32**, 541.

CHAPTER 11

Medical Applications of Hyperpolarized and Inert Gases in MR Imaging and NMR Spectroscopy

MARCUS J. COUCH,[a,b] MATTHEW S. FOX,[c,d] BARBARA BLASIAK,[e] ALEXEI V. OURIADOV,[d] KRISTA M. DOWHOS,[a,b] BOGUSLAW TOMANEK[b,e,f] AND MITCHELL S. ALBERT*[a,b]

[a] Lakehead University, 955 Oliver Road, Thunder Bay, Ontario P7B 5E1, Canada; [b] Thunder Bay Regional Research Institute, 980 Oliver Rd, Thunder Bay, Ontario P7B 6V4, Canada; [c] Department of Medical Biophysics, Western University, London, Ontario N6A 5C1, Canada; [d] Robarts Research Institute, Western University, London, Ontario N6A 5B7, Canada; [e] Institute of Nuclear Physics, Polish Academy of Sciences, Radzikowskiego 152, 31-342 Krakow, Poland; [f] University of Alberta, Department of Oncology, 11560 University Avenue, Edmonton, Alberta T6G 1Z2, Canada

*Email: albertmi@tbh.net

11.1 Introduction

The available imaging modalities used for the diagnosis of human disease include planar X-ray, computed tomography (CT), ultrasound (US), magnetic resonance imaging (MRI), optical coherence tomography (OCT),

New Developments in NMR No. 6
Gas Phase NMR
Edited by Karol Jackowski and Michał Jaszuński

Published by the Royal Society of Chemistry, www.rsc.org

single-photon emission computed tomography (SPECT), positron emission tomography (PET), and optical imaging. In particular, MRI is a non-invasive and non-ionizing technique that is able to provide images with a high spatial resolution and excellent contrast-to-noise ratio (CNR) for soft tissues. Considering its advantages, MRI has developed as a critical research and diagnostic tool since its discovery in the early 1970s.[1] A potential limitation of MRI is that its sensitivity is lower than that of PET or SPECT. MR signal arises from a net magnetization due to the small population difference between the nuclear Zeeman energy levels of spin-$\frac{1}{2}$ nuclei in an external magnetic field. The polarization of the spin-$\frac{1}{2}$ nuclei is defined as the fraction of excess nuclei in the lower energy level.[2] Conventional MRI detects hydrogen nuclei (^1H) and typical ^1H polarization values at body temperature and in clinical magnetic field strengths are 4.9×10^{-6} at 1.5 T and 9.9×10^{-6} at 3 T. Since polarization scales linearly with the magnetic field strength, sensitivity and thus signal-to-noise ratio (SNR) of conventional MRI can be improved by increasing the magnetic field strength. The strongest magnet used for whole body human MRI research is currently 9.4 T (Max Planck Institute, Tübingen, Germany).[3] The increase in magnetic field strength comes with additional challenges, including a higher specific absorption rate (SAR), and also interactions between the radio frequency (RF) field and tissue, which can cause image inhomogeneities.[4]

In general, sensitivity is the main prerequisite for early diagnosis of almost any disease. For example, Wu *et al.* reported a study of lung cancer detection using ^1H MRI in asymptomatic individuals, and it was determined that the expected detection threshold in MRI is approximately 0.3 cm for contrast enhanced MRI and 0.5 cm for non-contrast enhanced MRI.[5] In other words, malignant tumours that are smaller than 0.3 cm in diameter may go undetected with present MRI sensitivity. Improvements in diseased cell detection can be achieved by the application of contrast agents, including endogenous, exogenous non-targeted, and exogenous targeted contrast agents.[6] Iron particles that are naturally occurring in the body can serve as endogenous MRI contrast agents. Since these particles have high magnetic moments, they decrease the T_2 and $T_2{}^*$ relaxation times of surrounding protons, leading to a contrast generating decrease in MR signal for surrounding tissues in T_2 and $T_2{}^*$-weighted images.[7] Changes in brain iron distribution have been associated with neurodegenerative diseases, including multiple sclerosis, Alzheimer's disease and Parkinson's disease.[8,9] Exogenous non-targeted MRI contrast agents generally use Gadolinium (III) compounds, which have a large electron magnetic moment that tends to reduce the T_1 relaxation time of the surrounding water protons. Gadolinium-based contrast agents are expected to accumulate in diseased regions that have increased vascularity.[10] Targeted exogenous MRI contrast agents are uniquely tailored to take advantage of the distinctive properties of a target cell of interest, such as a unique pattern of protein expression. For example, superparamagnetic nanoparticles (*e.g.* Fe_3O_4) can be conjugated with organic compounds, such as single domain antibodies (sdAb) that are specific for proteins overexpressed on the surface of the cells.[11,12]

In essence, conventional MRI contrast agents produce image contrast by reducing the relaxation times of the water protons in a tissue of interest. These methods are limited since they indirectly detect the presence of the contrast (*i.e. via* ^{1}H relaxation) and a background MR signal from the surrounding organs and tissues is always present, thereby limiting potential increases in CNR. Furthermore, additional images (such as a control image without contrast) and post-processing techniques are often required for image interpretation and the extraction of physiological parameters. On the other hand, hyperpolarized (HP) agents have been developed for MR imaging, and they have the potential to overcome some of the limitations of conventional MRI contrast agents.[13,14] HP techniques generally use NMR-sensitive nuclei other than ^{1}H (*e.g.* ^{3}He, ^{13}C, ^{129}Xe), as they can be polarized to enhance their net magnetization by a factor of up to 100 000 times above thermal equilibrium levels. Therefore, an HP agent can be directly imaged in an organ of interest at low concentrations and with no background signal. In this chapter, current progress in HP gas lung imaging using ^{3}He and ^{129}Xe is discussed, along with inert fluorinated gas MRI, which is a new inert gas lung imaging technique. Although inert fluorinated gas MRI does not use HP agents, it will likely be one important future direction for the HP gas lung imaging community. Applications and future prospects of HP ^{129}Xe brain imaging and HP ^{129}Xe biosensors are also discussed in this chapter. Overall, HP agents have the potential to vastly improve the sensitivity of MRI procedures and to assess a variety of diseases.

11.2 Hyperpolarized ^{3}He and ^{129}Xe Lung MRI

11.2.1 Overview of HP Gas MRI

Conventional ^{1}H MR imaging of the lungs is very challenging due to a number of reasons. There is a low proton density in lung tissue compared to other body organs, which leads to an inherently low magnetization that is available for imaging. Furthermore, air-tissue interfaces within the lung create significant magnetic susceptibility differences, leading to a short T_2* relaxation time in lung tissue and significant image distortions. The use of optimized ultra-short echo time (UTE) imaging techniques[15] can help to mitigate the issues associated with short T_2* relaxation; however, the resulting images still lack important functional information. Furthermore, the respiratory and cardiac cycles can cause motion artifacts. The use of respiratory gating and breath-hold imaging can help to alleviate these issues, but only to some extent.[16] A number of recent efforts have investigated functional lung imaging using ^{1}H-based techniques, including O_2-enhanced MRI,[17] Fourier-decomposition MRI,[18] and UTE imaging.[15] Although these techniques can be used on any MRI scanner, it should be noted that they are only indirectly sensitive to pulmonary ventilation.

HP noble gas MRI, using helium-3 (^{3}He) or xenon-129 (^{129}Xe), can overcome the limitations of conventional ^{1}H MRI by providing high-quality images of the lung that are directly sensitive to ventilation and gas exchange.[19] HP noble gas MR imaging of the lungs was first demonstrated in 1994 by Albert *et al.*, and ^{129}Xe MR images of excised mouse lungs were reported.[13] HP gases that have been polarized through spin-exchange optical pumping (SEOP) or metastability exchange optical pumping (MEOP) have a nuclear spin polarization that is up to 100 000 times higher than thermal equilibrium levels, thereby allowing for direct MR imaging of the distribution of the noble gas inside the lung airspaces.[20] A detailed description of hyperpolarization physics can be found in Chapter 10 of this book. Following these discoveries, there has been an increasing interest in the use of HP ^{3}He and ^{129}Xe for obtaining structural and functional information from the lungs and other organs, such as the brain.[21–25]

11.2.2 Static Breath-hold Imaging

It is well known that HP gas MRI can be used to obtain high-quality images of the lungs, and that imaging phenotypes can be identified and associated with various respiratory diseases. For the past 20 years, HP gas imaging has been performed in the lungs of animals[26–28] and humans[22,29–32] by various research groups around the world. Figure 11.1 shows examples of static breath-hold HP ^{3}He MR images that were obtained in our lab from a healthy volunteer and three patients with pulmonary diseases: asthma, moderate chronic obstructive pulmonary disease (COPD), and severe COPD.[33]

Signal voids can be observed in patients with pulmonary diseases, known as ventilation defects, indicating that there is an obstruction preventing the HP gas from ventilating that region of the lung. By co-registering the HP ^{3}He MR image with a conventional ^{1}H lung image, the total ventilated volume (VV) and ventilation defect volume (VDV) can be measured.[34] Various studies have shown that this technique has a high degree of reproducibility. O'Sullivan *et al.* showed that HP ^{3}He MRI is repeatable in stable cystic fibrosis patients over a four-week period.[35] Similarly, de Lange *et al.* showed

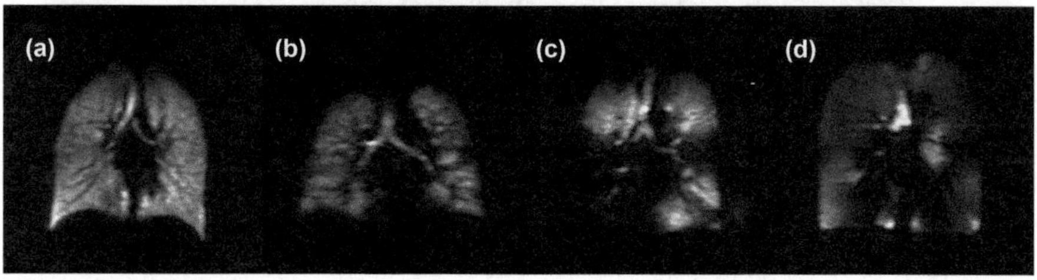

Figure 11.1 Representative static breath-hold HP ^{3}He MR images obtained from (a) a healthy volunteer and three patients with pulmonary diseases: (b) asthma, (c) moderate COPD, and (d) severe COPD.
Images were reproduced with permission from Couch *et al.*[33]

that ventilation defects in asthma are relatively persistent, and this finding was consistent regardless of disease severity and asthma medications.[36]

The quantity, size, location, and heterogeneity of ventilation defects can provide insight into the severity of the disease, making HP ^{3}He MRI a very sensitive technique.[37] Figure 11.2 shows static breath-hold HP ^{3}He images acquired in our lab from one subject in a study of mild-moderate asthmatics.[38] In this study, HP ^{3}He images were acquired at baseline (left) and post-methacholine challenge (right). This procedure was performed at day 1 (top) and again 45 days later (bottom).

For the images acquired at baseline, there was a 75% recurrence in ventilation defects. For the images acquired following a methacholine challenge, there was a 96% recurrence in ventilation defects (as noted by the arrows and circles), which shows that ventilation defects can persist over weeks and months. These data may suggest that asthma is a disease localized to specific airways in individual subjects, as opposed to a global diffuse disease as previously thought.[39] In addition to new information regarding asthma pathophysiology, HP ^{3}He MRI can potentially be used to detect localized improvements in lung function following treatment with bronchial thermoplasty.[40] Preliminary studies have shown encouraging results, as there is a trend towards improved ventilation (*i.e.* a smaller VDV) following treatment.[41]

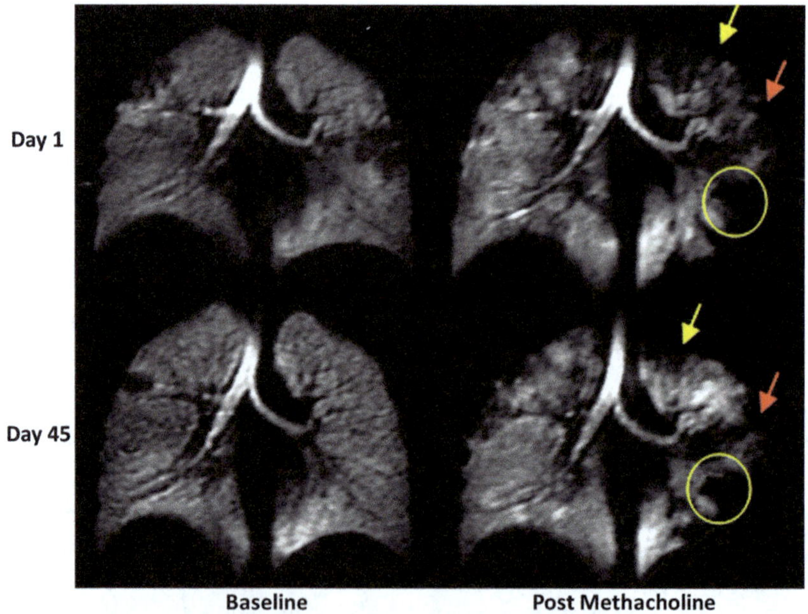

Figure 11.2 Static breath-hold HP ^{3}He MR images obtained in a mild-moderate asthmatic subject. Left: Pre Mch challenge; Right: Post Mch challenge. Top: Day 1; Bottom: Day 45.
Images were reproduced with permission from Liu *et al.*[38]

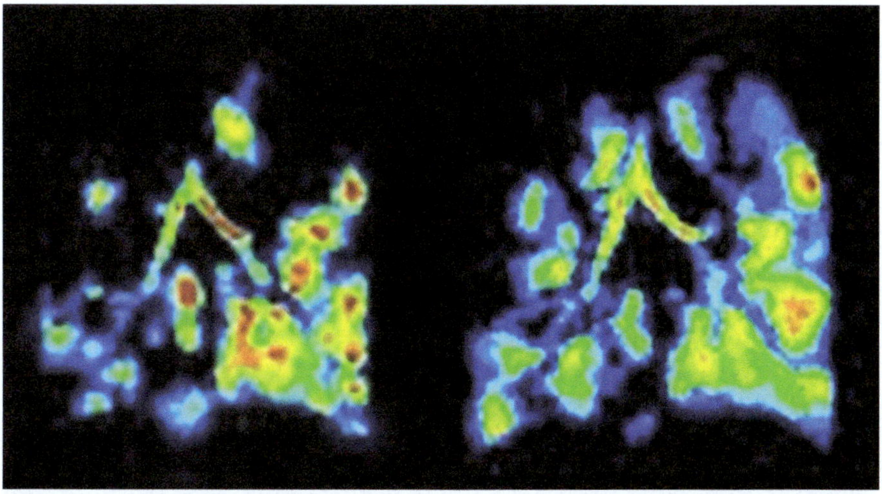

Figure 11.3 Color-coded ventilation maps obtained using HP ^{3}He MRI in a 17-year-old male cystic fibrosis patient (left) at baseline and (right) following treatment with antibiotics, hypertonic saline, chest physiotherapy, and rhDNase.
Images were reproduced with permission from Sun *et al.*[42]

Since HP gas MRI is a non-invasive and non-ionizing technique, it can be performed longitudinally to assess disease progression and response to treatment. For example, HP gas MRI has shown promise in the quantitative evaluation of treatment efficacy in cystic fibrosis patients.[42] Figure 11.3 shows examples of color-coded ventilation maps obtained using HP ^{3}He MRI in a 17-year-old male cystic fibrosis patient at baseline and following treatment with antibiotics, hypertonic saline, chest physiotherapy, and recombinant human DNase (rhDNase).

Overall, the HP ^{3}He MR images obtained in this patient following treatment had a 25% improvement in the total VV, and this result agreed well with spirometric indices. In particular, these images show a substantial improvement in ventilation in the upper portions of the lung, and this regional information cannot be obtained from spirometry. HP ^{3}He MRI has been recently used to demonstrate the efficacy of a new investigational drug for the treatment of cystic fibrosis, Ivacaftor, which improves the defective cystic fibrosis transmembrane conductance regulator (CFTR) protein function in patients with the G551D mutation.[43] HP ^{3}He MRI has also shown promise in pre-clinical mouse studies for detecting lung metastases following an injection of iron oxide nanoparticles that have been functionalized with cancer-binding ligands.[44] Beyond static breath-hold imaging, a variety of functional lung measurements can be performed using HP ^{3}He MRI, such as ventilation dynamics[23,45] and measurements of the regional alveolar partial pressure of oxygen (P_AO_2).[46,47]

HP ^{3}He MR images have historically been superior to HP ^{129}Xe imaging due to the high gyromagnetic ratio of ^{3}He compared to ^{129}Xe (^{3}He γ is 2.7 times greater than that of ^{129}Xe), and the available polarizer technology. Unfortunately ^{3}He is an extremely scarce and expensive isotope, and this is partly due to government sequestering of ^{3}He for use in neutron detectors for national security.[48,49] With recent improvements in polarizer technology,[50,51] the use of HP ^{129}Xe for lung MRI is gaining traction in the literature. Recent efforts have focused on validating HP ^{129}Xe for lung imaging with a direct comparison to HP ^{3}He imaging in the same subjects.[52,53] Figure 11.4 shows examples of HP ^{3}He and ^{129}Xe MR images obtained in two subjects: a healthy volunteer and a patient with COPD.[53] Qualitatively, these images appear very similar; however, measurements of VDV in COPD patients were significantly greater in HP ^{129}Xe MR images.[53] Naturally, the physical properties of the gas need to be taken into account when interpreting HP gas images, as the higher density and lower diffusivity of ^{129}Xe may lead to slower filling in the periphery of the lung, and hence larger ventilation defects.

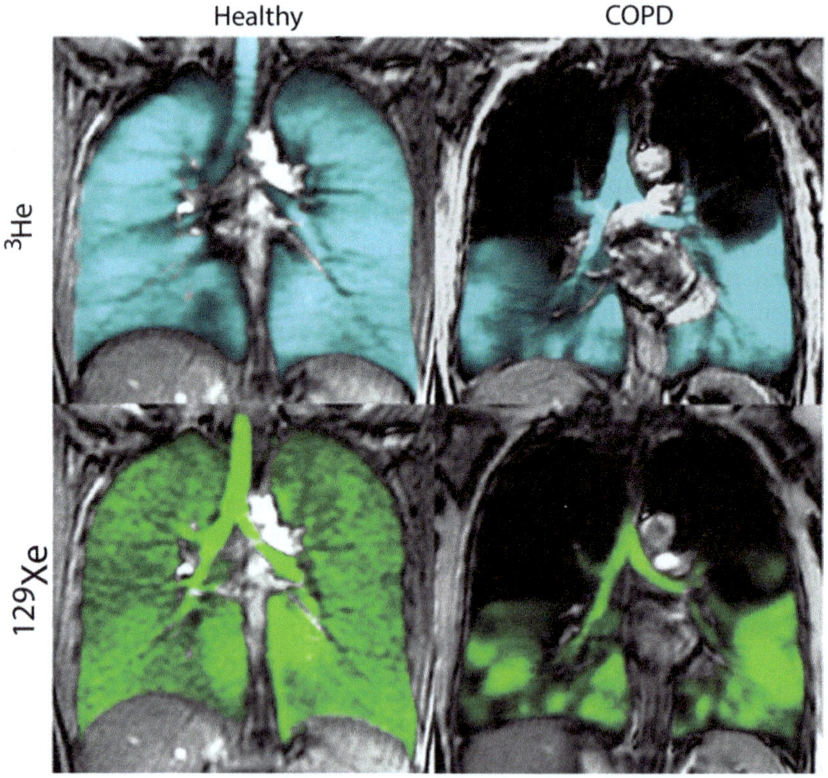

Figure 11.4 Representative HP ^{3}He and ^{129}Xe static breath-hold MR images obtained from the same healthy volunteer, and the same COPD subject. Courtesy of Grace Parraga *et al.*

11.2.3 Diffusion Imaging

In addition to static breath-hold imaging, it is possible to probe the lung microstructure using diffusion-weighted HP gas MR imaging.[23,53] With the application of diffusion-sensitizing gradients, the diffusion of gas atoms due to Brownian motion can be measured, and this technique is similar to diffusion-weighted imaging (DWI) of water in conventional ^{1}H MRI. It should be noted that the diffusion coefficients of ^{3}He and ^{129}Xe in air are 0.9 $cm^2\ s^{-1}$ and 0.14 $cm^2\ s^{-1}$, respectively,[54] and these values are four orders of magnitude larger than the diffusion coefficient of water.[55] With a typical diffusion time of a few milliseconds, the inhaled HP ^{3}He or ^{129}Xe is able to probe a diffusion length scale that is on the order of the alveolar diameter ($\sim$200 μm).[56] Therefore, elevated apparent diffusion coefficients (ADCs), due to alveolar wall destruction, can be detected in patients with emphysema.[23]

Since ^{129}Xe has been receiving increased attention recently, the use of HP ^{129}Xe for diffusion measurements requires validation with HP ^{3}He, especially considering the different physical properties of the two gases. Figure 11.5

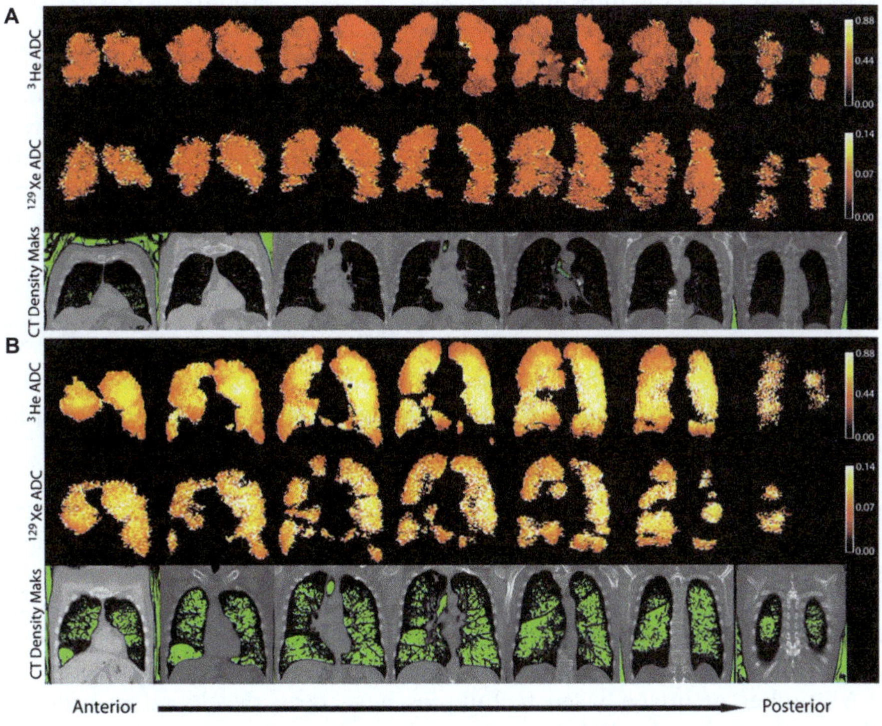

Figure 11.5 HP ^{3}He and ^{129}Xe ADC maps for two representative COPD subjects. The CT density mask is also shown in green to highlight the areas with attenuation ≤950 HU. (A) 78-year-old male with moderate COPD. (B) 73-year-old male with very severe COPD.
Images were reproduced with permission from Kirby *et al.*[52]

shows an example of a comparison of HP ^{3}He and ^{129}Xe ADC maps that were acquired in two representative COPD subjects: a 78-year-old male with moderate COPD, and a 73-year-old male with very severe COPD.[52]

As expected, the mean ^{3}He and ^{129}Xe ADC values were greater for these COPD subjects compared to healthy volunteers.[53] Of the two subjects, the latter subject with severe COPD had greater ADC values, and the CT density masks confirm the presence of substantial emphysema. For the subject with moderate COPD, the HP ^{3}He and ^{129}Xe ADC maps appear to be qualitatively similar, as both HP gases appear to be probing the same regions of the lung. However, ADC maps from the subject with severe COPD are visually different, as greater ventilation defects are visible in the HP ^{129}Xe ADC map. It has been hypothesized that ^{3}He can more easily access emphysematous regions of the lung due to its high diffusivity and low density, leading to the appearance of greater ventilation defects in ^{129}Xe ADC mapping.[52]

11.2.4 Probing Dissolved-phase ^{129}Xe

Since ^{3}He is virtually insoluble in blood and tissues, HP ^{3}He MRI is only sensitive to gas in the lung airspaces and, therefore, probing gas exchange is difficult. However, by measuring regional P_AO_2 with HP ^{3}He, it is possible to indirectly measure the ventilation/perfusion ratio (V/Q), which is a sensitive biomarker for gas exchange.[57] Since xenon readily dissolves into blood and tissue, HP ^{129}Xe MRI is a potentially suitable technique for performing direct measurements of parameters related to gas exchange.[58] ^{129}Xe is known to have a very broad chemical shift range of over 200 ppm in such studies, and as many as four spectral peaks appear in the lungs, where each peak is associated with a physically different compartment.[59] ^{129}Xe peaks appearing in the lungs include the following: one peak from ^{129}Xe in the alveoli, two overlapping peaks from ^{129}Xe dissolved in plasma and alveolar septum, one peak from ^{129}Xe dissolved in red blood cells (RBCs), and one peak from ^{129}Xe dissolved in fat/lipids. The temporal dynamics of the ^{129}Xe dissolved phase peaks can be probed and fitted to physiological models in order to extract information such as surface to volume ratio (S/V) and gas transfer times.[60–63] This information can potentially be very important for assessment of radiation induced lung injury (RILI) due to lung cancer or breast cancer radiotherapy, as inflammation associated with the radiation dose should significantly increase the xenon transfer time in the lung septum.[64] Unfortunately, these measurements currently lack regional information, as they typically use whole lung spectroscopy in order to acquire time-resolved dissolved phase information with sufficient SNR.

A number of image-based techniques have been developed in order to obtain ^{129}Xe dissolved-phase information. Once such technique, called xenon polarization transfer contrast (XTC), uses an approach similar to magnetization transfer contrast (MTC) to selectively destroy the ^{129}Xe dissolved-phase signal.[65] Due to the ongoing exchange between the gas and dissolved phases, a reduction in the gas phase signal can be observed, and fractional depolarization maps can be calculated. Decreased fractional depolarization has been observed in emphysematous regions of COPD subjects, indicating that less gas exchange is occurring in those regions of the lung.[66] Although this technique can potentially obtain meaningful physiological information, it is disadvantaged because it is only indirectly sensitive to the dissolved phase dynamics.

Taking inspiration from ^{1}H-based methods of water and fat separation, such as "iterative decomposition of water and fat with echo asymmetry and least squares estimation" (IDEAL),[67] a number of recent efforts have focused on simultaneous and direct imaging of ^{129}Xe in the gas and dissolved phases.[58,68] Figure 11.6 shows an example of isotropic and inherently co-registered ^{129}Xe gas and dissolved-phase MR images that were acquired in a healthy volunteer during a single breath-hold.[68]

By normalizing the dissolved-phase signal to the gas-phase, a ^{129}Xe gas-transfer map can be generated. In this healthy volunteer, a mean ratio of 1 indicates a good "matching" between the gas and dissolved phases, possibly suggesting that this measurement could be a surrogate for V/Q. Further decomposition of the dissolved phase ^{129}Xe signal allows for the generation of tissue and RBC images, which can be used to calculate additional ratio maps, such as tissue-to-gas, RBC-to-gas, and RBC-to-tissue ratios.[58] Figure 11.7 shows examples of ^{129}Xe gas, tissue, and RBC images from one healthy volunteer, and three patients with severe COPD, asthma, and cystic fibrosis.

As expected, the healthy volunteer had relatively homogenous images and ratio maps. The COPD subject had significant ventilation defects, and all ratio maps were heterogeneous and low. The asthma subject showed predominantly high RBC-to-tissue ratios, low tissue-to-gas ratios, and normal RBC-to-gas ratios. The subject with cystic fibrosis showed very high tissue-to-gas and RBC-to-gas ratios with heterogeneous distribution, but the RBC-to-tissue ratios were similar to the healthy volunteer. Therefore, ^{129}Xe dissolved phase imaging can potentially be a very useful clinical tool for the assesment and management of lung deseases such as COPD, asthma, and cystic fibrosis.[69]

11.3 ^{19}F Lung Imaging

Fluorine-19 (^{19}F) MRI of the lungs using inhaled inert fluorinated gases is a new pulmonary imaging modality that has recently received increased attention, as this technique may be able to provide images and functional information that are similar to HP noble gas MRI.[33] Although inert fluorinated gases do not need to be hyperpolarized prior to their use in MRI, this technique will be of great interest to the lung imaging community as these gases are non-toxic, abundant, and inexpensive compared to HP gases. This technique uses gases such as tetrafluoromethane (CF_4), sulfur hexafluoride (SF_6), hexafluoroethane (C_2F_6), and perfluoropropane (C_3F_8 or PFP) as inhaled signal sources. Efficient ^{19}F MR imaging can be accomplished

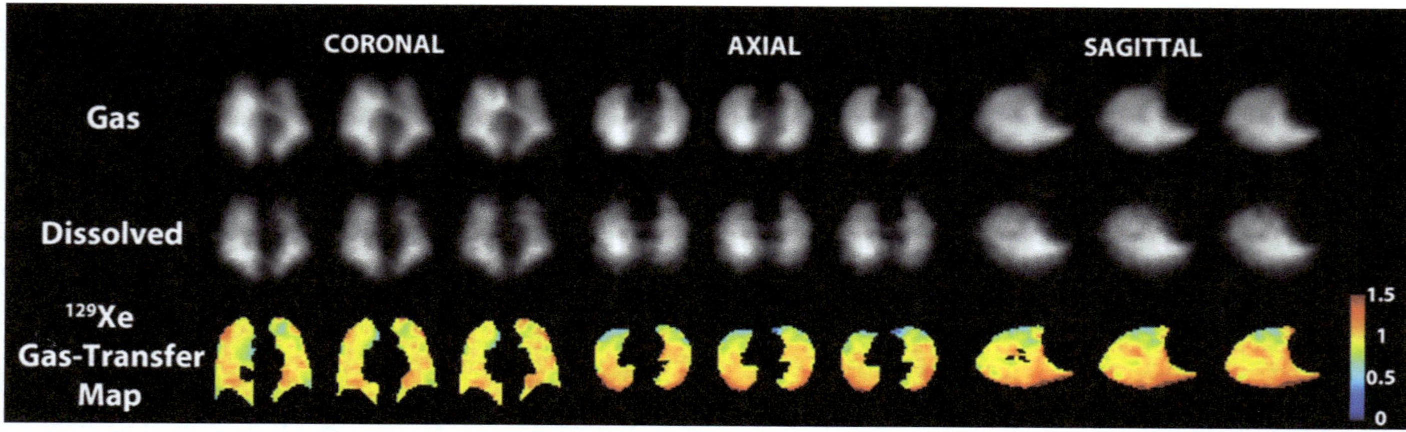

Figure 11.6 Single-breath, isotropic and inherently co-registered gas and dissolved-phase HP ^{129}Xe images, acquired using a 3D radial pulse sequence. The acquisition strategy enables the dissolved-phase signal to be normalized by the gas phase, to generate the ^{129}Xe gas-transfer map.
Images were reproduced with permission from Kaushik *et al.*[68]

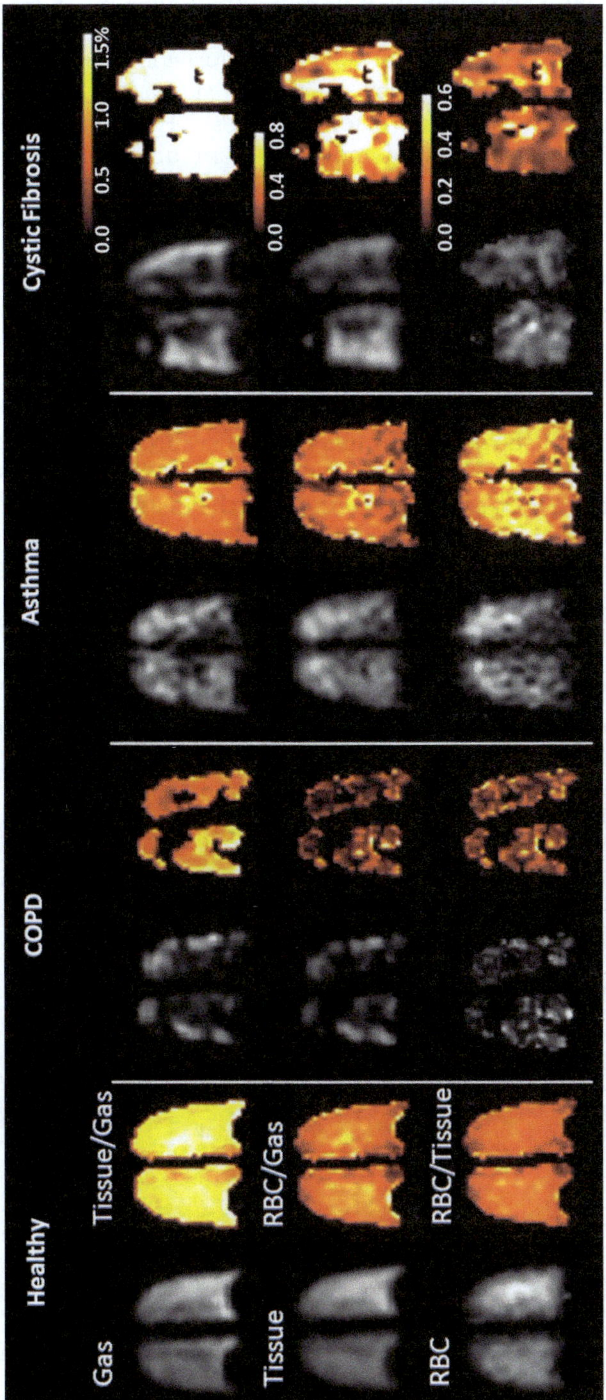

Figure 11.7 Representative HP ^{129}Xe MR images (^{129}Xe in the alveoli, ^{129}Xe dissolved in tissue, and ^{129}Xe dissolved in RBCs) and ratio maps obtained in one healthy volunteer, and three patients with severe COPD, asthma, and cystic fibrosis. Courtesy of John Mugler III *et al.*

considering several factors: the high natural abundance of ^{19}F (100%); the high gyromagnetic ratio of ^{19}F (251.662 MHz rad T^{-1}), which leads to a high thermal polarization (9.3×10^{-6} at 3 T); the short longitudinal relaxation times of fluorinated gases,[70] which allows for signal averaging within a single breath-hold; and the lack of endogenous ^{19}F signal sources. Inert fluorinated gas MRI was first demonstrated in animal lungs by Paul Lauterbur and colleagues in the 1980s,[71] and since that time a number of investigators have explored the technique in animals with high resolution 3D imaging,[72] ADCs,[73] ventilation mapping,[74] and V/Q ratios.[75]

Inert fluorinated gas MRI has been reported in healthy volunteers[76] and patients with pulmonary diseases, such as COPD, asthma, and post lung transplantation,[77] using a mixture of 79% PFP and 21% O_2. Due to the short T_2^* of inert fluorinated gases (~2 ms in the lungs at 3 T), this technique is best performed by the application of pulse sequences with a short TE, such as UTE.[76] Figure 11.8 shows an example of 12 slices from a ^{19}F 3D UTE MR image that was acquired in a healthy volunteer at 3 T during a 15-second breath-hold of the PFP/O_2 mixture. As expected, the distribution of ^{19}F signal was fairly homogenous in the lungs of this healthy volunteer.

Similarly, Figure 11.9 shows 12 slices from a ^{19}F 3D UTE MR image obtained in the axial plane from another healthy volunteer. These images were obtained under breath-hold conditions, following several breaths of the PFP/O_2 mixture.

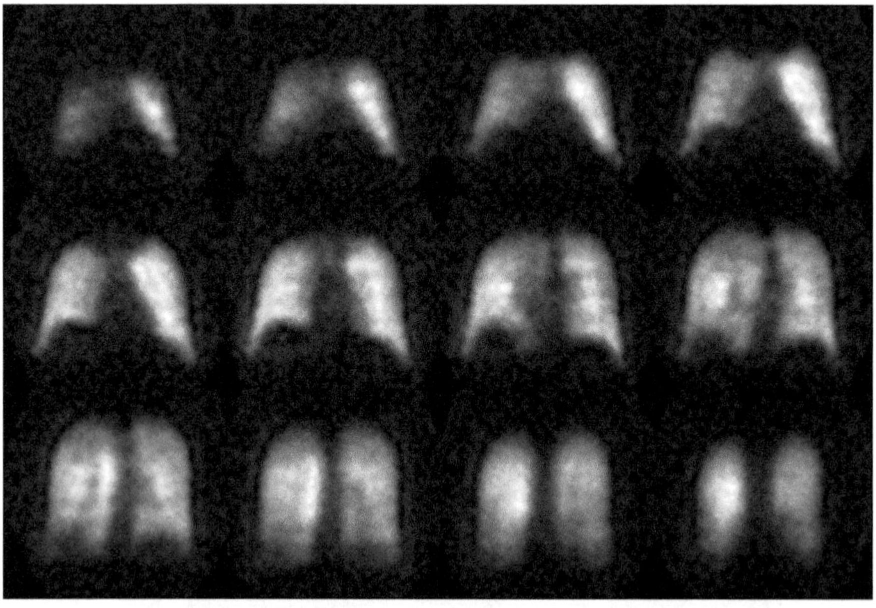

Figure 11.8 Coronal pulmonary ^{19}F 3D UTE MR images obtained in a healthy volunteer during a 15 s breath-hold after continuous breathing of a mixture of 79% PFP and 21% O_2.
Images were reproduced with permission from Couch *et al.*[76]

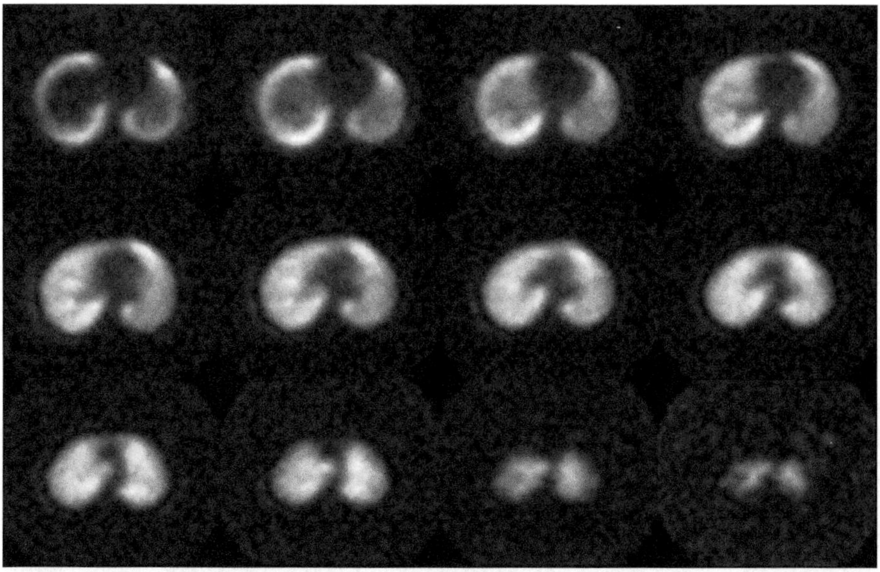

Figure 11.9 Axial pulmonary ^{19}F 3D UTE MR images obtained in a healthy volunteer during a 15 s breath-hold after continuous breathing of a mixture of 79% PFP and 21% O_2.
Images were reproduced with permission from Couch *et al.*[76]

In contrast to HP gases, inert fluorinated gases can be mixed with O_2 without significantly sacrificing image quality. Therefore, patients can safely breathe the gas mixture continuously in order to wash out residual air from the lungs and to maximize the available magnetization for imaging. Figure 11.10(A) shows a ^{19}F 3D gradient echo MR image that was acquired in a COPD patient with emphysema. Similar to HP gas MRI, these images show substantial signal voids and ventilation defects.

Figure 11.10(B) shows the same ^{19}F 3D gradient echo image overlaid on a conventional ^{1}H localizer image, and this approach helps to facilitate analysis of the VV and VDV.

Dynamic imaging using a wash-in/wash-out approach can provide a more physiologically meaningful measurement than static breath-hold imaging by quantifying regional gas replacement and gas trapping. A number of animal studies have explored dynamic imaging of inert fluorinated gases,[74,78–80] and these results have been validated with a comparison to respiratory gas analysis.[81] Halaweish *et al.* recently investigated the wash-in and wash-out characteristics of PFP in healthy volunteers and patients with respiratory diseases,[82] and preliminary efforts to quantify wash-in and wash-out time constants are currently underway.[83]

Due to the high cost of enriched noble gas isotopes and the limited availability of polarizer technology, the pulmonary MR imaging community is interested in developing alternative techniques that can yield information similar to HP gas MRI. The image quality of ^{19}F lung MRI using inert

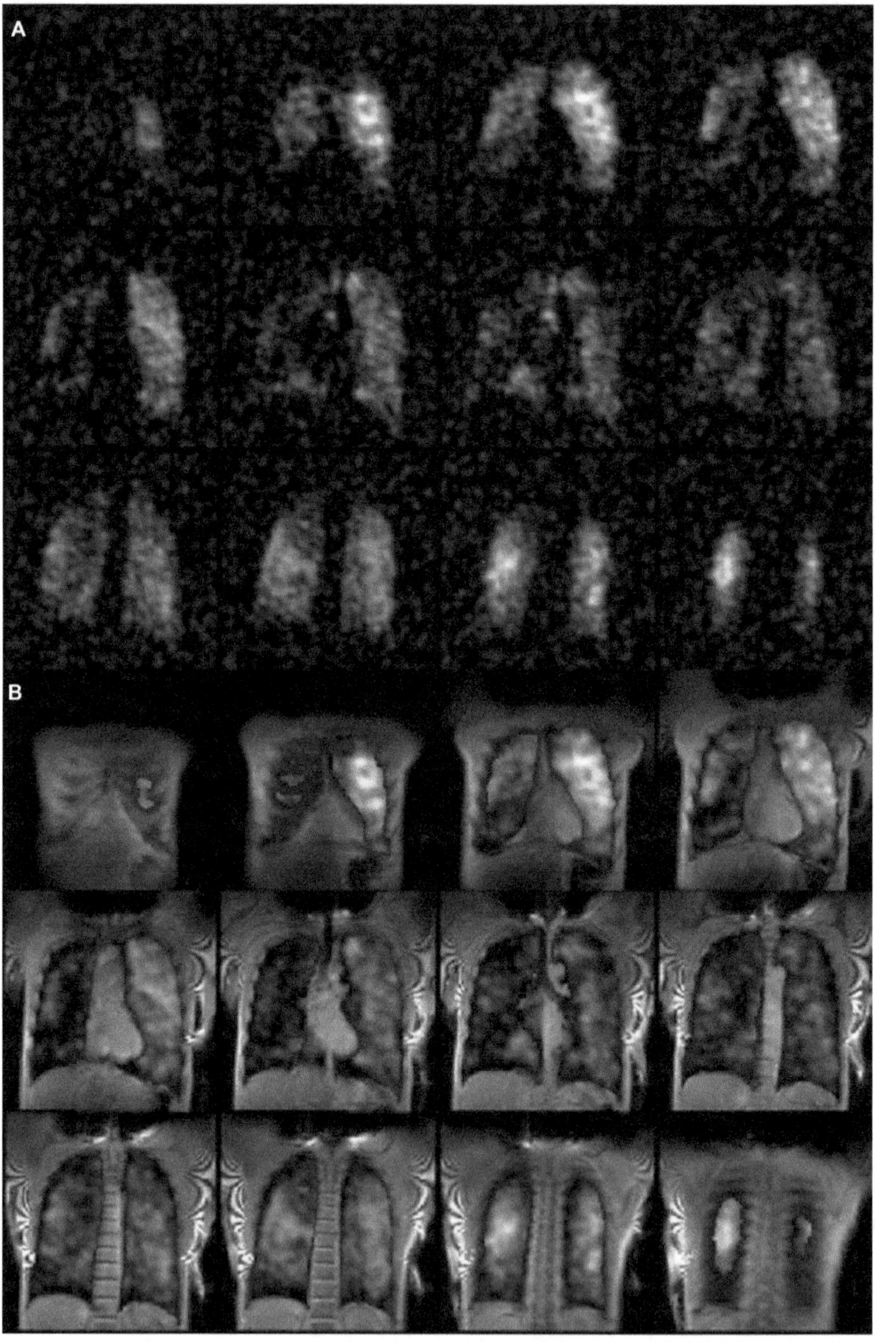

Figure 11.10 (A) ^{19}F 3D MR lung images obtained in the coronal plane from a COPD patient with emphysema following inhalation of a mixture of 79% PFP and 21% O_2. (B) ^{19}F MR lung images co-registered with conventional ^{1}H MR images.
Images were reproduced with permission from Halaweish *et al.*[77]

fluorinated gases is currently lower than HP gas lung MRI, and efforts to develop optimized acquisition strategies are ongoing. One of the most important advantages of ^{19}F lung imaging is perhaps its low cost, as the technique can be performed on any MRI scanner with broadband capability and dedicated ^{19}F RF coils, and without the need for expensive polarizer technology. Overall, ^{19}F MRI of inert fluorinated gases may be a viable clinical imaging modality that can provide useful information for the diagnosis of chronic respiratory diseases.

11.4 ^{129}Xe MRI of the Brain

HP ^{129}Xe is a potentially valuable MR tracer for functional brain imaging due to its high solubility, its ability to readily cross the blood–brain barrier, and its large chemical shift range. The first *in vivo* human brain ^{129}Xe spectrum was reported in 1997 by Mugler *et al.*[24] Swanson *et al.* investigated *in vivo* ^{129}Xe brain signals in animal models, and demonstrated the first *in vivo* brain spectra in animals.[84] Albert *et al.* and Wolber *et al.* investigated the longitudinal relaxation properties of ^{129}Xe dissolved in deoxygenated blood and oxygenated blood.[85,86] For HP gases in the lungs, the presence of paramagnetic O_2 tends to decrease the T_1 of ^{129}Xe in the alveoli; however, deoxygenated hemoglobin is paramagnetic, and an increased concentration of oxygen reduces the amount of deoxygenated hemoglobin in arterial blood, which leads to a longer T_1. Norquay *et al.* recently measured the T_1 of ^{129}Xe in RBCs over a wide range of blood oxygenation levels and found a linear relationship between T_1 and blood oxygenation.[87] Kilian *et al.* demonstrated the first 2D chemical shift image (CSI) map of ^{129}Xe distribution in the human brain.[88] The same group also proposed a theoretical model to describe ^{129}Xe transit to the brain. Zhou *et al.* reported the T_1 of HP ^{129}Xe in the rat brain to be $\sim$15 s, with slight variations depending on the acquisition method.[89] In humans, Kilian *et al.* measured the T_1 of HP ^{129}Xe at 2.94 T for grey and white matter, and reported T_1 values of 14 s and 8 s, respectively.[90]

Building on previous work in animals, Wakai *et al.* were able to observe multiple peaks in the rat brain over a large range of chemical shifts (185–210 ppm), by allowing the animal to continuously breathe a ^{129}Xe gas mixture during extensive averaging in the head region.[91] Using an animal model involving external carotid artery and pterygopalatine artery ligation, Kershaw *et al.* found that two peaks belonged to brain tissue and assigned them grey (193–197 ppm) and white matter (191–194 ppm).[92] Finally, Zhou *et al.*[93] and Mazzanti *et al.*[94] published the first results using ^{129}Xe brain imaging techniques for the measurement of cerebral ischemia (stroke model) and cortical brain function in animals, respectively. Figure 11.11 shows an example of HP ^{129}Xe MR imaging of a stroke model in rats (middle cerebral artery occlusion), where HP ^{129}Xe signal voids were observed in regions of the ischemic core, and this result was verified by ^{1}H diffusion weighted imaging and histology.[93]

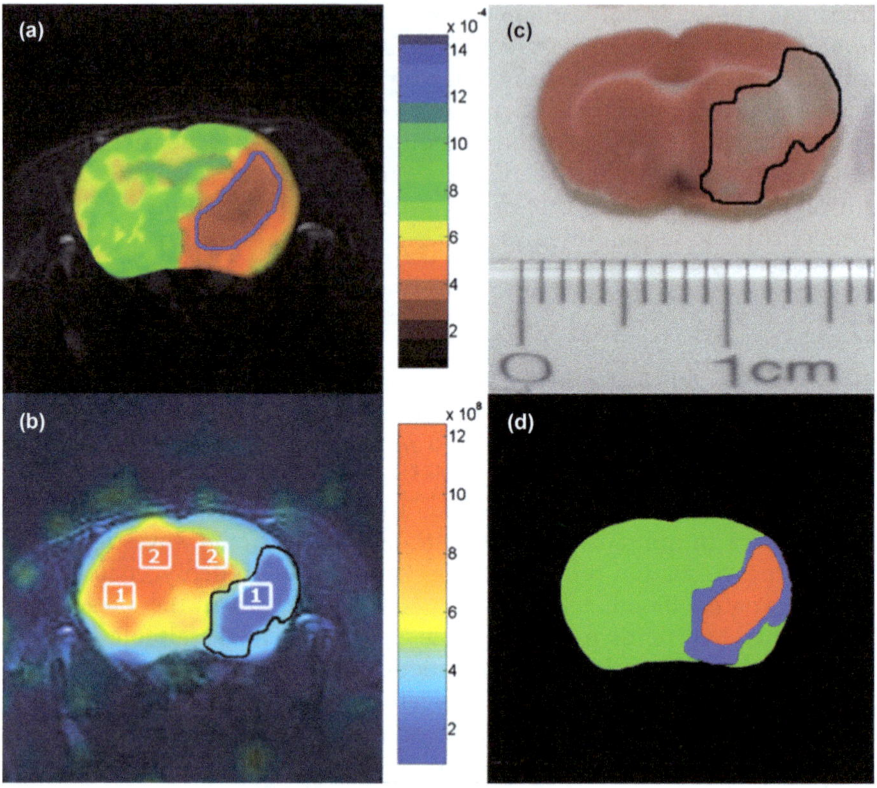

Figure 11.11 ^{1}H and HP ^{129}Xe MR imaging of a model of ischemic stroke in the rat brain. (a) ^{1}H diffusion weighted imaging indicating a region of poor perfusion (ischemic core). (b) Corresponding ^{129}Xe CSI map depicting signal void in the region of the ischemic core. (c) Histological confirmation of stroke model related ischemia. (d) Tri-colour map depicting normal tissue (green), ischemic core (red), and penumbra (blue). Images were reproduced with permission from Zhou *et al.*[93]

Figure 11.12 shows an example of HP ^{129}Xe MR imaging of brain function in rats that received an injection of capsaicin into the forepaw as a pain stimulus.[94] ^{129}Xe CSI maps were acquired at baseline and post-stimulus, and increased ^{129}Xe signal was observed in regions of the brain responsible for processing pain response.

CSI is in general a slower imaging technique than gradient echo imaging, and efforts to optimize faster image acquisition techniques that can achieve a high SNR are ongoing. Nouls *et al.* reported fast, high-resolution, isotropic images of healthy rat brains using a 3D radial acquisition method.[95] It is clear that valuable groundwork has already established the potential of this technique, and dynamic ^{129}Xe spectroscopy of the human brain is only now being revisited.[96] Further refinements will be required in order to translate these techniques to clinical human use.

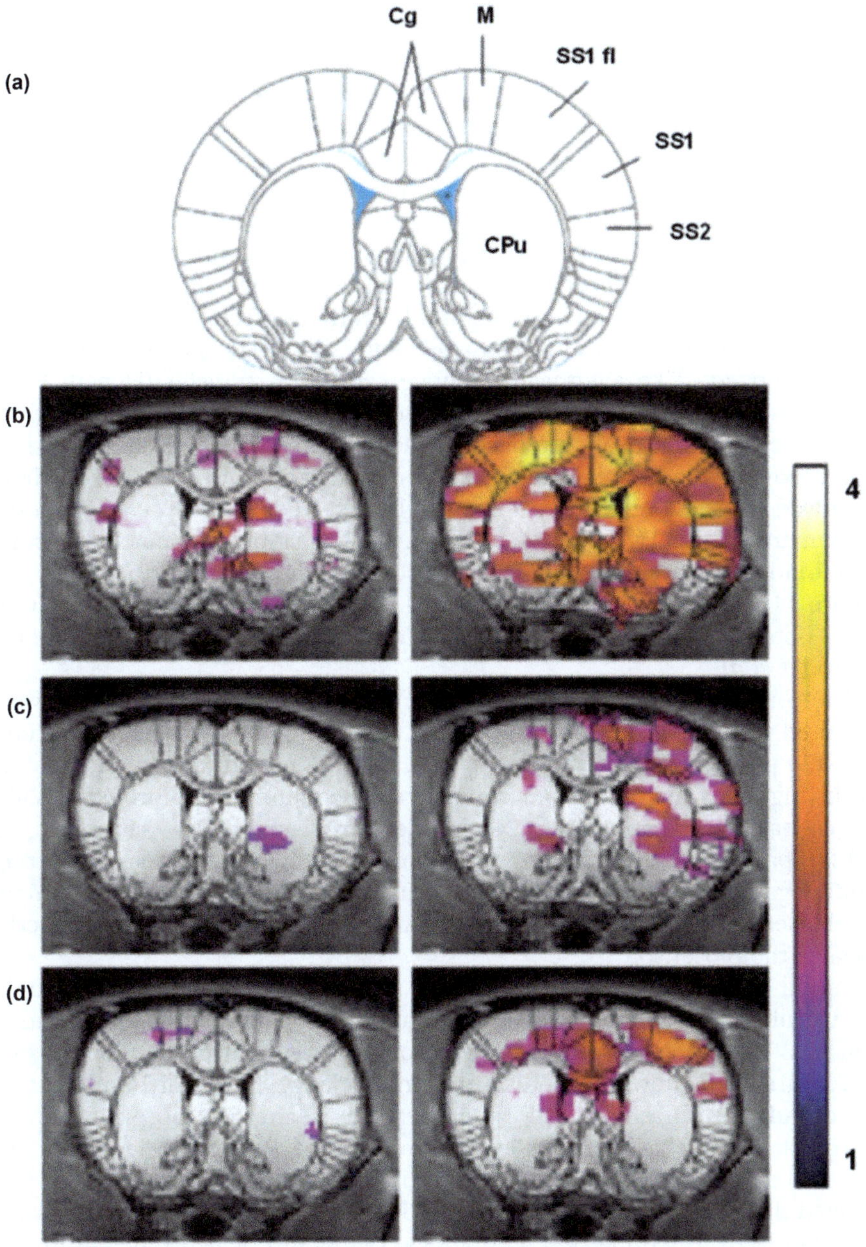

Figure 11.12 ^{1}H and HP ^{129}Xe MR imaging of a model of pain response in the rat brain. (a) Rat brain atlas indicating several important brain regions such as the cingulate cortex (Cg), primary somatosensory cortex (SS1), and secondary somatosensory cortex (SS2). (b–d) depict rat brain CSI maps registered to their corresponding ^{1}H brain images. The left column shows data obtained at baseline and the right column are maps obtained post injection of capsaicin. Each set of maps taken in (b) through (d) are obtained within the same animal, hence $n=3$. Images were reproduced with permission from Mazzanti *et al.*[94]

11.5 Hyperpolarized ^{129}Xe Biosensors

Although noble gases have no specificity for biological receptors, they can be delivered to a target by means of dedicated molecular systems that can encapsulate the noble gas and bind to the biological sites of interest.[97,98] An example of such a molecular system is a cryptophane-A cage, which is a large organic cage molecule with a hydrophobic interior that can reversibly bind to ^{129}Xe *via* van der Waals forces. Cryptophane-A cages can be functionalized with a linker as well as a targeting moiety, such as an antibody or a ligand, which enables detection of a specific biomarker.[99] Cage-encapsulated ^{129}Xe resonates at an NMR frequency distinct from that of free ^{129}Xe, and ^{129}Xe is exchanged between its encapsulated and dissolved states. Therefore, hyperpolarized ^{129}Xe chemical exchange saturation transfer (Hyper-CEST) can be used to indirectly detect the presence of the ^{129}Xe biosensor with a high specificity and near-zero background. That is, by selectively saturating and destroying cage-encapsulated ^{129}Xe magnetization, a reduction in free dissolved-phase ^{129}Xe magnetization can be detected. Schröder *et al.* recently demonstrated the Hyper-CEST technique using cryptophane-A biosensors that bind to avidin-functionalized agarose beads *via* a biotin moiety.[100] Figure 11.13 demonstrates the Hyper-CEST technique with a series of CSI maps obtained in a two-compartment avidin-agarose bead phantom used by Schröder *et al.*

The control image of free ^{129}Xe in the avidin-agarose bead medium was acquired with off-resonance saturation (*i.e.* continuous wave (cw) saturation at +128 ppm with respect to free ^{129}Xe), and the resulting image is unable to distinguish which phantom compartment contains the biosensor. With on-resonance saturation of ^{129}Xe in the biosensor (*i.e.* cw saturation at −128 ppm with respect to free ^{129}Xe), an image of free ^{129}Xe signal in the avidin-agarose bead medium shows signal depletion in the area containing the biosensor. Finally, a subtraction image can be calculated to indirectly localize the biosensor.

Recent work has focused on improving the Hyper-CEST technique through the use of pulsed saturation as opposed to more conventional cw saturation, and this approach has been shown to improve control over selectivity (*i.e.* saturation bandwidth) and saturation efficiency.[101,102] Furthermore, the use of pulsed saturation makes the Hyper-CEST technique more convenient to translate to *in vivo* molecular imaging, as the RF power deposition is substantially lower than in the cw saturation method. Recent work in our lab applied the pulsed saturation Hyper-CEST approach to a cryptophane-A cage that had been functionalized with a PK11195 ligand, which targets inflammation sites in the body.[103] Further contributing to the translational capability of the Hyper-CEST technique, this study was performed on a 3 T clinical whole-body MR scanner. Figure 11.14(left) shows an example of a ^{129}Xe spectrum of PK11195 functionalized cryptophane-A, and a 77% depletion in the dissolved-phase ^{129}Xe SNR was detected following the presaturation pulses.[104]

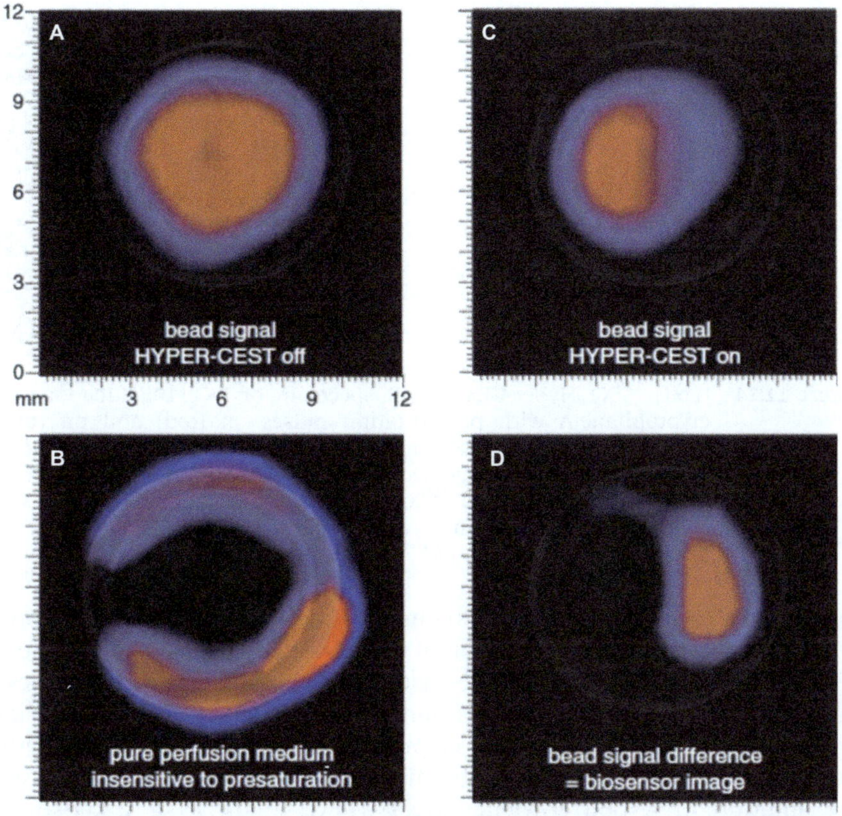

Figure 11.13 Chemical shift imaging of a two-compartment phantom demonstrating the Hyper-CEST effect. (A) Free ^{129}Xe in an avidin-agarose bead medium (off-resonance saturation). (B) Free ^{129}Xe in water. (C) On-resonance saturation of ^{129}Xe in the biosensor leads to a depletion in free ^{129}Xe signal in the avidin-agarose bead medium. (D) Subtraction image that localizes the phantom compartment containing the biosensor. Images were reproduced with permission from Schröder *et al.*[100]

This method was also capable of producing ^{129}Xe MR images at 3 T with Hyper-CEST contrast. Figure 11.14(right) shows a ^{129}Xe Hyper-CEST saturation map that spatially locates the PK11195-cryptophane molecules in solution. This map was obtained by subtracting a ^{129}Xe image acquired with pre-saturation pulses on and off of the encapsulated ^{129}Xe resonance, and then segmenting the resulting subtraction image from the surrounding background noise. Future *in vivo* work will determine the potential for using Hyper-CEST, in combination with PK11195-cryptophane molecules, to detect inflammation sites in the body caused by diseases such as COPD and arthritis.

Some emerging applications of ^{129}Xe Hyper-CEST that have been investigated include the use of gas vesicles as genetically encoded reporters of gene

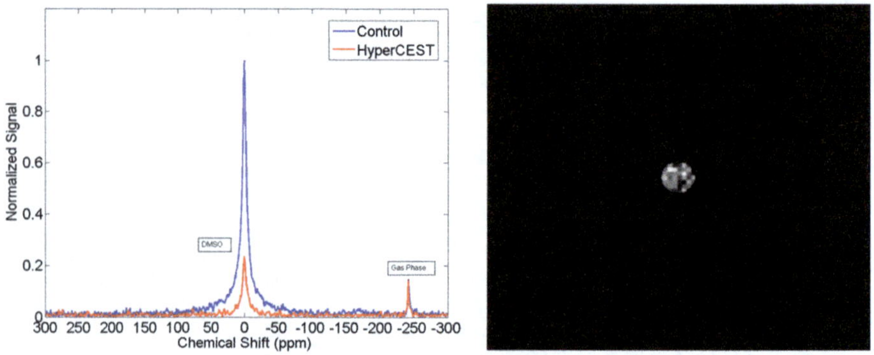

Figure 11.14 (Left) ^{129}Xe Hyper-CEST NMR spectrum of PK11195 functionalized cryptophane-A with presaturation pulses on (red) and off (blue). (Right) ^{129}Xe Hyper-CEST saturation map showing the difference in signal between an image acquired with presaturation pulses on and off of the encapsulated ^{129}Xe resonance, which spatially locates the PK11195-cryptophane molecules in solution. Adapted from Couch *et al.*[104]

expression,[105] as well as cell tracking using lipophilic cryptophanes in order to detect the cellular internalization of encapsulated ^{129}Xe.[106] Another focus of ongoing work with cryptophane-based biosensors is related to the development of fast and efficient acquisition sequences, which may be especially important for detecting very low cryptophane concentrations.[107,108] Other work has focused on improving the biocompatibility of these sensors with the addition of a poly(ethylene glycol) (PEG) chain as a water-soluble moiety,[97] and also the development of novel biosensors for molecular and cellular imaging of cancer.[98] Stevens *et al.* recently extended the ^{129}Xe Hyper-CEST technique to perfluorooctyl bromide (PFOB) nanoemulsions,[109] which have a Xe gas solubility that is approximately 10 times higher than Xe in water.[110] PFOB nanoemulsions were detected *in vitro* at sub-picomolar concentrations. Since it is possible to use PFC nanoemulsions as vehicles for localized drug delivery,[111] the use of Hyper-CEST can potentially enable high-sensitivity drug tracking and molecular imaging.[109]

11.6 Conclusions

The potential usefulness of HP and inert gases in MRI has been investigated for the past 20 years, and many potential diagnostic applications have been demonstrated by the research community. Although many of these applications are promising, continued research efforts are required in order to bring these techniques to the clinic and routine practice. The use of non-invasive and non-ionizing diagnostic methods, as offered by HP MRI, is especially important in younger populations where the risk of cancer induced by medical radiation is high. HP noble gas MRI is able to provide high-quality images of the lungs, as well as measure a wide range of functional biomarkers that are sensitive to disease, as well as targeted treatments. Due to the expense and scarcity of ^{3}He isotopes, a continued focus on using HP ^{129}Xe may help to make clinical translation a reality. Although the image quality from inert fluorinated gas MRI is currently not as high as HP gas MRI, it may be sufficient for clinical applications. Developments are ongoing as more and more groups are taking an interest in investigating and improving inert fluorinated gas MRI. Some advantages of inert fluorinated gas MRI are that it is inexpensive and it can be widely implemented without the need for expensive polarizer technology. Improvements in HP ^{129}Xe brain imaging are also ongoing and future work will certainly explore a wider range of animal models and human imaging applications. HP ^{129}Xe biosensors have many possible applications, as they can be tailored to target specific biological receptors, and the technique has the potential to provide the sensitivity of PET molecular imaging with the high spatial resolution of MRI. Overall, there is a potentially wide range of applications for HP media in the diagnosis and treatment of human diseases, and these techniques will provide new insights into disease pathophysiology.

Acknowledgements

Portions of this chapter were reproduced with permission from Couch *et al.*, *Mol. Imaging Biol.*, 2015, **17**(2), 149. This work was funded in part by the Natural Sciences and Engineering Research Council of Canada (NSERC) and the Thunder Bay Regional Research Institute (TBRRI). MJC was supported by an NSERC Canada Graduate Scholarship (CGS). Thanks to Bastiaan Driehuys, Sivaram Kaushik, Grace Parraga, Damien Pike, John Mugler III, and Kun Qing for generously providing figures, to Brenton DeBoef for synthesizing cryptophanes, to Ralph Hashoian for making the RF coils and interface electronics, and to the Thunder Bay Regional Health Sciences Centre (TBRHSC) MR technologists for their time and assistance with MR scanning of volunteers.

References

1. P. C. Lauterbur, *Nature*, 1973, **242**, 190.
2. J. C. Leawoods, D. A. Yablonskiy, B. Saam, D. S. Gierada and M. S. Conradi, *Concepts Magn. Res.*, 2001, **13**, 277.
3. G. L. Chadzynski, R. Pohmann, G. Shajan, R. Kolb, S. Bisdas, U. Klose and K. Scheffler, *Magn. Reson. Mater. Phys., Biol. Med.*, 2015, **28**, 239.
4. D. I. Hoult and D. Phil, *J. Magn. Reson. Imaging*, 2000, **12**, 46.
5. N. Y. Wu, H. C. Cheng, J. S. Ko, Y. C. Cheng, P. W. Lin, W. C. Lin, C. Y. Chang and D. M. Liou, *BMC Cancer*, 2011, **11**, 242.
6. J. V. Frangioni, *J. Clin. Oncol.*, 2008, **26**, 4012.

7. B. Blasiak, S. Barnes, T. Foniok, D. Rushforth, J. Matyas, D. Ponjevic, W. P. Weglarz, R. Tyson, U. Iqbal, A. Abulrob, G. R. Sutherland, A. Obenaus and B. Tomanek, *BMC Med. Imaging*, 2013, **13**, 20.
8. F. Bagnato, S. Hametner, B. Yao, P. van Gelderen, H. Merkle, F. K. Cantor, H. Lassmann and J. H. Duyn, *Brain*, 2011, **134**, 3602.
9. J. F. Schenck, in *Medicinal Inorganic Chemistry*, ed. J. Sessler, S. Doctrow, T. McMurry and S. Lippard, American Chemical Society, 2005, pp. 44–61.
10. E. J. Werner, A. Datta, C. J. Jocher and K. N. Raymond, *Angew. Chem. Int. Ed.*, 2008, **47**, 8568.
11. B. Blasiak, F. C. J. M. van Veggel and B. Tomanek, *J. Nanomater.*, 2013, **2013**, 148578.
12. B. Tomanek, U. Iqbal, B. Blasiak, A. Abulrob, H. Albaghdadi, J. R. Matyas, D. Ponjevic and G. R. Sutherland, *Neuro Oncol.*, 2012, **14**, 53.
13. M. S. Albert, G. D. Cates, B. Driehuys, W. Happer, B. Saam, C. S. Springer Jr. and A. Wishnia, *Nature*, 1994, **370**, 199.
14. A. Bifone, Y. Q. Song, R. Seydoux, R. E. Taylor, B. M. Goodson, T. Pietrass, T. F. Budinger, G. Navon and A. Pines, *Proc. Natl Acad. Sci. U. S. A.*, 1996, **93**, 12932.
15. K. M. Johnson, S. B. Fain, M. L. Schiebler and S. Nagle, *Magn. Reson. Med.*, 2013, **70**, 1241.
16. S. Fain, M. L. Schiebler, D. G. McCormack and G. Parraga, *J. Magn. Reson. Imaging*, 2010, **32**, 1398.
17. R. C. Sa, M. V. Cronin, A. C. Henderson, S. Holverda, R. J. Theilmann, T. J. Arai, D. J. Dubowitz, S. R. Hopkins, R. B. Buxton and G. K. Prisk, *J. Appl. Physiol.*, 2010, **109**, 1950.
18. G. Bauman, M. Puderbach, M. Deimling, V. Jellus, C. Chefd'hotel, J. Dinkel, C. Hintze, H. U. Kauczor and L. R. Schad, *Magn. Reson. Med.*, 2009, **62**, 656.
19. T. A. Altes and M. Salerno, *J. Thorac. Imaging*, 2004, **19**, 250.
20. T. G. Walker and W. Happer, *Rev. Mod. Phys.*, 1997, **69**, 629.
21. B. M. Goodson, *J. Magn. Reson.*, 2002, **155**, 157.
22. Z. I. Cleveland, G. P. Cofer, G. Metz, D. Beaver, J. Nouls, S. S. Kaushik, M. Kraft, J. Wolber, K. T. Kelly, H. P. McAdams and B. Driehuys, *PLoS One*, 2010, **5**, e12192.
23. S. S. Kaushik, Z. I. Cleveland, G. P. Cofer, G. Metz, D. Beaver, J. Nouls, M. Kraft, W. Auffermann, J. Wolber, H. P. McAdams and B. Driehuys, *Magn. Reson. Med.*, 2011, **65**, 1154.
24. J. P. Mugler, 3rd, B. Driehuys, J. R. Brookeman, G. D. Cates, S. S. Berr, R. G. Bryant, T. M. Daniel, E. E. de Lange, J. H. Downs, 3rd, C. J. Erickson, W. Happer, D. P. Hinton, N. F. Kassel, T. Maier, C. D. Phillips, B. T. Saam, K. L. Sauer and M. E. Wagshul, *Magn. Reson. Med.*, 1997, **37**, 809.
25. X. Zhou, in *Advances in Brain Imaging*, ed. D. V. Chaudhary, InTech, Shanghai, 2012.

26. M. J. Couch, A. Ouriadov and G. E. Santyr, *Magn. Reson. Med.*, 2012, **68**, 1623.
27. A. C. Thomas, E. N. Potts, B. T. Chen, D. M. Slipetz, W. M. Foster and B. Driehuys, *NMR Biomed.*, 2009, **22**, 502.
28. K. Emami, S. J. Kadlecek, J. M. Woodburn, J. Zhu, J. Yu, V. Vahdat, S. Pickup, M. Ishii and R. R. Rizi, *Magn. Reson. Med.*, 2010, **63**, 137.
29. G. Parraga, A. Ouriadov, A. Evans, S. McKay, W. W. Lam, A. Fenster, R. Etemad-Rezai, D. McCormack and G. Santyr, *Invest. Radiol.*, 2007, **42**, 384.
30. J. M. Wild, S. Fichele, N. Woodhouse, M. N. Paley, L. Kasuboski and E. J. van Beek, *Magn. Reson. Med.*, 2005, **53**, 1055.
31. J. C. Woods, C. K. Choong, D. A. Yablonskiy, J. Bentley, J. Wong, J. A. Pierce, J. D. Cooper, P. T. Macklem, M. S. Conradi and J. C. Hogg, *Magn. Reson. Med.*, 2006, **56**, 1293.
32. N. J. Tustison, T. A. Altes, G. Song, E. E. de Lange, J. P. Mugler, 3rd and J. C. Gee, *Magn. Reson. Med.*, 2010, **63**, 1448.
33. M. J. Couch, I. K. Ball, T. Li, M. S. Fox, A. V. Ouriadov, B. Biman and M. S. Albert, *NMR Biomed.*, 2014, **27**, 1525.
34. M. Kirby, M. Heydarian, S. Svenningsen, A. Wheatley, D. G. McCormack, R. Etemad-Rezai and G. Parraga, *Acad. Radiol.*, 2012, **19**, 141.
35. B. O'Sullivan, M. Couch, J. P. Roche, R. Walvick, S. Zheng, D. Baker, M. Johnson, M. Botfield and M. S. Albert, *Acad. Radiol.*, 2014, **21**, 1524.
36. E. E. de Lange, T. A. Altes, J. T. Patrie, J. J. Battiston, A. P. Juersivich, J. P. Mugler, 3rd and T. A. Platts-Mills, *Radiology*, 2009, **250**, 567.
37. E. Y. Lee, Y. Sun, D. Zurakowski, H. Hatabu, U. Khatwa and M. Albert, *J. Thorac. Imaging*, 2009, **24**, 110.
38. Z. Liu, T. Araki, Y. Okajima, M. Albert and H. Hatabu, *Eur. J. Radiol.*, 2014, **83**, 1282.
39. W. G. Teague, N. J. Tustison and T. A. Altes, *J. Asthma*, 2014, **51**, 677.
40. M. E. Wechsler, M. Laviolette, A. S. Rubin, J. Fiterman, J. R. Lapa e Silva, P. L. Shah, E. Fiss, R. Olivenstein, N. C. Thomson, R. M. Niven, I. D. Pavord, M. Simoff, J. B. Hales, C. McEvoy, D. J. Slebos, M. Holmes, M. J. Phillips, S. C. Erzurum, N. A. Hanania, K. Sumino, M. Kraft, G. Cox, D. H. Sterman, K. Hogarth, J. N. Kline, A. H. Mansur, B. E. Louie, W. M. Leeds, R. G. Barbers, J. H. Austin, N. S. Shargill, J. Quiring, B. Armstrong, M. Castro and G. Asthma Intervention Research 2 Trial Study, *J. Allergy Clin. Immunol.*, 2013, **132**, 1295.
41. A. Sheshadri, R. Thomen, J. Kozlowski, D. S. Gierada, J. Tarsi, T. Koch, D. Burgdorf, G. Sajol, H. Yin-Declue, B. Wilson, K. Schechtman, J. Quirk, J. Woods and M. Castro, *Am. J. Respir. Crit. Care Med.*, 2013, **187**, A5443.
42. Y. Sun, B. P. O'Sullivan, J. P. Roche, R. Walvick, A. Reno, D. Baker, J. K. Mansour and M. S. Albert, *J. Magn. Reson. Imaging*, 2011, **34**, 1206.
43. T. A. Altes, M. Johnson, J. P. Mugler, D. Froh, L. Flors, G. W. Miller, J. H. Mata, C. L. Salinas, N. Tustison, P. S. Lee, T. Song, K. Yen,

E. E. d. Lange and M. C. Botfield, *Proc. Intl. Soc. Mag. Reson. Med.*, 2012, 1359.

44. R. T. Branca, Z. I. Cleveland, B. Fubara, C. S. Kumar, R. R. Maronpot, C. Leuschner, W. S. Warren and B. Driehuys, *Proc. Natl Acad. Sci. U. S. A.*, 2010, **107**, 3693.
45. J. H. Holmes, R. L. O'Halloran, E. K. Brodsky, Y. Jung, W. F. Block and S. B. Fain, *Magn. Reson. Med.*, 2008, **59**, 1062.
46. J. Yu, M. Law, S. Kadlecek, K. Emami, M. Ishii, M. Stephen, J. M. Woodburn, V. Vahdat and R. R. Rizi, *Magn. Reson. Med.*, 2009, **61**, 1015.
47. H. Hamedani, S. J. Kadlecek, K. Emami, N. N. Kuzma, Y. Xu, Y. Xin, P. Mongkolwisetwara, J. Rajaei, A. Barulic, G. Wilson Miller, M. Rossman, M. Ishii and R. R. Rizi, *Magn. Reson. Med.*, 2012, **67**, 1332.
48. J. C. Woods, *J. Appl. Physiol.*, 2013, **114**, 705.
49. R. T. Kouzes, J. H. Ely, L. E. Erikson, W. J. Kernan, A. T. Lintereur, E. R. Siciliano, D. L. Stephens, D. C. Stromswold, R. M. Van Ginhoven and M. L. Woodring, *Nucl. Instrum. Methods Phys. Res., Sect. A*, 2010, **623**, 1035.
50. P. Nikolaou, A. M. Coffey, L. L. Walkup, B. M. Gust, N. Whiting, H. Newton, S. Barcus, I. Muradyan, M. Dabaghyan, G. D. Moroz, M. S. Rosen, S. Patz, M. J. Barlow, E. Y. Chekmenev and B. M. Goodson, *Proc. Natl Acad. Sci. U. S. A.*, 2013, **110**, 14150.
51. F. W. Hersman, I. C. Ruset, S. Ketel, I. Muradian, S. D. Covrig, J. Distelbrink, W. Porter, D. Watt, J. Ketel, J. Brackett, A. Hope and S. Patz, *Acad. Radiol.*, 2008, **15**, 683.
52. M. Kirby, S. Svenningsen, N. Kanhere, A. Owrangi, A. Wheatley, H. O. Coxson, G. E. Santyr, N. A. Paterson, D. G. McCormack and G. Parraga, *J. Appl. Physiol.*, 2013, **114**, 707.
53. M. Kirby, S. Svenningsen, A. Owrangi, A. Wheatley, A. Farag, A. Ouriadov, G. E. Santyr, R. Etemad-Rezai, H. O. Coxson, D. G. McCormack and G. Parraga, *Radiology*, 2012, **265**, 600.
54. X. J. Chen, H. E. Moller, M. S. Chawla, G. P. Cofer, B. Driehuys, L. W. Hedlund and G. A. Johnson, *Magn. Reson. Med.*, 1999, **42**, 721.
55. P. J. Basser, J. Mattiello and D. Lebihan, *J. Magn. Reson., Ser. B*, 1994, **103**, 247.
56. J. E. Hansen and E. P. Ampaya, *J. Appl. Physiol.*, 1975, **38**, 990.
57. R. R. Rizi, J. E. Baumgardner, M. Ishii, Z. Z. Spector, J. M. Edvinsson, A. Jalali, J. Yu, M. Itkin, D. A. Lipson and W. Gefter, *Magn. Reson. Med.*, 2004, **52**, 65.
58. K. Qing, K. Ruppert, Y. Jiang, J. F. Mata, G. W. Miller, Y. M. Shim, C. Wang, I. C. Ruset, F. W. Hersman, T. A. Altes and J. P. Mugler, 3rd, *J. Magn. Reson. Imaging*, 2014, **39**, 346.
59. K. Sakai, A. M. Bilek, E. Oteiza, R. L. Walsworth, D. Balamore, F. A. Jolesz and M. S. Albert, *J. Magn. Reson., Ser. B*, 1996, **111**, 300.
60. K. Ruppert, J. R. Brookeman, K. D. Hagspiel, B. Driehuys and J. P. Mugler, 3rd, *NMR Biomed.*, 2000, **13**, 220.

61. N. Abdeen, A. Cross, G. Cron, S. White, T. Rand, D. Miller and G. Santyr, *Magn. Reson. Med.*, 2006, **56**, 255.
62. S. Patz, I. Muradian, M. I. Hrovat, I. C. Ruset, G. Topulos, S. D. Covrig, E. Frederick, H. Hatabu, F. W. Hersman and J. P. Butler, *Acad. Radiol.*, 2008, **15**, 713.
63. Y. V. Chang, J. D. Quirk, I. C. Ruset, J. J. Atkinson, F. W. Hersman and J. C. Woods, *Magn. Reson. Med.*, 2014, **71**, 339.
64. M. S. Fox, A. Ouriadov, K. Thind, E. Hegarty, E. Wong, A. Hope and G. E. Santyr, *Med. Phys.*, 2014, **41**, 072302.
65. K. Ruppert, J. R. Brookeman, K. D. Hagspiel and J. P. Mugler, 3rd, *Magn. Reson. Med.*, 2000, **44**, 349.
66. I. Dregely, J. P. Mugler, 3rd, I. C. Ruset, T. A. Altes, J. F. Mata, G. W. Miller, J. Ketel, S. Ketel, J. Distelbrink, F. W. Hersman and K. Ruppert, *J. Magn. Reson. Imaging*, 2011, **33**, 1052.
67. S. B. Reeder, C. A. McKenzie, A. R. Pineda, H. Yu, A. Shimakawa, A. C. Brau, B. A. Hargreaves, G. E. Gold and J. H. Brittain, *J. Magn. Reson. Imaging*, 2007, **25**, 644.
68. S. S. Kaushik, M. S. Freeman, Z. I. Cleveland, J. Davies, J. Stiles, R. S. Virgincar, S. H. Robertson, M. He, K. T. Kelly, W. M. Foster, H. P. McAdams and B. Driehuys, *J. Appl. Physiol.*, 2013, **115**, 850.
69. J. P. Mugler, 3rd and T. A. Altes, *J. Magn. Reson. Imaging*, 2013, **37**, 313.
70. Y. V. Chang and M. S. Conradi, *J. Magn. Reson.*, 2006, **181**, 191.
71. P. A. Rinck, S. B. Petersen and P. C. Lauterbur, *Fortschr. Röntgenstr.*, 1984, **140**, 239.
72. D. O. Kuethe, A. Caprihan, E. Fukushima and R. A. Waggoner, *Magn. Reson. Med.*, 1998, **39**, 85.
73. L. Carrero-Gonzalez, T. Kaulisch and D. Stiller, *Magn. Reson. Med.*, 2013, **70**, 1761.
74. A. V. Ouriadov, M. S. Fox, M. J. Couch, T. Li, I. K. Ball and M. S. Albert, *Magn. Reson. Med.*, 2015, **74**, 550.
75. D. O. Kuethe, A. Caprihan, H. M. Gach, I. J. Lowe and E. Fukushima, *J. Appl. Physiol.*, 2000, **88**, 2279.
76. M. J. Couch, I. K. Ball, T. Li, M. S. Fox, S. L. Littlefield, B. Biman and M. S. Albert, *Radiology*, 2013, **269**, 903.
77. A. F. Halaweish, R. E. Moon, W. M. Foster, B. J. Soher, H. P. McAdams, J. R. MacFall, M. D. Ainslie, N. R. MacIntyre and H. C. Charles, *Chest*, 2013, **144**, 1300.
78. W. G. Schreiber, B. Eberle, S. Laukemper-Ostendorf, K. Markstaller, N. Weiler, A. Scholz, K. Burger, C. P. Heussel, M. Thelen and H. U. Kauczor, *Magn. Reson. Med.*, 2001, **45**, 605.
79. U. Wolf, A. Scholz, C. P. Heussel, K. Markstaller and W. G. Schreiber, *Magn. Reson. Med.*, 2006, **55**, 948.
80. U. Wolf, A. Scholz, M. Terekhov, R. Koebrich, M. David and L. M. Schreiber, *Magn. Reson. Med.*, 2010, **64**, 1478.
81. A. W. Scholz, U. Wolf, M. Fabel, N. Weiler, C. P. Heussel, B. Eberle, M. David and W. G. Schreiber, *Magn. Reson. Imaging*, 2009, **27**, 549.

82. A. F. Halaweish, W. M. Foster, R. E. Moon, N. R. MacIntyre, J. R. MacFall and H. C. Charles, *Proc. Intl. Soc. Mag. Reson. Med.*, 2013, 4111.
83. H. C. Charles, N. R. MacIntyre, R. Moon, W. M. Foster, H. P. McAdams and M. Ainslie, *Am. Respir. Crit. Care Med.*, 2014, **189**, A4348.
84. S. D. Swanson, M. S. Rosen, B. W. Agranoff, K. P. Coulter, R. C. Welsh and T. E. Chupp, *Magn. Reson. Med.*, 1997, **38**, 695.
85. M. S. Albert, D. F. Kacher, D. Balamore, A. K. Venkatesh and F. A. Jolesz, *J. Magn. Reson.*, 1999, **140**, 264.
86. J. Wolber, A. Cherubini, M. O. Leach and A. Bifone, *NMR Biomed.*, 2000, **13**, 234.
87. G. Norquay, G. Leung, N. J. Stewart, G. M. Tozer, J. Wolber and J. M. Wild, *Magn. Reson. Med.*, 2015, **74**, 303.
88. W. Kilian, F. Seifert and H. Rinneberg, *Proceedings of the International Society for Magnetic Resonance in Medicine*, Honolulu, Hawaii, 2002, p. 758.
89. X. Zhou, M. L. Mazzanti, J. J. Chen, Y. S. Tzeng, J. K. Mansour, J. D. Gereige, A. K. Venkatesh, Y. Sun, R. V. Mulkern and M. S. Albert, *NMR Biomed.*, 2008, **21**, 217.
90. W. Kilian, F. Seifert and H. Rinneberg, *Magn. Reson. Med.*, 2004, **51**, 843.
91. A. Wakai, K. Nakamura, J. Kershaw and I. Kanno, *Int. Congr. Ser.*, 2004, **1265**, 139.
92. J. Kershaw, K. Nakamura, Y. Kondoh, A. Wakai, N. Suzuki and I. Kanno, *Magn. Reson. Med.*, 2007, **57**, 791.
93. X. Zhou, Y. Sun, M. Mazzanti, N. Henninger, J. Mansour, M. Fisher and M. Albert, *NMR Biomed.*, 2010, **24**, 170.
94. M. L. Mazzanti, R. P. Walvick, X. Zhou, Y. Sun, N. Shah, J. Mansour, J. Gereige and M. S. Albert, *PLoS One*, 2011, **6**, e21607.
95. J. Nouls, Z. I. Cleveland, M. S. Freeman, H. E. Moeller, L. W. Hedlund and B. Driehuys, *Proceedings of the International Society for Magnetic Resonance in Medicine*, Montréal, Canada, 2011, p. 879.
96. M. R. Rao, N. J. Stewart, G. Norquay and J. M. Wild, *Proc. Intl. Soc. Mag. Reson. Med.*, 2014, 3532.
97. L. Delacour, N. Kotera, T. Traore, S. Garcia-Argote, C. Puente, F. Leteurtre, E. Gravel, N. Tassali, C. Boutin, E. Leonce, Y. Boulard, P. Berthault and B. Rousseau, *Chemistry*, 2013, **19**, 6089.
98. K. K. Palaniappan, R. M. Ramirez, V. S. Bajaj, D. E. Wemmer, A. Pines and M. B. Francis, *Angew. Chem. Int. Ed.*, 2013, **52**, 4849.
99. K. Bartik, M. Luhmer, J.-P. Dutasta, A. Collet and J. Reisse, *J. Am. Chem. Soc.*, 1998, **120**, 784.
100. L. Schröder, T. J. Lowery, C. Hilty, D. E. Wemmer and A. Pines, *Science*, 2006, **314**, 446.
101. T. Meldrum, V. S. Bajaj, D. E. Wemmer and A. Pines, *J. Magn. Reson.*, 2011, **213**, 14.
102. T. K. Stevens, K. K. Palaniappan, R. M. Ramirez, M. B. Francis, D. E. Wemmer and A. Pines, *Magn. Reson. Med.*, 2013, **69**, 1245.

103. K. M. Dowhos, M. S. Fox, I. K. Ball, T. Li, G. Gajawada, J. Wentzell, B. DeBoef and M. S. Albert, *Proceedings of the International Society for Magnetic Resonance in Medicine*, Milan, Italy, 2014, p. 3537.
104. M. J. Couch, B. Blasiak, B. Tomanek, A. V. Ouriadov, M. S. Fox, K. M. Dowhos and M. S. Albert, *Mol. Imaging Biol.*, 2015, **17**, 149.
105. M. G. Shapiro, R. M. Ramirez, L. J. Sperling, G. Sun, J. Sun, A. Pines, D. V. Schaffer and V. S. Bajaj, *Nat. Chem.*, 2014, **6**, 629.
106. S. Klippel, J. Dopfert, J. Jayapaul, M. Kunth, F. Rossella, M. Schnurr, C. Witte, C. Freund and L. Schroder, *Angew. Chem. Int. Ed.*, 2014, **53**, 493.
107. C. Boutin, E. Leonce, T. Brotin, A. Jerschow and P. Berthault, *J. Phys. Chem. Lett.*, 2013, **4**, 4172.
108. J. Dopfert, C. Witte, M. Kunth and L. Schroder, *Contrast Media Mol. Imaging*, 2014, **9**, 100.
109. T. K. Stevens, R. M. Ramirez and A. Pines, *J. Am. Chem. Soc.*, 2013, **135**, 9576.
110. J. Wolber, I. J. Rowland, M. O. Leach and A. Bifone, *Magn. Reson. Med.*, 1999, **41**, 442.
111. D. Bartusik and B. Tomanek, *Adv. Drug Delivery Rev.*, 2013, **65**, 1056.

Subject Index

References to figures are given in *italic* type. References to tables are given in **bold** type.

ab initio methods *see* quantum chemistry
acetaldehyde, 171–2, **171**
acetone, **113**, 141, **171**, 246
acetonitrile, **113**
acetyl chloride, **171**
acetylene, 21
ADF, 211
adiabatic Longitudinal Transport After Dissociation Engenders Net Alignment (ALTADENA), 305–6, 307–9
ammonia, 65–6, 175, 254
ammonium nitrate, 186–7
apparent diffusion coefficients (ADC), 371–2
argon, 37
aromatic molecules, 231, 232–3, 242, *243*
axicon, 359

basis sets, 240–1
benzene, 114
Berry pseudorotation, 40
biopolymers, 120
Bok globule B335, 163
Born–Oppenheimer approximation, 218–19, 269–70
Breit–Pauli perturbation theory (BPPT), 194, 241
Breitfussin A and B, 290
bromine, **108**
bulk susceptibility correction, 115

cancer, 384
carbon diselenide, 12
carbon monoxide, 163, **166**
carbon tetrafluoride, 5
carbon-13
  nuclear magnetic moment, **108**, 114–15
  quadrupolar interaction, 73–5
  shielding scale, 62–3, 119–20
    vibrational corrections, 231–2
  shielding surfaces, 12
carbonyl sulfide, **166**, 168
carboxylic acid, 138
Carr–Purcell–Meiboom–Gill (CPMG) sequence, 306
CFOUR, 210–11
Chapman–Enskog procedure, 25
chemical exchange, 39
chemical reactions see reaction monitoring
chemical shift, 3–4, *117*, **166**
  density coefficient, 4–6
  isotope effects, 173–5
  limitations, 116
  shielding constants and, relativistic effects, 289–95
  temperature dependence, 9–10
chemical shift anisotropy (CSA), 29–30
chlorine, **108**

chlorine-35, 107
chlorine-37, 107
chloroform, **113**
chronic obstructive pulmonary disease (COPD), 373
clathrates, 120
complete basis set limit, 240–1
condensation, 171–2
conformational dynamics, 38–41, 256–7
contact shifts, 8
continuous transformation of the origin of current density leading to formal annihilation of diamagnetic contribution (CTOCD-DZ), 191
coupled cluster methods
  singles-and-doubles (CCSD), 11–12, 17–18, 66, 204–5, 208–9
  singles-and-doubles (triplets) (CCSD(T)), 11–12, 192–3
    shielding constants, 196–7, **197**, 199–207, **234**
    vibrational corrections, 241
coupling constant, 175
cycloadditions, 138
cyclohexane (deuterated), **113**
cyclopropene, **249**

density functional theory (DFT), 7, 83–4, 160, 203–4, 255–6
  absolute errors, 17
  relativistic effects, 274
  spin-rotation and magnetic shielding constants, 72–3
  spin–spin coupling vibrational corrections, 255–6
deuterated solvents, 113–15
deuteron, 102
  magnetic moment, **106**
di-*t*-butyl peroxide (DTBP), 139–43
diatomic molecules
  electric field gradient, 77
  hyperfine Hamiltonian, 54–6, *57*
  shielding, 61, 203–4
  shielding surface rovibrational averaging, 15–16
differential cross-section (DCS), 35
diffusion-ordered spectroscopy (DOSY), 147–8
diffusion-weighted imaging, 371–2
difluoroethane, 236
dimethyl ethers, 163, 165
N,N-dimethylformamide, 39
dimethylsulfoxide (DMSO), **113**
dipolar coupling constant, 79
DIRAC, 211
Dirac comb function, 323–4
Dirac equation, 271
Dirac matrices, 270
Dirac spinors, 270
Dirac–Coulomb–Breit Hamiltonian, 274
Douglas–Kroll (DK) method, 275–6
Douglas–Kroll–Hess (DKH) method, 276, 281
dynamic nuclear polarization (DNP), 305

echo time, 311
effective geometry, 223–4
electric field gradient (EFG), 73–4, 74–5
electron correlation, 238–9
ethane, 21, **249**
  as decomposition product, 141
ethanol, **113**
ethene, **249**
ethyne, 244, **249**
exact two-component (X2C) approach, 279, 291
EXSY, 143
external field-dependent unitary transformation (EFUT), 66, 284

Fermi-contact (FC) mechanism, 19–20
fluorinated ethanes, 147
fluorine monoxide, 171, **171**
fluorine-19, 68, 82
  isotope effects, 174–5
  magnetic shielding, 68–9

fluorine-19 (*continued*)
  MRI using, 373–9
  nuclear magnetic moment, **108**
  shielding corrections, 235–6
  shielding scale, 63–5, *64*
fluoromethanes, 235–6
fluoropropane, 236
Flygare's formula, 189
Flygare's relations, 293
Foldy–Wouthuysen transformation, 284
full configuration interaction (FCI), 11, 254
full field-dependent unitary transformation at matrix level (FFUTm), 66
fullerenes, 119, 246
furan, **171**

gas samples, 96–7
  high-pressure, 97–8
gas-to-liquid parameter, 171–2
gauge problem, 160
gauge-including atomic orbitals (GIAOs), 191, 286
germane, 107
germanium-73, nuclear magnetic moment, 107
Gordon, Roy, 23–4
gradient-compensated stimulated echo (GCSTE), 147

halosilanes, 258–9
Hartree–Fock (HF) methods, 238–9
heavy atom on light atom (HALA) effect, 206, 268
helium-3, 18, 305
  compression, 355–6
  metastability exchange optical pumping, 352–5
  MRI using, 366–71
  nuclear magnetic moment, **108**, 110
  as reference standard, *112*
  as reference material, 111–13
  shielding, 111, 120
hexafluorooethane, 236
hexafluoropropylene oxide, 136–8
hexyne, 309–10, 325, **326**
high-resolution spectroscopy, 147
hydrogen chloride
  to 2-methylpropene, 136
  to propene, 136
hydrogen fluoride, 53–4, 68, **248**
  splitting, 82
hydrogen halides, 70–1, 290–1
hydrogen (molecular), **248**
  para isomer, 307
  spin relaxation, 37–8
hydrogen-1
  isotope effects, 174–5
  shielding, vibrational corrections, 230–2
hydrogen-2 *see* deuterium
hydronium ion, 254
hyper-CEST, 382–4
hyperfine structure, 54, 63–4

INEPT, 158
infinite-order regular approximation (IORA), 279
intermolecular interactions, 3–4, 170
  shielding function, 6–8, 196–7
intramolecular potential energy surface, 2
iron-57, **110**
isomerization reactions, 138
isotope effects
  chemical shift, 173–5
  rovibrational effects, 226–7, 245–6, 260
isotope labelling, 156–7, 178
isotope shifts, 14
isotopic mass, 2

J-coupling, 313
J-spectroscopy, 310–17
  interpretation of partial spectra, 317
  in PHIP, 319–27
  technical considerations, 317–20
J-surface, 19–23, 82–4
J-tensor, 82–4

Karplus, Martin, 2
krypton-83, 86–7, **108**

Lamb-dip technique, 69, 164
Larmor frequency, 338
laser diode arrays, 351
laser polarization, 305
linear molecules, 61–3, 230–1
lungs, 336–7, 367, 373–9

magic angle spinning (MAS), 119–20
magnetic balance, 281
magnetic moment, 95–6, 101–2
    deuteron, 102, **106**
    experimentally determined, 105–9
    measurement, 85–8
    nuclear, **108**, 210
        undetermined, **110**
    proton, 102–5, **106**
    tritium, 107
magnetic resonance imaging (MRI), 304, 336–7, 364–7
    breath-hold imaging, 367–71
    diffusion imaging, 371–2
    dissolved xenon, 372–3
    fluorine-19 lung imaging, 373–9
    hyperpolarized, 366–7
    xenon-129 biosensors, 382–4
magnetization transfer contrast, 372–3
magnetogyric ratio, 86
MCSCF, 20
metastability exchange optical pumping (MEOP), 352–5
    helium-3, 353–4
    high magnetic field and pressure, 356–8
methane, **248**
methanol, **113**, *177*
methyl acetate, **171**
methyl ether, **171**
methyl ether ketone, 114
methyl formate, **171**
microwave spectroscopy, 74, 82–3
molecular spectroscopy, 80
    quadrupolar tensors from, 75–8
Møller–Plesset second-order perturbation theory (MP2), 52, 160, 192
Morse–Morse–Spline–van der Waals PES, 36
MRCC, 210
multiconfigurational self-consistent field (MCSCF), 160, 163, 203

neon-21, nuclear magnetic moment, **108**
nitrogen-14, nuclear magnetic moment, **108**
nitrogen-15
    nuclear magnetic moment, **108**
    shielding corrections, 233–4
nitromethane, **113**
nitrous oxide, **166**, 178
non-collinear approach, 274
normalized elimination of the small component (NESC), 277
nuclear magnetic moment *see* magnetic moment
nuclear magnetic shielding *see* shielding
nuclear quadrupole moment *see* quadrupole moment
nuclear site effect, 5–6
nuclear spin polarization, 338–9
nuclear spin-rotation, 54–7

orbital decomposition (ODA), 66–7, 286
oxygen, 69, 153–4
    chemical shifts and shielding parameters, **166**
    isotopic abundance, 154
    isotopic labelling, 156–7
    NMR properties, **156**
    nuclear magnetic moment, **108**
    shielding corrections, 239
    spin–spin coupling, 175–80

oxygen-17
isotope labelling, 156–8
shielding in small molecules, 164–7
spin–spin coupling, 175–80

parahydrogen, 307
parahydrogen induced polarization (PHIP), 305–6
hydrogenation, 309–11
J-spectroscopy in, 319–27
experimental results, 324–7
gases, 327–31
theoretical basis, 319–24
parahydrogen and synthesis allow dramatically enhanced nuclear alignment (PASADENA), 305, 307–9, 320, 327–9
paramagnetic gases, 8
Penning collisions, 356
Penning trap, 104
perfluoro-2-butenes, 136
perfluoro-2-pentenes, 136
perfluoropropane, 373
perturbation theory, 160–1, 283
alternative expansions, 227–8
relativistic effects, 268
rovibrational corrections, 219–20
spin–spin coupling, 251–3, 256–9
vibrational second-order (VPT2), 220, 250, 252
phosphine, 107
phosphorus-31, 205
nuclear magnetic moment, 107, **108**, 114
vibrational corrections, 236–7
photochemical reactions, 138–9
PK11195-cryptophane, 383
polytetrafluoroethylene (PTFE), 130
potential energy surface, 13–14, 35, 218–19
pressure, 39, 127–9, **128**
pressurized samples, 97–8
propane, *329*, 330
propionaldehyde, **171**
proton, magnetic moment, 102–5, 103–4, **106**
proton exchange, 175
pulse field gradient (PFG) coil, 134

quadrupolar coupling constant, 74–6, **76**
quadrupolar spin relaxation, 24
quadrupole moment, 73–5
quantum chemistry
coupled cluster methods *see* coupled cluster methods
electron configuration, 11
J-tensors, 82
MP2, 52, 160, 192
shielding, 6–7, 32–3, 65, 191–6, 192–4
basis sets, 191–2, 240–1
electron correlation effects, 202–3, 238–9
gauge origin, 160
relativistic effects, 193–4
spin-rotation and, 71–3, 87–8
software, 210–11
spin-rotation tensors, 71–3, 87–8
spin–spin coupling, 162–3

radiation induced lung injury (RILI), 372
radiation trapping, 344
Ramsey, N. F., 53
Ramsey–Flygare measurement, 60–73
linear molecules, 61–3
non-linear molecules, 63–6
relativistic, 66–7
Ramsey's formula, 189
rare gases, 5
Raynes, W. T., 152
reaction monitoring, 126–7
acquisition parameters, 134–5
di-*t*-butyl peroxide (DTBP) decomposition, 139–43

experimental considerations
concentration and
pressure, 127–9
temperature, 130
hydrogen chloride to propene, 136
phase and frequency drift, 134
probe design, 133–4
published studies, 135–9
reaction product characterization, 143–8
reaction product characterization, 143–8
COSY, 144–7
DOSY, 147–9
see also chemical reactions
relativistic effects, 267–9, 288–95
spin–spin coupling, indirect, 295–7
theoretical models
four-component Hamiltonians, 269–74
two-component Hamiltonians, 274–9
resonance frequency, isolated molecules, 98–100
ReSpect, 211
restricted magnetic balance3, 284–5
Rice–Ramsperger–Kassel–Marcus (RRKM) theory, 39–40
rotational spectroscopy, 64
rovibrational effects, 218–19, 246–7
basis set effects, 240–1
expansion term analysis, 242–3
isotope effects, 226–7, 245–6
parameter calculation, 228–9
perturbation theory, 219–20
relativistic effects, 241–2
rotational contributions, 224–5
shielding, 13–17
spin–spin coupling, 21–3
temperature and, 225–6, 245
transferability, 237–8
vibrational corrections, 221–3
carbon shielding, 231–2
effective geometry approach, 223–4
fluorine shielding, 235–6
fluorine shieldings, 235–6
hydrogen shielding, 230–1
nitrogen shieldings, 233–4
oxygen shieldings, 234–5
phosphorus shielding, 236–7
phosphorus and transition metals, 236–7
RRKM theory, 39–40
rubidium, 340, *341*, 342–3

sample preparation, 96–7
second-order polarization propagator approximation with coupled cluster singles and doubles amplitudes (SOPPA(CCDS)), 252
selenium-77, 12
self-consistent field (SCF), 11
shielding, 2, 157, 158–60
ab initio methods, 191–6
electron correlation effects, 192–3, 198–205
intermolecular interactions, 208–9
relativistic effects, 193–4, 205–7, 279–88
zero-point vibrational effects, 207
absolute scales, 17–18, 292–5
bulk susceptibility contribution, 157–8
bulk susceptibility correction, 115
carbon monoxide as reference source, 163–4
carbon-13 scale, 119–20
carbonyl sulfide, 168
chemical shift and, relativistic effects, 289–95

shielding (*continued*)
  computational determination, 71–3
    gauge origin, 160
    verification, 118
  condensation effect, 171–3
  condensation effects, 171–3
  diamagnetic contribution, 59, 160
  diamagnetic term, 160
  experimentally derived scales, 68–73
  Hamiltonian, 58
  intermolecular interactions, 6–8, 170, 195–6, 196–7, 208–9
  intramolecular effects, 8–18
    rovibrational averaging, 13–17
    shielding surface for di- and polyatomic molecules, 10–13
  isotope effects, 119
  measurement
    direct, 108–10, 109–11
    from spin–spin interaction, 60–1
    linear molecules, 61–3
    non-linear molecules, 63–6
    Ramsey–Flygare method, 60–73
    referencing, 111–15
  NMR spectrum standardization, 115–18
  non-linear molecules, 64–6
  non-relativistic formulation, 188–9
  paramagnetic contribution, 59, 160
  paramagnetic term, 160
  Ramsey's method, 53, 161
  reference materials
    carbon monoxide, 163
    water, 164
  relativistic effects, 66–8, 72, 193–4
    perturbation approach, 67–8
  second virial coefficient, 4–5
  sulfur hexafluoride, 168–9
  verification, 118–20
  vibrational correction, 229–33
silane, 107, **248**
silicon-29
  magnetic moment, 107
  nuclear magnetic moment, **108**
single-photon emission computed tomography (SPECT), 365
sodium-23, 10
software, 210–11
solid-phase measurements, 119–20
solvents, **113**
  rovibrational effects, 246–7
SOPPA(CCSD), 20
spin echoes, 306
spin exchange optical pumping (SEOP), 339–40
  alkali metals, 341–5
    noble gas nuclei and, 345–9
  helium-3, 351–2
  xenon, 349–51
spin relaxation, 23–4, 23–8, 134–5
  chemical shift anisotropy, 29–30
  cross-sections, 26
  effect addition, 30
  hydrogen molecule, 37–8
  intermolecular mechanisms, 30–3
    classical cross-sections, 35–7
    nuclear spin dipole electron spin dipole mechanism, 33–5
  intramolecular dipolar, 29
  noble gas polarization, 349–52
  quadrupolar mechanism, 28–9

spin-rotation
  derivation of magnetic shielding from, 62–3
  mechanism, 27–8
  nuclear, 54–7
  relaxation values, **27**, 28
spin–spin coupling, 18–19, 78–80, 157, 162–3, 175–80
  density coefficient, 19, **176**
  indirect
    calculation, 82–4
    relativistic effects, 295–7
  indirect tensor characterization, 80–2
  isotope effects, 22–3, 173–5
  J surface, 19–23
  molecules with quadrupolar nuclei, 178
  vibrational corrections, 229, 247–62
    density functional theory, 255–60
    general trends, 261–2
    isotope effects, 260–1
    perturbation theory, 251–3
    relativistic effects, 260
    variational methods, 259–61
Stern, Otto, 86
Stern–Gerlach experiment, 102–4
sulfur, 153–4, 155
sulfur difluoride, 63–4
sulfur hexafluoride, 168–9, 373
  spin–spin coupling, **180**
sulfur oxides, 153, 169–70
sulfur tetrafluoride, 40–1
sulfur-33, 69, 155
  isotopic labelling, 157
  NMR properties, **156**
  nuclear magnetic moment, **108**
  shielding, carbonyl sulfide as reference, 168
  spin–spin coupling, 175–80

temperature, 2, 225–6, 245
  calibration, 129–30
  chemical shift and, at zero density, 9–10
  rovibrational effects and, 225–6
tetrafluoroethene, 138
tetrafluoromethane, 373
tetramethylsilane (TMS), 116
tetramethyltin, 71
thallium halides, 84, **84**
thermotropic liquid crystals, 178–9
toluene, **113**
2,2,3-trifluoro-3-(trifluoromethyl)oxirane, 136–8
trifluoroethane, 139
N-trifluoroacetamide, 40
trimethylene oxide, **171**
tritium, magnetic moment, 107
tubes, **128**

ultra-short echo time, 367
unrestricted kinetically balanced basis set (UKB), 285

vanadium, 236–7
variational methods, 259–61
ventilation/perfusion ratio (V/Q), 372
vessel design, 130–3
vibrational second-order perturbation theory (VPT2), 220, 250, 252
virial expansion, 3–4, 157, 208

water, 154–5, 164
  deuterated, **113**
  gas-to-liquid shift, **171**, 172
  J-surface, 21
  shielding surfaces, 12
  spin–spin coupling constant, 176–8, **176**, **248**

xenon difluoride, 13, 30
xenon fluorides, 205–6, **206**
xenon gas, 4, 5
  intermolecular shielding function, 6–7

xenon polarization transfer contrast (XTC), 372
xenon-129, 120, 241–2, 349–50
  biosensors, 382–3
  chemical shift density coefficient, 4–5
  intermolecular shielding function, 6–7
  MRI using, 305, 367, 372–3
  nuclear magnetic moment, **108**
  nuclear spin relaxation, 34–5
  spin-exchange optical pumping, 349–50
xenon-131, **108**
zeolites, 120
zero-point vibration (ZPV), 167, 237–8, 239–40, 247
zero-point vibrational corrections (ZPVC), 230
  carbon, 231–2
  fluorine, 235–6
  hydrogen, 230–1, 237–8
  nitrogen, 233–4, 239
  oxygen, 234–5, 239
  phosophorus, 236–7, 239
zeroth-order regular approximation (ZORA), 260, 279, 281, 291